RAINFALL

Modeling, Measurement and Applications

RAINFALL
Modeling, Measurement and Applications

Edited by

Renato Morbidelli

Department of Civil and Environmental Engineering,
University of Perugia, Perugia, Italy

Elsevier
Radarweg 29, PO Box 211, 1000 AE Amsterdam, Netherlands
The Boulevard, Langford Lane, Kidlington, Oxford OX5 1GB, United Kingdom
50 Hampshire Street, 5th Floor, Cambridge, MA 02139, United States

British Library Cataloguing-in-Publication Data
A catalogue record for this book is available from the British Library

Library of Congress Cataloging-in-Publication Data
A catalog record for this book is available from the Library of Congress

ISBN: 978-0-12-822544-8

Publisher: Candice Janco
Acquisitions Editor: Louisa Munro
Editorial Project Manager: Jose Paolo Valeroso
Production Project Manager: Bharatwaj Varatharajan
Cover Designer: Miles Hitchen

Typeset by Aptara, New Delhi, India

Contents

Contributors

Haider Ali
School of Engineering, Newcastle University, United Kingdom

Ali Alnahit
Department of Civil Engineering, King Saud University, Riyadh, Saudi Arabia

Marco Borga
Department of Land Environment Agriculture and Forestry, University of Padova, Padova, Italy

Maria Teresa Brunetti
CNR IRPI - Italian National Research Council, Research Institute for the Geo-Hydrological Protection, Perugia, Italy

O. Castro-Orgaz
University of Córdoba, Department of Agronomy, Córdoba, Spain

Arianna Cauteruccio
University of Genova, Dept. of Civil, Chemical and Environmental Engineering, Genoa, Italy; WMO/CIMO Lead Centre "B. Castelli" on Precipitation Intensity, Italy

Pierluigi Claps
Department of Environment, Land and Infrastructure Engineering, Politecnico di Torino, Torino, Italy

Corrado Corradini
Department of Civil and Environmental Engineering, University of Perugia, Perugia, Italy

Jacopo Dari
National Research Council, Research Institute for Geo-Hydrological Protection, Perugia, Italy; Department of Civil and Environmental Engineering, University of Perugia, Perugia, Italy

Alessia Flammini
Department of Civil and Environmental Engineering, University of Perugia, Perugia, Italy

Hayley J. Fowler
School of Engineering, Newcastle University, United Kingdom

Marco Gabella
Meteoswiss, Locarno Monti, Switzerland

Daniele Ganora
Department of Environment, Land and Infrastructure Engineering, Politecnico di Torino, Torino, Italy

Stefano Luigi Gariano
CNR IRPI - Italian National Research Council, Research Institute for the Geo-Hydrological Protection, Perugia, Italy

J.V. Giráldez
University of Córdoba, Department of Agronomy, Córdoba, Spain; Institute for Sustainable Agriculture, CSIC, Department of Agronomy, Córdoba, Spain

Rao S. Govindaraju
Lyles School of Civil Engineering, Purdue University, West Lafayette, IN, United States

Abhishek Goyal
Lyles School of Civil Engineering, Purdue University, West Lafayette, IN, United States

Wojciech W. Grabowski
Mesoscale and Microscale Meteorology Laboratory, NCAR, Boulder, CO, United States

Fausto Guzzetti
Civil Protection Department, Office of the Prime Minister, Rome, Italy

J.A. Gómez
Institute for Sustainable Agriculture, CSIC, Department of Agronomy, Córdoba, Spain

Theano Iliopoulou
Department of Water Resources and Environmental Engineering, National Technical University of Athens, Zographou, Greece

Christopher Kidd
Earth System Science Interdisciplinary Center, University of Maryland, MD, United States; NASA Goddard Space Flight Center, Greenbelt, MD, United States

Demetris Koutsoyiannis
Department of Water Resources and Environmental Engineering, National Technical University of Athens, Zographou, Greece

A.M. Laguna
University of Córdoba, Department of Applied Physics, Córdoba, Spain

Luca G. Lanza
University of Genova, Dept. of Civil, Chemical and Environmental Engineering, Genoa, Italy; WMO/CIMO Lead Centre "B. Castelli" on Precipitation Intensity, Italy

Vincenzo Levizzani
Institute of Atmospheric Sciences and Climate, National Research Council (CNR-ISAC), Bologna, Italy

Francesco Marra
Institute of Atmospheric Sciences and Climate, National Research Council (CNR-ISAC), Bologna, Italy

Paola Mazzoglio
Department of Environment, Land and Infrastructure Engineering, Politecnico di Torino, Torino, Italy

Greg M. McFarquhar
Cooperative Institute for Mesoscale Meteorological Studies, University of Oklahoma, Norman, OK, United States

Massimo Melillo
CNR IRPI - Italian National Research Council, Research Institute for the Geo-Hydrological Protection, Perugia, Italy

Ashok Mishra
Glenn Department of Civil Engineering, Clemson University, South Carolina, United States

Renato Morbidelli
Department of Civil and Environmental Engineering, University of Perugia, Perugia, Italy

Sourav Mukherjee
Glenn Department of Civil Engineering, Clemson University, South Carolina, United States

Silvia Peruccacci
CNR IRPI - Italian National Research Council, Research Institute for the Geo-Hydrological Protection, Perugia, Italy

Carla Saltalippi
Department of Civil and Environmental Engineering, University of Perugia, Perugia, Italy

Mattia Stagnaro
University of Genova, Dept. of Civil, Chemical and Environmental Engineering, Genoa, Italy; WMO/CIMO Lead Centre "B. Castelli" on Precipitation Intensity, Italy

Ramesh S.V. Teegavarapu
Department of Civil, Environmental and Geomatics Engineering, Florida Atlantic University, Boca Raton, Florida, United States

Preface

Even though different types of precipitation can occur, this book is substantially related to rainfall formed from condensation or sublimation of water vapor over condensation nuclei with successive coagulation and precipitation at the ground surface. Rainfall, characterized by a droplet size distribution, is a major component of the water cycle and gives a crucial contribution to the fresh water on the Earth. It produces suitable conditions for many types of ecosystems and is useful for hydroelectric power plants and crop irrigation.

This book integrates different rainfall perspectives from microphysics and modeling developments to experimental measurements and their analysis also in the context of applications in surface and subsurface hydrology. It is mainly directed to postgraduate readers studying meteorology, civil and environmental engineering, geophysics, agronomy and natural science, as well as practitioners working in the fields of hydrology, hydrogeology, agronomy, and water resources management. Each chapter provides an updated representation of the involved subject with relative open problems.

Chapter 1 describes elements of microphysics inside and beneath clouds producing rainfall at Earth's surface, highlighting discoveries of the last 100 years that have led to significant improvement of our knowledge. The chapter describes processes associated to both warm and cold rain.

In Chapter 2, the production of precipitation through the development of vertical motions in the atmosphere is considered. The lifting of humid air mass due to frontal disturbances, orographic chains, convective systems, and humid air convergence is discussed. In this context, the spatio-temporal distribution of rainfall is widely analyzed for frontal systems influenced by the interaction of the above lifting mechanisms.

Chapter 3 discusses the progress in the last decades regarding two fundamental elements for rainfall modeling: formulation of moist air dynamics and representation of formation and fallout of precipitation.

Chapter 4 describes and discusses the catching and noncatching instruments of rainfall measurements (raingauges) and their main characteristics. Standard calibration methods are reported for catching-type gauges. Optimal correction algorithms for the interpretation of tipping-bucket raingauge records are presented, together with correction methods for both tipping-bucket and weighing gauges. The impact of wind on rainfall measurements is discussed on the basis of the outer shape of the gauge body, and suitable correction curves are reported for cylindrical gauges. The relevance of measurement accuracy and quality in rainfall monitoring is highlighted and a brief section on the design of monitoring networks is included.

Chapter 5 provides an outline of the principles of precipitation estimation by means of weather radar, with coverage of the main measurement techniques and methods used to generate rainfall products starting from weather radar observations.

Chapter 6 points out that the use of conventional instruments (gauge or radar) to map global precipitation is essentially limited to land areas and thus satellite observations must be used to provide estimates of global precipitation. Many satellite sensors operating in the last 50 years provided data for a range of techniques, algorithms, and schemes developed to obtain quantitative precipitation estimates. Space-time limitations of current satellite-based precipitation products are described. This chapter outlines the basis of satellite precipitation estimation, satellites and sensors types, and techniques and schemes used to generate the precipitation products.

In Chapter 7, the role of a limited and not homogeneous temporal resolution of rain gauge data in the analysis of commonly available historical series is discussed to provide evidence of possible errors in hydrological investigations. Particular emphasis is given to the effects on the analysis of extreme rainfalls that have a crucial role in designing hydraulic structures. Simple equations to improve the determination of extreme values are also provided.

Mean areal rainfall estimate using deterministic and stochastic methods is presented in Chapter 8. Conceptually weighting methods that use raingauge-based observations and gridded rainfall data from radar and satellite-based sources are described.

Chapter 9 presents the typical form of mathematical relationships linking maximum rainfall intensity of different durations to the return period, also known as intensity-duration-frequency curves, along with its merits and limitations. A modeling framework to overcome the limitations is also described. Two variants of the model are presented: a full version valid over time scales and a simplified relationship applicable over fine scales of the order of common applications, i.e., sub-hourly to daily.

In Chapter 10, the main factors influencing the rainfall areal reduction factors (ARFs) are described. The main empirical and analytical approaches available in the scientific literature to estimate ARFs are presented and critically discussed. The crucial issue of the transposition and applicability of ARFs developed for a certain area to other regions is also deepened by presenting the results of several studies.

In Chapter 11, recent advances in studying the extreme rainfall through recorded quantities available from measurements on a sub-daily/multi-day time scale are described. Future changes of rainfall extremes are discussed on the basis of climate model outputs. This is achieved by examining different available models and understanding the relationships between rainfall extremes and temperature.

Chapter 12 summarizes the state of the art of regionalization techniques applied to rainfall data. First, current problems in data availability are identified. Then, differences between traditional and more innovative approaches aimed to provide intensity-duration-frequency curves everywhere in a large area are highlighted. Furthermore, this chapter explores the advantages of interpolation methods over the homogeneous region paradigm, addressing in particular the objective of valorization of the local information deriving from short records.

Chapter 13 first deals with the formation and separation of the flood hydrograph through the effective hyetograph associated to a specific rainfall-runoff event. On this basis the main structure of typical rainfall-runoff models for simulating single flood events is highlighted in general terms. Then, the specific structure of an adaptive rainfall-runoff model for real-time flood forecasting is also examined. Finally, through a synthetic statistical analysis of extreme rainfalls, a classical procedure for determining the design hydrograph of hydraulic structures is presented.

In Chapter 14, the rainfall pattern role in determining the infiltration process is examined. In this context, a quantitative representation of the rainfall infiltration process at different spatial scales is provided considering also erratic spatio-temporal rainfall distributions. Artificial rainfall systems useful for determining the main soil properties are also synthetically presented.

In Chapter 15, the exploration of the main features of soil erosion controlled by rainfall has been carried out; starting from this analysis some relevant aspects that might deserve more attention by the research in the near future can be detected.

In Chapter 16, a grid-based slope stability model for the spatial and temporal prediction of rainfall-induced landslides is described after a general characterization of physically-based and empirical approaches. A particular emphasis has been placed on a widely used empirical method for the prediction of landslide initiation, i.e., rainfall threshold.

Chapter 17 highlights the importance of rainfall in drought assessment. An overview of the role of rainfall in drought evaluation is provided and the most common precipitation-based drought indices are pointed out. The various challenges and limitations associated with quantifying the evolution of drought using the rainfall-based drought indices are emphasized.

CHAPTER 1

Rainfall microphysics

Greg M. McFarquhar

Cooperative Institute for Mesoscale Meteorological Studies, University of Oklahoma, Norman, OK, United States

1.1 Introduction

Water is crucial for life. Even though Earth can be inhabited only because rain is part of the water cycle, destructive impacts of excessive or inadequate rainfall abound. Rain is essential for agriculture, replenishes the water table that is the main source of drinking water, provides the water source for hydroelectricity, has helped shape Earth's topographical features, and through associated phase changes redistributes heat in the atmosphere. The absence of rain can have devastating impacts through loss of live, livelihood, and other social and socioeconomic impacts. But, the generation of too much rain in a short interval of time can be problematic due to flooding that leads to loss of life and damage of property. Better prediction of rain on short time scales (i.e., nowcasts or forecasts), subseasonal to seasonal time scales, and long-term climatic time scales is critically needed to help society take advantage of, prepare for, and adapt to rain.

Knowledge on what controls the spatial and temporal distribution of precipitation, and its intensity and phase (i.e., solid versus liquid) is critical for generating quantitative precipitation forecasts and for assessing how rainfall distributions will change in a warming and more polluted environment. Although precipitation is only possible when the appropriate synoptic and mesoscale conditions are present, knowledge of small-scale microphysical processes occurring within and below cloud determine when rain will occur, its intensity, phase, and spatial distribution. Even though the temporal and spatial scales of cloud microphysical processes are substantially smaller than the scales of any rain producing weather system, the accompanying release or absorption of latent heat is so large that heat is redistributed both vertically and horizontally in the atmosphere, affecting the evolution of the weather system.

There are two mechanisms by which rain forms: the warm rain process and the cold rain process. Rising motion in the atmosphere initiates both processes. Air ascent can be initiated different ways, including orographic lifting, frontal lifting associated with weather systems, and convection associated with instability in the atmosphere (Corradini et al., 2022). As a parcel rises, it cools because the kinetic energy of the molecules is converted to work to expand the parcel. Since the vapor

RAINFALL. Modeling, Measurement and Applications. DOI: https://doi.org/10.1016/B978-0-12-822544-8.00009-3

mixing ratio of a rising unsaturated air parcel remains constant, the parcel eventually becomes saturated, and subsequently supersaturated.

In the warm rain process, a cloud droplet is said to be nucleated when the supersaturation is large enough to grow a sufficiently large deliquesced aerosol so that the reduction of Gibbs free energy associated with the creation of the higher order liquid phase is greater than the energy barrier associated with the formation of the new water surface. Thereafter the rate of droplet growth is governed by a balance between the heat added by the condensation of water vapor and the heat advected away from the droplet by conduction. Because condensational growth cannot explain the development of rain on the time scales in which rain develops, additional mechanisms must be at work. Small cloud drops collide and coalesce with each other, forming larger size drops, which ultimately attain sufficient terminal velocities to fall out of cloud. During their descent, collisions with other drops continue leading to coalescence and collision-induced breakup. Combined with evaporation, these processes control the distribution of raindrop sizes observed at the ground.

Although the cold rain process is more complex, it was discovered before warm rain. The key difference between the warm and cold rain process is that ice crystals play a role in the development of precipitation in the latter. The cold rain process starts with the nucleation of an ice crystal, which involves the creation of a new higher-order lower-energy ice surface through one of several primary nucleation mechanisms. Thereafter, the crystal grows by vapor deposition, accretion of supercooled water drops, and aggregation with other ice crystals. In some conditions, more ice crystals are produced by secondary mechanisms. When they have sufficient fall speeds, ice crystals or the resultant snowflakes, graupel, or hail particles fall out of the cloud. They then either evaporate, melt to rain, or fall to the ground some other phase.

The remainder of this chapter discusses warm and cold rain process in more detail, citing both historical and more recent studies. It is noted that even though the basic mechanisms of rain formation have been known for over 70 years, there are still significant uncertainties in the understanding of both warm and cold rain. These uncertainties are highlighted.

1.2 The warm rain process

1.2.1 Importance of warm rain

Warm rain is defined by the American Meteorological Society Glossary of Meteorology as rain forming in clouds with temperatures greater than 0°C. The warm rain process refers to the production of rain from droplet coalescence, with growth limited by drop breakup. The key distinction between warm rain and cold rain processes is ice particles have no influence on the precipitation process in warm rain. There can be some ice particles or supercooled drops in cloud provided they are not playing a role in the production of rain. The warm rain process is frequently active in clouds with top temperatures as low as −4°C or −5°C. Textbooks giving fundamental information on the warm rain process include Mason (1971), Rogers and Yau (1989), Young (1993), Pruppacher and Klett (1996), Lamb and Verlinde (2011), and Lohmann et al. (2016). Other review papers include those of Kreidenweis et al. (2020) who overview progress in cloud physics research over the last 100 years, including a description of the warm rain mechanism, Beard and Ochs (1993) who provide an overview of the understanding of microphysical processes acting in warm rain based on studies conducted before 1993, and McFarquhar (2010) who outlines factors that affect the evolution of raindrop size distributions.

The existence of the warm rain process was hypothesized after the cold rain process had been described. Riehl et al. (1951), Byers and Hall (1955), and Battan and Braham (1956) were among the first to observe that rain could be produced in clouds with tops entirely below the freezing level. Prior to these observations, it was felt that the influence of ice was needed to grow precipitation sized drops because condensation on liquid drops alone is not able to describe the development of rain in the approximately 30-minute period in which rain is observed to develop (Saunders, 1965; Rauber et al., 2007). Warm rain is an important component of the hydrological cycle as Nuijens et al., 2017 showed that 10% to 50% of clouds over the oceans are warm clouds, and between 20% and 40% of these warm clouds were shown to produce rain using spaceborne radar data. Warm rain is most important in the tropics where the typical location of the freezing level is 4 to 5 km above ground so that up to 80% of clouds do not penetrate above it (Squires, 1956). These clouds, devoid of ice particles, frequently produce precipitation. Fig. 1.1 summarizes the basic physical mechanisms associated with the warm rain process. Fig. 1.1.

1.2.2 Nucleation of cloud drops

In rising unsaturated air, the saturation ratio and relative humidity increase until the lifting condensation level is reached. Aerosols, suspensions of fine solid or liquid particles, are swept upwards. During the ascent hydrophilic aerosol particles

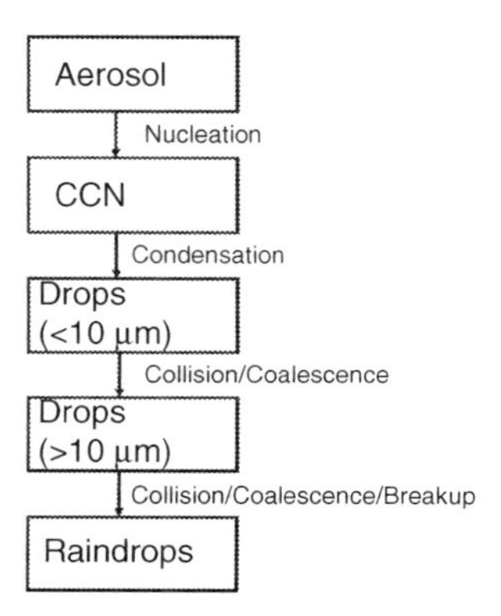

FIG. 1.1 Overview of the warm rain process.

deliquesce, meaning that liquid condenses upon them so that they become dilute solutions of water. At a given humidity, they grow or shrink until there is a balance between the number of microscopic evaporation and condensation events, at which point the vapor pressure over the surface of the dilute solution drop is equal to the ambient vapor pressure. Physically, the vapor pressure over the surface of a dilute solution drop is determined by two terms: the curvature or Kelvin effect that describes the increase in saturation vapor pressure over a curved surface relative to that over a plane surface of water; and the Raoult or soluble substance effect that describes the reduction in vapor pressure over a dilute solution compared to that of pure water. These two effects are combined in the Kelvin-Kohler-Junge equation, which describes the vapor pressure over a dilute drop with radius r (e_r') compared to the saturation vapor pressure over a plane surface of water at the same temperature (e_s),

$$\frac{e_r'}{e_s} = \left[1 + \frac{2\sigma}{\rho_w R_v T r}\right]\left[1 - \frac{i m_0 M_w}{\frac{4}{3}\pi r^3 \rho_w M_0}\right] \tag{1.1}$$

where σ is the surface tension of water, ρ_w the density of water, R_v the gas constant for water vapor, T the temperature, $m_0 = 4/3\pi r_0^3$ the mass of the aerosol with radius r_0, M_w the molecular mass of water, M_0 the molecular mass of the aerosol and i the Van't Hoff disassociation factor, which describes the number of ions an aerosol dissolves into when in water. For simplicity constants a and b are used to represent the Kelvin and Raoult terms, respectively where:

$$a = \frac{2\sigma}{\rho_w R_v T} \tag{1.2}$$

and

$$b = \frac{i m_0 M_w}{\frac{4}{3}\pi \rho_w M_0} \tag{1.3}$$

Thus, for a given saturation ratio $s = e_r'/e_s$ in the atmosphere, the size r to which a solution drop would grow can be computed. Alternatively, for a given r, the s at which the particle would be at equilibrium (i.e., not evaporating or growing) can be computed.

There are important implications of Eq. (1.1). First, without the influence of the Raoult term due to soluble substances, e_r', the vapor pressure over the surface of the liquid drop would be much larger and thus larger ambient humidity would be required for the drop to grow. Given typical conditions in the troposphere, statistical mechanics suggests that 41 vapor molecules could converge as an embryo. But the corresponding drop size of 6.7×10^{-10} m for such an embryo would have a saturation vapor pressure at its surface of $e_r' = 5\ e_s$ if there was no influence from dilute aerosols. This means that a supersaturation of 400% would have to be present to support the continued growth of the drop. As such supersaturations do not occur in the atmosphere, it is clear the formation of water droplets requires the presence of aerosols to reduce e_r' to match a vapor pressure that can be reached in a rising cloud. Taking the derivative of Eq. (1.1) with respect to r shows that there is a critical value of s beyond which further increases of r lead to a reduction in the saturation ratio. This means that once that supersaturation is exceeded in the atmosphere in the rising updraft, the particle is nucleated and will further grow. This is equivalent to saying nucleation has occurred as the new phase with lower thermodynamic energy has formed, with the supersaturation allowing the energy barrier between forced and spontaneous growth to be overcome. A consequence of this argument is that to determine if an aerosol particle will be nucleated into a cloud drop (and hence called a cloud condensation nuclei, CCN) requires not only information about the size and composition of the aerosol, but also the supersaturation (S) of the environment, where $S = s - 1$. Finally, for nucleation to occur, the ambient vapor pressure must exceed e_r'. This means the environment must be supersaturated.

In the natural atmosphere where aerosols with varying sizes and compositions exist, those whose critical supersaturation is lowest nucleate first. Thereafter, as a parcel continues to rise and S increases, more aerosols are nucleated. At some point, S ceases to increase because the decrease of S due to the reduction of water vapor from condensational growth of the nucleated cloud drops equals the increase of S due to the reduction of temperature and corresponding decrease of e_s in the rising air parcel. Because the rate at which the mass of cloud droplets grows by condensation increases with r, S starts to slightly decrease in the rising air parcel at some point. This usually occurs within tens of meters of cloud base, and at this location the nucleation is mainly finished, and the number of cloud drops ceases to increase.

The number of CCN, N_{CCN}, thus varies as a function of supersaturation. The dependence of N_{CCN} on S is typically represented as:

$$N_{CCN} = aS^k \tag{1.4}$$

where a and k are coefficients determined by performing fits of N_{CCN} measured at different S using CCN counters. Because N_{CCN} plays a critical role in determining the microphysical properties of the cloud such as the total cloud drop concentration, extinction, effective radius and ultimately the development of precipitation and radiative impacts of the cloud, N_{CCN} is an important parameter to determine and measure.

Studies measuring CCN prior to the 1993 were reviewed by Hudson (1993). Since that time there have been many more measurements of CCN in a wide variety of environments, including in pristine environments over the Southern Ocean (Hudson et al., 1998; McFarquhar et al., 2021), in pristine and polluted environments over the Indian Ocean (Twohy et al., 2001; Hudson and Yum, 2002; Nair et al., 2020), in biomass burning over the Amazon (Vestin et al., 2007) in more polluted environments such as the Saharan Air Layer (Haarig et al., 2019) and over China (Leng et al., 2014; Zhang et al., 2014), and in many other locations. Other studies have focused on the role of specific particles as CCN. For example, Steiner et al. (2015) looked at the activity of pollen and Bauer et al. (2003) at the role of airborne bacteria as CCN, whereas other studies examined how the mixing of soluble and insoluble particles, and of organics and black carbon affect CCN activation (Dalirian et al., 2015, 2018; Miyazaki et al., 2016).

Studies focusing on measuring sea salt have particular interest because of the potential role of ultragiant nuclei in warm rain initiation. For example, Collins et al. (2013) examined the impact of marine biogeochemical processes on the composition and mixing state of sea salt particles, and Jensen and Nugent (2017) looked at CCN forming from giant sea-salt aerosol particles. Although it was thought at one time that sea spray aerosol constituted a significant fraction of CCN over oceans, Quinn et al. (2017) showed that with the exception of the high southern latitudes sea spray aerosol make up less than 30% of the oceanic CCN. However, the potential role of ultragiant sea salt on warm rain initiation is still uncertain.

Recent studies have shown bimodality in CCN spectra (Hudson et al., 2015) which could be associated with in-cloud processing (Hudson and Noble 2020). Although it is commonly thought that particles in the accumulation mode of aerosols mainly serve as CCN because their larger sizes means nucleation at lower S, Fan et al. (2018) raised the possibility that ultrafine aerosols with diameters smaller than 50 nm could serve as CCN over the Amazon where excessive S could develop in deep convective clouds in low aerosol environments. Further, it is possible these ultrafine aerosols could be nucleated higher above cloud base than those in the accumulation mode. It is clear more details about the size and composition of particles that serve as CCN in different environments are required to better understand the nucleation of drops, and ultimately the development of precipitation. Much work on the role of anthropogenic aerosols as CCN is also required. Such particles may impact projections of climate change not only from their direct effect on radiation, but also their indirect effect on radiation through modification of cloud properties and response of the clouds to these modifications.

1.2.3 Condensational growth of cloud drops

After cloud drops are nucleated, they continue to increase in size due to condensation of water vapor on their surface. The condensation growth rate is determined by a balance between the heat added to the drop from the release of latent heat of vaporization and the heat advected away from the droplet surface due to the gradient between the drop's surface temperature and the ambient environment. Equations developed to describe this growth rate assume the growth is isotropic, and that all drops grow independently from each other. The rate at which the vapor that condenses on the drop is transported towards the drop is determined by Fick's law that states the net diffusion rate of a gas across a surface is equal to the product of the area of the surface, the gradient across the surface and a constant, which is the diffusivity of water vapor in air (D_v) when looking at the diffusion of water vapor. Analogous to Fick's law, Fourier's law of heat conduction determines the heat flux away from the droplet as the product of the gradient in temperature across the surface, the surface area, and the thermal conductivity of air (K). Multiplying the mass growth rate by the latent heat of vaporization (L_v) gives the latent heat, which is equal to the heat of conduction, and thus the mass growth equation for a single drop can be written as

$$4\pi K\left[T_r - T\right] = L_v 4\pi r D_v\left[\rho_v - \rho_{vr}\right] \tag{1.5}$$

where T_r is the temperature at the surface of the drop, ρ_{vr} the vapor pressure at the surface of the drop, T the ambient temperature, and ρ_v the ambient vapor density. Using the Kelvin-Kohler-Junge equation, the ideal gas law, and the Clausius-Clapeyron equation to describe the change in vapor pressure with temperature, this equation is simplified to give the explicit growth equation

$$r\frac{dr}{dt} = \frac{D_v}{\rho_w R_v T}\left[e - e_s(T)exp\left(\frac{L_v \delta}{R_v t}\right)\left(1 + \frac{a}{r} - \frac{b}{r^3}\right)\right] \tag{1.6}$$

where $T_r = T(1 + \delta)$ and $\delta << 1$.

The solution to the explicit growth equation is non-trivial because δ is not known and must be determined using either a numerical or iterative approach. Neiburger and Chien (1960) solved this equation iteratively, giving sample growth rates and temperature differences between the environment and droplet. An easier to use approximate analytic equation giving the growth rate was derived by Mason (1971) as

$$r\frac{dr}{dt} = \frac{S-1}{\left[\left(\frac{L_v}{R_v T} - 1\right)\frac{L_v \rho_w}{KT} + \frac{\rho_w R_v T}{D_v e_s}\right]} \tag{1.7}$$

where terms for the effects of soluble substances and curvature have been removed because they are negligible given the sizes of drops growing by condensation. The first term in the denominator is frequently referred to as F_k, the thermodynamic term associated with heat conduction, and the second term F_d, the term associated with vapor diffusion.

Although Eq. (1.7) can be used to show that the mass growth rate increases as r increases, it also shows the radial growth rate decreases as r increases. Thus, considering only condensation, drop size distributions become narrower as parcels rise above cloud base and drops grow. Although narrow size distributions are seen near cloud base (Fitzgerald, 1972), observations show that distributions typically broaden as parcels rise above cloud base (Warner, 1969). Multiple observations of broad drop size distributions exist in cumulus (Geoffroy et al., 2014). Calculations using Eq. (1.7) show it would take approximately 1 h for a drop to reach r of 60 μm based on condensation alone, yet rain develops from isolated cumuli below the freezing layer in less than 30 min. Thus, it is clear another mechanism acts to develop drizzle and rain in subfreezing clouds.

1.2.4 Growth due to collision and coalescence

Drops of different sizes fall at different speeds and collide with each other. This can be represented by a collection kernel $K(m_R,m_r)$ (Berry, 1967) that describes the rate that a volume is swept out by a large drop of radius R and mass m_R collecting a small drop of radius r with mass m_r as

$$K\left(m_R, m_r\right) = \pi(R+r)^2\left[V_t(R) - V_t(r)\right]E(R,r) \tag{1.8}$$

where V_t is the terminal fall speed of a drop and $E(R,r)$ describes the collection efficiency of the larger drop R collecting the smaller drop r. The $E(R,r)$ is the product of a collision efficiency typically determined by theoretical modeling and a coalescence efficiency. Since the collection efficiency is typically measured in lab experiments (Beard and Ochs, 1984), the coalescence efficiency is determined by dividing the collection efficiency by the collision efficiency. Terminal fall speeds are determined from laboratory observations (Gunn and Kinzer, 1949), field observations (Bringi et al., 2018; Das et al., 2020) or theoretical studies depending on the flow regime. Fall speeds also exhibit a dependence on pressure (Locatelli and Hobbs, 1974).

Continuous and stochastic models can be used to determine the growth rate due to collision and coalescence. In the continuous model, all drops of the same size grow continuously at the same rate assuming that the drops being collected are evenly distributed in the cloud with a specified liquid water content, LWC, which is the number of grams of liquid water per cubic meter. In the stochastic model, drops collide with each other in a statistical manner and different drops grow at different rates. Drops grow quicker and size distributions become broader in the stochastic model, which better represents the way in which drops grow in nature. However, the stochastic model requires more computational time and is difficult to implement because drops may lose their identities in coalescence events.

For the collection kernel (Eq. 1.8) to be large enough to promote the development of rain in the approximately 30-min period in which rain can develop in cumulus entirely beneath the freezing layer (Rauber et al., 2007), the colliding drops must have sufficiently different sizes so that the difference in terminal velocities is large. A vexing problem is then that condensational growth, which describes the initial development of drops, leads to a narrowing of the droplet size distribution that gives small values for the collection kernel. If there are a few large drops, this is sufficient to initiate the warm rain process in a cascading process in a quick time frame (Telford, 1955). Therefore, the pressing question is what causes the initial broadening of the drop size distribution that allows the collision and coalescence process to lead to the development of rain. This has yet to be well solved.

There have been several explanations for this broadening using numerical models (Grabowski and Wang, 2009) and observational studies (Rauber et al., 2007). Proposed explanations include the presence of ultragiant nuclei (Woodcock, 1953; Johnson, 1982; Szumowski et al., 1999; Lasher-Trapp et al., 2002; Blyth et al., 2003), the impact of surfactants or film-forming compounds (Feingold and Chuang, 2002), a radiative effect on particle growth (Zeng, 2018), thermal radiative cooling

(Barekzai and Mayer, 2020), mixing favoring the growth of larger particles (Baker and Latham, 1979; Cooper et al., 1986), turbulence (Pinsky and Khain, 1997, 2002; Xue et al., 2008; Franklin et al., 2014; Chen et al., 2016), inhomogeneous mixing (Baker et al., 1980; Hoffman et al., 2019), and clustering of drops (Bodenschatz et al., 2010; Madival, 2019). It is likely that different explanations or combinations of explanations hold depending on the meteorological, surface and aerosol conditions, and combinations of theoretical, modeling and observational studies are required to further address this.

The rate at which the number distribution function, $n(m)$, of drops of a given mass m are changing with time is given by:

$$\frac{dn(m)}{dt} = \int_0^{m/2} n(\mu)n(m-\mu)C(\mu, m-\mu)d\mu - n(m)\int_0^{\infty} n(\mu)C(m,\mu)d\mu \tag{1.9}$$

where $n(m)$ represents the number of drops per unit volume per bin width in mass coordinate dm so that $N(m, m + dm)$ is the total concentration of drops with masses between m and $m + dm$. Thus, $dn(m)/dt\ dm$ represents the probability that a drop with mass between m and $m + dm$ is produced or lost during interactions of drops with these masses with any other drops.

1.2.5 Collision-induced breakup

Although there are observations of very large-sized raindrops up to 8 mm in diameter (Hobbs and Rangno, 2004) and drops as large as 10 mm are aerodynamically stable and have been observed floating in cloud wind tunnels (Pruppacher and Pitter, 1971), there must be some mechanism that limits the size growth of raindrops. Although Langmuir (1948) and Villermaux and Bossa (2009) postulated that aerodynamic or spontaneous breakup could limit the growth of raindrop size, McFarquhar (2010) reviews a multitude of laboratory studies and evidence (Pruppacher and Pitter, 1971; McTaggart-Cowan and List, 1975; Srivastava, 1978) that shows the collision-induced breakup of raindrops is the principal factor limiting the size growth of raindrops. Although fragment size distributions generated by binary collisions of raindrops have recently been observed in field studies (Testik and Rahman, 2017) and predicted by theoretical models (Schlottke et al., 2010; Straub et al., 2010), most modeling studies examining the evolution of raindrop size distributions due to collision-induced breakup (Brown, 1987; McFarquhar and List, 1991) have used results of laboratory collisions between specific pairs of raindrops (Low and List, 1982a) that were extended in a parameterization to predict results between any colliding pair of raindrops (Low and List, 1982b; McFarquhar 2004).

Raindrop breakup, combined with other factors such as evaporation (Hu and Srivastava, 1995), and size sorting due to varying fall speeds (McFarquhar and List, 1991a) and wind shear (Dawson II et al., 2015) can be incorporated into more detailed models to predict the temporal and spatial evolution of raindrop size distributions. Box models, where particles falling out of a volume element are immediately reinserted at the top, allow a focus on the impact of collision-induced breakup. These models suggest an equilibrium distribution independent of initial conditions (Valdez and Young, 1985; List et al., 1987) is generated regardless of initial conditions provided there are sufficient collisional interactions (McFarquhar and List, 1991b). However, observations show that such an equilibrium distribution rarely occurs (McFarquhar et al., 1996; D'Adderio et al., 2018) except possibly under conditions of very heavy rainfall (Garcia-Garcia and Gonzalez, 2000). Nevertheless, observations of raindrop size distributions provide a framework to interpret microphysical processes that occur in natural rain. Examples of observational studies include those that distinguish between size distributions in convective and stratiform rain (Tokay and Short, 1995) and made by ground probes (Tokay et al., 2008; Chen et al., 2017; Dolan et al., 2018; D'Adderio et al. 2015) and aircraft probes (Willis, 1984; Yuter and Houze, 1997; Testud et al., 2000) in multiple geographic locations. Fits of such observations to analytic functions (Marshall and Palmer, 1948; Willis, 1984; Haddad et al., 1996; Smith 2003; Handwerker and Straub, 2010; McFarquhar et al., 2015) aid in interpretation of measured raindrop size distributions and in their representation in model parameterization schemes. Knowledge of raindrop size distributions is important for interpretation of remote sensing signals, development of parameterizations for numerical models, and for hydrological applications. Future work should concentrate on further understanding how raindrop size distributions vary according to meteorological and aerosol conditions, the detailed processes that affect their evolution, and implications of the assumed form of raindrop size distribution for model and remote sensing parameterization schemes.

1.3 Cold rain process

1.3.1 Overview and Importance

Although the cold rain process is more complicated than the warm rain process, the role of ice crystals in rain formation was postulated before the warm rain process was discovered. The cold rain process, also termed the Wegener-Bergeron-Findeisen process after those who discovered it, is first attributed to Wegener (1911) who theorized that ice crystals would grow at the expense of water droplets from observations of frost. Bergeron (1935) then hypothesized that the growth of ice crystals at the expense of water droplets would lead to rain, with Findeisen (1938) further extending the work. In essence, the key feature is that the rapid growth of ice crystals is possible in water saturated conditions because of the difference in vapor pressures over ice and water, giving a range of particle sizes that can start the collection process. Further, ice crystals grow at the expense of water drops and can even be growing when water drops are evaporating. Fig. 1.2 gives a simplified diagram of different microphysical processes that occur in the cold rain process.

Cold rain is important meteorologically and for understanding distributions of rain. Using satellite remote sensing observations from CloudSat satellite instruments, Field and Heymsfield (2015) showed that about 50% of all global surface precipitation

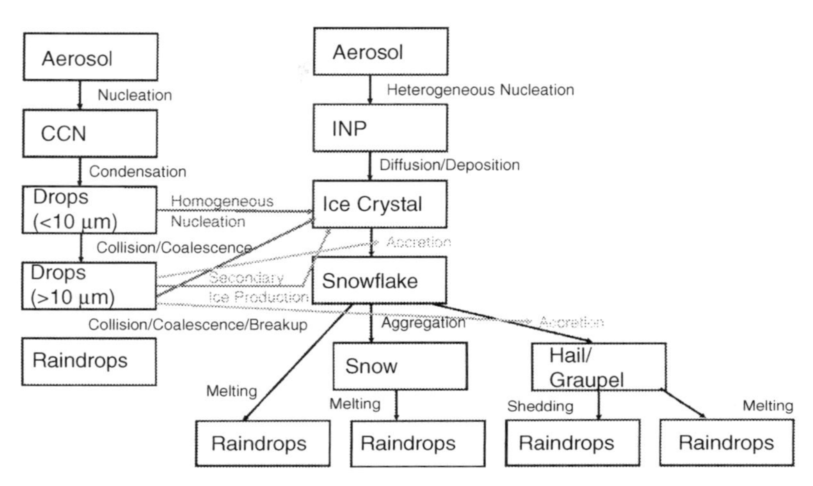

FIG. 1.2 Simplified overview of the cold rain process.

involves the ice-phase, with this percentage increasing to 80% and 90% for the mid-latitude and polar regions respectively. Although many concepts developed for the warm rain process can be extended to understand cold rain, the more complicated and variety of ice crystal and snow shapes, as well as the co-existence of three phases of water, makes its description and understanding more complex. Fig. 1.2.

1.3.2 Nucleation of ice phase

The nucleation of ice involves the creation of a lower-energy higher-order state from either water vapor of liquid. As with the formation of the liquid phase, an energy barrier to the formation of the ice phase must be overcome to create the new ice surface. As with vapor to liquid transitions, there can be no direct transition from vapor to ice without the aid of a foreign object, namely an aerosol. However, ice can be created from liquid water either with or without the aid of an aerosol, and hence both heterogeneous and homogeneous nucleation of ice is possible. Due to ambiguities in the terminology used to describe ice nucleation, Vali et al. (2015) standardized the terminology with the term ice nucleating particle (INP) used to describe the agent responsible for heterogeneous nucleation. Vali et al. (2015) also clarified other terms used to describe both ice-phase and liquid-phase nucleation.

Supercooled water has been observed in the atmosphere at temperatures as low as −38°C (Schaefer, 1962; Heymsfield and Sabin, 1989). Further, laboratory studies have shown that the lowest temperature to which pure water can be supercooled is approximately −39C for 1 μm drops, −36°C for 10 μm drops, and −33°C for 1 mm drops (Jacobi, 1954). Mason (1952) showed theoretically that the rate of homogeneous ice nucleation greatly increased at temperatures around −40°C. Thus, if liquid drops in rising parcels have not frozen due to a heterogeneous mechanism when they reach a level with temperature –40°C, it can be safely assumed that they will freeze homogeneously at that level.

But most clouds have ice at temperatures greater than –40°C, and further, most clouds are totally glaciated for temperatures less than about –20°C. Thus, heterogeneous ice nucleation must be occurring in the atmosphere. Studies have shown that aerosols that have a large size (Hiranuma et al., 2019), are water insoluble (Pruppacher and Klett, 1996), have a lattice structure resembling that of ice (Vonnegut, 1947), and a history of preactivation possess typical characteristics of INPs (Mossop, 1956). Early studies, mainly based on laboratory investigations, suggested that INPs could be mixtures of minerals (Kumai, 1951), organic and inorganic substances (Soulage, 1955) and provided some information about how the INP concentration varied with temperature (Fletcher, 1966). However, INP concentration can also vary considerably with the number of aerosol particles larger than 0.5 μm (DeMott et al., 2010) and humidity (López and Ávilla, 2016). But even when those parameters and temperature are constrained, variations in INP concentration are still noted (Hoose and Möhler, 2012). Thus, there can be large variations in the concentration and composition of INPs with time, space and height in the atmosphere. Traditionally the measurement of INPs has been difficult because their concentrations are much lower than those of aerosols or CCN, especially at higher temperatures near freezing. In the last 15 years, there has been a resurgence of ice nucleation research in both laboratory and field studies. This work is summarized elsewhere (Vali, 1996; Kreidenweis et al., 2020).

In addition to a lack of knowledge on the composition and concentration of INPs, there is uncertainty on the mechanism by which ice nucleates heterogeneously. Vali et al. (2015) and Kanji et al. (2017) summarize different primary heterogeneous nucleation pathways including condensation freezing, immersion freezing, contact freezing, deposition nucleation, and evaporation freezing. Deposition nucleation involves direct nucleation of vapor onto an INP akin to the heterogeneous nucleation of water vapor onto a CCN for liquid particles. Contact freezing occurs when a supercooled droplet collides with an INP; the contact can also take place on the inside of a drop when an INP touches the surface from within. Immersion freezing happens when an INP immersed in an aqueous solution is activated. Although it has not been truly established whether condensation freezing is different from immersion freezing, the conventional definition is that freezing occurs at the same time as liquid forms on a CCN at supercooled temperatures. In general, deposition freezing occurs when there is less humidity (i.e., water subsaturated conditions), and the other mechanisms occur when there are greater amounts of vapor. Immersion freezing typically occurs over longer time scales than the other mechanisms.

1.3.3 Depositional growth

After nucleation, ice crystals grow by deposition of water vapor onto their surface. Although somewhat analogous to the condensational growth of drops, there are important differences. First, the growth rate is typically much quicker because much higher supersaturations with respect to ice than with respect to water occur in the troposphere. Second, the humidity, temperature, and imperfections on the crystal lattice determine which crystal habit preferentially grows under those conditions (Magono and Lee, 1966; Bailey and Hallett, 2004). Thus, the non-spherical nature of ice crystals and their varying shapes precludes the use of a radius r to describe ice crystal size in Eq. (1.5). However, r can be replaced by a capacitance C, where C for a number of idealized ice crystal shapes is given by values used in electrostatics since the equation for current flowing to an object through a medium given a potential difference is analogous to the equation for mass diffusing towards the particle given a difference in vapor density. The equation for heat transfer is similarly written given a temperature difference. Capacitances for other common shapes of ice crystals, such as bullet rosettes (Chiruta and Wang, 2003), have also been derived.

Therefore, Eq. (1.7) describing the mass growth equation for water droplets can be rewritten for ice crystals as:

$$\frac{dM}{dt} = \frac{4\pi C\left(S_i - 1\right)}{\left[\left(\frac{L_s}{R_v T} - 1\right)\frac{L_s}{KT} + \frac{R_v T}{e_i(T) D_v}\right]} \tag{1.10}$$

where M is the mass of the ice crystal, L_s the latent heat of sublimation and e_i the saturation vapor pressure over ice at temperature T. Corrections to Eq. (1.10) have been made by incorporating ventilation effects into the conduction term (first term in the denominator) (Cheng et al., 2014) and by modifying the vapor diffusion term (second term in the denominator) (Srivastava and Coen, 1992; Zhang and Harrington, 2014) to ensure that vapor molecules are added into the crystal structure. Knowledge of the capacitance and the ventilation and diffusion terms are important for representation in numerical models where growth processes are simulated.

1.3.4 Growth by accretion and aggregation

Even though diffusional growth occurs at faster rates than the condensational growth of water drops (e.g., a 100 μm hexagonal plate at −12°C and water saturation would grow at a rate of about 3 μm s^{-1}), other processes in cold clouds further promote particle growth. These include accretion and aggregation. The AMS Glossary of Meteorology defines accretion as the growth of ice particles during collisions with supercooled water droplets and aggregation as the clumping together of snow or ice crystals due to collisions.

An equation that describes accretional growth is similar to one developed for the growth of cloud droplets by collection, so that the mass growth rate of an ice particle is written as:

$$\frac{dM}{dt} = E_{coll}\pi\left(R_i + r\right)^2\left(V_{Ri} - V_r\right)LWC \tag{1.11}$$

where E_{coll} is the collection efficiency of an ice crystal collecting supercooled drops of radius r, V_{Ri} and V_r the fall speeds of the ice crystal and supercooled drop, and R_i the ice crystal dimension related to the cross-sectional area of the falling ice crystal.

Because ice crystals typically fall with their major axis oriented normal to the ground as seen through both *in-situ* observations (Magono, 1953) and satellite polarized radiance measurements (Noel and Chepfer, 2004), R_i is typically represented as the maximum dimension (D_{max}) of an ice crystal divided by 2. Although Eq. (1.11) is conceptually simple, its application is fraught with uncertainties because the collection efficiency, fall speed and even D_{max} of ice crystals are not well known and highly variable given the large numbers of shapes and sizes of ice crystals that exist.

Collection efficiencies have been derived theoretically (Langmuir and Blodgett, 1946; Lew et al., 1985), measured in laboratory studies (Macklin and Bailey, 1968; Makkonen and Stallabrass, 1987) and determined numerically (Wang and Ji, 2000), but are highly uncertain. More studies to better constrain these values through use of laboratory facilities or appropriate instrumentation should be a priority. Databases for crystal fall velocities as a function of size and shape exist (Locatelli and Hobbs, 1974; Mitchell, 1996). But the fact that many ice crystals are irregular (Korolev et al., 1999) and frequently do not match the idealized shapes assumed in libraries of ice crystal properties leads to large uncertainties. Although radars measure Doppler velocity, this is not a direct measurement of fall speed because a combination of air motion and particle fall speed is detected. Additional assumptions need to be made to estimate fall speeds, such as estimating that updrafts and downdrafts balance over sufficiently long distances (Rosenow et al., 2014). Continuing efforts need to concentrate on making direct measurements of fall speed in-situ (Schmidt et al., 2019), in laboratories (Heymsfield and Westbrook 2010), through interpretation of remote sensing data (Protat and Williams, 2011), and theoretical studies (Dunnavan, 2020). Uncertainties in crystal orientation due to vibrations and fluctuations (Jayawera and Mason, 1965) and in how maximum dimension is defined (Wu and McFarquhar, 2016) impact the estimate of D_{max}. Care must be taken to account for these uncertainties in calculations of accretion rate, which in turn affects the representation of accretion in numerical models.

Because ice crystals are non-spherical, there is an extra complication in converting mass growth rates into dimensional growth rates. To do this, the mass m of a non-spherical particle is typically written as a function of D_{max} using a relation of the form $m = aD_{max}{}^{b}$, where a and b are coefficients that are derived by forcing agreement between measures of m and D_{max} of individual particles (Locatelli and Hobbs, 1974) or between bulk mass measured in-situ or retrieved radar reflectivity and equivalent quantities observed by integrating different moments of ice particle size distributions (Brown and Francis, 1995; McFarquhar et al., 2007). Because past studies have derived very different a and b coefficients that vary with meteorological conditions and even within the same meteorological conditions, Finlon et al. (2019) proposed that a surface of equally realizable coefficients in (a, b) phase space could represent the coefficients and their variability. However, work still needs to be performed to determine how these surfaces vary with meteorological conditions, to better constrain uncertainties, and to determine if the instruments used in the derivation of the coefficients affect the results.

The density of the rimed particles that grow by accretion must be known for process studies and model parameterization schemes. When supercooled drops freeze on impact with ice crystals, rimed particles with a porous structure are formed because the drops freeze in place and maintain crystal shape. This is called dry growth. However, when the drops do not freeze on impact a layer of supercooled water forms on the ice particle surface. This is termed wet growth and results in the formation of denser structures like hail. Relative contributions of terms such as latent heating from freezing, sensible heating of drops, cooling by sublimation and evaporation, and heat loss by conduction to the environment to the energy balance determines whether wet or dry growth occurs. In general, the rime density decreases with supercooling, and increases with droplet size and impact velocity (Macklin, 1962; Cober and List, 1993). Knowledge of the density is important as it affects how fast the particles fall, accretional growth rates, and is used in model parameterization schemes (Jin et al., 2019). Future work should concentrate on further refining controls of accretional density.

An aggregation growth equation is written similarly to that for accretion, but instead uses the collection efficiencies of colliding ice crystals and the fall speeds of the two ice crystals. As with accretion, there is considerable uncertainty in the aggregation efficiency. Aggregation efficiencies have been measured in laboratory studies (Hallgreen and Hosler, 1960; Keith and Saunders, 1989) or deduced based on observations (Passarelli, 1978). *In-situ* observational studies can examine aggregation by using Lagrangian descents (Lo and Pasarelli, 1982) of aircraft, where the same population of ice crystals is tracked by descending at typical crystal fall speeds, to assess how the size distributions are changing (Field and Heymsfield, 2003; McFarquhar et al., 2007). In general, the study of aggregation is complicated not only by the fact there are so many primary ice crystal shapes, but also by the fact there is also wide variability in aggregate shape (Hobbs et al., 1974). Aggregation can occur due to different mechanisms, such as contact freezing when there is a liquid layer on the ice particle surface (Hosler et al., 1957), cementing of crystals by rime (Rauber, 1987), and mechanical interlocking of complex shapes (Um and McFarquhar, 2007). In general, aggregation also depends on temperature, with the maximum size of aggregates occurring near 0°C (Magono, 1953) due to a marked increase in aggregation efficiency near −3°C (Pruppacher and Klett, 1996). Ongoing work is needed to further observe aggregates to better quantify the size, shape, density and fall speed of aggregates, to understand controls on their formation and evolution in context of the meteorological systems in which they occur, to use parcel models evaluated against observations to better quantify the aggregation process (Grim et al., 2009), and to develop new and improved aggregation schemes for use in models (Sulia et al., 2020).

1.3.5 Secondary ice crystal production

A significant issue in understanding and representing the cold rain process is caused by the fact that in some conditions the number of ice crystals is substantially greater than the number of INPs (Mossop et al., 1974; Hobbs and Rangno, 1985; Mossop, 1985; DeMott et al., 2016). The concept of secondary ice production has been proposed to explain this discrepancy. This concept establishes that although ice crystals initially form from INP, the majority of crystals form by secondary processes in which primary crystals are multiplied. Field et al. (2017) reviews the current understanding of secondary ice crystal production and remaining issues to be resolved.

To understand mechanisms promoting the secondary production of ice, conditions under which the concentration of ice crystals approximately matches the concentration of INPs must first be established. For example, Mossop (1985) showed that the number of ice crystals approximately equaled the number of INP when there were no supercooled drops with $r > 25$ μm and temperatures were greater than −10°C. This suggests large supercooled drops at temperatures just less than freezing have a role in secondary ice crystal production. Based on similar observations Hallett and Mossop (1974) proposed a potential mechanism to explain secondary production, namely the rime splintering process now often referred to as the Hallett-Mossop process. The rime splintering mechanism itself was originally proposed by Findeisen and Findeisen (1943). Here many splinters are produced when supercooled drops collide with ice, which subsequently serve as ice nuclei. Other secondary production mechanisms proposed include collision fragmentation (Vardiman, 1978; Fridlind et al., 2007), droplet shattering (Leisner et al., 2014) and sublimation fragmentation (Bacon et al., 1998). Recent work has noted the presence of spicules, or bubbles bursting from the surface of frozen droplets, in both laboratory (Lauber et al., 2018) and aircraft observations (Lawson et al., 2015). It is likely that a single mechanism cannot explain observed ice crystal concentrations in all environmental conditions, so combinations of mechanisms acting, even in the same cloud, might be needed to explain the observations (Sotiropoulou et al., 2020).

Much activity is ongoing to explain secondary ice crystal production using laboratory, field observational, and numerical modeling studies. A useful approach has been to implement multiple mechanisms and microphysical parameterization schemes in models, comparing observed and modeled fields with and without specific mechanisms acting to identify their importance (Qu et al., 2018). One complication in such studies is the quality of the observations, and in particular those of the total concentration of ice crystals which is dominated by small particles. Small crystals are difficult to measure in-situ because common aircraft probes have small and size-dependent depths of field for particles with $D_{max} < 150$ μm (Baumgardner and Korolev, 1997) and because the shattering of large crystals on probe tips can artificially amplify small ice crystal concentration (Korolev et al., 2011; Jackson et al., 2014). Due to the development of shatter-mitigating probe tips (Korolev et al., 2013) and the development of algorithms to remove shattered artifacts in the processing of probe data (Field et al., 2003, 2006), better quality observations are now available. It is anticipated that this will lead to a better understanding of secondary ice production processes.

1.3.6 Formation of rain, melting and ground precipitation type

The final step understanding the cold rain process is determining the evolution of particles as they fall from cloud, melt, and reach the ground. The vertical profile of temperature and humidity is important in determining how much evaporation occurs below cloud base and hence whether precipitation reaches the ground. It is also possible that the shedding of liquid from the surface of hail can provide more source of drops for accretion, or for collision, coalescence, and breakup of rain. The temperature profile and presence of inversions determines precipitation type (rain, snow, sleet, freezing rain), which is important for forecasting, the issuance of watches and warnings that can save lives and property, and reduction of economic damage from storms. Studies of melting snow first occurred over 100 years ago (Horton, 1915) and combinations of in-situ observations (Stewart et al., 1984; Fujiyoshi, 1986; Heymsfield et al., 2015) and remote sensing data (Battan and Bohren, 1982; Vivekanandan et al., 1990) have been used to continue these studies. In addition to understanding the amount of phase and precipitation at the surface, the process of melting affects the evolution of cloud systems as the associated cooling can generate cold pools (Biggerstaff and Houze, 1991), downbursts (Fujita, 1985; Wakimoto, 1985), frontogenesis (Szeto and Stewart, 1997), and other meteorological phenomena. Thus, it is very important to understand the small-scale characteristics of the precipitation so that the implications of melting can be better understood and modeled.

1.4 Conclusion

This chapter provided an overview of microphysical processes that lead to rain formation, focusing on the warm and cold rain processes. Although much has been learned in the past 100 years about the microphysical processes that govern precipitation formation, there are still significant uncertainties and unknowns that must be resolved to better understand how microphysical processes affect the evolution of meteorological systems that produce rainfall (Corradini et al., 2022). Further, knowledge about these processes is important for the development of parameterization schemes for numerical models that are used to model rainfall and their impacts (Grabowski, 2022). Information about the microphysical properties of clouds and the rain itself are also needed to interpret signals received from radars (Borga et al., 2022) and to evaluate retrievals of rainfall from satellites (Kidd and Levizzani, 2022). Knowledge of the rainfall size distributions arriving at the ground, which are controlled by microphysical processes, are also important for hydrological applications (Saltalippi et al., 2022; Govindaraju and Goyal, 2022), erosion (Giraldez et al., 2022), and even the development of landslides (Guzzetti et al., 2022) due to the importance of rainfall intensity. Thus, an improved understanding of the microphysical processes producing rain are also important for rainfall applications.

Acknowledgments

This work was supported by the Cooperative Institute for Mesoscale Meteorological Studies at the University of Oklahoma.

References

Bacon, N.J., Swanson, B.D., Baker, M.D., et al., 1998. Breakup of levitated frost particles. J. Geophys. Res. 103, 763–775.

Bailey, M., Hallett, J., 2004. Growth rates and habits of ice crystals between −20° and −70°C. J. Atmos. Sci. 61, 514–544.

Baker, M.B., Latham, J., 1979. The evolution of droplet spectra and rate of production of embryonic raindrops in small cumulus clouds. J. Atmos. Sci. 36, 1612–1615.

Baker, M., Corbin, R., Latham, J., 1980. The influence of entrainment on the evolution of cloud droplet spectra: I. a model of inhomogeneous mixing. Quart. J. Roy. Meteor. Soc. 106, 581–598.

Barekzai, M., Mayer, B., 2020. Broadening of the cloud droplet size distribution due to thermal radiative cooling: turbulent parcel simulations. J. Atmos. Sci. 77, 1993–2010.

Battan, L.J., Bohren, C.F., 1982. Radar backscattering by melting snowflakes. J. Appl. Meteor. Climatol. 21, 1937–1938.

Battan, L.J., Braham, R.R., 1956. A study of convective precipitation based on cloud and radar observations. J. Meteorol. 13, 587–591.

Bauer, H., Giebl, H., Hitzenberger, R., et al., 2003. Airborne bacteria as cloud condensation nuclei. J. Geophys. Res. 108. doi:10.1029/2003JD003545.

Baumgardner, D., Korolev, A., 1997. Airspeed corrections for optical array probe sample volumes. J. Atmos. Oceanic Technol. 14, 1224–1229.

Beard, K.V., III Ochs, H.T., 1984. Collection and coalescence efficiencies for accretion. J. Geophys. Res. 89, 7165–7169.

Beard, K.V., III Ochs, H.T., 1993. Warm-rain initiation: an overview of microphysical mechanisms. J. Appl. Meteor. 32, 608–625.

Bergeron, T., 1935. On the physics of cloud and precipitation, Proc. 5th Assembly UGGI. Lisbon, 2, 156.

Berry, E.X., 1967. Cloud droplet growth by collection. J. Atmos. Sci. 24, 688–701.

Biggerstaff, M.I., Houze Jr, R.A., 1991. Kinematic and precipitation structure of the 10-11 June 1985 squall line. Mon. Wea. Rev. 119, 3034–3065.

Blyth, A., Lasher-Trapp, S.G., Cooper, W.A., Knight, C.A., Latham, J., 2003. The role of giant and ultragiant nuclei in the formation of early radar echoes in warm c clouds. J. Atmos. Sci. 60, 2557–2572.

Bodenschatz, E., Malinowski, S.P., Shaw, R.A., et al., 2010. Can we understand clouds without turbulence? Science 327, 970–971.

Borga, M., Marra, F., Gabella, M., 2022. Rainfall Estimation by Weather Radar. In: Morbidelli, R. (Ed.), Rainfall. Modeling, Measurement and Applications. Elsevier, Amsterdam, pp. 109–134. doi:10.1016/C2019-0-04937-0.

Bringi, V., Thurai, M., Baumgardner, D., 2018. 2018: raindrop fall velocities from an optical array probe and 2-D video disdrometer. Atmos. Meas. Tech 11, 1377–1384.

Brown, P.R.A., Francis, P.N., 1995. Improved measurements of the ice water content in cirrus using a total-water probe. J. Atmos. Oceanic. Technol. 12, 410–414.

Brown, P.S., 1987. Parameterization of drop-spectrum evolution due to coalescence and breakup. J. Atmos. Sci. 44, 242–249.

Byers, H.R., Hall, R.K., 1955. A census of cumulus-cloud height versus precipitation in the vicinity of Puerto Rico during the winter and spring of 1953-1954. J. Atmos. Sci. 12, 176–178.

Chen, B., Hu, Z., Liu, L., Zhang, G., 2017. Raindrop size distribution measurements at 4,500 m on the Tibetan plateau during TIPEX-III. J. Geophys. Res. 122, 11092–11106.

Chen, S., Bartello, P., Yau, M.K., et al., 2016. Cloud droplet collisions in turbulent environment: collision statistics and parameterization. J. Atmos. Sci. 73, 621–636.

Cheng, K.Y., Wang, P.K., Wang, C.K., 2014. A numerical study on the ventilation coefficients of falling hailstones. J. Atmos. Sci. 71, 2625–2634.

Chiruta, M., Wang, P.K., 2003. The capacitance of rosette ice crystals. J. Atmos. Sci. 60, 836–846.

Cober, S.G., List, R., 1993. Measurements of the heat and mass transfer parameters characterizing conical graupel growth. J. Atmos. Sci. 50, 1591–1609.
Collins, D.B., Ault, A.P., Moffet, R.C., et al., 2013. Impact of marine biogeochemistry on the chemical mixing state and cloud forming ability of nascent sea spray aerosol. J. Geophys. Res. 118, 8553–8565.
Cooper, W.A., Baumgardner, D., Dye, J.E., 1986. Evolution of the Droplet Spectra in Hawaiian Orographic Clouds, 9th Conf. on Cloud Physics, 52–55, Amer. Meteor. Soc. Snowmass, CO.
Corradini, C., Morbidelli, R., Saltalippi, C., Flammini, A., 2022. Meteorological Systems Producing Rainfall. In: Morbidelli, R. (Ed.), Rainfall. Modeling, Measurement and Applications. Elsevier, Amsterdam, pp. 27–48. doi:10.1016/C2019-0-04937-0.
D'Adderio, L.P., Porcù, F., Tokay, A., 2015. Identification and analysis of collisional breakup in natural rain. J. Atmos. Sci. 72, 3404–3416.
D'Adderio, L.P., Porcu, F., Tokay, A., 2018. Evolution of drop size distribution in natural rain. Atmos. Res. 200, 70–76.
Dalirian, M., Keskinen, H., Ahlm, L., Ylisirnio, A., Romakkaniemi, S., Laaksonen, A., Virtanen, A., Riipinen, I., 2015. CCN activation of fumed silica aerosols mixed with soluble pollutants. Atmos. Chem. Phys. 15, 3815–3829.
Dalirian, M., Ylisirniö, A., Buchholz, A., Schlesinger, D., Ström, J., Virtanen, A., Riipinen, I., 2018. Cloud droplet activation of black carbon particles coated with organic compounds of varying solubility. Atmos. Chem. Phys. 18, 12477–12489.
Das, S.K., Simon, S., Kolte, S.Y.K., et al., 2020. Investigation of raindrops fall velocity during different monsoon seasons over the Western Ghats. India. Earth Space Sci.
Dawson II, D.T., E.R., Mansell, M.R., Kumjian, 2015. Does wind shear cause hydrometeor size sorting? J. Atmos. Sci. 72, 340–348.
DeMott, P.J., Prenni, A.J., Liu, X., et al., 2010. Predicting global atmospheric ice nuclei distributions and their impacts on climateProc. Natl. Acad. Sci. U. S. A.107, 11217–11222.
DeMott, P.J., Thomas, C.J., Hill, C.S., et al., 2016. Sea spray aerosol as a unique source of ice nucleating particles, Proc. Natl. Acad. Sci. U. S. A.113, 5797–5803.
Dolan, B., Fuchs, B., Rutledge, S.M., Barnes, E.A., 2018. Primary modes of global drop size distributions. J. Atmos. Sci. 75, 1453–1476.
Dunnavan, E.L., 2020. How snow aggregate ellipsoid shape and orientation variability affects fall speed and self-aggregation rates. J. Atmos. Sci. 78, 51–73.
Fan, J., Rosenfeld, D., Zhang, Y., et al., 2018. Substantial convection and precipitation enhancement by ultrafine aerosol particles. Science 359, 411–418.
Feingold, G., Chuang, P.Y., 2002. Analysis of the influence of film-forming compounds on droplet growth: Implications for cloud microphysical processes and climate. J. Atmos. Sci. 59, 2006–2018.
Field, P.R., Heymsfield, A.J., 2003. Aggregation and scaling of ice crystal size distributions. J. Atmos. Sci. 60, 544–560.
Field, P.R., Heymsfield, A.J., 2015. Importance of snow to global precipitation. Geophys. Res. Lett. 42, 9512–9520.
Field, P.R., Wood, R., Brown, P.R.A., et al., 2003. Ice particle interarrival times measured with a fast FSSP. J. Atmos. Oceanic. Technol. 20, 249–261.
Field, P.R., Heymsfield, A.J., Bansemer, A., 2006. Shattering and particle interarrival times measured by optical array probes in ice clouds. J. Atmos. Oceanic Technol. 23, 1357–1371.
Field, P.R., Lawson, R.P., Brown, P.R.A., et al., 2017. Secondary ice production: current state of the science and recommendations for the future. Amer. Meteor. Soc. Monographs 58, 7.1–7.20.

Findeisen, W., 1938. Die kolloidmeteorologischen Vorgänge bei der Niederschlagsbildung (colloidal meteorological processes in the formation of precipitation). Met. Z. 55, 1218.
Findeisen, W., Findeisen, E., 1943. Untersuchungen uber die eissplitterbildung an reifschichten (ein beitrag zur frage der entstehung der gewitterelektrizitat und zur mikrostruktur der cumulonimben). Met. Z. 60, 145–154.
Finlon, J.A., McFarquhar, G.M., Nesbitt, S.W., et al., 2019. A novel approach for characterizing the variability in mass-dimension relationships: results from MC3E. Atmos. Chem. Phys. 19, 3621–3643.
Fitzgerald, J.W., 1972. A study of the initial phase of cloud droplet growth by condensation: comparison between theory and observation. Department of Geophysical Sciences, University of Chicago, Chicago, p. 144.
Fletcher, N.H., 1966. The Physics of Rainclouds. Cambridge, England, p. 390.
Franklin, C.N., 2014. The effects of turbulent collision-coalescence on precipitation formation and precipitation-dynamical feedbacks in simulations of stratocumulus and shallow cumulus convection. Atmos. Chem. Phys. 14, 6557–6570.
Fridlind, A.M., Ackerman, A.S., McFarquhar, G., et al., 2007. Ice properties of single-layer stratocumulus during the mixed-phase arctic cloud experiment: 2. Model results. J. Geophys. Res. 112, D24202.
Fujita, T.T., 1985. The downburst, microburst and microburst. SMRP Research Paper 210, 122.
Fujiyoshi, Y., 1986. Melting snowflakes. J. Atmos. Sci. 43, 307–311.
Garcia-Garcia, F., Gonzalez, J.E., 2000. Raindrop spectra observations form convective showers in the valley of Mexico, Proceedings of the 13th International Conference on Clouds and Precipitation. Reno, NV.
Geoffroy, O., Siebesma, A.P., Burnet, F., 2014. Characteristics of the raindrop distributions in RICO shallow cumulus. Atmos. Chem. Phys. 14, 10897–10909.
Giraldez, J.V., Castro-Orgaz, O., Gómez, J.A., Laguna, A.M., 2022. Rainfall and Erosion/Sediment Transport. In: Morbidelli, R. (Ed.), Rainfall. Modeling, Measurement and Applications. Elsevier, Amsterdam, pp. 397–426. doi:10.1016/C2019-0-04937-0.
Govindaraju, R.S., Goyal, A., 2022. Rainfall and Infiltration. In: Morbidelli, R. (Ed.), Rainfall. Modeling, Measurement and Applications. Elsevier, Amsterdam, pp. 367–396. doi:10.1016/C2019-0-04937-0.
Grabowski, W., 2022. Rainfall Modeling. In: Morbidelli, R. (Ed.), Rainfall. Modeling, Measurement and Applications. Elsevier, Amsterdam, pp. 49–76. doi:10.1016/C2019-0-04937-0.
Grabowski, W.W., Wang, L.P., 2009. Diffusional and accretional growth of water drops in a rising adiabatic parcel: effects of the turbulent collision kernel. Atmos. Chem. Phys. 9, 2335–2353.
Grim, J., McFarquhar, G.M., Rauber, R.M., et al., 2009. Microphysical and thermodynamic structure and evolution of the trailing stratiform regions of mesoscale convective systems during BAMEX. Part II: column model simulations. Mon. Wea. Rev. 137, 1186–1205.
Gunn, R, Kinzer, G.D., 1949. The terminal velocity of fall for water drops in stagnant air. J. Meteor. 6, 243–248.
Guzzetti, F., Gariano, S.L., Peruccacci, S., Brunetti, M.T., Melillo, M., 2022. Rainfall and Landslide Initiation. In: Morbidelli, R. (Ed.), Rainfall. Modeling, Measurement and Applications. Elsevier, Amsterdam, pp. 427–450. doi:10.1016/C2019-0-04937-0.
Haarig, M., Walser, A., Ansmann, A., et al., 2019. Profiles of cloud condensation nuclei, dust mass concentration, and ice-nucleating-particle-relevant aerosol properties in the Saharan Air Layer over Barbados from polarization lidar and in situ measurements. Atmos. Chem. Phys. 19, 13773–13788. doi:10.5194/acp-2019-466.

Haddad, Z.S., Durden, S.L., Im, E., 1996. Parameterizing the raindrop size distribution. J. Appl. Meteor. 35, 3–13.

Hallett, J., Mossop, S.C., 1974. Production of secondary ice particles during the riming process. Nature 249, 26–28.

Hallgreen, R.E., Hosler, C.L., 1960. Ice crystal aggregation. Geophys. Monogr. 5, 251–260.

Handwerker, J., Straub, W., 2010. Optical determination of parameters for gamma-type drop size distributions based on moments. J. Atmos. Ocean. Tech. 28, 513–529.

Heymsfield, A.J., Sabin, R.M., 1989. Cirrus crystal nucleation by homogeneous freezing of solution droplets. J. Atmos. Sci. 46, 2252–2264.

Heymsfield, A.J., Westbrook, C.D., 2010. Advances in the estimation of ice particle fall speeds using laboratory and field measurements. J. Atmos. Sci. 67, 2469–2482.

Heymsfield, A.J., Bansemer, A., Poellot, M.R., et al., 2015. Observations of ice microphysics through the melting layer. J. Atmos. Sci. 72, 2902–2928.

Hiranuma, N., Adachi, K., Bell, D.M., et al., 2019. A comprehensive characterization of ice nucleation by three different types of cellulose particles immersed in water. Atmos. Chem. Phys. 19, 4823–4849.

Hobbs, P.V., Rangno, A.L., 1985. Ice particle concentrations in clouds. J. Atmos. Sci. 42, 2523–2549.

Hobbs, P.V., Rangno, A.L., 2004. Super-large raindrops. Geophys. Res. Lett. 31, L13102.

Hobbs, P.V., Chang, S., Locatelli, J.D., 1974. The dimensions and aggregation of ice crystals in natural clouds. J. Geophys. Res. 79, 2199–2206.

Hoffman, F., Yamaguchi, T., Feingold, G., 2019. Inhomogeneous mixing in Lagrangian cloud models: effects on the production of precipitation embryos. J. Atmos. Sci 76, 113–133.

Hoose, C., Möhler, O., 2012. Heterogeneous ice nucleation on atmospheric aerosols: a review of results from laboratory experiments. Atmos. Chem. Phys. 12, 9817–9854.

Horton, R.E., 1915. The melting of snow. Mon. Wea. Rev. 43, 599–605.

Hosler, C.L., Jensen, D.C., Goldshlak, L., 1957. On the aggregation of ice crystals to form snow. J. Atmos. Sci. 14, 415–420.

Hu, Z., Srivastava, R.C., 1995. Evaporation of raindrop size distribution by coalescence, breakup and evaporation. J. Atmos. Sci. 52, 1761–1783.

Hudson, J.G., 1993. Cloud condensation nuclei. J. Appl. Meteor. Climatol. 32, 596–607.

Hudson, J.G., Noble, S., Tabor, S., 2015. Cloud supersaturations from CCN spectra Hoppel minima. J. Geophys. Res. 120, 3436–3452.

Hudson, J.G., Noble, S., 2020. CCN spectral shape and cumulus cloud and drizzle microphysics. J. Geophys. Res. 125, e2019JD031141.

Hudson, J.G., Yum, S.S, 2002. Cloud condensation nuclei spectra and polluted and clean clouds over the Indian Ocean. J. Geophys. Res. 107, 8022.

Hudson, J.G., Xie, Y., Yum, S.S., 1998. Vertical distributions of cloud condensation nuclei spectra over the summertime Southern Ocean. J. Geophys. Res. 103, 16609–16624.

Jackson, R.C., McFarquhar, G.M., Stith, J., et al., 2014. An assessment of the impact of anti-shattering tips and artifact removal techniques on cloud ice size distributions measured by the 2D cloud probe. J. Atmos. Oceanic. Technol. 31, 2567–2590.

Jacobi, W., 1954. Homogeneous nucleation in supercooled water. J. Meteor. 12, 408–409.

Jayawera, K.O.L., Mason, B.J., 1965. The behaviour of freely falling cylinders and cones in a viscous fluid. J. Fluid. Mech. 22, 709–720.

Jensen, J.B., Nugent, A.D., 2017. Condensational growth of drops formed on giant sea-salt aerosol particles. J. Atmos. Sci. 74, 679–697.

Ji, W., Wang, P.K., 1999. Ventilation coefficients for falling ice crystals in the atmosphere at low-intermediate Reynolds numbers. J. Atmos. Sci. 56, 829–836.

Jin, H.G., Lee, H., Baik, J.J., 2019. A new parameterization of the accretion of cloud water by graupel and its evaluation through cloud and precipitation simulations. J. Atmos. Sci. 76, 381–400.
Johnson, D.B., 1982. The role of giant and ultragiant aerosol particles in warm rain initiation. J. Atmos. Sci. 38, 448–460.
Kanji, Z.A., Ladino, L.A., Wex, H., et al., 2017. Overview of ice nucleating particles. Meteor. Mono. 58, 1.1–1.33.
Keith, W.D., Saunders, C.P.R., 1989. The collection efficiency of a cylindrical target for ice crystals. Atmos. Res. 23, 83–95.
Kidd, C., Levizzani, V., 2022. Satellite Rainfall Estimation. In: Morbidelli, R. (Ed.), Rainfall. Modeling, Measurement and Applications. Elsevier, Amsterdam, pp. 135–170. doi:10.1016/C2019-0-04937-0.
Korolev, A.V., Isaac, G.A., Hallett, J., 1999. Ice particle habits in Arctic clouds. Geophys. Res. Lett. 26, 1299–1302.
Korolev, A.V., Emery, E.F., Strapp, J.W., et al., 2011. Small ice particles in tropospheric clouds: fact or artifact? Bull. Amer. Meteor. Soc. 92, 967–973.
Korolev, A.V., Emery, E., Creelman, K., 2013a. Modification and tests of particle probe tips to mitigate effects of ice shattering. J. Atmos. Oceanic. Technol. 30, 690–708.
Kreidenweiss, S., Petters, M., Lohmann, U., 2020. 100 years of progress in cloud physics, aerosols, and aerosol chemistry research. Amer. Meteor. Soc. Monographs 59, 11.1–11.72.
Kumai, M., 1951. Electron-microscope study of snow-crystal nuclei. J. Atmos. Sci. 8, 151–156.
Lamb, D., Verlinde, J., 2011. Physics and Chemistry of Clouds. Cambridge University Press, p. 600.
Langmuir, I., Blodgett, K.B., 1946. A mathematical investigation of water droplet trajectories, Collected Works of Irving Langmuir, Pergamon Press, New York, 10, 348–393.
Langmuir, I., 1948. The production of rain by a chain reaction in cumulus clouds at temperatures above freezing. J. Atmos. Sci. 5, 175–192.
Lasher-Trapp, S.G., Cooper, W.A., Blyth, A.M., 2002. Measurements of ultragiant aerosol particles in the atmosphere from the small cumulus microphysics study. J. Atmos. Ocean. Tech. 19, 402–408.
Lauber, A., Kiselev, A., Pander, T., et al., 2018. Secondary ice formation during freezing of levitated droplets. J. Atmos. Sci. 75, 2815–2826.
Lawson, R.P., Woods, S., Morrison, H., 2015. The microphysics of ice and precipitation development in tropical cumulus clouds. J. Atmos. Sci. 72, 2429–2445.
Leisner, T., Pander, T., Handmann, P., et al., 2014. Secondary ice processes upon heterogeneous freezing of cloud droplets, 14th Conference on Cloud Physics and Atmospheric Radiation, 2.3. Amer Meteor Soc, Boston, MA.
Leng, C., Zhang, Q., Zhang, D., et al., 2014. Variations of cloud condensation nuclei (CCN) and aerosol activity during fog-haze episode: a case study from Shanghai. Atmos. Chem. Phys. 14, 12499–12512.
Lew, J.K., Kingsmill, D.E., Montague, D.C., 1985. A theoretical study of the collision efficiency of small planar ice crystals colliding with large supercooled water drops. J. Atmos. Sci. 42, 857–862.
List, R., Donaldson, N.R., Stewart, R.E., 1987. Temporal evolution of drop spectra to collisional equilibrium in steady and pulsating rain. J. Atmos. Sci. 44, 362–371.
Lo, K.K., Passarelli Jr, R.E., 1982. Growth of snow in winter storms: an airborne observational study. J. Atmos. Sci. 39, 697–706.

Locatelli, J.D., Hobbs, P.V., 1974. Fall speeds and masses of solid precipitation particles. J. Geophys. Res 79, 2185–2197.

Lohmann, U., Lüöand, F., Mahrt, F., 2016. An Introduction to Clouds: From the Microscale to Climate. Cambridge University Press, p. 391.

López, M.L., Ávilla, E.E., 2016. Influence of ambient humidity on the concentration of natural deposition-mode ice-nucleating particles. Atmos. Chem. Phys. 16, 927–932.

Low, T.B., List, R., 1982a. Collision, coalescence and breakup of raindrops. Part I: experimentally established coalescence efficiencies and fragment size distributions in breakup. J. Atmos. Sci. 39, 1591–1606.

Low, T.B., List, R., 1982b. Collision, coalescence and breakup of raindrops. Part II: parameterization of fragment size distributions. J. Atmos. Sci. 39, 1607–1618.

Macklin, W.C., 1962. The density and structure of ice formed by accretion. Quart. J. Roy. Meteor. Soc. 88, 30–50.

Macklin, W.C., Bailey, I.H., 1968. The collection efficiencies of hailstones. Quart. J. Roy. Meteor. Soc. 94, 393–396.

Madival, D.G., 2019. A model study of the effect of clustering on the last stage of drizzle formation. Quart. J. Roy. Meteor. Soc. 145, 3790–3800.

Magono, C., 1953. On the growth of snowflakes and graupel. Sci. Rep. Yokohama Natl. Univ., Ser. 1 2, 18–40.

Magono, C., Lee, C.W., 1996. Meteorological classification of natural snow crystals. J. Fac. Sci. Hoddaido Univ. Ser. 2, 321–335.

Makkonen, L., Stallabrass, J.R., 1987. Experiments on the cloud droplet collision efficiency of cylinders. J. Climate Appl. Meteor. 26, 1406–1411.

Marshall, J.S., Palmer, W.M., 1948. The distribution of raindrops with size. J. Meteor. 5, 165–166.

Mason, B.J., 1952. The spontaneous crystallization of supercooled water. Quart. J. Roy. Meteor. Soc. 78, 52–57.

Mason, B.J., 1971. The Physics of Clouds. Oxford, p. 671.

McFarquhar, G.M., 2004. A new representation of collision-induced breakup of raindrops and its implication for the shape of raindrop size distributions. J. Atmos. Sci. 61, 777–794.

McFarquhar, G.M., 2010. Raindrop size distribution and evolution. Rainfall: state of the science. Geophysical Monograph Series 191. American Geophysical Union. doi:10.1029/2010GM000971.

McFarquhar, G.M., List, R., 1991a. The evolution of three-peak raindrop size distributions in one-dimensional shaft models. Part II: multiple pulse rain. J. Atmos. Sci. 48, 1587–1595.

McFarquhar, G.M., List, R., 1991b. The raindrop mean free path and collision rate dependence on rainrate for three-peak equilibrium and Marshall-Palmer distributions. J. Atmos. Sci. 48, 1999–2004.

McFarquhar, G.M., List, R., Hudak, D.R., et al., 1996. Flux measurements of pulsating rain packages with disdrometers and Doppler radar made during phase ii of the joint tropical rain experiment in Penang, Malaysia. J. Appl. Meteor. 35, 859–874.

McFarquhar, G.M., Timlin, M.S., Rauber, R.M., et al., 2007. Vertical variability of cloud hydrometeors in the stratiform region of mesoscale convective systems and bow echoes. Mon. Wea. Rev. 135, 3405–3428.

McFarquhar, G.M., Hsieh, T.L., Freer, M., et al., 2015. The characterization of ice hydrometeor gamma size distributions as volumes in $N_0/\lambda/\mu$ phase space: implications for microphysical process modeling. J. Atmos. Sci. 72, 892–909.

McFarquhar, G.M., Bretherton, C., Marchand, R., et al., 2021. Unique observations of clouds, aerosols and precipitation over the Southern Ocean: an overview of Capricorn, Marcus, Micre and Socrates. Bull. Amer. Meteor. Soc. doi:10.1175/BAMS-D-20-0132.1.

McTaggart-Cowan, J.D., List, R., 1975. An acceleration system for water drops. J. Atmos. Sci. 32, 1395–1400.

Mitchell, D.L., 1996. Use of mass- and area-dimensional power laws for determining precipitation particle terminal velocities. J. Atmos. Sci. 53, 1710–1723.

Miyazaki, Y., Coburn, S., Ono, K., et al., 2016. Contribution of dissolved organic matter to submicron water-soluble organic aerosols in the marine boundary layer over the eastern equatorial Pacific. Atmos. Chem. Phys. 16, 7695–7707.

Mossop, S.C., 1956. Sublimation nuclei. P. Phys. Soc. Lond. B. 69, 161–164.

Mossop, S.C., Brownscombe, J.J., Collins, G.J., 1974. The production of secondary ice particles during riming. Quart. J. Roy. Meteor. Soc. 100, 427–436.

Mossop, S.C., 1985. Secondary ice particle production during rime growth. The effect of drop size distribution and rimer velocity. Quart. J. Roy. Meteor. Soc. 111, 1113–1124.

Nair, V.S., Jayachandran, V.N., Kompalli, S.K., et al., 2020. Cloud condensation nuclei properties of South Asian outflow over the northern Indian Ocean during winter. Atmos. Chem. Phys. 20, 3135–3149.

Neiburger, M., Chien, C.W., 1960. Computations of the growth of cloud drops by condensation using an electronic digital computer. Meteor. Mono 5, 191–210.

Noel, V., Chepfer, H., 2004. Study of ice crystal orientation in cirrus clouds based on satellite polarized radiance measurements. J. Atmos. Sci. 16, 2073–2081.

Nuijens, L., Emanuel, K., Masunaga, H., et al., 2017. Implications for warm rain in shallow cumulus and congestus clouds for large-scale circulations. Surv. Geophys. 38, 1257–1282.

Quinn, P.K., Coffman, D.J., Johnson, J.E., et al., 2017. Small fraction of marine cloud condensation nuclei made up of sea spray aerosol. Nat. Geosci. 10, 674–681.

Passarelli Jr, R.E., 1978. Theoretical and observational study of snow-size spectra and snowflake aggregation efficiencies. J. Atmos. Sci. 35, 882–889.

Pinsky, M.B., Khain, A.P., 1997. Turbulence effects on the collision kernel. I: formation of velocity deviations of drops falling within a turbulent three-dimensional flow. Quart. J. Roy. Meteor. Soc. 123, 1517–1542.

Pinsky, M.B., Khain, A.P., 2002. Effects of in-cloud nucleation and turbulence on droplet spectrum formation in cumulus clouds. Quart. J. Roy. Meteor. Soc. 128, 501–533.

Protat, A., Williams, C.R., 2011. The accuracy of radar estimates of ice terminal fall speed from vertically pointing Doppler radar measurements. J. Appl. Meteor. Clim. 50, 2020–2038.

Pruppacher, H.R., Klett, J.D., 1996. Microphysics of Clouds and Precipitation. Springer, p. 976.

Pruppacher, H.R., Pitter, R.L., 1971. A semi-empirical determination of the shape of cloud and raindrops. J Atmos Sci 28, 86–94.

Qu, Z., Barker, H.W., Korolev, A.V., et al., 2018. Evaluation of a high-resolution numerical weather prediction model's simulated clouds using observations from CloudSat, GOES-13 and in situ aircraft. Q. J. R. Meteorol. Soc. 144, 1681–1694.

Rauber, R.M., 1987. Characteristics of cloud ice and precipitation during wintertime storms over the mountains of Northern Colorado. J. Appl. Meteor. Climatol. 26, 488–524.

Rauber, R.M., Stevens, B., Ochs III, H.T., 2007. Rain in shallow cumulus over the ocean: The RICO campaign. Bull. Amer. Meteor. Soc. 87, 1912–1928.

Riehl, H., Yeh, T.C., Malkus, J.S., et al., 1951. The northeast trades of the Pacific Ocean. Quart. J. Roy. Meteor. Soc. 72, 598–626.

Rogers, R.R., Yau, M.K., 1989. A Short Course in Cloud Physics, *3rd Ed.* Butterworth-Heinemann, p. 304.
Rosenow, A., Plummer, D.M., Rauber, R.M., et al., 2014. Vertical velocity and physical structure of generating cells and convection in the comma head region of continental winter cyclones. J. Atmos. Sci. 71, 1538–1558.
Saltalippi, C., Corradini, C., Dari, J., Morbidelli, R., Flammini, A., 2022. Rainfall and Development of Floods. In: Morbidelli, R. (Ed.), Rainfall. Modeling, Measurement and Applications. Elsevier, Amsterdam, pp. 351–366. doi:10.1016/C2019-0-04937-0.
Saunders, P.M., 1965. Some characteristics of tropical marine showers. J. Atmos. Sci. 22, 167–175.
Schaefer, V.J., 1962. Condensed water in the free atmosphere in air colder than −40°C. J. Appl. Meteor. 1, 481–488.
Schlottke, J., Straub, W., Beheng, K.D., et al., 2010. Numerical investigation of collision-induced breakup of raindrops. Part I: Methodology and dependencies on collision energy and eccentricity. J. Atmos. Sci. 67, 557–575.
Schmidt, C.G., Sulia, K., Lebo, Z.J., et al., 2019. The fall speed variability of similarly sized ice particle aggregates. J. Appl. Meteor. Clim. 58, 1751–1761.
Smith, P.L., 2003. Raindrop size distributions: exponential or gamma-does the difference matter? J. Appl. Meteor. Climatol. 42, 1031–1034.
Sotiropoulou, G., Sullivan, S., Savre, J., et al., 2020. The impact of secondary ice production on Arctic stratocumulus. Atmos. Chem. Phys. 20, 1301–1316.
Soulage, G., 1955. Les noyaux de congélations de l'atmosphère. Ann. Geophys. 13, 103–133.
Squires, P.K., 1956. The micro-structure of cumuli in maritime and continental clouds. Tellus 8, 443–444.
Srivastava, R.C., 1978. Size distribution of raindrops generated by their breakup and coalescence. J. Atmos. Sci. 28, 410–415.
Srivastava, R.C., Coen, J.L., 1992. New explicit equations for the accurate calculation of the growth and evaporation of hydrometeors by the diffusion of water vapor. J. Atmos. Sci. 48, 1643–1651.
Steiner, A.L., Brooks, S.D., Deng, C., et al., 2015. Pollen as atmospheric cloud condensation nuclei. Geophys. Res. Lett. 42, 3596–3602.
Stewart, R.E., Marwitz, J.D., Pace, J.C., et al., 1984. Characteristics through the melting layer of stratiform clouds. J. Atmos. Sci. 41, 3237–3237.
Straub, W., Beheng, K.D., Seifert, A., et al., 2010. Numerical investigation of collision-induced breakup of raindrops. Part II: Parameterizations of coalescence efficiencies and fragment size distributions. J. Atmos. Sci. 67, 576–588.
Sulia, K.J., Lebo, Z.J., Przybylo, V.M., et al., 2020. A new method for ice-ice aggregation in the adaptive habit model. J. Atmos. Sci. 78, 133–154.
Szeto, K.K., Stewart, R.E., 1997. Effects of melting on frontogenesis. J. Atmos. Sci. 54, 689–702.
Szumowski, M.J., Rauber, R.M., Ochs III, H.T., 1999. The microphysical structure and evolution of Hawaiian rainband clouds. Part III: a test of the ultragiant nuclei hypothesis. J. Atmos. Sci. 56, 1980–2003.
Telford, J.W., 1955. A new aspect of coalescence theory. J. Meteor. 12, 436–444.
Testik, F.Y., Rahman, M.K., 2017. First in situ observations of binary raindrop collisions. Geophys. Res. Lett. 44, 1175–1181.
Testud, J., Oury, S., Black, R.A., et al., 2000. The concept of "normalized" distribution to describe raindrop spectra: a tool for cloud physics and cloud remote sensing. J. Appl. Meteor. Clim. 40, 1118–1140.

Tokay, A., Short, D.A., 1995. Evidence from tropical raindrop spectra at the origin of rain from stratiform versus convective clouds. J. Appl. Meteor. Climatology. 35, 355–371.
Tokay, A., Bashor, P.G., Habib, E., et al., 2008. Raindrop size distribution measurements in tropical cyclones. Mon. Wea. Rev. 136, 1669–1685.
Twohy, C.H., Hudson, J.G., Yum, S.S., et al., 2001. Characteristics of cloud-nucleating aerosols in the Indian Ocean region. J. Geophys. Res. 106, 28699–28710.
Um, J., McFarquhar, G.M., 2007. Single-scattering properties of aggregates of bullet rosettes in cirrus. J. Appl. Meteor. Clim. 46, 757–775.
Vali, G., 1996. Ice nucleation- a review. Nucleation and Atmospheric Aerosols, Pergamon, 271–279. ISBN 9780080420301, . https://doi.org/10.1016/B978-008042030-1/50066-4.
Vali, G., DeMott, P.J., Möhler, O., et al., 2015. Technical note: a proposal for ice nucleation terminology. Atmos. Chem. Phys. 15, 10263–10270.
Vardiman, L., 1978. The generation of secondary ice particles in clouds by crystal–crystal collisions. J. Atmos. Sci. 35, 2168–2180.
Vestin, A., Rissler, J., Swietlicki, E., et al., 2007. Cloud-nucleating properties of the Amazonian biomass burning aerosol: cloud condensation nuclei measurements and modeling. J. Geophys. Res. 112. doi:10.1029/2006JD008104.
Villermaux, E., Bossa, B., 2009. Single drop fragmentation determines size distribution of raindrops. Nat. Phys. 5, 697–702.
Vivekanandan, J., Ellis, S.M., Oye, R., et al., 1990. Cloud microphysics retrieval using S-band dualpolarization radar measurements. J. Am. Meteorol. Soc. 80, 381–388.
Vonnegut, B., 1947. The nucleation of ice formation by silver iodide. J. Appl. Phys. 18, 593–595.
Wakimoto, R.M., 1985. Forecasting dry microburst activity over the high plains. Mon. Wea. Rev. 113, 1131–1143.
Wang, P.K., Ji, W., 2000. Collision efficiencies of ice crystals at low-intermediate Reynolds numbers colliding with supercooled droplets: a numerical study. J. Atmos. Sci. 57, 1001–1009.
Warner, J., 1969. The microstructure of cumulus cloud. Part I. General features of the droplet spectrum. J. Atmos. Sci. 26, 1049–1059.
Wegener, A., 1911. Thermodynamik der Atmospäre. Barth, Leipzig.
Willis, P.T., 1984. Functional fits of observed drop size distributions and parameterization of rain. J. Atmos. Sci. 41, 1648–1661.
Woodcock, H.A., 1953. Salt nuclei in marine air as a function of altitude and wind force. J. Appl. Meteor. 10, 362–371.
Wu, W., McFarquhar, G.M., 2016. On the practical definitions of maximum dimension for non-spherical particles recorded by 2D probes. J. Atmos. Ocean. Tech. 33, 1057–1072.
Xue, Y., Wang, L.P., Grabowski, W.W., 2008. Growth of cloud droplets by turbulent collision-coalescence. J. Atmos. Sci. 65, 331–356.
Young, K., 1993. Microphysical Processes in Clouds. Oxford University Press, p. 448.
Yuter, S.E., Houze Jr, R.A., 1997. Measurements of raindrop size distributions over the Pacific warm pool and implications for Z-R relations. J. Appl. Meteor. 36, 847–867.
Zeng, X., 2018. Modeling the effect of radiation on warm rain initiation. J. Geophys. Res. 123, 6896–6906.
Zhang, C., Harrington, J.Y., 2014. Including surface kinetic effects in simple models of ice vapor diffusion. J. Atmos. Sci. 71, 372–390.
Zhang, F., Li, Y., Li, Z., et al., 2014. Aerosol hygroscopicity and cloud condensation nuclei activity during the AC3Exp campaign: implications for cloud condensation nuclei parameterization. Atmos. Chem. Phys. 14, 13423–13437.

CHAPTER

Meteorological systems producing rainfall

2

Corrado Corradini, Renato Morbidelli, Carla Saltalippi, Alessia Flammini
Department of Civil and Environmental Engineering, University of Perugia, Perugia, Italy

2.1 Introduction

Frontal and convective systems give a major contribution to rainfall production in most geographic areas. The spatiotemporal structure of the associated rainfall fields is rather complex and difficult to represent properly by theoretical investigations and frequently even by experimental measurements. Its understanding relies upon elements of microphysics (McFarquhar, 2022) and dynamics of atmosphere (Grabowski, 2022).

The substantial differences characterizing the aforementioned systems can be roughly caught on through the typical rainfall extension and duration. At a given time, frontal rainfalls spread over areas larger than tens of thousands square kilometers while convective rainfalls over smaller areas up to 100 to 200 km^2. At a given site, frontal rainfalls persist many hours while convective rainfalls have a short duration typically up to 2 h.

Rainfall at ground is characterized by its intensity and the droplet size distribution that, under conditions of supersaturation determined by air uplift, first develops with air temperature greater than 0°C by the process of water vapor condensation over condensation nuclei. This process (Visconti, 2016) commonly allows to have droplets with radius up to about 20 μm. Then, a significant fall of droplets occurs and their dimensions increase by coalescence up to a radius of few millimeters limited by a droplet disruption process. Furthermore, considering that air temperature generally decreases with height, if air supersaturation develops above the 0°C level it occurs sublimation of water vapor over the condensation nuclei, with formation of snowflakes that evolve through the same mechanisms described for the droplets until they reach the 0°C level. Then the snowflakes, in an atmospheric layer of about 100 to 200 m, experience a melting process and get to ground as liquid water. It is important to note that in the case the duration of air supersaturation stage is too short the droplets or snowflakes do not reach dimensions such to produce a significant falling and there is only formation of not precipitant clouds.

Formation of clouds and precipitation requires air supersaturation that is generated through the development of air motion with a vertical velocity component. This implies uplift of air (Roe and Baker, 2006; Cheng and Yu, 2019) that, as a first approximation, experiences a thermodynamic transformation of adiabatic type

RAINFALL. Modeling, Measurement and Applications. DOI: https://doi.org/10.1016/B978-0-12-822544-8.00011-1

because the surrounding air is not a good heat conductor and the atmospheric pressure decreases with height. During the ascent motion the saturation ratio increases and can reach values greater than one, which means supersaturation with absence of thermodynamic equilibrium that determines water vapor condensation/sublimation over the nuclei with lower critical supersaturation (activated nuclei). Continuing air ascent supersaturation still increases, and other nuclei become active, until a maximum is reached when its growth trend associated with air uplift becomes balanced by the vapor depletion due to the growing particles.

Air uplift can be produced by different meteorological systems linked with air convergency, convection, orography and fronts. Air convergency takes place at the confluence of valleys or in centers of low pressure at ground. Convective motions originate from strong differences of temperature over adjacent areas that produce air instability. Mountain chains can generate an air mass uplift under propitious conditions of wind (speed and direction) and air stability (James and Houze, 2005). Frontal systems derive from the contact of two extended air masses characterized by pronounced differences in the basic meteorological quantities. They give a substantial contribution to rainfall at mid latitudes.

In the frontal systems there is interaction with the other aforementioned mechanisms causing uplift of air masses. Primarily, their interaction with orography has been widely investigated using both experimental results and conceptual physical/mathematical modeling. Numerical solutions of complex models based on a coupling of equations for dynamics and microphysics of atmosphere have been also proposed (Pielke, 2013; Hou et al., 2020).

In this context, the main objective of this chapter is to highlight the structure of the rainfall fields produced by frontal disturbances, considering also their fundamental role in the development of hydrological processes at different spatiotemporal scales.

Another objective is to provide a quantitative evidence of the crucial physical process involved in the interaction between frontal systems and orography that is commonly designated as seeder-feeder process (Smith et al., 2016; Rauber et al., 2019; Smith et al., 2019). For the sake of simplicity the last topic is treated by a diagnostic approach in the absence of convective processes.

2.2 Schematization of frontal systems

As earlier specified the supersaturation condition for the formation of clouds and rainfalls requires the presence of air motions with a vertical velocity component. Air uplift can be produced by frontal systems that frequently are influenced by orographic and convective effects as well as by air convergency. Frontal systems develop through the confluence of two air masses with characteristics determined by different heat-transfer and humidification, with formation of a narrow separation zone denoted as front. In this zone there is a large horizontal variation of temperature, humidity, wind and atmospheric pressure while their variations in each air mass are substantially rather limited. Frontal systems are classed as warm fronts, cold fronts, cyclones and occluded fronts. The last type, that is not much spread, is a combination of a warm and a cold front. A huge number of meteorological measurements carried out for many years in most geographic areas have shown that the frontal zone can be schematized by an inclined surface along the system propagation direction with an extension of hundreds kilometres. Schematically Fig. 2.1a shows the vertical section of a warm frontal system with the warmer air upon the colder one. The frontal zone has a slope α typically in the range $1/300 < tg\alpha < 1/100$. The front position at the ground level is represented in the geographic maps by the weather charts using the symbology highlighted in Fig. 2.1b. By a similar approach the vertical section of a cold front system is shown in Fig. 2.2a. It typically involves a larger slope with $1/150 < tg\alpha < 1/50$ and an extension somewhat similar to that of warm frontal systems. The symbology adopted for representing the cold front position at the ground level by the weather charts is highlighted in Fig. 2.2b. As it can be observed in Figs. 2.1a and 2.2a both the typologies involve warm air above the front that ascends and produces precipitations trough adiabatic transformations, however it is important to note that there are also cold fronts with descending warm air (see later). Each system is characterized by an overall propagation velocity denoted as front velocity, while in each air mass there are important motions with respect to that of the frontal

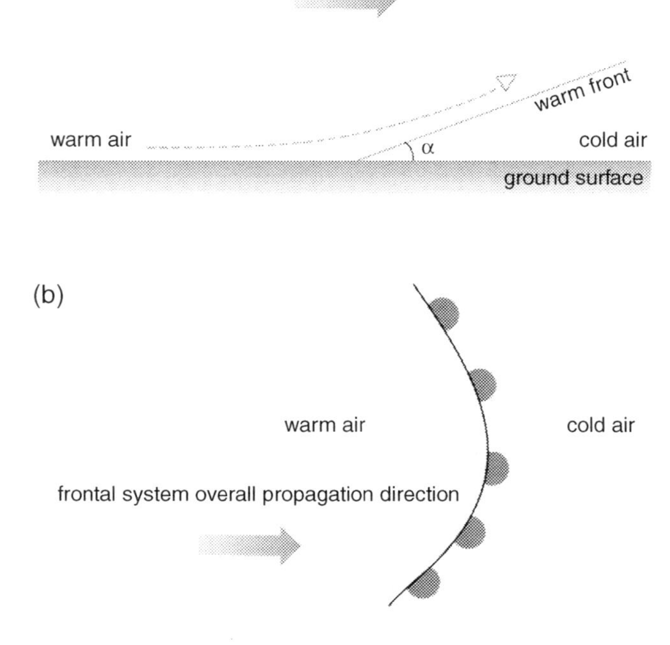

FIG. 2.1

(a) Schematic representation of the vertical section of a warm frontal system.
(b) Symbolical representation of a warm front at the ground level on the weather charts.

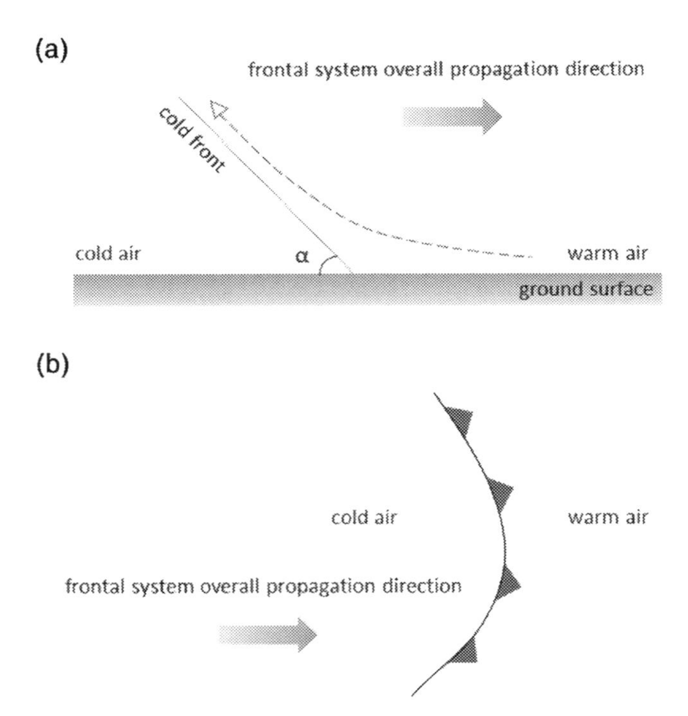

FIG. 2.2

(a) Schematic representation of the vertical section of a cold frontal system. The warm air ascent above the front refers to ana-cold fronts (see later). (b) Symbolical representation of a cold front at the ground level on the weather charts.

surface. The passage of a warm front at the ground surface occurs when warmer air substitutes colder air, and vice versa for a cold front, can be easily observed by a classical meteorological station placed at the soil surface.

A meteorological system widely observed, f.i. at mid latitudes, and designated as a cyclone consists of a combination of a warm front with a cold front that converge in a low pressure center. Fig. 2.3 shows its representation by the weather charts at the ground level and points out the existence of a warm sector between the two fronts. Fig. 2.4 shows a representation of an occluded front by the weather charts at the ground level.

2.3 Rainfall fields associated to mid-latitude frontal systems

Knowledge of the main features of rainfalls produced at different spatiotemporal scales by warm frontal disturbances, cold frontal disturbances and cyclonic systems is of utmost importance for dealing with a variety of experimental and theoretical issues. In this section is not presented a review of the existing investigations

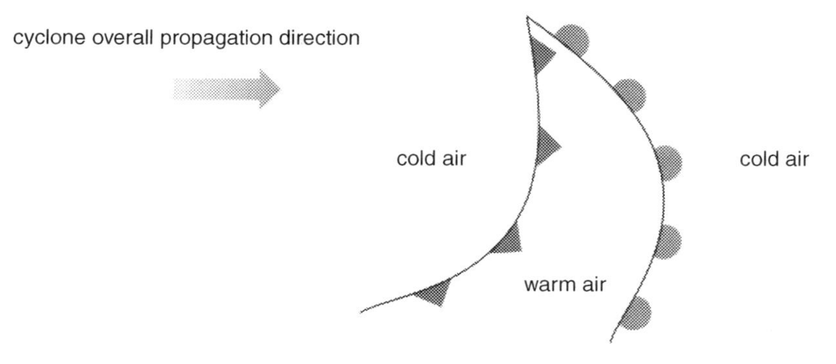

FIG. 2.3

Symbolic representation of a cyclone at the ground level on the weather charts.

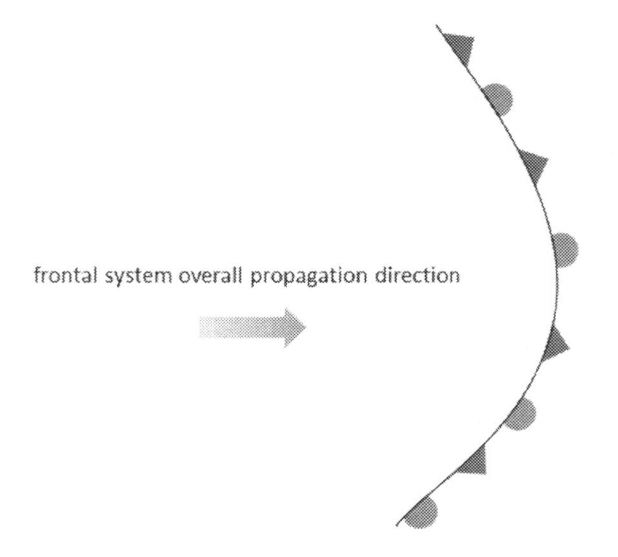

FIG. 2.4

Symbolic representation of an occluded front at the ground level on the weather charts.

addressed to specific aspects of the topic, but it is chosen to provide a short analysis of mid-latitude frontal rainfalls mainly focused on well-grounded and exhaustive studies even though performed a few tens of years ago. These studies are still used as benchmarks (Berne at al., 2009; Pielke, 2013; Dorsi et al., 2015; Plummer et al., 2015; Murphy et al., 2017; Hou et al., 2020), because of the clear interpretation of the meteorological processes based on synoptic situation analyses, data of weather radars and rain gauge networks, and satellite information. Many valuable studies on the rainfall field structures were carried out in Britain (Browning and Harrold, 1970; Browning et al., 1974; Browning et al., 1975; Hill and Browning, 1979; Hill et al., 1981), particularly for precipitation in warm sectors of cyclonic disturbances and in cold fronts, and in USA with major emphasis to precipitations in Pacific extratropical cyclones (Austin and Houze, 1972; Herzegh and Hobbs, 1980; Houze et al., 1981; Hobbs et al., 1980; Hobbs and Persson, 1982; Parsons and Hobbs, 1983; Rutledge and Hobbs, 1983; Wang et al., 1983). These studies indicated that a precipitation area associated to cold front systems and cyclones (of dimension larger than 10^4 km^2, with synoptic features) has generally one or several smaller-scale areas with more heavy precipitation embedded within it (Fig. 2.5). The precipitation areas within a synoptic disturbance are characterized by a wide variety of sizes and can be classified (Harrold and Austin, 1974) as large mesoscale precipitation areas (typically of 10^3–10^4 km^2), Small Mesoscale precipitation Areas (SMSA, typically of 50–1000 km^2) and rain cells of area typically less than 50 km^2. This classification also distinguishes between large areas of stratiform precipitation (LMSA) and embedded clusters of convective cells that are often aligned in bands or lines of precipitation. Sometimes the mesoscale precipitations areas have an irregular distribution with a wide spectrum of sizes and are simply classified as MPAs. The mesoscale variability of precipitation is linked with convective instabilities that in Britain and in northern USA were found to be usually associated to the presence of a layer of potential instability in the middle troposphere (Browning et al., 1974; Hobbs and Locatelli, 1978). As to the characterization of warm frontal systems, there are not widespread

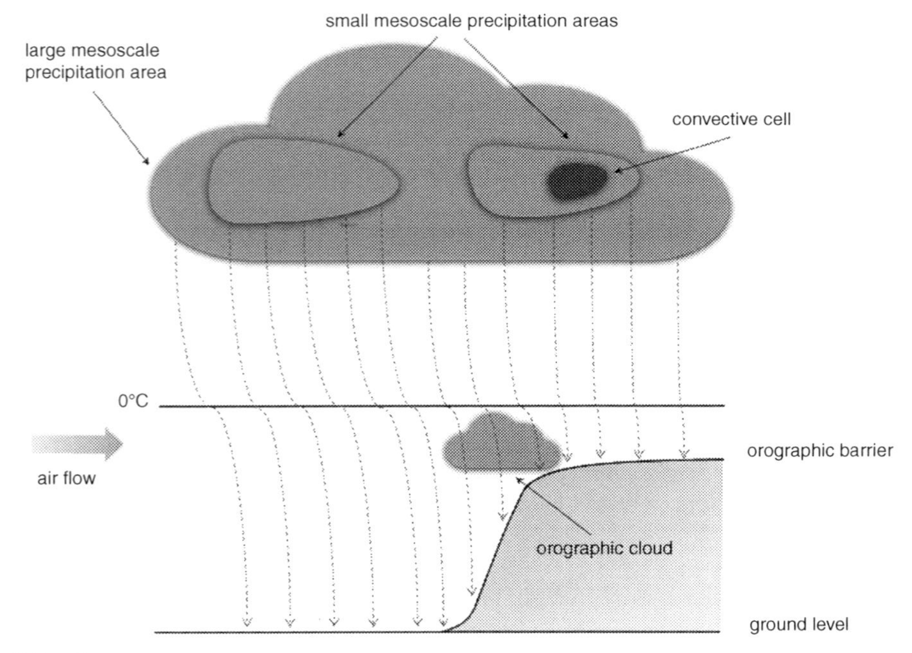

FIG. 2.5

Schematic vertical section of surface rainfall over an orographic barrier, with an orographic cloud produced by air uplift during the passage of a frontal disturbance involving convection aloft.

experimental analyses carried out to explain the mechanisms producing the surface rainfall pattern. Rainfall spatial distributions ahead of a few Mediterranean type warm fronts were investigated by Corradini (1985) and two representative cases are detailed at the end of this chapter. Furthermore, an analysis in depth was carried out by Hill and Browning (1979) even though with the warm front as a component of a cyclonic disturbance (see later). In any case, it is widely recognized that even for a given typology of frontal system the spatiotemporal distribution of surface rainfall is highly affected by origin of meteorological disturbance (Hofstätter et al., 2016) and orography (Smith, 1979)

Most of the above studies ascribe a primary role to the interaction between atmospheric disturbance and orography. A significant amount of rainfall can be generated through a stable slantwise ascent over a mountain-barrier leading to a low-level cloud often with droplets that have insufficient time to reach precipitation sizes. However, in the presence of a precipitation from above (seeder cloud) the not precipitant cloud (feeder cloud) can be washed out increasing surface rainfall (Fig. 2.5). In the case of airstream with potential instability at low levels the forced uplift can trigger convective processes (Kirshbaum et al., 2018) that increase the magnitude of the orographic effects on surface rainfall. Techniques for separation of stratiform precipitation produced by widespread lifting in frontal systems and convective precipitation were discussed by Houze (2014).

2.3.1 Cold frontal systems

Cold frontal systems are still commonly classed as ana- and kata-cold fronts (Sansom, 1951; Browning et al., 1975). The associated time evolution of rain is rather different. Roughly, in ana-cold fronts the warm air ascends above the slantwise frontal surface (Fig. 2.2a) and at a given site most of the rain is observed during and after the front passage at the ground level, with a trend usually represented by a burst of rain followed by a longer period of lighter rain. On the other hand, in the kata-cold fronts the warm air above the frontal surface descends and the rain is mainly observed ahead or during the front passage at the surface. From the investigations performed by Browning and his collaborators in the British Isles important insights on the rainfall fields associated to wintertime cold frontal systems can be drawn. Two representative meteorological disturbances observed on November 11, 1970 and October 29, 1972 are first shortly examined. Both included ahead of the front warm air with a low-level jet at 850 mb from between SSW and WSW so that existing convective rain areas approached the study regions (Wales and SE England) without experiencing prior orographic change upwind. The first front (Browning and Pardoe, 1973; Browning et al., 1975) was a typical ana-cold front characterized by an initial burst of heavy rain at the surface cold front, which was associated to a continuous thin band of convective areas, followed by a period of moderate rain. Significant changes between the surface rainfall (ahead of, on and behind the surface front) observed in the coast and hills of south Wales were not found. This result was ascribed to the non-existent conditions of a strong (not descending) flow of warm and moist air at low levels ahead of and behind the surface front, while the absence of orographic enhancement during the passage of the surface cold front was due to the fact that the vertical motion linked with the dynamically induced low-level convection was much greater than that generated by the forced orographic ascent. The second front was substantially a classical kata-cold front involving rain mainly ahead of the surface cold front, and during its passage when a band of small rain cells was observed. Heavy orographic rain was found over parts of the south Wales hills with average rate greater than 4 mmh^{-1} while most low-lying areas had a little quantity of rain with average rate less than 0.5 mmh^{-1}. This large orographic effect was found in the absence of seeding rain from middle tropospheric levels and the seeding droplets were produced in the orographic cloud (thickness 3 km) formed by air uplift of low-level layers. This was supported by the fact that ahead of the cold front, as in many other cold fronts in the British Isles, in the conveyor belt (Harrold, 1973) there was a low-level jet of moist air.

The arrangement of the bands of convective areas at surface cold fronts was examined on the basis of a line convection that was frequently observed by radar over the British Isles (James and Browning, 1979) and the western seaboard of the USA (Houze et al., 1976). James and Browning (1979) considered a large number of ana-cold fronts involving the line convection. This line was really found to be usually broken up into line elements variable in both length and lifetime approximately by an order of magnitude. Nevertheless these elements were largely unaffected by topography and had some common features that allowed to follow their travel in a reasonable manner. The minor role of topography was found to be a surprising result considering that line convection is a low-level phenomenon that in the investigated cases was limited to the lowest 2–3 km. It is important to note that the uniform line convection associated to the above front of November 11, 1970 was considered to represent an extreme case.

2.3.2 Cyclonic systems

Rainfall patterns in mid-latitude cyclones are shortly examined on the basis of experimental results obtained in the hilly western parts of the British Isles (Browning et al., 1974; Hill and Browning, 1979; Hill et al., 1981).

Browning et al. (1974) selected a cyclone observed on December 5, 1972 to investigate the structure upwind, over and downwind of the south Wales hills of the precipitation associated to the warm sector. Warm and moist air with strong winds between SW and WSW and impinging on the SW-facing hills of south Wales was involved. The warm sector was characterized by an airstream with potential instability at low levels as well as by a layer with remarkable instability in the middle troposphere. Over the sea pre-existing scattered MPAs in the middle troposphere were strengthen 100 km upwind of the hills by new convection areas developed between them, suggesting that in this tropospheric region the orographic uplift occurred far upwind of the hills. Convective motions developed also in the low levels during the forced orographic ascent generating the feeder cloud washed out by the seeding from aloft. Rain substantially occurred ahead of the surface warm front and during the passage of the warm sector. It was intermittent in many sites but continuous over the hills. A comparison of the total rain fallen ahead of the surface warm front, in the warm sector and close to the surface cold front in the south Wales coast, in the south Wales hills and in a downwind region was performed. With respect to the upwind coastal area, in the warm sector there was a rainfall increase of a factor of six over the hills while in the warm and cold fronts the corresponding increase was of a factor of three and four, respectively. A significant rainfall decrease was observed downwind for both warm front and warm sector.

Hill and Browning (1979) investigated another cyclone, occurred on February 12, 1976, which was moving eastwards producing rainfall over Ireland, Wales and England. The main objective was to examine the changes experienced by MPAs observed in the warm sector in their motion through two regions of high land separated by a stretch of sea. The MPAs were associated with middle-level potential instability in the layer 800 to 600 mb within the warm sector, characterized by moist air and strong winds. From the distribution of the total rainfall observed ahead of the warm front, in the warm sector and behind the cold front it was possible to deduce significant insights useful for clarifying the mechanisms producing surface rainfall and also as a support for application to rainfall forecast. The total rainfall associated to these three stages of the cyclone was distributed in the quantities 12, 4 and 2 mm, respectively, on the west coast of Wales, while the corresponding quantities over the hills increased to 22, 14, and 4 mm. The observed increase of rainfall over the hills, both in extension and intensity, was ascribed to both the invigoration of the MPAs after crossing the sea between Ireland and the mainland of England and Wales and, probably to a great extent, the development of low-level clouds due to orographic ascent. In any case, a larger quantity of rain fell in a broad region of continuous precipitation ahead of the warm front with respect to that in the warm sector. Nevertheless, the MPAs can play an important role in many hydrometeorological studies. In this context the possibility to track them over great distances even in the presence of a large orographic modulation can represent a significant support.

A few cyclones were also investigated in depth by Hill et al. (1981) in their passage on the 20 km wide Glamorgan Hills in the south Wales. They involved winds from the SW quadrant below 700 mb so that the study region was not sheltered by other hilly areas. In this chapter the cyclonic disturbances observed on November 28, 1976 (case 1) and February 3, 1977, both substantially associated with rainfall ahead of the surface warm fronts and in the associated warm sectors, are discussed. Both the cyclones produced an increase of the rainfall rate from the sea to the hills, where a significant contribution was associated with the passage of precipitation areas that were previously identified upwind by radar. In the middle troposphere MPAs were observed over the sea as well as over the hills with minor orographic enhancement in the travel from sea to hills. However, a great increase of the maximum and the mean rainfall rate over the hills, computed through the duration of surface rain, was generated in the low levels. In particular for the first cyclone the maximum increased from about 3 to 9 mmh^{-1} in the lowest 1.5 km above the hills and the corresponding enhancement for the other case was from about 1.5 to 4 mmh^{-1}. The stronger low-level wind speed was used to explain the different enhancement between the two cases, even though both involved strong airflow with mean wind speed at 600 m asl greater than 25 m/s. These conditions together with both the high mean relative humidity below 2 km and the existence of raindrops from upper clouds indicated that the seeder-feeder mechanism in the low levels had a primary role in the observed orographic enhancement of the surface rainfall. The impact of the potential instability occurred in two layers placed in the middle troposphere and between 1 and 2 km provides important insights. The upper layer with weak potential instability produced convective processes and seeder clouds that were not affected by important changes due to orography, while the lower layer was sufficiently moist to determine low-level convection during forced uplift over the hills.

2.4 Interaction between frontal systems and orographic barriers in the low levels of troposphere

2.4.1 Some remarks on the adiabatic transformation of humid air

Solutions of the adiabatic transformation equations (Visconti, 2016) are requested to estimate the seeder-feeder contribution to rainfall at the ground level.

For an infinitesimal volume of humid air, using the state law of ideal gases the first thermodynamic principle can be expressed by:

$$m\left(c_p dT - RT\frac{dP}{P}\right) = \delta Q \tag{2.1}$$

where m is the air mass, c_p is the specific heat at constant pressure, R is the specific gas constant, T and P are absolute air temperature and pressure, respectively, and δQ is heat provided from outside. Considering an adiabatic ascent of the air parcel in the atmosphere under unsaturated conditions, starting from an initial state with temperature T_0 and pressure P_0 up to a final state with corresponding values T and P Eq. (2.1), with $\delta Q = 0$, becomes:

$$\int_{T_0}^{T} c_p dT - \int_{P_0}^{P} RT\frac{dP}{P} = 0; \quad e < e_s \tag{2.2}$$

where e is the real vapor pressure and e_s denotes its saturated value (vapor tension). Eq. (2.2) gives:

$$\frac{T}{T_0} = \left(\frac{P}{P_0}\right)^{\frac{R}{c_p}} \tag{2.3}$$

with $\frac{R}{c_p} \cong \frac{R_d}{c_{pd}}$ where the subscript d stands for quantities referred to dry air. During the parcel uplift e decreases with P while the variation of T obtained by (3) determines a more marked decrease of e_s that for a flat surface of liquid water is expressed by the Clausius-Clapeyron exponential relation as:

$$e_s = \gamma \, exp\left[-\frac{L_{lv}}{R_v T}\right] \tag{2.4}$$

where γ is a constant, L_{lv} is the latent heat of evaporation and R_v is the specific constant for water vapor. For solid water Eq. (2.4) is still applicable with L_{lv} substituted by the latent heat of sublimation.

Therefore, at a given uplift height the saturation ratio $S_r = \frac{e}{e_s}$ becomes equal to 1. The successive uplift, if any, implies $S_r > 1$ (with supersaturation $S_r - 1 > 0$) and requires to adopt a saturated adiabatic transformation that can be expresses by Eq. (2.1) rewritten as:

$$\left(m_d + m_{vs}\right)\left[c_p dT - RT\frac{dP}{P}\right] = -L_{lv} dm_{vs} \; ; S_r > 1 \tag{2.5}$$

where m_d and m_{vs} are the dry air and saturated vapor masses, respectively, of the parcel and $-L_{lv}dm_{vs}$ is the heat added to the parcel by condensation/sublimation of water vapor. Eq. (2.5) can be also rewritten in terms of saturation mixing ratio, $q_s = \frac{m_{vs}}{m_d}$, as:

$$\left(1 + q_s\right)\left[c_p\frac{dT}{T} - R\frac{dP}{P}\right] = -L_{lv}\frac{dq_s}{T} \tag{2.6}$$

By expressing q_s as a function of T and P (that is $q_s = 0.622\frac{e_s}{P}$) Eq. (2.6) can be solved to obtain the variation of T associated to that of P.

The first stage of cloud formation should be described by coupling Eq. (2.6) with an equation describing the growth of droplets/snowflakes by condensation/sublimation. The last equation would be required to derive correctly the droplet size spectrum that is influenced by the values of S_r larger than the critical values associated to the available condensation nuclei.

2.4.2 A parameterized seeder-feeder model for estimating orographic rainfall

A diagnostic conceptual model formulated by Corradini (1985) (see also Bell, 1978; Bader and Roach, 1977; Collier, 1975) for simulating the rainfall spatial distribution of widespread rainfall at the meso-local scale is presented. All the involved altitudes are referred to the sea level, and the geographic area upwind of the study region is considered to be flat. This means that meteorological data observed by vertical sounding in the flat region can be directly used as inputs. The model does not represent convective processes and could be therefore more suitable for applications to rainfalls occurring before the passage of warm fronts at the ground level. In any case, it incorporates the essential physics of the processes that determine the orographic effects on surface rainfalls. The model, formulated considering precipitation as rainfall, relies upon the knowledge of the distortion of air fluxes produced by orography that primarily implies the development of a vertical velocity component, W_T, and a change of relative humidity due to the adiabatic ascent. Orographic effects are assumed to be mainly produced in the lower region of the troposphere, with air uplift influencing the levels, Z, up to $Z_{max} = 2000\ m$, while the rainfall at higher levels is generated by large-scale processes. The last rainfall, P_{ex}, represents a model input and is that falling over the highest distorted air layer. In a saturated underlying atmosphere it is assumed equal to the rainfall measured at the soil surface in flat areas, far from ridges. In the case of unsaturated lower layers the above observed rainfall requires a correction by a trial-and-error procedure to allow for the evaporation process of droplets.

The rainfall at the ground surface is derived by computing the variation of P_{ex} beneath Z_{max} through a seeder-feeder process. The following quantitative formulation can be applied to estimate the orographic effects.

A regular horizontal grid of dimensions ΔX, ΔY allowing to properly schematize (see later a case study) the orographic features is chosen, while in the vertical the air mass between 0 and 2000 m is divided into layers of thickness ΔZ, with associated variation of the atmospheric pressure ΔP. The air uplift at each grid point (X, Y) and at each level Z is expressed by the relation for the vertical displacement, S, as:

$$S = K_D H \tag{2.7}$$

where H is the height of the terrain and K_D is a dimensionless parameter taken as:

$$K_D = 1 - 0.4\frac{Z}{Z_{max}} \tag{2.8}$$

Fig. 2.6 displays, as an example, the trend of the flow lines provided by Eqs. (2.7) and (2.8).

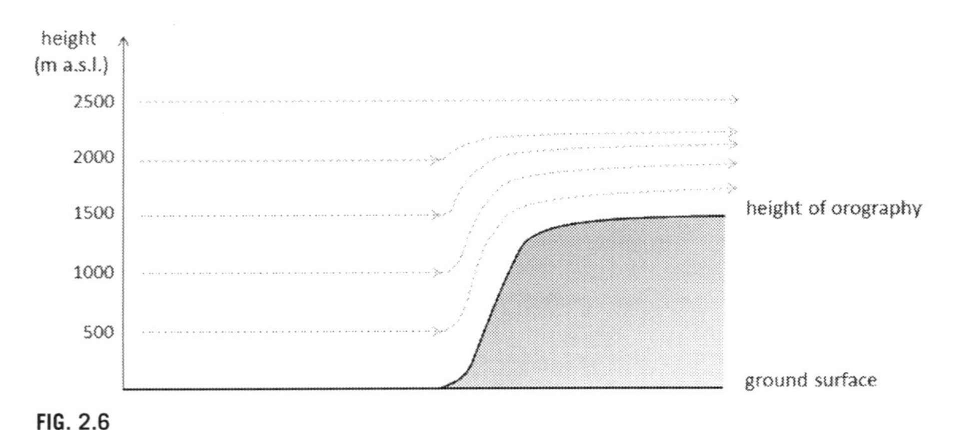

FIG. 2.6

Trend of the flow lines of an air mass over a hill (vertical profile).

The air uplift is associated with a vertical velocity, W_T, given by:

$$W_T = K_D \vec{V} \cdot \overrightarrow{\nabla H} \quad (2.9)$$

with $\vec{V} \cdot \overrightarrow{\nabla H}$ scalar product of horizontal velocity, $\vec{V}$, and local topographic gradient, $\overrightarrow{\nabla H}$.

During the displacement with vertical velocity W_T from the position in the absence of orography to the final position determined by Eq. (2.7) the air mass experiences adiabatic transformation and, using the continuity equations for dry air and water vapor together with thermodynamic elements, the rainfall per unit volume and unit time, I, reaching directly the ground surface is expressed as:

$$I = -K_1 K_2 \frac{dq_s}{dt} \rho \quad (2.10)$$

with

$$\frac{dq_s}{dt} = -\frac{q_s}{R}\left(\frac{L_{lv} R - c_p R_v T}{c_p R_v T^2 + q_s L_{lv}^2}\right) g W_T \quad (2.11)$$

where, in the absence of a droplet growth equation (Cotton et al., 2011) q_s is approximated as the value associated to a flat surface of pure water; t is the time; g is the magnitude of gravity; K_1 and K_2 are parameters assumed as:

$$K_1 = \begin{cases} 0 & \text{if } W_T \le 0 \text{ or the air is unsaturated after the ascent} \\ 1 & \text{if } W_T > 0 \end{cases} \quad (2.12)$$

$$K_2 = \begin{cases} 0 & \text{if } t_g \le 300 \\ 1 & \text{if } t_g \ge 1200 \text{ or the air is saturated before ascending} \\ \frac{(t_g - 300)}{900} & \text{if } 300 < t_g < 1200 \end{cases} \quad (2.13)$$

where t_g denotes the duration of the supersaturation period available for the droplet growth, given by:

$$t_g = \frac{(S - S_s)}{W_T} \quad (2.14)$$

with S_s vertical displacement for reaching air saturation. Eq. (2.13) implies that, with $W_T > 0$, if $t_g > 1200$ all the droplets reach sufficient sizes to fall directly, while for lower t_g values there are suspended particles that can be caught up by larger raindrops generated in the upper layers and producing a washout process by the coalescence between them. The last seeder-feeder mechanism is quantified by (Bader and Roach, 1977):

$$C = c\sum_{r} \pi r^2 N_r V_r E_r \Delta r \tag{2.15}$$

where C is the rainfall per unit volume per unit time produced by coalescence; c is the suspended liquid water content; r is the raindrop radius and N_r is the corresponding raindrop concentration per unit radius, that is expressed by the Best distribution (Mason, 1971) through the rainfall rate; V_r is the fall speed of raindrops of radius r and E_r is the corresponding collection efficiency with the suspended droplets characterized by a representative radius of 10 μm (Mason, 1971). The quantity c is computed by:

$$c = 0.1\rho\left(q - 0.9q_s\right)K_3 \tag{2.16}$$

with q mixing ratio and $K_3 = 0$ if $K_1 = 0$ or $K_1K_2 = 1$ in Eq. (2.10) while in different conditions $K_3 = 1$. The quantity q is estimated before the orographic uplift, while $q_s(T,P)$ after the ascent through Eqs. (2.4) and (2.6) for given $P(Z)$.

The evaporation process in a layer of unsaturated air is treated through the aforementioned drop-size Best distribution.

The precipitation drift, linked with the fall speed of precipitation (5ms^{-1} as representative value for raindrops) and wind speed and direction, is incorporated to derive the rainfall rate at the ground level. The following procedure is adopted. The three-dimensional matrix $\hat{I}$ relative to the quantity I is first computed in the absence of drift. Afterwards the precipitation trajectories are estimated and a new matrix $\widehat{I'}$, in which all the I' elements of a column contribute to the surface rainfall rate at a given point (X,Y), is derived. The same procedure is used for the matrix of both the relative humidity, $\hat{F}$ after the ascent and $\hat{c}$, so obtaining $\widehat{F'}$ and $\widehat{c'}$. Then the precipitation rate at each level, from the highest to the lowest, is computed. Finally, the surface rainfall rate is smoothed by applying the operator 1-2-1 in both the X and Y-direction; such an approach is primarily suggested by the discretization scheme involved in the model.

The precipitation formed in a layer, $\left(\overline{I'} + \overline{C'}\right)\Delta Z$, is subjected to the constraint:

$$\overline{I'} + \overline{C'} \le \overline{I'}_{max} \tag{2.17}$$

where $\overline{I'}$ is the mean value of I' between the levels delimiting the layer and $\overline{I'}_{max}$ is the same mean referred to the I' values obtained for $K_1K_2 = 1$; the $\overline{C'}$ value is computed by Eq. (2.15) using the mean value of c' in the layer.

2.5 Study cases

Two case studies taken out from Corradini (1985) and related to rainfalls occurred ahead of surface warm fronts are reported to highlight the reliability of the diagnostic model described in the previous sub-section. A three-dimensional view of the involved geographic area is shown in Fig. 2.7, while Fig. 2.8 highlights the 250-, 550- and 850-m contours. The locations of the operative rain gauge stations are also indicated. The main features of topography can be synthetized by a flat region at 250 m and a contiguous SW-facing hilly region where a significant percentage of the land ranges in altitude from 550 to 850 m. The schematization of orography in a physical/mathematical model is a complex factor (Elvidge, 2019). For representing the terrain in the model, a smoothing procedure (see also Hill et al., 1981) was applied to isolate peaks and narrow valleys obtaining the simplified schematization of orography displaced in Fig. 2.9. The study area belongs to the Upper Tiber River basin, central Italy, where floods generally occur during the autumn-winter period and are caused by widespread frontal rainfalls. The rain gauge network consisted

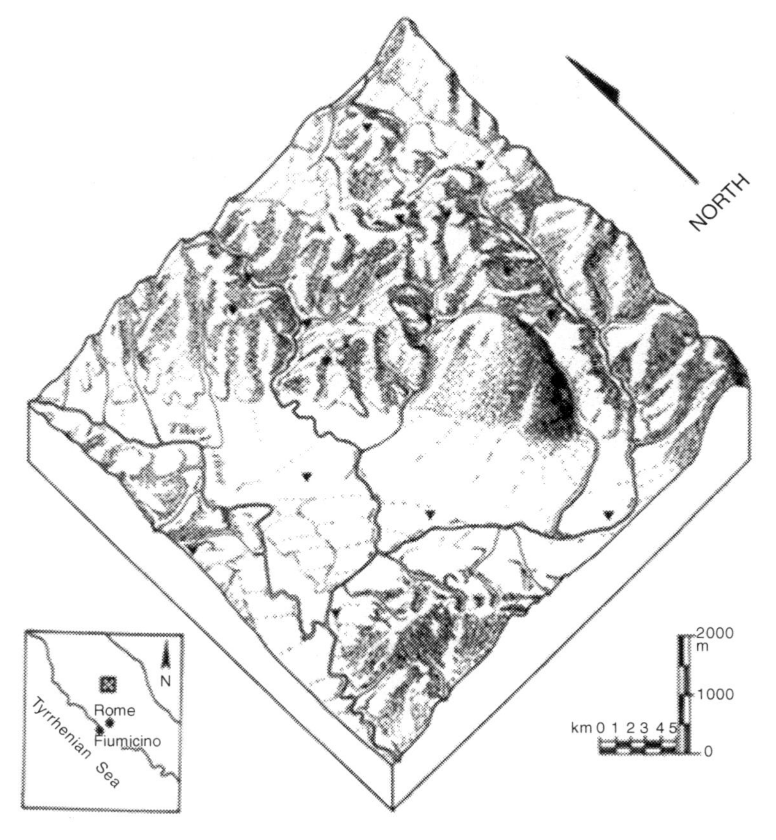

FIG. 2.7

Three-dimensional view of the model area with indication of the rain gauge positions (black triangles). (Modified from Corradini, 1982). The geographic location (Central Italy) is also shown.

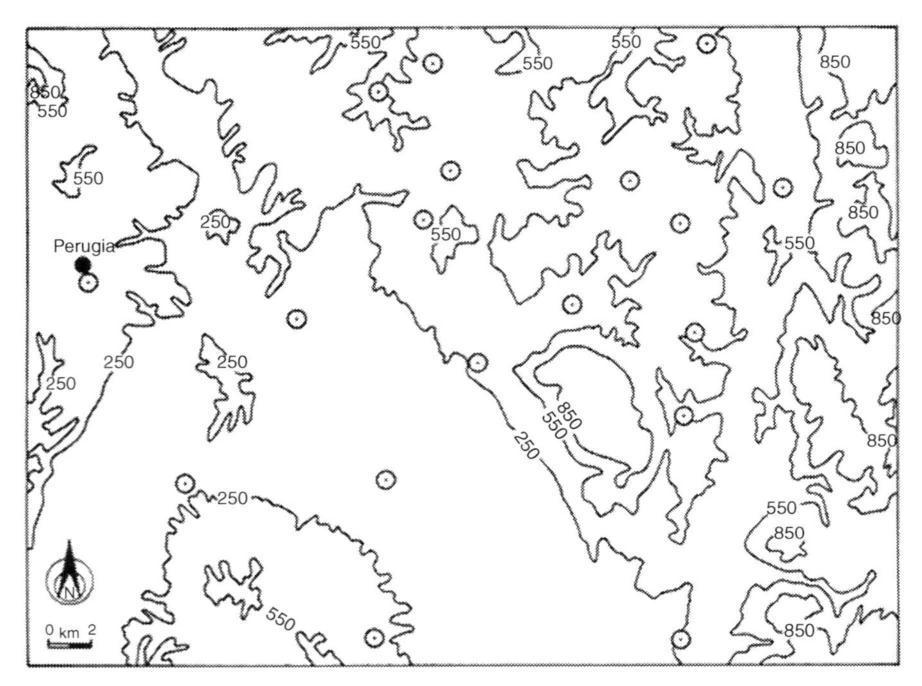

FIG. 2.8

Map of the model area. The 250-, 550-, and 850-m contours are shown together with the rain gauge positions (⊙). (From Corradini, 1985).

of eighteen rain gauges with continuous data logging. The model area was also equipped with meteorological sensors of air temperature and humidity, wind speed and direction, all located in the flat region far from ridges.

Two warm frontal systems came from SW and the investigated surface rainfalls were observed in the periods 12^h00^m–15^h00^m GMT on February 11, 1981 and 12^h00^m–15^h00^m GMT on March 2, 1981. During each rainfall event the temperature, relative humidity, wind speed, wind direction and rainfall rate at the soil surface were practically constant. Therefore, the input data to the model were assumed to be time invariant. For the first study case (Table 2.1) they were deduced from experimental data as specified below.

1. The values of T, F, wind speed and direction at the $Z = 0$ level were derived from the data observed by the available meteorological stations in the flat region. In particular, T and F were adjusted to obtain the observed saturation condition after the adiabatic uplift from $Z = 0$ to $Z = 250$ m.
2. The values of T, F, wind speed and direction at levels $Z > 0$ and atmospheric pressure at $Z = 0$ were taken from the Fiumicino vertical sounding performed at 12 GMT by the Air Force Meteorological Service of Italy. It is noted that the

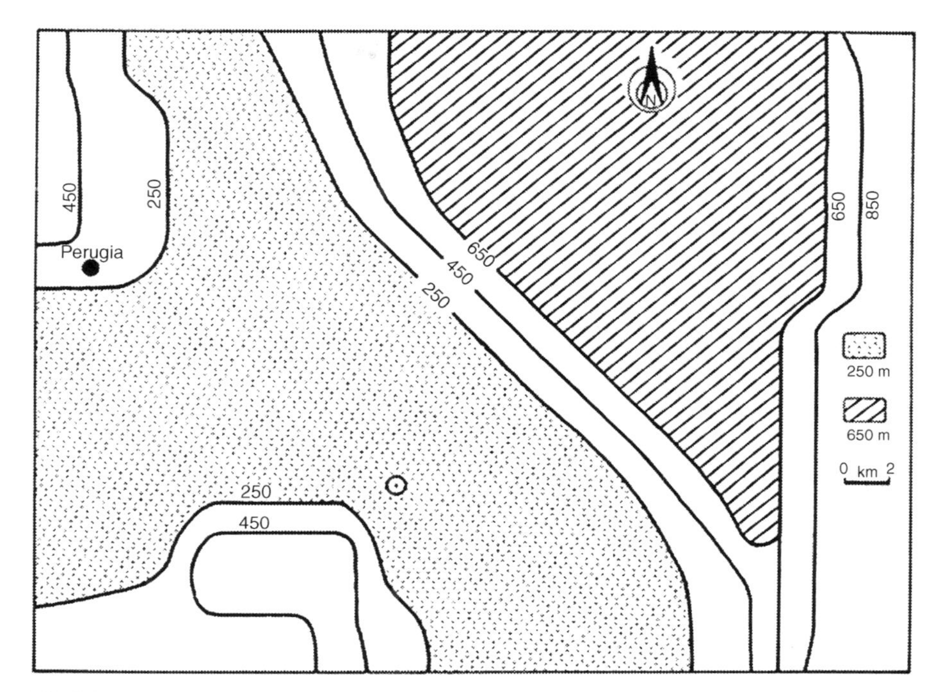

FIG. 2.9

Smoothed representation of the terrain in the model. The 250-, 450-, 650-, and 850-m contours are shown together with the location of the rain gauge (ʘ) used to estimate the rainfall falling above the atmospheric layers influenced by orography. (From Corradini, 1985).

Table 2.1 Model input meteorological data.

Altitude a.s.l. (m)	Absolute temperature (K)	Relative humidity (%)	Wind speed (ms^{-1})	Wind direction (deg)
(a) Period 12^h00^m–15^h00^m GMT on February 11, 1981 (P = 1017 mbar; P_{ex} = 0.43 g m^{-2} s^{-1})				
0	283	87	2.0	175
500	282	89	10.5	175
1000	280	88	8.4	180
1500	277	88	8.4	205
2000	274	87	14.2	225
(b) Period 12^h00^m–15^h00^m GMT on March 2, 1981 (P = 1014 mbar; P_{ex} = 0.30 g m^{-2} s^{-1})				
0	285	87	1.7	150
500	281	87	15.5	170
1000	279	80	14.0	195
1500	276	81	12.5	195
2000	273	82	11.0	190

P, atmospheric pressure at zero altitude; P_{ex}, rainfall rate at the 2000 m level.

T and F values led by the adiabatic ascent to nearly saturated air at each level over the flat surface at $Z = 250$ m.

3. Considering the nearly saturated air condition beneath $Z = 2000$ m, the rainfall rate P_{ex} was assumed equal to the value measured in the low-lying flat zone.

The computed and observed rainfall spatial distribution associated to the event occurred on February 11, 1981 are shown in Fig. 2.10a. Many experimental features are reliably reproduced by the model, in particular in the central region defined by the zone of orographic slope, with $W_T > 0$, and by the continuous downwind zone, which is influenced by the precipitation drift process, the rainfall field is fairly well simulated. The precipitation drift determined rather limited displacements, up to ~ 4000 m, because beneath $Z = 2000$ m the precipitation is in the rain form. These two zones are henceforth designated as CS zone and DS zone. Over the DS zone far from ridges the model produces rainfall rates similar to the ones computed over the low-lying zone at $Z = 250$ m. This is due to both the value of $W_T = 0$ over the last two zones and the nearly saturated air mass before the orographic uplift above 250 m. These theoretical results compare reasonably well with the experimental ones, except for a part of the DS region where the observed rainfall rates are larger than the values computed by the model. Quantitatively, using for each measurement the computed rainfall at the nearest grid point, the errors, ϵ, in the simulation of the mean rainfall rates have values $\epsilon = -4\%$ for the entire study area, $\epsilon_1 = -2$ for the CS zone, $\epsilon_2 = -15\%$ for the DS zone and $\epsilon_3 = -5\%$ for the remaining part of the model area. Furthermore 75% of the point estimates have an error in magnitude, $\epsilon_{(75)}$, less than 20%.

For the second study case the model input data (Table 2.1) were selected with a similar approach. The theoretical and experimental results are compared in Fig. 2.10b. As it can be seen also in this case the model provides a useful simulation of the rainfall field. In particular, ϵ, ϵ_1, ϵ_2, ϵ_3, and $\epsilon_{(75)}$ have values of 1%, -5%, 18%, 5% and less than 30%, respectively.

An overall analysis of the procedure adopted to apply the diagnostic seeder-feeder model to the study cases highlights a questionable element. The use of the vertical soundings carried out at the Fiumicino station does not represent an optimal solution because of the orography existing between this station (Fig. 2.7) and the study area, even though there are not high and extended barriers. A more appropriate and simple model application would have required the availability of a radiosonde station in the low-lying flat zone of the study area.

2.6 Conclusions

The knowledge of the mechanisms that contribute to the production of rainfalls in the different stages characterizing the frontal disturbances is of utmost importance in the formulation of prediction models for their distribution at different spatiotemporal scales. In this context, investigations based on integrated experimental systems with rain gauges, radiosondes, radars and satellite observations, even though carried out a few decades ago, provided crucial insights that should be further improved with

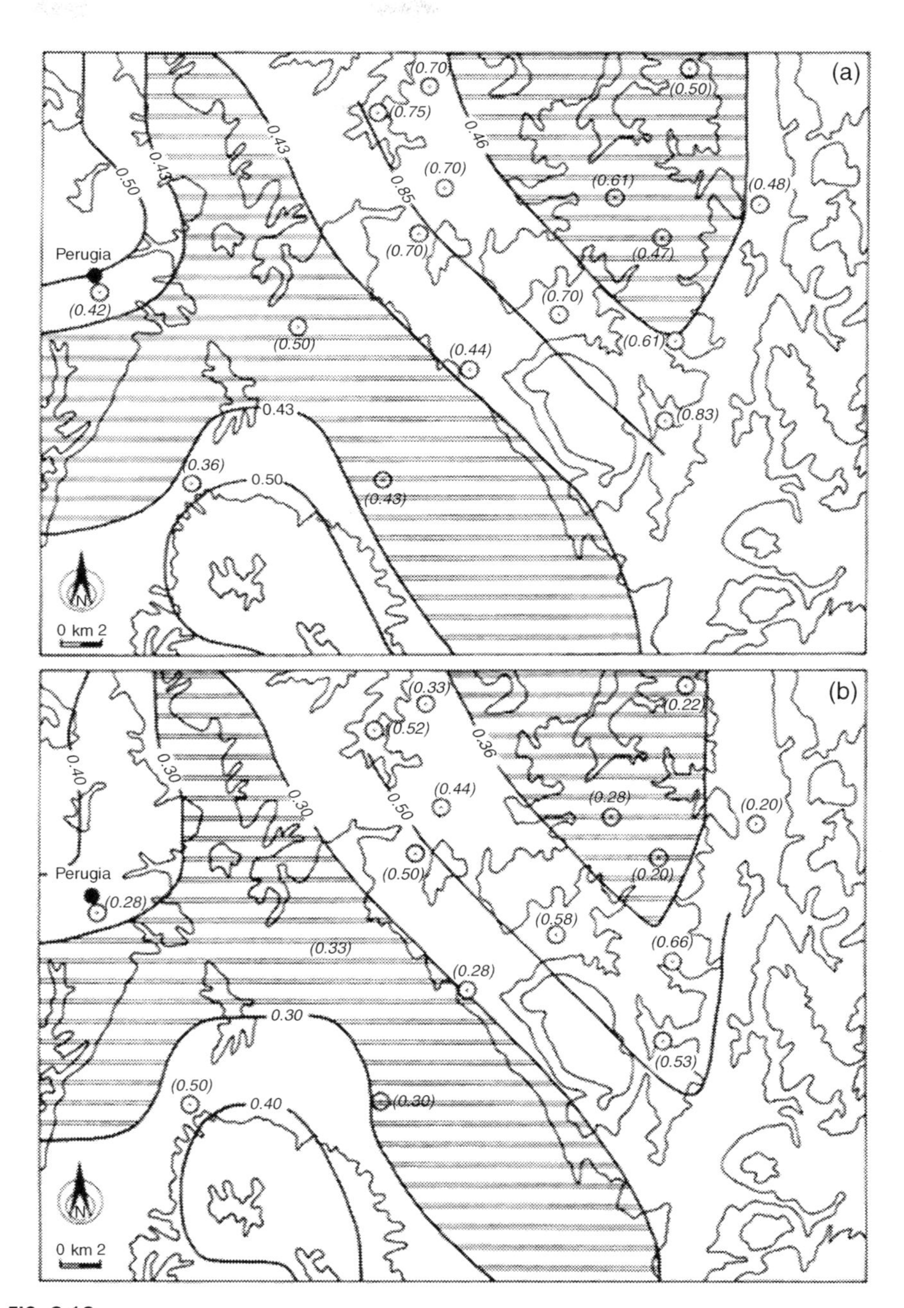

FIG. 2.10

Computed average rainfall spatial distribution ahead of warm fronts at soil surface, in g m^{-2} s^{-1}. Hatched areas indicate constant rainfall rate. Numbers enclosed in parentheses represent rain gauge (⊙) derived rainfall rates in g m^{-2} s^{-1}. The 250-, 550-, and 850-m contours are also shown. (a) 12^h00^m–15^h00^m GMT on February 11, 1981. (b) 12^h00^m–15^h00^m GMT on March 2, 1981. (From Corradini, 1985).

using new technologies (Borga et al., 2022; Kidd and Levizzani, 2022; Lanza et al., 2022). In this chapter primary importance has been given to studies performed in the British Isles because the study areas, with Atlantic frontal systems followed from seaboard to the first downwind hills and beyond, allowed to mark out the role of the mechanisms involved at different spatial scales.

A few open issues to be better understood for improving models for rainfall simulations concern the representation of orography (Smith et al., 2016; Elvidge et al., 2019), the development of moist orographic convection due to blocked and unblocked airflow by orographic barriers (Kirshbaum et al., 2018), the development of convective cells due to potential instability in the middle troposphere and the interaction between microphysics and dynamics of the atmosphere (Grabowski, 2022). Finally, it is noted that the presented diagnostic model could be also applied using input data provided by larger scale meteorological models.

References

Austin, P.M., Houze, R.A.Jr., 1972. Analysis of the structure of precipitation patterns in New England. J. Appl. Meteorol. 11 (6), 926–935.

Bader, M.J., Roach, W.T., 1977. Orographic rainfall in warm sectors of depressions. Q. J. Roy. Meteor. Soc. 103 (436), 269–280.

Bell, R.S., 1978. The forecasting of orographically enhanced rainfall accumulations using 10-level model data. Met. Mag. 107 (April 1978), 113–124.

Berne, A., Delrieu, G., Boudevillain, B., 2009. Variability of the spatial structure of intense Mediterranean precipitation. Adv. Water. Resour. 32 (7), 1031–1042.

Borga, M., Marra, F., Gabella, M., 2022. Rainfall Estimation by Weather Radar. In: Morbidelli, R. (Ed.), Rainfall. Modeling, Measurement and Applications. Elsevier, Amsterdam, pp. 109–134. doi:10.1016/C2019-0-04937-0.

Browning, K.A., Harrold, T.W., 1970. Air motion and precipitation growth at a cold front. Q. J. Roy. Meteor. Soc. 96 (409), 369–389.

Browning, K.A., Pardoe, C.W., 1973. Structure of low-level jet streams ahead of mid-latitude cold fronts. Q. J. Roy. Meteor. Soc. 99 (422), 619–638.

Browning, K.A., Hill, F.F., Pardoe, C.W., 1974. Structure and mechanism of precipitation and the effect of orography in a wintertime warm sector. Q. J. Roy. Meteor. Soc. 100 (425), 309–330.

Browning, K.A., Pardoe, C.W., Hill, F.F., 1975. The nature of orographic rain at wintertime cold fronts. Q. J. Roy. Meteor. Soc. 101 (428), 333–352.

Cheng, L.W., Yu, C.K., 2019. Investigation of orographic precipitation over an isolated, three-dimensional complex topography with a dense gauge network, radar observations, and upslope model. J. Atmos. Sci. 76 (11), 3387–3409.

Collier, C.G., 1975. A representation of the effects of topography on surface rainfall within moving baroclinic disturbances. Q. J. Roy. Meteor. Soc. 101 (429), 407–422.

Corradini, C., 1982. Analysis of widespread rainfall spatial distribution by a parameterized numerical model. Riv. Meteorol. Aeronaut. XLII (4), 383–389.

Corradini, C., 1985. Analysis of the effects of orography on surface rainfall by a parameterized numerical model. J. Hydrol. 77 (1–4), 19–30.

Cotton, W.R., Bryan, G., van den Heever, S.C., 2011. The Parameterization or modeling of microphysical processes in clouds. In: Cotton, W., Bryan, G., van den Heever, S.C. (Eds.),

Storm and Cloud Dynamics, International Geophysics, 99. Academic Press, Elsevier, USA, pp. 87–142.

Dorsi, S.W., Shupe, M.D., Persson, P.O.G., Kingsmill, D.E., 2015. Phase-specific characteristics of wintertime clouds across a midlatitude mountain range. Mon. Weather Rev. 143 (10), 4181–4197.

Elvidge, A.D., Sandu, I., Wedi, N., Vosper, S.B., Zadra, A., Boussetta, S., Bouyssel, F., van Niekerk, A., Tolstykh, M.A., Ujiie, M., 2019. Uncertainty in the representation of orography in weather and climate models and implications for parameterized drag. J. Adv. Model. Earth Sy. 11 (8), 2567–2585.

Grabowski, W., 2022. Rainfall Modeling. In: Morbidelli, R. (Ed.), Rainfall. Modeling, Measurement and Applications. Elsevier, Amsterdam, pp. 49–76. doi:10.1016/C2019-0-04937-0.

Harrold, T.W., 1973. Mechanisms influencing the distribution of precipitation within baroclinic disturbances. Q. J. Roy. Meteor. Soc. 99 (420), 232–251.

Harrold, T.W., Austin, P.M., 1974. The structure of precipitation systems – a review. J. Rech. Atmos 8 (1–2), 41–57.

Herzegh, P.H., Hobbs, P.V., 1980. The mesoscale and microscale structure and organization of clouds and precipitation in midlatitude cyclones. II: Warm-frontal clouds. J. Atmos. Sci. 37 (3), 597–611.

Hill, F.F., Browning, K.A., 1979. Persistence and orographic modulation of mesoscale precipitation areas in a potentially unstable warm sector. Q. J. Roy. Meteor. Soc. 105 (443), 57–70.

Hill, F.F., Browning, K.A., Bader, M.J., 1981. Radar and raingauge observations of orographic rain over South Wales. Q. J. Roy. Meteor. Soc. 107 (453), 643–670.

Hobbs, P.V., Locatelli, J.D, 1978. Rainbands, precipitation cores and generating cells in a cyclonic storm. J. Atmos. Sci. 35 (2), 230–241.

Hobbs, P.V., Persson, P.O.G., 1982. The mesoscale and microscale structure and organization of clouds and precipitation in midlatitude cyclones. Part V: The substructure of narrow cold-frontal rainbands. J. Atmos. Sci. 39 (2), 280–295.

Hobbs, P.V., Matejka, T.J., Herzegh, P.H., Locatelli, J.D., Houze, R.A.Jr., 1980. The mesoscale and microscale structure and organization of clouds and precipitation in midlatitude cyclones. I: a case study of a cold front. J. Atmos. Sci. 37 (3), 568–596.

Hofstätter, M., Chimani, B., Lexer, A., Blöschl, G., 2016. A new classification scheme of European cyclone tracks with relevance to precipitation. Water Resour. Res. 52 (9), 7086–7104.

Hou, T., Lei, H., Hu, Z., Yang, J., Li, X., 2020. Simulations of microphysics and precipitation in a stratiform cloud case over Northern China: comparison of two microphysics schemes. Adv. Atmos. Sci. 37 (1), 117–129.

Houze R.A., 2014. Nimbostratus and the separation of convective and stratiform precipitation. In: Houze R.A., (Ed.) Cloud Dynamics, International Geophysics 104. Academic Press, Elsevier, Oxfordx, UK, pp. 141–163.

Houze, R.A.Jr., Locatelli, J.D., Hobbs, P.V., 1976. Dynamics and cloud microphysics of the rainbands in an occluded frontal system. J. Atmos. Sci. 33 (10), 1921–1936.

Houze, R.A.Jr., Rutledge, S.A., Matejka, T.J., Hobbs, P.V., 1981. The mesoscale and microscale structure and organization of clouds and precipitation in midlatitude cyclones. III: air motions and precipitation growth in a warm-frontal rainband. J. Atmos. Sci. 38 (3), 639–649.

James, P.K., Browning, K.A., 1979. Mesoscale structure of line convection at surface cold fronts. Q. J. Roy. Meteor. Soc. 105 (444), 371–382.

James, C.N., Houze, R.A.Jr., 2005. Modification of precipitation by coastal orography in storms crossing Northern California. Mon. Weather Rev. 133 (11), 3110–3131.

Kidd, C., Levizzani, V., 2022. Satellite Rainfall Estimation. In: Morbidelli, R. (Ed.), Rainfall. Modeling, Measurement and Applications. Elsevier, Amsterdam, pp. 135–170. doi:10.1016/C2019-0-04937-0.

Kirshbaum, D.J., Adler, B., Kalthoff, N., Barthlott, C., Serafin, S., 2018. Moist orographic convection: physical mechanisms and links to surface-exchange processes. Atmos. Basel 9 (3), 80.

Lanza, L.G., Cauteruccio, A., Stagnaro, M., 2022. Rain Gauge Measurements. In: Morbidelli, R. (Ed.), Rainfall. Modeling, Measurement and Applications. Elsevier, Amsterdam, pp. 77–108. doi:10.1016/C2019-0-04937-0.

Mason, B.J., 1971. The Physics of Clouds. Oxford University Press, London, New York.

McFarquhar, G., 2022. Rainfall Microphysics. In: Morbidelli, R. (Ed.), Rainfall. Modeling, Measurement and Applications. Elsevier, Amsterdam, pp. 1–26. doi:10.1016/C2019-0-04937-0.

Murphy, A.M., Rauber, R.M., McFarquhar, G.M., Finlon, J.A., Plummer, D.M., Rosenow, A.A., Jewett, B.F., 2017. A microphysical analysis of elevated convection in the comma head region of continental winter cyclones. J. Atmos. Sci. 74 (1), 69–91.

Parsons, D.B., Hobbs, P.V., 1983. The mesoscale and microscale structure and organization of clouds and precipitation in midlatitude cyclones. IX: Some effects of orography on rainbands. J. Atmos. Sci. 40 (8), 1930–1949.

Pielke, R.A.Sr., 2013. Mesoscale meteorological modeling. In: International Geophysics 98. Academic Press, Elsevier, USA.

Plummer, D.M., McFarquhar, G.M., Rauber, R.M., Jewett, B.F., Leon, D.C., 2015. Microphysical properties of convectively generated fall streaks within the stratiform comma head region of continental winter cyclones. Mon. Weather Rev. 72 (6), 2465–2483.

Rauber, R.M., Geerts, B., Xue, L., French, J., Friedrich, K., Rasmussen, R.M., Tessendorf, S.A., Blestrud, D.R., Kunkel, M.L., Parkinson, S., 2019. Wintertime orographic cloud seeding—a review. J. Appl. Meteorol. Clim. 58 (10), 2117–2140.

Roe, G.H., Baker, M.B., 2006. Microphysical and geometrical controls on the pattern of orographic precipitation. J. Atmos. Sci. 63 (3), 861–880.

Rutledge, S.A., Hobbs, P.V., 1983. The mesoscale and microscale structure and organization of clouds and precipitation in midlatitude cyclones. VIII: A model for the "seeder-feeder" process in warm-frontal rainbands. J. Atmos. Sci. 40 (5), 1185–1206.

Sansom, H.W., 1951. A study of cold fronts over the British Isles. Q. J. Roy. Meteor. Soc. 77 (331), 96–120.

Smith, R.B., 1979. The influence of mountains on the atmosphere. In: Advances in Geophysics 21. Academic Press, Elsevier, New York, 87–230.

Smith, S.A., Field, P.R., Vosper, S.B., Shipway, B.J., Hill, A.A., 2016. A parametrization of subgrid orographic rain enhancement via the seeder-feeder effect. Q. J. Roy. Meteor. Soc. 142 (694), 132–142.

Smith, S.A., Field, P.R., Vosper, S.B., Derbyshire, S.H., 2019. Verification of a seeder–feeder orographic precipitation enhancement scheme accounting for low-level blocking. Q. J. Roy. Meteor. Soc. 145 (724), 2909–2932.

Visconti, G., 2016. Fundamentals of Physics and Chemistry of the Atmospheres. Springer International Publishing, Switzerland.

Wang, P.Y., Parsons, D.B., Hobbs, P.V., 1983. The mesoscale and microscale structure and organization of clouds and precipitation in midlatitude cyclones. VI. Wavelike rainbands associated with a cold-frontal zone. J. Atmos. Sci. 40 (3), 543–558.

CHAPTER

Rainfall modeling

3

Wojciech W. Grabowski[1]

Mesoscale and Microscale Meteorology Laboratory, NCAR, Boulder, CO, United States

[1]*National Center for Atmospheric Research is sponsored by the National Science Foundation*

3.1 Introduction

Clouds, especially those that produce significant precipitation, form because of the rising motion in the stratified atmosphere. The rising motion may be forced by environmental conditions (e.g., flow over topography or a frontal circulation) or may come from small-scale or mesoscale dynamics as in the case of moist convection and mesoscale convective systems (Corradini et al., 2022). Rising motion is accompanied by adiabatic expansion that lowers the air temperature and can lead to the condensation and formation of cloud droplets, the first step in precipitation formation. Rainfall modeling involves two separate aspects: simulation of the airflow (e.g., modeling of a synoptic or mesoscale weather system, a convective cloud, or a flow over topography) and simulation of microphysical processes that lead to formation and fallout of precipitation. Simulation of the airflow is the key element of the numerical weather prediction (NWP) as well as climate modeling. It involves solving air flow equations that include the impact of the water substance phase changes. Modeling cloud microphysics, on the other hand, has to do with representation of formation, transformation, and eventual fallout of cloud and precipitation particles that accompany air flow simulation. The two aspects are discussed in the following two sections emphasizing the progress made over the last several decades. Section 3.4 provides a brief discussion of two simulation methodologies for the moist nonhydrostatic atmospheric dynamics, contrasting their applications in small-scale and global modeling. A brief discussion of the outlook in section 3.5 concludes the chapter.

3.2 Simulation of precipitating system dynamics

Numerical models applied to simulate moist atmospheric dynamics are typically cast in the Eulerian frame with the fluid flow equations solved on a predefined mesh. Such a methodology is convenient for geophysical applications such as NWP because it has numerical benefits, for instance, from the mass and energy conservation point of

RAINFALL. Modeling, Measurement and Applications. DOI: https://doi.org/10.1016/B978-0-12-822544-8.00010-X

view. In the past, evolution of numerical models for moist atmospheric flow simulation proceeded in two separate paths. At large scales, such as in NWP and climate models, the necessity to apply large horizontal domains required large horizontal grid lengths, from a few ten kms for limited-area and global NWP models to several hundred kms for global climate models. In addition, such models needed to be run for extended periods, days and weeks for NWP and decades or longer for the climate simulation. The limited-area models (LAMs) applied computational domains covering a small fraction of the planet (say, Europe or North America), whereas climate models need to cover the entire Earth and are typically referred to as the atmospheric general circulation models (AGCMs). Because of the large horizontal grid lengths, hydrostatic flow equations were typically used. Hydrostatic approximation reduces the vertical momentum equation to the diagnostic relationship between the pressure and air density profiles within the column. Loosely speaking, the pressure comes from the weight of the air above a given level within the atmospheric column. The hydrostatic pressure is subsequently used in the horizontal flow equations, and the vertical velocity is diagnosed from the horizontal velocity components applying the air mass continuity. Hydrostatic pressure can be applied as the vertical coordinate because this simplifies model equations. Such an equation set is typically referred to as the primitive equations (Holton, 2004).

Because hydrostatic models cannot resolve nonhydrostatic small-scale motions, such as dry and moist convection or short gravity (buoyancy) waves, a critical component of such models is the representation of unresolved processes, typically referred to as "parameterizations." Examples of such parameterizations include the boundary layer parameterization (i.e., the representation of processes controlling the exchange of heat, moisture, and momentum between the Earth surface and the atmosphere), dry convection (to represent the impact of vertical mixing in regions where the atmospheric becomes statically unstable, that is, the temperature lapse rate becomes larger that the dry-adiabatic), and—most importantly for precipitation processes—the parameterization of moist shallow and deep convection. Since moist convection, shallow and deep, is inherently nonhydrostatic and requires order 1 km or smaller horizontal grid lengths, the moist convection parameterization is the only way to include convective processes in large-scale models based on primitive equations. Perhaps more importantly, moist convective instability is dramatically different from other forms of classical fluid dynamics instabilities, such as the Kelvin-Helmholtz or Rayleigh-Taylor instabilities that can be studied applying small-perturbation theory and linearized fluid flow equations (Chandrasekhar, 1961). Moist convective instability is in essence a finite-amplitude instability, where a sufficiently large vertical displacement is needed to bring the air to saturation and to allow the release of latent heat of condensation that provides positive buoyancy driving a convective updraft. It is thus not surprising that moist convection parameterization emerged as the key element of weather and climate models and it has occupied the modeling community from early days of the large-scale atmospheric modeling (Arakawa, 2004 and references therein). It remains the key component of AGCMs applied in climate modeling today. To improve the representation of near-surface processes, large-scale models often apply stretched vertical grid, with vertical grid length as small as

~100 meters near the surface and increasing to about 1 km or more in the middle and upper troposphere. Such models are capable in resolving large-scale, synoptic-scale, and some mesoscale dynamical processes related to precipitation formation, such as the frontal circulations in midlatitudes. Precipitation simulated by such models is typically divided between the one resolved by the simulated flow, the so-called large-scale precipitation, and the one represented by shallow and deep convection parameterizations. Examples of such hydrostatic limited area models applied towards the end of the 20th century in research and operational NWP include the US Mesoscale Model Version 4 (MM4; Anthes et al., 2021) and the European consortium applying the HIRLAM (High Resolution Limited Area Model) model (see http://hirlam.org/index.php/hirlam-programme-53/welcome-to-hirlam). Up to the end of the 20th century, such models were the workhorse of national weather centers. For the global NWP, it is impossible not to mention the European Center for Medium-Range Weather Forecasts (ECMWF, https://www.ecmwf.int/) and its Integrated Forecasting System (IFS) that is in continues development and improvement since mid-1970ies providing global NWP at increasing resolutions (at about 9 km horizontal resolutions and 137 vertical levels up to 0.01 hPa at the time of this writing).

Models simulating small-scale nonhydrostatic dynamics emerged in the 1960s (Ogura, 1962, 1963; Lilly, 1962; Orville, 1965) and became a viable small-scale research tool in the following decades. Such models do not rely on the hydrostatic approximation because they apply nonhydrostatic equations, either anelastic (Schlesinger, 1975; Clark, 1977; 1979; Lipps and Hemler, 1982) or compressible (Tapp and White, 1976; Klemp and Wilhelmson, 1978) and are capable of simulations targeting dry and moist convective processes. Small-scale nonhydrostatic models were initially applied in relatively short process-study-type simulations of moist convection, say, up to a few hours as in the above references, in different geometries (2D versus 3D) and applying various lateral boundary conditions, either open or periodic. Open lateral boundaries allow partial mitigation of the limited horizontal domain extent (Klemp and Wilhelmson, 1978; Lilly, 1981, and references therein), but they provide a rather imperfect representation of the interaction between simulated convective system and its far environment. Periodic lateral boundaries, on the other hand, do not allow any such interactions and are appropriate to study a closed system, especially when the horizontal domain is sufficiently large to include many clouds (Soong and Ogura, 1980; Soong and Tao, 1980; Tao and Soong, 1986; Tao et al., 1987; Krueger, 1988; Xu et al., 1992; Robe and Emanuel, 1996; see a review by Krueger, 2000). Such applications featured horizontal grid lengths of around 1 km and were referred to as the cloud-resolving models, or—perhaps more appropriately—convection-permitting models. This is because spatial resolutions significantly higher than ~1 km horizontal grid length are needed to resolve key elements of convective dynamics, such as entrainment (Bryan et al., 2003).

In the early 1990ies, the computational technology advanced sufficiently to allow extended-time (e.g., several days) limited-area convection-permitting simulations, that is, applying nonhydrostatic equations with a horizontal grid length ~1 km and computational domains covering horizontal areas up to ~million km^2. For the

NWP, examples of nonhydrostatic LAMs include the US Mesoscale Model Version 5 (https://a.atmos.washington.edu/~ovens/newwebpage/mm5-home.html) and its successor, the Weather Research and Forecasting (WRF) model (https://www.mmm.ucar.edu/weather-research-and-forecasting-model), the family of German nonhydrostatic models for mesoscale modeling in complex terrain (see review in Schluenzen, 1994), German Lokal Modell (LM) introduced in 1999 as a nonhydrostatic operational LAM for German Weather Service (Steppeler et al., 2003), and the French Meso-NH model (http://mesonh.aero.obs-mip.fr/mesonh51). For climate-related applications, nonhydrostatic limited-area simulations were driven by either idealized forcing (e.g., convective–radiative quasi-equilibrium as in Robe and Emanuel, 1996 or Tompkins and Craig, 1998) or applying observed evolving large-scale conditions obtained from field campaigns, such as GATE (Global Atmospheric Research Program [GARP] Atlantic Tropical Experiment) as in Grabowski et al. (1996, 1998) and Xu and Randall (1996). The periodic (in 2D) or doubly periodic (in 3D) horizontal domains were typically of the size of a AGCM grid box and allowed a simulation of cloud ensembles rather than a single cloud as typically applied in early cloud-scale simulations. The main motivation of such studies was to improve the understanding of processes that required parameterizations in hydrostatic models, moist convection in particular. These simulations led to some improvements of traditional parametrizations, for instance, the impact of convection on the surface exchange (Redelsperger et al., 2000) or cloud-radiation interactions (Xu and Randall, 1995). One of the key benefits of those cloud-resolving simulations was a demonstration that a 2D cloud-scale model can provide useful information about unresolved processes and can replace most of subgrid-scale parameterizations, deep convection in particular (Grabowski and Smolarkiewicz, 1999; Grabowski, 2001; Khairoutdinov and Randall, 2001; Randall et al., 2003). Such an approach, often referred to as the "super-parameterization", pushed the climate modeling community in a new direction. Cloud-resolving simulations continue to be used in process-studies today (e.g., in the discussion concerning possible convective invigoration in polluted environments, see the Fan and Khain (2021), Grabowski and Morrison (2021) and references therein, or in an investigation into subgrid-scale transport processes in the boundary layer as in Moeng et al., 2009, reaching even higher spatial resolutions (e.g., the large-eddy simulation or LES) to document the required resolution convergence (Bryan et al., 2003; Khairotdinov et al., 2009).

For the NWP, limited-area simulations (for instance applying the WRF model) usually apply lateral boundary conditions obtained for larger-scale models (e.g., AGCMs) and use a nesting technique with a small number (say, two or three) computational domains that cover smaller areas but feature increasing horizontal (and sometimes vertical) resolutions. Such an approach allows studying details of weather processes (including precipitation) with increasing fidelity, albeit over a small area only. The outer domain applies lateral boundary conditions interpolated from a larger-scale model and provides lateral boundary conditions for the inner higher-resolution domains. Inner domains can move to follow the specific weather event, like a tropical cyclone. The outer domain grid length is often not sufficiently high to simulate convective motions, and thus convection parameterization has to be used. However, the innermost domain features horizontal resolutions high enough so the convection parameterization is not needed. Such convection-permitting NWP systems are today's workhorses of national weather prediction centers of many countries. The increasing horizontal resolution of such modeling systems allows better representation of land-surface features (topography, coastlines, land features, etc.) and brings them closer to LES-type simulations.

Positive experience with convection-permitting limited-area models without convection parameterization motivated the development and applications of global nonhydrostatic convection-permitting models. The first one was the Japanese model NICAM (Satoh et al., 2014; see http://nicam.jp/hiki/). A similar more recent development is the German nonhydrostatic global Icosahedral Nonhydrostatic (ICON) model, an effort supported by the Max Planck Institute for Meteorology and the German Weather Service (Zängl et al., 2014; see http://www.mpimet.mpg.de/en/science/models/icon.html) and the NCAR's Model for Prediction Across Scales (MPAS; https://ncar.ucar.edu/what-we-offer/models/model-prediction-across-scales-mpas). Together with other convection permitting AGCMs, these models are involved in a DYAMOND (DYnamics of the Atmospheric general circulation Modeled On Nonhydrostatic Domains) initiative (https://www.esiwace.eu/services/dyamond), see preliminary results in Stevens et al. (2019). There is no doubt that this is the future of weather and climate modeling at the global scale.

What is the spatial resolution limit of weather and climate modeling in the future? One may argue that such a limit is the so-called large eddy simulation (LES). LES have been applied in the past to cloud modeling (Grabowski and Clark, 1993; Carpenter et al., 1998; Brown et al., 2002; Siebesma et al., 2003; Stevens et al., 2005) and more recently deep convective clouds (Bryan et al., 2003; Khairoutdinov et al., 2009; Lebo and Morrison, 2015). The entire premise of the LES approach is that the mean features of the turbulent flow are determined by the behavior of large (energy-containing) scales of motion, typically tens to hundreds of meters in the atmosphere, with smaller scales (down to the cloud microscale) slaved to the large eddies. Although such models typically feature quite sophisticated and physically well-posed subgrid-scale schemes to represent the impact of unresolved flow on the dynamics and transport, the effects of unresolved flow features on the growth of cloud particles are typically not considered. We will return to this issue towards the end of the next section.

In summary, modeling of the moist atmospheric dynamics, either for process-studies or numerical weather prediction, evolved over the last several decades from applying the hydrostatic equations with convection parameterization to convection-permitting nonhydrostatic models. The trend of applying higher spatial resolution towards LES in nonhydrostatic NWP will likely continue as higher horizontal resolution allow a better representation of surface features and such atmospheric flow features as fronts, boundary-layer circulations, gravity waves, shallow and deep convection, etc. The nonhydrostatic climate modeling will likely follow as the computational technology evolves.

3.3 Modeling of cloud microphysics

The growth of individual cloud and precipitation particles involves processes at sub-centimeter scales, typically referred to as the cloud microscale (these processes are discussed in considerable detail in McFarquhar (2022)). For instance, diffusional growth of a cloud droplet is driven by the molecular exchange of the water vapor and thermal energy between the droplet and its immediate environment. Similarly, collisions between cloud droplets to produce a raindrop or between small ice crystals to form a larger snowflake take place at the cloud microscale. At the same time, one cubic meter of cloudy air contains several tens of millions to a few billions cloud droplets. Although at smaller concentrations, the number of precipitation particles (raindrops, snowflakes, graupel particles, etc.) is still enormous for a typical grid volume of an NWP numerical model. It follows that such models cannot consider individual cloud and precipitation particles because of their sheer number. There are two possibilities to cope with this problem. First, one can use selected parameters of the particle population within a grid cell, for instance, their size distribution or the total mass, and predict spatial and temporal evolution of those parameters applying fluid flow predicted by the dynamic model. Second, one can select a small but representative subset of the particle population, predict their time and space evolution, and scale-up the subset to represent all particles within the grid cell volume.

The first approach is typically referred to as the Eulerian method because the time and space evolution of particle population characteristics is predicted in the same way as other continuous dynamic and thermodynamic variables in the Eulerian fluid flow model, such as the momentum, air density, temperature, etc. The variables applied are density-like (e.g., the total mass or number of particles per unit volume of air) or more often the so-called mixing ratios, that is, the mass or number of particles per unit mass of dry air. The mixing ratio is a convenient variable because it does not change when the air density changes due to air expansion as, for example, during air parcel rise in the atmosphere. The Eulerian methodology has been the workhorse of cloud-scale modeling from its early days (Kessler, 1969; Liu and Orville, 1969; Murray, 1970; Schlesinger 1973; Klemp and Wilhelmson, 1978; Clark 1979) and also in large-scale NWP and climate models. The discussion below focuses on the Eulerian methodology, starting with simple approaches, and reviewing the evolution of their complexity dictated by details of microphysical processes that need to be represented to simulate precipitation formation in different cloud systems. The evolution of the microphysical model complexity is illustrated in Fig. 3.1 taken from Grabowski et al., (2019). The second approach is referred to as the Lagrangian microphysics and because of its novelty it has only been applied in idealized simulations. However, its benefits are significant, and this is why its short discussion is included towards the end of this section. Fig. 3.1

A casual observation of natural clouds suggests a simple partitioning of cloud populations into precipitating and nonprecipitating, see Fig. 3.2 for a simple illustration. Moreover, clouds often form as nonprecipitating, and they develop precipitation at some point later during their lifecycle. For instance, cumulus clouds start raining after

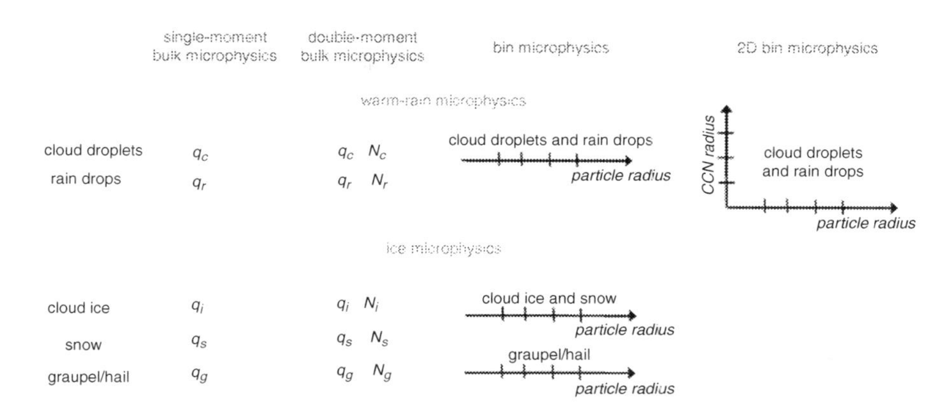

FIG. 3.1 An illustration of the increasing complexity of Eulerian microphysics schemes. Single-moment bulk microphysical schemes predict only mass mixing ratios q of various cloud and precipitation categories (cloud water, rain, cloud ice, snow, graupel/hail). Double¬moment bulk schemes add the corresponding number mixing ratio N that together with q allows estimation of the mean particle size. Bin microphysical schemes represent a size distribution of each category. If additional detail is needed (e.g., chemical composition of drops due to CCN, or ice crystal habits) then 2D (or even more dimensional) size distributions are required. Reproduced with permission from Grabowski et al. (2019).

they become sufficiently deep, that is, during a later stage of their natural lifecycle. The classical lifecycle of a precipitating convective cell, from developing, through mature, to dissipating, was formulated based on observations during the Thunderstorm Project in the US shortly after World War II (Byers and Braham, 1948). Convective cells are building blocks of precipitating convection, such as multicellular thunderstorms or mesoscale convective systems (e.g., squall lines). The partitioning between precipitating and nonprecipitating clouds provides a simple methodology of cloud and precipitation modeling. The idea, well-posed for clouds that develop in environments above freezing (i.e., ice-free clouds), is that clouds form through water vapor condensation to form cloud droplets once the air reaches saturation with respect to the plain water surface, see discussion in McFarquhar (2022). In reality, some supersaturation is needed to form and grow cloud droplets, but the ubiquity of cloud condensation nuclei (CCN) on which cloud droplets form limits the supersaturation to relatively small values, say, of the order of 1%. Because ice-free clouds are typically close to water saturation (however, this is not always the case), a very convenient and numerically simple technique to model such clouds is the so-called saturation adjustment. Saturation adjustment assumes that the air is always at saturation by either condensing the excess of water vapor to form or grow cloud droplets or evaporating the available cloud water to bring the water vapor back to saturation. This is the simplest approach to model the phase change from water vapor to cloud water. The key variable is the mass that changes from water vapor to the cloud water because it determines the accompanying release of the latent heat of condensation. Fig. 3.2

FIG. 3.2 A simple depiction of (left) nonprecipitating and (right) precipitating cloud.

See text for a discussion. The air temperature together with the mass of water vapor and the condensed cloud water determine the density of the air volume. The differences in the air density due to changes of the temperature, water vapor, and cloud water affect the cloud buoyancy, the excess of the air density over its environmental (reference) value. The buoyancy drives the air vertical motion in the nonhydrostatic model. Because cloud droplets are typically small (with radii of the order of 10 micron), their fall velocity is small (around 1 cm/sec) and thus can be neglected considering vertical velocities in precipitating weather systems. Similar argument applies to ice-bearing clouds as small ice crystals fall with similar velocities. This explains why, from the distance, clouds look like suspended in the air as shown in the left cloud. As a result, the cloud water mixing ratio (as well as the cloud ice mixing ratio) can be assumed to follow the air motion. When compared to cloud droplets, drizzle and rain drops have significantly larger fall velocities (i.e., around 1 m/s for drop radius of 0.5 mm) and thus can fall out of the cloud as shown in the schematic of a precipitating cloud.

Once sufficient mass of cloud water is produced through condensation, one can expect formation of drizzle- or rain-size drops, that is, drops larger than about 100 microns (i.e., 0.1 mm). Drizzle and rain drops form through droplet and drop collisions or by melting of ice particles falling from above through the melting level. Direct formation of drizzle-size drops through the diffusion of water vapor is only possible on extremely large CCN particles, referred to as giant CCN, with dry sizes larger than 1 micron. Modeling of droplet collisions leading to drizzle and rain formation is possible by applying the so-called bin microphysics, that is, an approach in which evolution of the particle size distribution is predicted (this modeling technique will be discussed later in this section). The drizzle/rain category can be modeled in a similar way to the cloud water (i.e., by an additional density or mixing ratio), with the only exception being its fall velocity relative to the air. The average fall velocity of the drizzle/rain mass can be formulated assuming a size distribution of raindrops and applying a relationship between the drop fall velocity and its radius. The former is typically assumed to be exponential following pioneering observations by Marshall and Palmer (1948) for raindrops and Marshall and Gunn (1952) for snowflakes. Formulation of fall velocities for the entire size range of liquid particles (droplets and drops) is quite involved. Cloud droplets fall as solid Stokes particles with their fall (terminal) velocity increasing with the square of their radii. This is valid for droplets with radii smaller than about 30 microns. Formulation of the fall velocity for larger droplets, drizzle and rain drops is more involved. Raindrops are deformed as they fall, and large raindrops oscillate because of the complex unsteady air flow around them. They beak either spontaneously or – much more often –because of collision with other drops, see discussion in McFarquhar (2022). The comprehensive formulation of cloud droplet and raindrop fall velocities that is used by modeling community to this day was compiled by Beard (1976) For ice particles, the formulation typically involves a combination of theoretical formulations (e.g., similar to Beard's publication) as well as numerous field and laboratory observations (see references in Pruppacher and Klett, 1996). All these are the key ingredients of modern cloud physics modeling.

The initial formation of drizzle/rain mixing ratio out of the cloud water is referred to as the autoconversion. This is the key process affected by the details of the droplet size distribution. Observations suggest that the mean cloud droplet radius needs to exceed a threshold value, around 12 μm to develop significant drizzle or rain (Rosenfeld and Gutman, 1994; Rosenfeld, 2000; Pawlowska and Brenguier, 2003; Khain et al., 2013). This is because smaller droplets have insignificant inertia and tend to avoid collisions by following the air flow around a larger droplet, at least in the case of gravitational collision/coalescence. The tendency for droplets to avoid collisions is reflected in a dramatic increase of the collision efficiency from a fraction of 1% for collision between droplet with radii around 10 μm to close to 100% for drops of several tens of micros radii (see Table 1 in Hall, 1980). It follows that the mean radius is a better indication of a cloud susceptibility to develop drizzle/rain than the cloud water mixing ratio. Unfortunately, microphysics schemes predicting only the condensed water mass, referred to as single-moment bulk schemes (see Fig. 3.1), do not provide the mean droplet size information, and the autoconversion can only be formulated based on the cloud water mixing ratio (for instance, making the autoconversion proportional to the cloud water mixing ratio). When rain is already present, it falls through a cloud and collects cloud droplets. This process is simpler to represent because it involves fast-falling raindrops interacting with slowly falling cloud droplets, and the continuous collection model (see Fig. 1 in Berry, 1967 and references therein) provides an appropriate context. When rain falls outside a cloud, it can partially or completely evaporate in the subsaturated air before reaching the surface.

In summary, the basic equations describing moist thermodynamics with condensation and rain formation can be written as follows (Kessler, 1969; Klemp and Wilhelmson, 1978; Grabowski and Smolarkiewicz, 1996):

$$\frac{D\theta}{Dt} = \frac{L_v \theta}{c_p T}(C_d - EVAP)$$

$$\frac{Dq_v}{Dt} = -C_d + EVAP$$

$$\frac{Dq_c}{Dt} = C_d - AUT - ACC$$

$$\frac{Dq_r}{Dt} = \frac{1}{\rho}\frac{\partial(\rho q_r v_t)}{\partial z} + AUT + ACC - EVAP$$

$$\frac{D}{Dt} = \frac{\partial}{\partial t} + u \cdot \nabla$$

These equations provide the simplest form of the so-called single-moment bulk microphysics parameterization, see Fig. 3.1. The top equation describes the evolution of the potential temperature θ that is conserved in dry-adiabatic motions, but changes through the latent heat release. The latent heat of condensation comes from either conversion of water vapor into cloud water (the C_d term) or evaporation of rain (the $EVAP$ term). These two terms affect the water vapor mixing ratio, q_v. Cloud water and rainwater mixing ratios, q_c and q_r, respectively, change by condensation (C_d), conversion from cloud water to rain (AUT and ACC, separated into autoconversion and accretion), and rain evaporation ($EVAP$). The rain equation also includes the rain fallout term, the first term on the right-hand-side, the divergence of the vertical rain flux (ρ is the air density and v_t is the rain fall velocity). The material derivate (shown at the bottom of the above panel) is typically converted into the flux form (with help of the air mass continuity equation) to ensure conservation of the thermal energy and water mass in the numerical implementation of the model-resolved transport. Moreover, the above equations are typically supplemented by representations of processes beyond moist precipitating thermodynamics, such as the impact of surface fluxes of water vapor and thermal energy, radiative transfer, unresolved subgrid-scale processes or other parameterizations.

Modeling condensation through saturation adjustment does not require any information about droplet concentration because their total mass (or mixing ratio) is the only relevant variable. In contrast, formulation of physically-based autoconversion term AUT in the above equations requires information about droplet concentration as already mentioned above (Seifert and Beheng, 2001; Khairoutdinov and Kogan, 2000; Kogan 2013). Moreover, droplet concentration together with the total liquid mass provides estimate of the mean droplet radius, a parameter important for the interaction of clouds with solar radiation (e.g., the partitioning between transmitted and reflected radiative flux in the cloudy atmosphere). For that reason, incorporation of the number mixing ratios, in addition to the mass mixing ratios in the above equations, is the next logical step to expand sophistication of the bulk microphysics parameterization (see Fig. 3.1). Once the number mixing ratios are added to the above equations, such a scheme is typically referred to as the double-moment bulk scheme, that is, predicting two moments of the particle size distribution, the mean mass and the mean concentration (Meyers et al., 1997; Khairoutdinov and Kogan, 2000; Seifert and Beheng, 2001; Morrison and Grabowski, 2007). Model equations for the Eulerian double-moment warm-rain scheme are similar to those shown above, with mass and number mixing ratios for cloud water and drizzle/rain, and with appropriate formulation of fall velocities (i.e., mass-waited versus number-weighted for mass versus number mixing ratios, respectively), see, for instance, Morrison and Grabowski (2007). To predict the droplet concentration, one needs to consider processes that lead to the initial droplet formation, the CCN activation (see McFarquhar, 2022). This requires prediction of the supersaturation inside a cloud instead of applying saturation adjustment. However, microphysical schemes that assume saturation adjustment can also include droplet concentration prediction by parameterizing processes leading to CCN activation. For instance, such a parameterization can consider vertical velocity near the cloud base and simply represent the concentration of activated CCN as a function of the updraft speed near the cloud base (Abdul-Razzak and Ghan, 2000; Saleeby and Cotton, 2004). Changes in droplet concentration due to some processes can be modelled without much difficulty (e.g., droplet collisions), but other require some uncertain assumptions, like for example, the impact of the unsaturated environmental air entrainment that depending on the mixing assumptions can lead to a wide range of concentration changes (the so called homogeneous versus inhomogeneous mixing, see Jarecka et al. (2013) and references therein).

The most comprehensive and thus computationally most demanding is the spectral (bin) microphysics that aims at prediction of the droplet/drop spectra (Kogan, 1991; Khain and Sednev, 1995; Khain et al., 2004). The droplet/drop size distribution is represented by the spectral density function for which the evolution equation includes initial formation of the droplet (i.e., CCN activation), advection in the physical space, and transformations due to particle population changes (i.e., growth by condensation and by collision/coalescence). For the collision/coalescence, the changes of the spectral density function are represented by the Smoluchowski equation (Smoluchowski, 1916), referred to in the cloud physics as the stochastic collection equation. The spectral density function is discretized in a finite number of droplet size bins and interactions between all bins need to be consider to calculate growth by collision/coalescence (see examples of solutions in Berry and Reinhardt, 1974a, b, c). In simulations of natural clouds, each bin has to be advected in the physical space similarly to cloud water or rainwater categories in the above equations. With typically a few dozens of bins, the physical space advection of each bin is the most computationally intensive part of the bin scheme. Representing droplet growth and evaporation requires prediction of the in-cloud supersaturation that is numerically cumbersome, especially when compared to the simplicity of the saturation adjustment. This adds to the complexity of the model applying bin microphysics. Bin microphysics schemes come in different specific implementations. For instance, one can have either a single- or double-moment bin microphysics, that is, predicting either one variable per bin (i.e., either the mean mass or the mean number of droplets within a bin as, for example, in Kogan, 1991; Khain and Sednev, 1995; and Bott, 1998) or predicting both the mean mass and number (Tzivion et al., 1987, 1989; Feingold et al., 1988; Simmel et al., 2002). If evolution of the CCN spectrum due to droplet collisions and drop fallout (referred to as the CCN processing) is to be simulated, a 2D bin microphysics is required, that is, one dimension for the droplet size (as in 1D bin microphysics) and the other for the CCN mass within a drop of a given size (Lebo and Seinfeld, 2011). See Fig. 3.3 for an illustration of the 1D and 2D bin microphysics. Transporting all bins in the physical space makes the 2D bin microphysics extremely expensive. Fig. 3.3

Modeling ice processes is by far more complicated for several reasons. First, natural ice particles occur in variety of forms and shapes, referred to as habits, depending on the temperature and supersaturation conditions they grow in as well as on their growth history. Second, initial formation of ice in natural clouds (often referred to as cloud glaciation) is possible in variety of ways, such as a direct formation of

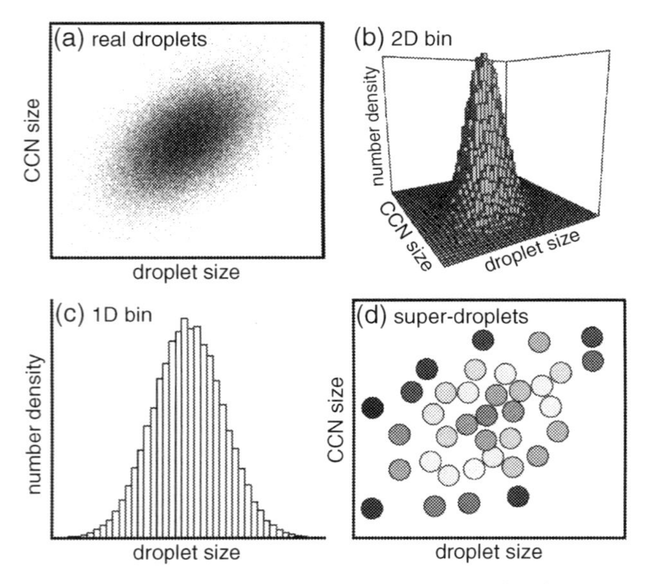

FIG. 3.3 Comparison between real droplets, super-droplets and bin microphysics.

Panel (a) shows cloud droplets with different sizes (horizontal axis) each containing a different CCN size (vertical axis). Such a droplet ensemble can be represented by a two-dimension number density function as shown in panel (b). If CCN is of no interest, the ensemble can be represented by a one-dimensional number density function as in panel (c). If used in a cloud model, each bin in (b) and (c) needs to be advected in physical space and all bin combinations have to be considered in collision-coalescence calculations. Panel (d) shows a super-droplet representation of the ensemble. Each symbol shows a single super-droplet on the same plane as in panel (a), with colors depicting increasing multiplicity from very low multiplicity (dark blue), through low to moderate multiplicity (green and yellow), to high multiplicity (red). Transport and growth of the real droplet ensemble in (a) is represented in a computationally tractable way by the orders-of-magnitude smaller super-droplet ensemble in (d). Reproduced with permission from Grabowski et al. (2018).

ice from the water vapor or freezing of already-present cloud droplets, see discussion in McFarquhar (2022). For temperatures warmer than about –40°C, initial formation of an ice particle requires presence of the ice forming nucleus (IFN), typically an insoluble organic or inorganic particle (e.g., mineral dust) featuring crystal lattice similar to ice. Only for temperatures colder that about –40°C, ice can form by homogeneous freezing of small liquid droplets. Finally, in contrast to liquid droplets and drops, growth of the ice phase can take place through variety of mechanisms, such as the diffusional growth of individual ice crystals, their collisions to produce large snowflakes, and collisions between ice particles and supercooled liquid droplets and drops that freeze upon contact with the ice particle, referred to as growth by riming. Moreover, the ice phase can grow in different ways for the same macroscopic cloud properties (such as the cloud updraft strength) depending on the microphysical cloud characteristics. For instance, when ice crystal concentration is high, water vapor can stay supersaturated with respect to ice and subsaturated with respect to water, without forming supercooled liquid water droplets. With the same updraft and a small ice crystal concentration, water vapor can reach water saturation and supercooled cloud droplets can form. All those factors affect the way ice processes can be modeled.

The simplest way (Dudhia, 1989; Grabowski, 1998) is to assume that cloud water and rain water mixing ratios in the bulk warm-rain equations above change into cloud ice and snow for cold temperatures, affecting the bulk mass growth rates and fall velocities. In addition, saturation adjustment that applies water-saturation in warm temperatures changes to the saturation with respect to ice in ambient temperatures way below freezing. In essence, the form of the equations and the number of mixing ratio variables stays the same, only the formulation of various terms changes depending on the ambient temperature. However, the simplest approach does not allow representation of various ice growth mechanisms, for instance, separating growth by the diffusion of water vapor (that forms the picturesque individual ice crystals) from the growth by riming. To represent ice particles grown by the two mechanisms, one can include two separate mass mixing ratios for precipitating ice: one mixing ratio to represent ice particles grown by the diffusion of water vapor and by collisions of such particles to produce large snowflakes, and the second category for ice particles grown by riming (i.e., graupel and possibly hail). As a result, one has three classes of ice: cloud ice (small ice crystals similar to cloud water, carried by the flow), snow (large ice crystals and their aggregates, snowflakes, with fall velocities around 1 m/s) and graupel or hail (large ice particles grown by riming with fall velocities up to several m/s). Hence, in addition to cloud water and rain water mixing ratios, such a set of model equations includes the cloud ice mixing ratio (similar to the cloud water equation above) and two precipitating ice mixing ratio equations, one for the snow and one for the graupel/hail. Such a set of equations includes more complicated expressions for the latent heat release (of sublimation/resublimation for water vapor to ice phase change and of freezing/melting for the liquid water to ice phase change) together with more involved representations of various transfer processes between various forms of vapor, liquid, and solid water substance. Such methodology was independently developed by two groups in the 1980ies (Lin et al., 1983; Rutledge and Hobbs, 1984) and with various modifications provide the most popular methodology used in numerical models today. Such an approach can be expanded to include prediction of the number mixing ration for some or all mass mixing ratios (Thompson et al., 2004; Lim and Hong, 2010). The schematic of microphysical processes involved in such schemes is shown in Fig. 3.4

The key aspect (and arguably an important drawback) of bulk approaches that divide the bulk ice mass into predefined categories of cloud ice, snow and graupel/hail is the need to formulate various conversion rates, from cloud ice to snow and from snow to graupel. The key problem is that it is difficult to decide when an ice crystal continuously growing by the diffusion of water vapor ceases to be "cloud ice" and begins to be "snow". Similarly, a gradual growth by riming of an ice particle makes it difficult to decide on the boundary between rimed snowflake (still in the snow category) and a graupel particle. This is in contrast to warm rain processes

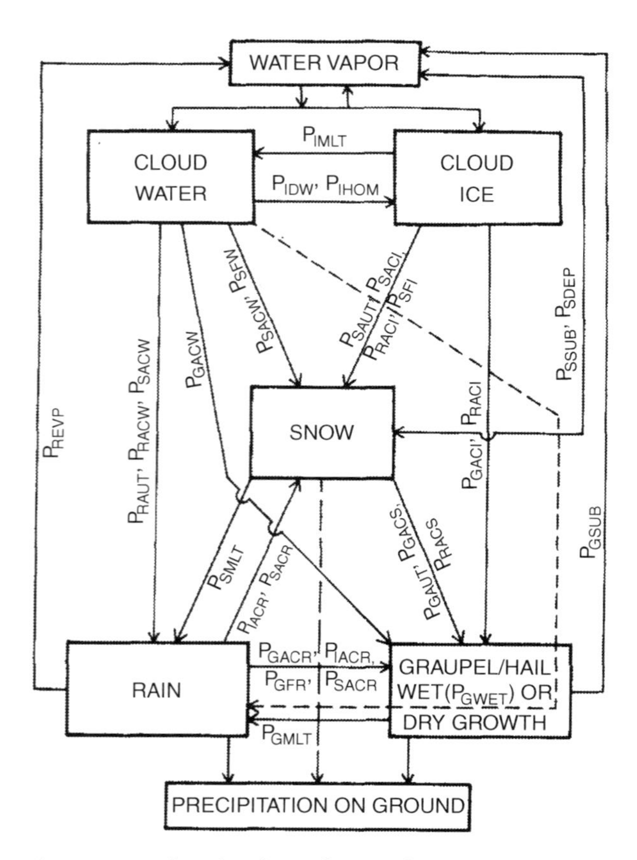

FIG. 3.4 Schematic representation of various microphysical processes that result in the formation of precipitation from the water vapor carried by the air.

Precipitation falls out of a cloud and may reach the surface. From Lin et al. (1983).

where separation between diffusional growth of cloud droplets and growth by collision/coalescence leading to drizzle/rain makes the partitioning between cloud water and drizzle/rainwater justified. Although including both the mass and the number mixing ratios for some or all those three classes of ice particles can provide a more physically based thresholds and conversion rates, the key problem remains. Adding more moments beyond mass and number mixing ratios can be beneficial for the overall scheme performance. For instance, Milbrandt and Yau (2005, 2006) discuss series of simulations with single-, double-, and triple-momnt bulk microphysics. They noticed that, in hail-producing thunderstorm simulations, the largest improvement is seen when a single-moment scheme is replaced with a double-moment scheme, with additional improvements possible when a triple-moment scheme is used. Finally, contemporary bin ice microphysics schemes follow the same logic of partitioning the ice field into particles that grow by the diffusion of water vapor and by aggregation (i.e., collision of ice crystals to form a snowflake) and particles that grow by riming to form graupel and hail (see Fig. 3.1; Khain et al., 2013; Lebo et al. 2012). Such complex schemes require representation of all ice processes together with controlling parameters (some poorly constrained by laboratory and field observations), and for that reason model results remain uncertain. This is highlighted in a recent comparison between three state-of-the-art bin schemes in a squall line simulation (Xue et al., 2017) that shows a surprising spread when compared to observed system.

The remove the artificial partitioning of the ice field into predefined ice categories, Morrison and Milbrandt (2015) developed a bulk ice scheme that predicts ice particle properties, the P3 (predicted particle properties) scheme. The key aspect is to predict time and space evolution of various bulk ice particle properties, such as the total mass and number mixing ratios, the rime mass and the rime volume, thus allowing physically based representation of various ice growth mechanisms within a bulk microphysical framework. One can argue that such an approach is conceptually similar to the Lagrangian methodology discussed below, with the key difference that the particle properties are evolved in the Eulerian frame. Despite having a single ice category (rather than three categories in bulk schemes described above), model simulations of an idealized squall line show realistically wide range of particle characteristics in different regions of the squall line (e.g., convective versus stratiform) in general agreement with observations. Morrison et al. (2015) discuss application of the P3 scheme to three-dimensional simulations of a squall line and an orographic precipitation event. Because of the conceptual simplicity of the P3 scheme, its simplified framework has been recently included in the ice scheme of the NCAR's Community Atmospheric Model version 5 (CAM5), the atmospheric component of the NCAR's climate model (Eidhammer et al., 2017).

We finish the discussion of the Eulerian microphysics with a general comment that the sensitivity of model results, including surface precipitation, to the representation of cloud microphysics is well appreciated by the NWP community (Weisman et al., 2008; Lean et al., 2008; Clark et al., 2012; Stein et al., 2015) and high-resolution regional climate modeling community (Liu et al., 2011 and references therein).

In contrast to the Eulerian approaches discussed above, the Lagrangian microphysics involves following in time and space a representative set of cloud particles and calculating their growth individually rather than predicting evolution of the population parameters as in the Eulerian scheme. Of course, one cannot follow all cloud and precipitation particles, so each computational particle represents a large ensemble of particles and can be referred to as a "super-particle" following the terminology introduced by Shima et al. (2009) for the case of the warm-rain microphysics, the "super-droplet method". Fig. 3.3 illustrates the conceptual difference between Lagrangian and Eulerian bin approaches. The Lagrangian scheme has so far been applied in research-like applications and focusing on warm-rain microphysics. Because of computational constraints, a typical number of super-particles per grid volume is between a few tens to several hundred (Arabas and Shima, 2013; Unterstrasser and Sölch, 2014; Grabowski et al. 2018). Examples of recent applications of Lagrangian microphysics to warm ice-free cloud simulations include Andrejczuk et al. (2008, 2010), Shima et al. (2009), Riechelmann et al. (2012), Arabas and Shima (2013), Hoffmann et al. (2015, 2019), Dziekan et al. (2019), Chandrakar et al. (2021). Grabowski et al. (2019) highlight advantages of the Lagrangian methodology for simulation of warm (ice-free) clouds when compared to the Eulerian approach reviewed earlier in this section. As far as ice processes are concerned, the Lagrangian ice microphysics has been applied in simulations of upper-tropospheric ice clouds such as cirrus and contrails (Sölch and Kärcher. 2010; Unterstrasser and Sölch, 2010). Only very recently a novel scheme was developed to simulate warm-rain and ice processes together, targeting deep convection (Shima et al., 2020). Such a comprehensive scheme can be applied in NWP simulations in the future. One of the key advantages of Lagrangian microphysics – in contrast to the Eulerian approach – is its capability to include physically-based representation of unresolved scales of motion. This is because super-particle motion and growth can be linked to parameterized subgrid-scale processes, such as vertical velocity fluctuations for the droplet growth (Grabowski and Abade, 2017) or supersaturation fluctuations resulting from mixing between a cloud and its environment (Hoffmann et al., 2019). This potentially can lead to a truly multiscale modeling of cloud microphysics in the future. Other advantages of the Lagrangian microphysics are discussed in Grabowski et al. (2019).

3.4 Convection-permitting modeling: anelastic versus compressible approach

As mentioned in section 3.2, there are two different approaches to model nonhydrostatic dynamics associated with convective motions in deep atmospheres. Anelastic approach (Ogura and Phillips, 1962; Lipps and Hemler, 1982; Clark 1977) is a methodology valid for low-Mach number flows (i.e., when the fluid velocity is much smaller than the speed of sound) and is in essence an extension of the incompressible dynamics to stratified deep atmospheres, that is, when the fluid depth is larger than the density scale height. In the anelastic system, the incompressible continuity equation div($\boldsymbol{u}$) = 0 (where $\boldsymbol{u}$ is the fluid velocity) replaced by div($\rho_0\, \boldsymbol{u}$) = 0, where $\rho_0(z)$ is the density profile. The key advantage of both incompressible and anelastic systems is the lack of sound waves that limit stability of the numerical algorithms because of their fast propagation speed. Sound waves play no role in climate- and weather-related low-Mach-number atmospheric flows and their presence in compressible Navier-Stokes equations creates a significant challenge for the numerical stability. An often-used technique in compressible atmospheric models, going back to the seminal work of Klemp and Wilhelmson (1978) and still used (with some modifications) in today's community models such as the Weather Research and Forecasting (WRF) model (https://www.mmm.ucar.edu/weather-research-and-forecasting-model) and Model for Prediction Across Scales (MPAS, https://ncar.ucar.edu/what-we-offer/models/model-prediction-across-scales-mpas, https://mpas-dev.github.io/) is to use the so-called split-explicit method, where the sound (acoustic) part of the governing equations is solved applying sub-stepping (i.e., with short time steps) to ensure numerical stability.

Applying anelastic and compressible models to deep convection simulations on the mesoscale leads to similar results. This has been documented in a study by Kurowski et al. (2014) who used the EULAG model (https://www2.mmm.ucar.edu/eulag/) in compressible and anelastic simulations of a supercell splitting following the Klemp and Wilhelmson (1978) and Weisman and Klemp (1982) case. The three-dimensional simulations used computational domain of 128 km x 128 km in the horizontal and 17.5 km in the vertical. The horizontal and vertical grid lengths were 2 km and 350 m, respectively. Fig. 3.5 illustrates some of the results from the 120 min long anelastic and compressible simulations. The figure shows the maximum velocity during the simulation as a function of the mixing length applied in the subgrid-scale scheme for the parameterized turbulent transport. The mixing length, one of the parameters of the scheme, determines the magnitude of the turbulent subgrid fluxes, with the magnitude of the fluxes increasing with the increase of the mixing length. The results in Fig. 3.5 can be summarized as follows. First, subgrid-scale mixing significantly affects the maximum updraft strength, with the strongest updrafts in simulations with no subgrid scheme. This is consistent with the parameterized transport reducing cloud buoyancy and mixing updraft momentum with the environment, both increasing with the mixing length increase. The key result is that both compressible and anelastic solutions agree relatively well, and this is true even for the maximum updraft speed of around 50 m/s, about 15% of the speed of sound, in simulations without subgrid scheme. However, an increase of small updraft strength differences between COMP and ANES with the reduction of the subgrid mixing is also apparent.

Applying the two computational approaches at the global scale exposes clear differences. This perhaps should not be surprising because baroclinic vorticity dynamics, the key driver of midlatitude weather systems on a rotating Earth, differs significantly between anelastic and compressible systems. Davies et al. (2003) highlight significant problems with the anelastic equations through the normal-mode analysis. Kurowski et al. (2015) document anelastic-compressible differences in a

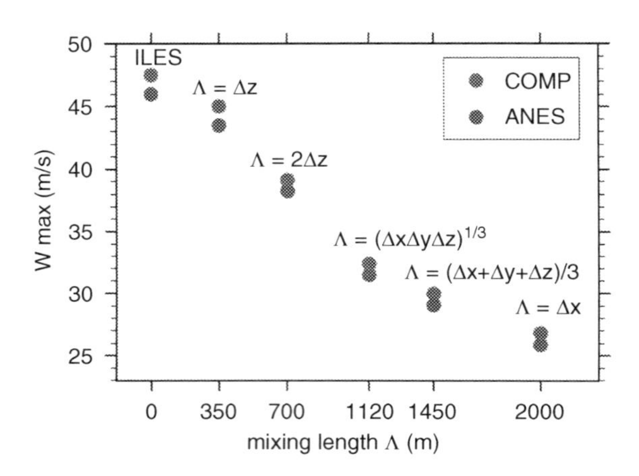

FIG. 3.5 Maximum vertical velocity for the supercell simulations applying anelastic (ANES) and compressible (COMP) versions of the EULAG model as a function of the mixing length in the subgrid-scale turbulence scheme.

Implicit large-eddy simulation (ILES) refers to simulations with no explicit subgrid-scale scheme. Reproduced with permission from Kurowski et al. (2014).

simple baroclinic wave development case of Jablonowski and Williamson (2006) applying the spherical version of the EULAG model. The horizontal mesh consists of 256 x 128 grid points on the regular longitude–latitude grid. The vertical domain of 23 km is covered with 48 uniformly distributed levels. Fig. 3.6 illustrates the differences by showing results from simulations at day 8 of the wave train development without condensation and precipitation in the left panels, and similar results from simulations with moist processes in the right panels. The ANES panels show results from the anelastic simulations that apply 5 min time step. The COMP and COMPe are compressible solutions. COMP applies a fully implicit solver that allows stable integrations with the same time step as ANES, 5 min (Smolarkiewicz et al., 2014). The explicit solver in COMPe requires 2 s time step for stable integrations. The solution obtained with the explicit solver, COMPe, can be considered the benchmark for ANES and COMPe. Fig. 3.6 illustrates the following points. First, anelastic solution show significantly smaller wave amplitude at day 8. This is because the wave develops slower as documented in Kurowski et al. (2015). Second, COMP and COMPe solutions are close providing a strong support for the accuracy and efficiency of the implicit solver in COMPe (see details in Smolarkiewicz et al., 2014).

In summary, the differences between anelastic and compressible approaches to modeling the dynamics of precipitating weather systems can be perhaps best summed up by the following statement from the abstract of Davies et al. (2003): "Whilst of key importance for small-scale theoretical studies and process modelling, the anelastic equations are not recommended for either operational numerical weather prediction or climate simulation at any scale".

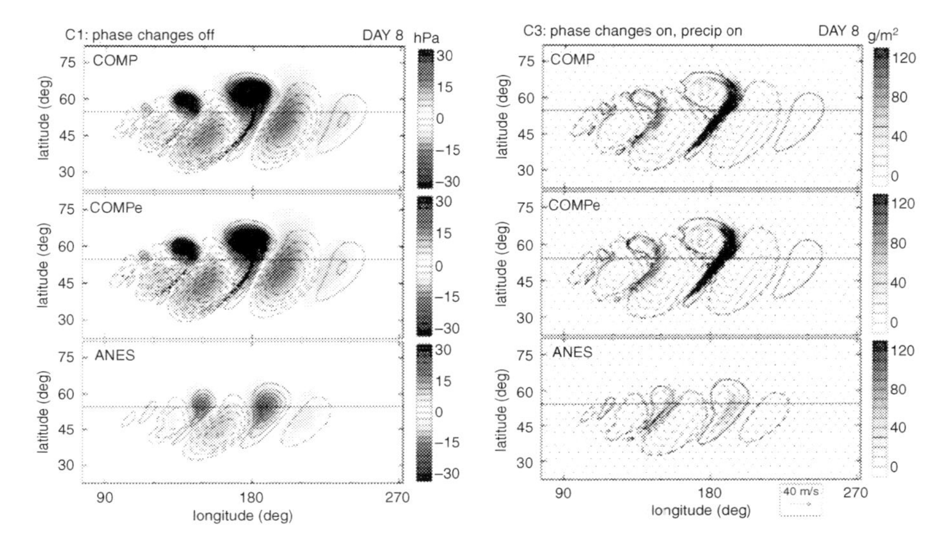

FIG. 3.6 Left panels: baroclinic instability at day 8 as simulated by COMP, COMPe, and ANES models.

Surface virtual potential temperature perturbations are shown by black isolines (contour interval 4 K; negative values dashed) and surface pressure perturbations are marked as colors (color scale to the right of the panels). Right panels: As left panels but with condensation and precipitation included. Cloud water path is shows as colors and vectors depict surface horizontal winds. Black contours show −1 K (dashed) and 1 K (solid) isolines of surface virtual potential temperature perturbations. Solid horizontal lines are drawn at 53oN in all panels to help comparing details of simulation results. Reproduced with permission from Kurowski et al. (2015) with changes.

3.5 Prospects for future rainfall modeling

The progress in the numerical weather prediction and in rainfall modeling over the last several decades has been impressive. Because of increasing computational resources, the spatial resolution of numerical models and the sophistication of microphysics parameterizations has been steadily increasing. This trend will likely continue for a foreseeable future. On the dynamics side, removing the need for convection parameterization in large-scale models (even global, e.g., DYAMOND initiative) puts modeling of precipitation processes on a solid footing. Continuing increase of the model spatial resolution not only allows representing convective processes directly, but it also improves representation of other precipitation formation mechanisms, like the flow over land-surface features (e.g., topography) or frontal circulations. Note that topography is smooth in low-resolution models and this significantly reduces the amount of the simulated orographic precipitation and the climate-warming signal (Rasmussen et al., 2011; Prein et al., 2013). For the moist convection, it is well appreciated that the convection-permitting model resolutions (i.e., ~1 km horizontal grid length) are still not sufficient and even higher (e.g., as in the large eddy simulation) resolutions are needed. One may expect that more comprehensive microphysics parameterizations should be used in high-resolution simulations, so increases of model resolution and microphysics sophistication should go hand in hand in the future. Although still in the exploratory phase today, the Lagrangian microphysics will likely have a significant role in future operational rainfall simulations as well.

Acknowledgements

This work was supported by the National Center for Atmospheric Research. Suggestions from NCAR's Jimmy Dudhia, DWD's Axel Seifert and ETH's Christoph Schär are greatly appreciated. National Center for Atmospheric Research is sponsored by the National Science Foundation.

References

Abdul-Razzak, H., Ghan, S.J., 2000. A parameterization of aerosol activation: 2. multiple aerosol types. J. Geophys. Res. 105, 6837–6844. doi:10.1029/1999JD901161.

Andrejczuk, M., Reisner, J.M., Henson, B., Dubey, M.K., Jeffery, C.A., 2008. The potential impacts of pollution on a nondrizzling stratus deck: does aerosol number matter more than type? J. Geophys. Res. 113, D19204. doi:10.1029/2007JD009445.

Andrejczuk, M., Grabowski, W.W., Reisner, J., Gadian, A., 2010. Cloud-aerosol interactions for boundary-layer stratocumulus in the Lagrangian cloud model. J. Geophys. Res. 115, D22214. doi:10.1029/2010JD014248.

Anthes, R.A., Hsie E.Y., Kuo Y.H., 1987. Description of the Penn State/NCAR Mesoscale Model: Version 4 (MM4) (No. NCAR/TN-282+STR). doi:10.5065/D64B2Z90.

Arabas, S., Shima, S., 2013. Large eddy simulations of trade-wind cumuli using particle-based microphysics with monte-carlo coalescence. J. Atmos. Sci. 70, 2768–2777.

Arakawa, A., 2004. The cumulus parameterization problem: past, present, and future. J. Climate 17, 2493–2525.

Beard, K.V., 1976. Terminal velocity and shape of cloud and precipitation drops aloft. J. Atmos. Sci. 33, 851–864.

Berry, E.X., 1967. Cloud droplet growth by collection. J. Atmos. Sci. 24, 688–701.

Berry, E.X., Reinhardt, R.L., 1974a. An analysis of cloud drop growth by collection part I. Double distributions. J. Atmos. Sci. 31, 1814–1824.

Berry, E.X., Reinhardt, R.L, 1974b. An analysis of cloud drop growth by collection part ii. single initial distributions. J. Atmos. Sci. 31, 1825–1831.

Berry, E.X., Reinhardt, R.L., 1974c. An analysis of cloud drop growth by collection: part III. Accretion and self-collection. J. Atmos. Sci. 31, 2118–2126.

Bott, A., 1998. A flux method for the numerical solution of the stochastic collection equation. J. Atmos. Sci. 55, 2284–2293.

Brown, A.R., Cederwall, R.T., Chlond, A., et al., 2002. Large-eddy simulation of the diurnal cycle of shallow cumulus convection over land. Q. J. R. Meteorol. Soc. 128, 1075–1093.

Bryan, G.H., Wyngaard, J.C., Fritsch, J.M, 2003. Resolution requirements for the simulation of deep moist convection. Mon. Wea. Rev. 131, 2394–2416.

Byers, H.R., Braham, R.R, 1948. Thunderstorm structure and circulation. J. Meteor. 5, 71–86.

Carpenter, R.L., Droegemeier, K.K., Blyth, A.M, 1998. Entrainment and detrainment in numerically simulated cumulus congestus clouds. Part I: general results. J. Atmos. Sci. 55, 3417–3432.

Chandrakar, K.K., Grabowski, W.W., Morrison, H., Bryan, G.H, 2021. Impact of entrainment-mixing and turbulent fluctuations on droplet size distributions in a cumulus cloud: An investigation using Lagrangian microphysics with a sub-grid-scale model. J. Atmos. Sci. in review.

Chandrasekhar, S, 1961. Hydrodynamic and Hydromagnetic Stability. Clarendon Press, p. 652.

Clark, A.J., Weiss, S.J., Kain, J.S., et al., 2012. An overview of the 2010 hazardous weather testbed experimental forecast program spring experiment. Bull. Amer. Meteor. Soc. 93, 55–74.

Clark, T.L., 1977. A small-scale dynamic model using a terrain-following coordinate transformation. J. Comp. Phys. 24, 186–215.

Clark, T.L, 1979. Numerical simulations with a three dimensional cloud model: Lateral boundary condition experiments and multicellular severe storm simulations. J. Atmos. Sci. 36, 2191–2215.

Corradini, C., Morbidelli, R., Saltalippi, C., Flammini, A., 2022. Meteorological Systems Producing Rainfall. In: Morbidelli, R. (Ed.), Rainfall. Modeling, Measurement and Applications. Elsevier, Amsterdam, pp. 27–48. doi:10.1016/C2019-0-04937-0.

Davies, T., Staniforth, A., Wood, N., Thuburn, J, 2003. Validity of anelastic and other equation sets as inferred from normal-mode analysis. Q. J. R. Meteorol. Soc. 129, 2761–2775.

Dudhia, J, 1989. Numerical study of convection observed during the winter monsoon experiment using a mesoscale two-dimensional model. J. Atmos. Sci. 46, 3077–3107.

Dziekan, P., Waruszewski, M., Pawlowska, H, 2019. University of Warsaw Lagrangian Cloud Model (UWLCM) 1.0: a modern large-eddy simulation tool for warm cloud modeling with Lagrangian microphysics. Geosci. Model Dev. 12, 2587–2606.

Eidhammer, T., Morrison, H., Mitchell, D., Gettelman, A., Erfani, E., 2017. Improvements in global climate model microphysics using a consistent representation of ice particle properties. J. Climate 30, 609–629.

Fan, J., Khain, A, 2021. Comments on "do ultrafine cloud condensation nuclei invigorate deep convection? J. Atmos. Sci. 78, 329–339.

Feingold, G., Tzivion, S., Levin, Z, 1988. Evolution of raindrop spectra. Part I: solution to the stochastic collection/breakup equation using the method of moments. J. Atmos. Sci. 45, 3387–3399.

Grabowski, W.W., Clark, T.L, 1993. Cloud-environment interface instability. Part II: extension to three spatial dimensions. J. Atmos. Sci. 50, 555–573.

Grabowski, W.W., Smolarkiewicz, P.K, 1996. On two-time-level semi-Lagrangian modeling of precipitating clouds. Mon. Wea. Rev. 124, 487–497.

Grabowski, W.W., Wu, X., Moncrieff, M.W., 1996. Cloud resolving modeling of tropical cloud systems during phase III of GATE. Part I: two–dimensional experiments. J. Atmos. Sci. 53, 3684–3709.

Grabowski, W.W, 1998. Toward cloud resolving modeling of large–scale tropical circulations: a simple cloud microphysics parameterization. J. Atmos. Sci. 55, 3283–3298.

Grabowski, W.W., Smolarkiewicz, P.K, 1999. CRCP: a cloud resolving convection parameterization for modeling the tropical convecting atmosphere. Physica D 133, 171–178.

Grabowski, W.W, 2001. Coupling cloud processes with the large-scale dynamics using the cloud-resolving convection parameterization (CRCP). J. Atmos. Sci. 58, 978–997.

Grabowski, W.W., Wu, X., Moncrieff, M.W., Hall, W.D, 1998. Cloud resolving modeling of tropical cloud systems during phase III of GATE. Part II: effects of resolution and the third spatial dimension. J. Atmos. Sci. 55, 3264–3282.

Grabowski, W.W., Abade, G.C., 2017. Broadening of cloud droplet spectra through eddy hopping: turbulent adiabatic parcel simulations. J. Atmos. Sci. 74, 1485–1493.

Grabowski, W.W., Dziekan, P., Pawlowska, H, 2018. Lagrangian condensation microphysics with Twomey CCN activation. Geosci. Model Dev. 11, 103–120.

Grabowski, W.W., Morrison, H., Shima, S., Abade, G.C., Dziekan, P., Pawlowska, H, 2019. Modeling of cloud microphysics: can we do better? Bull. Amer. Meteor. Soc. 100, 655–672. doi:10.1175/BAMS-D-18-0005.1.

Grabowski, W.W., Morrison, H, 2021. Reply to "Comments on 'Do ultrafine cloud condensation nuclei invigorate deep convection'". J. Atmos. Sci. 78, 341–350.

Hall, W.D, 1980. A detailed microphysical model within a two-dimensional dynamic framework: model description and preliminary results. J. Atmos. Sci. 37, 2486–2507.

Holton, J.R, 2004. An Introduction to Dynamic Meteorology, *4th edition*. Elsevier Academic Press, p. 535.

Hoffmann, F., Raasch, S., Noh, Y, 2015. Entrainment of aerosols and their activation in a shallow cumulus cloud studied with a coupled LCM–LES approach. Atmos. Res. 156, 43–57. doi:10.1016/j.atmosres.2014.12.008.

Hoffmann, F., Yamaguchi, T., Feingold, G, 2019. Inhomogeneous mixing in Lagrangian cloud models: effects on the production of precipitation embryos. J. Atmos. Sci. 76, 113. doi:10.1175/JAS-D-18-0087.1.

Jablonowski, C., Williamson, D.L, 2006. A baroclinic instability test case for atmospheric model dynamical cores. Quart. J. Roy. Meteor. Soc. 132, 2943–2975.

Jarecka, D., Grabowski, W.W., Morrison, H, Pawlowska, H., 2013. Homogeneity of the subgrid-scale turbulent mixing in large-eddy simulation of shallow convection. J. Atmos. Sci. 70, 2751–2767.

Kessler, E, 1969. On the distribution and continuity of water substance in atmospheric circulations. Meteor. Monogr. 10 (88).

Khain, A.P., Sednev, I, 1995. Simulation of hydrometeor size spectra evolution by water-water, ice-water and ice-ice interactions. Atmos. Res. 36, 107–138.

Khain, A.P., Pokrovsky, A., Pinsky, M., Seifert, A., Phillips, V., 2004. Simulation of effects of atmospheric aerosols on deep turbulent convective clouds using a spectral microphysics mixed-phase cumulus cloud model. Part I: model description and possible applications. J. Atmos. Sci. 61, 2963–2982.

Khain, A., Prabha, T.V., Benmoshe, N., Pandithurai, G., Ovchinnikov, M, 2013. The mechanism of first raindrops formation in deep convective clouds. J. Geophys. Res. Atmos. 118, 9123–9140. doi:10.1002/jgrd.50641.

Khairoutdinov, M., Kogan, Y, 2000. A new cloud physics parameterization in a large-eddy simulation model of marine stratocumulus. Mon. Wea. Rev. 128, 229–243 10.1175/1520-0493.

Khairoutdinov, M.F., Randall, D.A, 2001. A cloud resolving model as a cloud parameterization in the NCAR community climate system model: preliminary results. Geophys. Res. Lett. 28, 3617–3620.

Khairoutdinov, M.F., Krueger, S.K., Moeng, C.H., Bogenschutz, P., Randall, D.A, 2009. Large-eddy simulation of maritime deep tropical convection. J. Adv. Model. Earth Syst. 1. doi:10.3894/JAMES.2009.1.15.

Klemp, J.B., Wilhelmson, R.B, 1978. The simulation of three-dimensional convective storm dynamics. J. Atmos. Sci. 35, 1070–1096.

Kogan, Y.L, 1991. The simulation of a convective cloud in a 3-D model with explicit microphysics. Part I: model description and sensitivity experiments. J. Atmos. Sci. 48, 1160–1189.

Kogan, Y, 2013. A cumulus cloud microphysics parameterization for cloud-resolving models. J. Atmos. Sci. 70, 1423–1436. doi: 10.1175/JAS-D-12-0183.

Krueger, S.K, 1988. Numerical simulation of tropical cumulus clouds and their interaction with the subcloud layer. J. Atmos. Sci. 45, 2221–2250.

Krueger, S.K, 2000. Cloud system modeling. In: Randall, D.A. (Ed.), General Circulation Model Development. Academic Press, p. 803.

Kurowski, M.J., Grabowski, W.W., Smolarkiewicz, P.K, 2014. Anelastic and compressible simulation of moist deep convection. J. Atmos. Sci. 71, 3767–3787.

Kurowski, M.J., Grabowski, W.W., Smolarkiewicz, PK, 2015. Anelastic and compressible simulation of moist dynamics at planetary scales. J. Atmos. Sci. 72, 3975–3995.

Lean, H.W., Clark, P.A., Dixon, M., Roberts, N.M., Fitch, A., Forbes, R., Halliwell, C, 2008. Characteristics of high-resolution versions of the Met Office Unified Model for forecasting convection over the United Kingdom. Mon. Wea. Rev. 136, 3408–3424.

Lebo, Z.J., Seinfeld, J.H, 2011. A continuous spectral aerosol-droplet microphysics model. Atmos. Chem. Phys. 11, 12297–12316.

Lebo, Z.J., Morrison, H., Seinfeld, J.H, 2012. Are simulated aerosol-induced effects on deep convective clouds strongly dependent on saturation adjustment? Atmos. Chem. Phys. 12, 9941–9964.

Lebo, Z.J., Morrison, H, 2015. Effects of horizontal and vertical grid spacing on mixing in simulated squall lines and implications for convective strength and structure. Mon. Wea. Rev. 143, 4355–4375.

Lilly, D.K, 1962. On the numerical simulation of buoyant convection. Tellus 14, 148–172.

Lilly, D.K, 1981. Wave-permeable lateral boundary conditions for convective cloud and storm simulations. J. Atmos. Sci. 38, 1313–1316.

Lim, K.S., Hong, S., 2010. Development of an effective double-moment cloud microphysics scheme with prognostic cloud condensation nuclei (ccn) for weather and climate models. Mon. Wea. Rev. 138, 1587–1612.

Lin, Y.L., Farley, R.D., Orville, H.D., 1983. Bulk parameterization of the snow field in a cloud model. J. Climate Appl. Meteor. 22, 1065–1092.

Lipps, F.B., Hemler, R.S., 1982. A scale analysis of deep moist convection and some related numerical calculations. J Atmos Sci 39, 2192–2210.

Liu, C., Ikeda, K., Thompson, G., Rasmussen, R., Dudhia, J,, 2011. High-resolution simulations of wintertime precipitation in the colorado headwaters region: sensitivity to physics parameterizations. Mon. Weather Rev. 139, 3533–3553.

Liu, J.Y., Orville, H.D, 1969. Numerical modeling of precipitation and cloud shadow effects on mountain-induced cumuli. J. Atmos. Sci. 26, 1283–1298.

Marshall, J.S., Palmer, W.M.K, 1948. The distribution of raindrops with size. J. Atmos. Sci. 5, 165–166.
Marshall, J.S., Gunn, K.L.S, 1952. Measurement of snow parameters by radar. J. Atmos. Sci. 9, 322–327.
McFarquhar, G., 2022. Rainfall Microphysics. In: Morbidelli, R. (Ed.), Rainfall. Modeling, Measurement and Applications. Elsevier, Amsterdam, pp. 1–26. doi:10.1016/C2019-0-04937-0.
Meyers, M.P., Walko, R.L., Harrington, J.Y., Cotton, W.R, 1997. New RAMS cloud microphysics parameterization. Part II: the two-moment scheme. Atmos. Res. 45, 3–39.
Milbrandt, J.A., Yau, M.K., 2005. A multimoment bulk microphysics parameterization. Part II: a proposed three-moment closure and scheme description. J. Atmos. Sci. 62, 3065–3081.
Milbrandt, J.A., Yau, M.K, 2006. A multimoment bulk microphysics parameterization. Part IV: sensitivity experiments. J. Atmos. Sci. 63, 3137–3159.
Moeng, C.H., LeMone, M.A., Khairoutdinov, M.F., Krueger, S.K., Bogenschutz, P.A., Randall, D.A, 2009. The tropical marine boundary layer under a deep convection system: a large-eddy simulation study. J. Adv. Model. Earth Syst. 1, 16. doi:10.3894/JAMES.2009.1.16.
Morrison, H., Grabowski, W.W, 2007. Comparison of bulk and bin warm rain microphysics models using a kinematic framework. J. Atmos. Sci. 64, 2839–2861.
Morrison, H., Milbrandt, J.A, 2015. Parameterization of cloud microphysics based on the prediction of bulk ice particle properties. part i: scheme description and idealized tests. J. Atmos. Sci. 72, 287–311.
Morrison, H., Milbrandt, J.A., Bryan, G.H., Ikeda, K., Tessendorf, S.A., Thompson, G, 2015. Parameterization of cloud microphysics based on the prediction of bulk ice particle properties. Part II: case study comparisons with observations and other schemes. J. Atm. Sci. 72 (1), 312–339.
Murray, F.W, 1970. Numerical models of a tropical cumulus cloud with bilateral and axial symmetry. Mon. Wea. Rev. 98, 14–28.
Ogura, Y, 1962. Convection of isolated masses of a buoyant fluid: a numerical calculation. J. Atmos. Sci. 19, 492–502.
Ogura, Y, 1963. The evolution of a moist convective element in a shallow, conditionally unstable atmosphere: a numerical calculation. J. Atmos. Sci. 20, 407–424.
Ogura, Y., Phillips, N.A, 1962. Scale analysis of deep and shallow convection in the atmosphere. J Atmos Sci 19, 173–179.
Orville, H.D., 1965. A numerical study of the initiation of cumulus clouds over mountainous terrain. J. Atmos. Sci. 22, 684–699.
Pawlowska, H., Brenguier, J.L, 2003. An observational study of drizzle formation in stratocumulus clouds for general circulation model (GCM) parameterizations. J. Geophys. Res. 108, 8630. doi:10.1029/2002JD002679.
Prein, A.F., Holland, G.J., Rasmussen, R.M., Done, J., Ikeda, K., Clark, M.P., Liu, C.H., 2013. Importance of regional climate model grid spacing for the simulation of heavy precipitation in the colorado headwaters. J. Climate 26, 4848–4857.
Pruppacher, H.R., Klett, J.D, 1996. Microphysics of Clouds and Precipitation. Springer, p. 976.
Randall, D., Khairoutdinov, M., Arakawa, A., Grabowski, W.W, 2003. Breaking the cloud-parameterization deadlock. Bull. Amer. Meteor. Soc. 84, 1547–1564.

Rasmussen, R., Liu, C., Ikeda, K., et al., 2011. High-resolution coupled climate runoff simulations of seasonal snowfall over colorado: a process study of current and warmer climate. J. Climate 24, 3015–3048.

Redelsperger, J.L., Guichard, F., Mondon, S, 2000. A parameterization of mesoscale enhancement of surface fluxes for large-scale models. J. Climate 13, 402–421.

Riechelmann, T., Noh, Y., Raasch, S, 2012. A new method for large-eddy simulations of clouds with Lagrangian droplets including the effects of turbulent collision. New J. Phys. 14, 065008. doi:10.1088/1367-2630/14/6/065008.

Robe, F.R., Emanuel, K.A, 1996. Moist convective scaling: some inferences from three-dimensional cloud ensemble simulations. J. Atmos. Sci. 53, 3265–3275.

Rosenfeld, D, 2000. Suppression of rain and snow by urban and industrial air pollution. Science 287, 1793–1796.

Rosenfeld, D., Gutman, G, 1994. Retrieving microphysical properties near the tops of potential rain clouds by multispectral analysis of AVHRR data. Atmos. Res. 34, 259–283.

Rutledge, S.A., Hobbs, P.V., 1984. The mesoscale and microscale structure and organization of clouds and precipitation in midlatitude cyclones. XII: A diagnostic modeling study of precipitation development in narrow cold-frontal rainbands. J. Atmos. Sci. 41, 2949–2972 1984.

Saleeby, S.M., Cotton, W.R, 2004. A large-droplet mode and prognostic number concentration of cloud droplets in the colorado state university regional atmospheric modeling system (RAMS). Part I: module descriptions and supercell test simulations. J. Appl. Meteorol. 43, 182–195.

Satoh, M., Tomita, H., Yashiro, H., et al., 2014. The non-hydrostatic icosahedral atmospheric model: description and development. Prog. Earth. Planet. Sci. 1, 18. doi:10.1186/s40645-014-0018-1.

Schlesinger, R.E, 1973. A numerical model of deep moist convection: Part I. Comparative experiments for variable ambient moisture and wind shear. J. Atmos. Sci. 30, 835–856.

Schlesinger, R.E., 1975. A three-dimensional numerical model of an isolated deep convective cloud: preliminary results. J. Atmos. Sci. 32, 934–957.

Schluenzen, K.H, 1994. Mesoscale modelling in complex terrain, An overview on the German non-hydrostatic models. Contr. Atmosph. Phys. 67, 243–254.

Seifert, A., Beheng, K.D, 2001. A double-moment parameterization for simulating autoconversion, accretion and selfcollection. Atmos. Res. 59-60, 265–281.

Shima, S.I., Kusano, K., Kawano, A., Sugiyama, T., Kawahara, S, 2009. The super-droplet method for the numerical simulation of clouds and precipitation: A particle-based and probabilistic microphysics model coupled with a non-hydrostatic model. Q. J. Roy. Meteor. Soc. 135, 1307–1320.

Shima, S., Sato, Y., Hashimoto, A., Misumi, R, 2020. Predicting the morphology of ice particles in deep convection using the super-droplet method: development and evaluation of SCALE-SDM 0.2.5-2.2.0, -2.2.1, and -2.2.2. Geosci. Model Dev. 13, 4107–4157.

Siebesma, A.P., Bretherton, C.S., Brown, A., et al., 2003. A large eddy simulation intercomparison study of shallow cumulus convection. J. Atmos. Sci. 60, 1201–1219.

Simmel, M., Trautmann, T., Tetzlaff, G, 2002. Numerical solution of the stochastic collection equation - comparison of the Linear Discrete Method with other methods. Atmos. Res. 61, 135–148.

Smolarkiewicz, P.K., Kühnlein, C., Wedi, N.P, 2014. A consistent framework for discrete integrations of soundproof and compressible PDEs of atmospheric dynamics. J. Comp. Phys 263, 185–205.
Smoluchowski, M.V, 1916. Drei vortrage uber diffusion, brownsche bewegung und koagulation von kolloidteilchen. Physik. Zeit., 17, 557–585.
Soong, S., Ogura, Y., 1980. Response of tradewind cumuli to large-scale processes. J. Atmos. Sci. 37, 2035–2050.
Soong, S., Tao, W., 1980. Response of deep tropical cumulus clouds to mesoscale processes. J. Atmos. Sci. 37, 2016–2034.
Stein, T.H.M., Hogan, R.J., Clark, P.A., Halliwell, C.E., Hanley, K.E., Lean, H.W., Nicol, J.C., Plant, R.S ., 2015. The DYMECS project: a statistical approach for the evaluation of convective storms in high-resolution NWP models. Bull. Amer. Meteor. Soc. 96, 939–951.
Steppeler, J., Doms, G., Schattler, U., Bitzer, H.W., Gassmann, A., Damrath, U., Gregoric, G, 2003. Meso-gamma scale forecasts using the nonhydrostatic model LM. Meteorol. Atmospheric Phys. 82, 75–96. doi:10.1007/s00703-001-0592-9.
Stevens, B., Moeng, C.H., Ackerman, A.S., et al., 2005. Evaluation of large-eddy simulations via observations of nocturnal marine stratocumulus. Mon. Weather Rev. 133, 1443–1462.
Stevens, B, Satoh, M, Auger, L, et al., 2019. Dyamond: the dynamics of the atmospheric general circulation modeled on non-hydrostatic domains. Prog. Earth Planet. Sci. 6. doi:10.1186/s40645-019-0304-z.
Tao, W., Soong, S, 1986. A study of the response of deep tropical clouds to mesoscale processes: three-dimensional numerical experiments. J. Atmos. Sci. 43, 2653–2676.
Tao, W., Simpson, J., Soong, S, 1987. Statistical properties of a cloud ensemble: a numerical study. J. Atmos. Sci. 44, 3175–3187.
Tapp, M.C., White, P.W, 1976. A non-hydrostatic mesoscale model. Quart. J. Roy. Meteor. Soc. 102, 277–296.
Tompkins, A.M., Craig, G.C., 1998. Radiative–convective equilibrium in a three-dimensional cloud-ensemble model. Q. J. R. Meteorol. Soc. 124, 2073–2097.
Thompson, G., Rasmussen, R.M., Manning, K, 2004. Explicit forecasts of winter precipitation using an improved bulk microphysics scheme. Part I: description and sensitivity analysis. Mon. Wea. Rev. 132, 519–542.
Tzivion, S., Feingold, G., Levin, Z, 1987. An efficient numerical solution to the stochastic collection equation. J. Atmos. Sci. 44, 3139–3149.
Tzivion, S., Feingold, G., Levin, Z, 1989. The evolution of rain-drop spectra. Part II: collisional collection/breakup and evaporation in a rain shaft. J. Atmos. Sci. 46, 3312–3327.
Sölch, I., Kärcher, B, 2010. A large-eddy model for cirrus clouds with explicit aerosol and ice microphysics and Lagrangian ice particle tracking. Q. J. Roy. Meteor. Soc. 136, 2074–2093.
Unterstrasser, S, Sölch, I, 2010. Study of contrail microphysics in the vortex phase with a Lagrangian particle tracking model. Atmos. Chem. Phys. 10, 10003–10015. doi:10.5194/acp-10- 10003-2010.
Unterstrasser, S., Sölch, I, 2014. Optimisation of simulation particle number in a Lagrangian ice microphysical model. Geosci. Model Dev. 7, 695–709.
Weisman, M.L., Klemp, J.B, 1982. The dependence of numerically simulated convective storms on vertical wind shear and buoyancy. Mon. Weather Rev. 110, 504–520.

Weisman, M.L., Davis, C., Wang, W., Manning, K.W., Klemp, J.B, 2008. Experiences with 0-36-h convective forecasts with the WRF-ARW model. Weather Forecast 23, 407–436.
Xu, K., Arakawa, A., Krueger, S.K, 1992. The macroscopic behavior of cumulus ensembles simulated by a cumulus ensemble model. J. Atmos. Sci. 49, 2402–2420.
Xu, K., Randall, D.A., 1995. Impact of interactive radiative transfer on the macroscopic behavior of cumulus ensembles. Part II: mechanisms for cloud-radiation interactions. J. Atmos. Sci. 52, 800–817.
Xu, K., Randall, D.A., 1996. Explicit simulation of cumulus ensembles with the GATE phase III data: comparison with observations. J. Atmos. Sci. 53, 3710–3736.
Xue, L., Fan, J., Lebo, Z., Wu, W., et al., 2017. Idealized simulations of a squall line from the MC3E field campaign applying three bin microphysics schemes. Part I: dynamic and thermodynamic structure. Mon. Wea. Rev. 145, 4789–4812.
Zängl, G., Reinert, D., Ripodas, P., Baldauf, M, 2014. The ICON (ICOsahedral non-hydrostatic) model- ling framework of DWD and MPI-M: description of the non-hydrostatic dynamical core. Quart. J. Roy. Meteor. Soc. 141, 563–579.

CHAPTER

4 Rain gauge measurements

Luca G. Lanza[a,b], Arianna Cauteruccio[a,b], Mattia Stagnaro[a,b]

[a]*University of Genova, Dept. of Civil, Chemical and Environmental Engineering, Genoa, Italy*

[b]*WMO/CIMO Lead Centre "B. Castelli" on Precipitation Intensity, Italy*

4.1 Introduction

Rainfall was among the first environmental variables in history to require continuous and accurate measurements for immediate application in the everyday human life and activities. Initial efforts to quantify the amount of water made available by liquid precipitation for religious, agricultural, and even taxation purposes were based on simple graduated vessels exposed to the atmosphere and are documented in the literature since the very beginning of our civilization (Cauteruccio et al., 2021a; Strangeways, 2010).

Notwithstanding the extraordinary relevance of rainfall among the environmental variables, the development of accurate measurement instruments stalled for a long time, at least until the mid-19th century. It was only after understanding the role of rainfall intensity in hydrological processes that the need of developing automatic measurement instruments was recognized, to keep records of the variation of the rainfall depth with time.

Accurate measurement of the rainfall process remains a difficult task even under the current status of technological development and underlying scientific knowledge. Due to the associated technical difficulties, the role of accurate measurements is still understated in modern hydrological sciences and applications, and rainfall records still suffer from a lack of standardization and quality assurance procedures (Lanza and Stagi, 2008).

Well-known instrumental and environmental biases of rainfall measurements remain mostly ignored and uncorrected in the operation of many rainfall monitoring networks, and the historic information contained in the archived rainfall series continues to be affected by significant biases and uncertainties. Such errors propagate e.g., in the derivation of rainfall statistics related to the expected frequency of rainfall events, the calculation of design rainfall for engineering works (La Barbera et al., 2002; Molini et al., 2005), and in other hydrological applications (Biemans et al., 2009). The same is true in the field of climatology, where the quality and

RAINFALL. Modeling, Measurement and Applications. DOI: https://doi.org/10.1016/B978-0-12-822544-8.00002-0

homogeneity of historic rainfall records is imperative to ensure that the assessment of possible climatic trends is correctly substantiated (Morbidelli et al., 2022).

Further to a synthetic overview of the available technological solutions based on the various measuring principles, including the recently developed non-catching type instruments, the present chapter proposes a discussion of the most relevant rainfall measurement accuracy issues. Both instrumental and environmental biases are addressed with the aim of highlighting the most relevant calibration procedures, interpretation algorithms and correction methods that, once implemented, would ensure highly accurate rainfall measurements.

4.2 Principles and operation of rain gauges

The World Meteorological Organization (WMO), in the Guide to Meteorological Instruments and Methods of Observation (WMO, 2014), defines precipitation as "the liquid or solid products of the condensation of water vapour falling from clouds or deposited from air onto the ground." Precipitation intensity, usually expressed in mm h^{-1}, is defined as "the amount of precipitation collected per unit time interval" while the rainfall depth, in mm, is "the total amount of precipitation which reaches the ground in a stated period (…) expressed in terms of the vertical depth of water to which it would cover a horizontal projection of the Earth surface."

These quantities are measured using various principles and technological solutions, whose main characteristics are described below.

4.2.1 Volumetric method

The very first method historically used to measure rainfall consisted in collecting the rainwater in an open container exposed to the atmosphere and in measuring the total collected volume. The ratio between the collected volume and the surface area of the collector provided the measured rainfall depth.

4.2.1.1 *Storage gauge*

The storage gauge is still used today for manual measurements, typically by using a container with known geometry where the rainwater is collected, and the total accumulated water volume over a given period can be measured. An operator periodically performs the measurement by either annotating the water level inside the container or pouring the collected water into a reference storage volume. After each measurement, the container must be emptied.

Storage gauges are made of metal, glass, or plastic (see Fig. 4.1), and the material characteristics and thickness of the gauge body must avoid any deformation due to thermal effects. Transparent walls are preferable when the reading of the water level is manually performed, otherwise many metal gauges are provided with a transparent lateral window. Graduation marks are engraved on the container to allow simple reading of the water level.

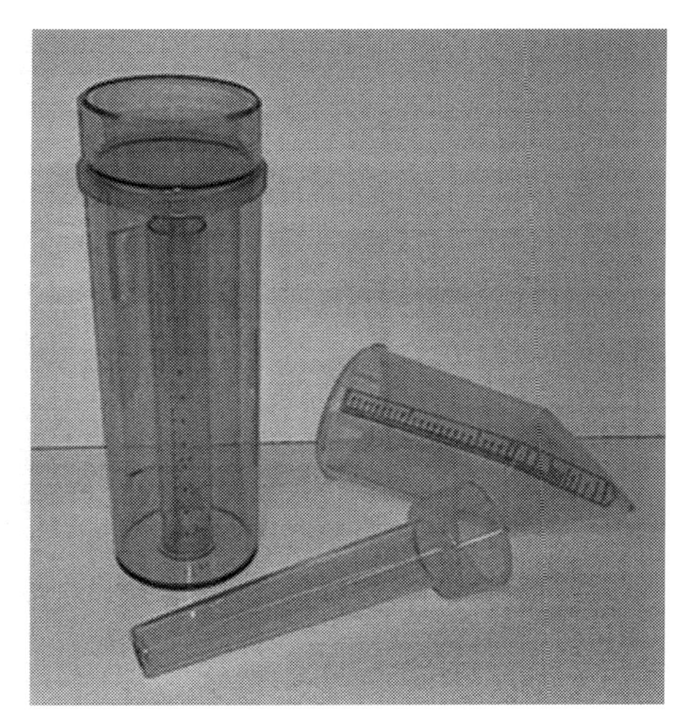

FIG. 4.1

Simple plastic storage gauges for manual reading of the accumulated rainwater.

The shape of the container is commonly cylindrical, but in some cases the bottom of the container has a narrower section area to enhance the readability of the water level when the accumulated rain amount is small. The size of the container may vary significantly to conform to the local climatology and the expected magnitude of rainfall events.

For manual measurements, the water volume is recorded once or twice a day, and every time the gauge is emptied by the operator. An improvement of this measuring system employs a float inside the container to detect the water level and transpose the reading onto a strip chart. Using a mechanical assembly to measure rainfall increases the time resolution of the records, although the measurement always depends on the operator for the reading of the data recorded on the charts and for the periodic emptying of the container.

Moving from manual to automatic precipitation gauges, the measurement of the water level collected in the container is performed by automatic sensors based on conductivity, acoustic distance, or hydrostatic pressure measurements. These methods, based on different measuring principles, allow continuous recording of the water level (and therefore of the volume collected by the gauge) with no need of operators to perform the measurement, while the need of periodically emptying the container remains.

4.2.1.2 *Storage gauge with automatic emptying principle*

One of the disadvantages of the storage methods consists in periodic maintenance needed to empty the container, especially during the wet season. To overcome this problem, some instruments have been equipped with automatic emptying system using syphon mechanisms.

Siphons consist of an inverted U-shaped tube which connects the inner side of the container with the outside part; when the water level inside the gauge reaches the top of the outer tube, water start flowing, and the container is emptied until the water level reaches the bottom inner tube of the siphon.

An alternative mechanism based on the siphon principle consists of a tilting siphon. When the water level inside the container reaches the top of the bucket it tilts over one side and activates the siphon to release the water outside.

The use of an emptying system allows this kind of gauges to operate automatically but introduces some underestimation when the siphon is activated during rain events.

4.2.2 Gravimetric methods

Instruments based on gravimetric methods use the effect of the force of gravity on the collected water to directly measure the weight of a water volume, to activate mechanical parts of the gauge or to release constant volumes of water.

4.2.2.1 *Tipping-bucket rain gauge*

Tipping-bucket rain gauges (TBRs) are worldwide employed in National Meteorological Services (NMSs) to measure the rain depth and intensity. The limited production cost, the ease of maintenance and their reliability are the main reasons behind the extensive use of this kind of instruments.

The mechanical principle of TBRs consists of a tilting balance with two compartments (buckets) having the same shape and volume (see Fig. 4.2). The balance has a horizontal rotation axis, and the two buckets are in an unstable equilibrium. Two stop screws under each bucket avoid complete tilting of the balance on either side and allow to regulate its inclination and therefore the volume required to trigger the rotation movement.

During precipitation events, the gauge funnel conveys the collected water into one of the twin buckets placed under the funnel nozzle. When the amount of water collected inside the bucket reaches the critical volume, the balance starts tilting, the first bucket is emptied, and the second bucket moves below the nozzle.

In ancient instruments, based only on mechanical principles, the rotation of the buckets activated the movement of a pen writing on a strip chart, while in modern automatic gauges the balance is provided with a magnet, which triggers a reed relay contact and produces an electrical impulse every time the bucket rotates. The output

FIG. 4.2

The tilting mechanism of a tipping-bucket rain gauge.

signal can be recorded by an impulse counter or a data logger; in the first case only the number of tips that occur in a chosen time interval (therefore the total rain volume) can be calculated, knowing the nominal volume of the two buckets. When a data logger is employed to acquire the output signal also the timestamp of each tip can be recorded, which is the finest information provided by this kind of gauges and allows the implementation of advanced algorithms based on the inter-tip time to calculate the rainfall intensity.

Instruments based on this principle need no manual interaction with the operator due to the automatic emptying principle and allow continuous measurements; however, due to the presence of rotating elements, they require periodic maintenance to clean the funnel area and avoid that dust or small particle may block the rotating movement of the buckets.

Usually, the gauges have a cylindrical shape (Fig. 4.3), but different outer shapes have been recently developed with the aim of reducing the impact of wind on the collection efficiency of the gauges, e.g., with a calix shaped external case.

The area of the collector and the nominal capacity of the bucket compartments provide the sensitivity of the gauge. The gauge size and shape vary with the manufacturer, although the WMO recommendations (WMO, 2014) suggest an instrument sensitivity between 0.1 and 0.5 mm (with 0.2 mm being the most typical figure for mid latitude climatology).

4.2.2.2 Weighing gauge

A weighing gauge (WG) consists of a bucket, usually made of plastic, used to collect and measure liquid and solid precipitation by means of a weighing principle.

FIG. 4.3

Tipping-bucket rain gauge with cylindrical outer shape installed in the field.

The bucket and the measuring principles are placed inside an external case, which often has a "chimney" shape (see Fig. 4.4), with a larger section at the bottom to allow collecting large volumes of water. This large capacity of the bucket allows to minimize the emptying interventions, which in many cases are operated manually. Those gauges equipped with an automatic emptying system have smaller size and a cylindrical outer shape. The typical emptying mechanism is based on the siphon principle, as described for the storage gauges. This technique does not allow to measure precipitation during the emptying period, leading to an underestimation error.

Precipitation is collected in the bucket and the water is periodically weighted, so that the precipitation rate can be calculated as the difference between the measured weight of the container in two consecutive measurements. In principle, the time resolution of these instruments can be very fine and the sensitivity of the weighing system very high, although the actual gauge sensitivity is generally limited by the filtering process needed to eliminate the noise of the raw signal.

The measurement can be obtained using different weighing principles, including a balance, a load cell, or a vibrating wire load sensor.

FIG. 4.4

A typical weighing gauge having a "chimney" shape.

A special WG, employing an automatic emptying principle was recently developed by making use of the tipping-bucket principle. In this case, the tilting buckets are used as a weighing container and an emptying mechanism at the same time, and the sensor weighs the water collected in each bucket, alternately. This rapid emptying principle allows the sensor to weigh a small volume of water and therefore to increase the resolution of the measurement. However, this leads to an underestimation of precipitation during the tilting time of the buckets (as is typical of the TBRs), and the presence of rotating parts requires additional maintenance.

4.2.2.3 Drop counting gauge

The catching-type drop counting gauge is provided with a funnel that collects rainfall and conveys the rainwater towards a calibrated nozzle, which generates a series of droplets. Below the nozzle, an optical sensor is positioned such that the passage of each released drop is detected by measuring the occlusion of the optical path. The size of the dispensed drops may vary with the gauge design and depends on the material and geometrical characteristics of the nozzle. The recent study by Stagnaro et al. (2018) revealed that also the rainfall intensity and the size of the collecting area affect the drop formation process, and their releasing frequency. Therefore, the drop volume varies with the drop frequency, or the rainfall rate, and must be accurately calibrated.

Knowing the size of the drops and the releasing frequency, the accumulated volume of rainfall over a given period can be obtained, as well as the rainfall rate.

The sensitivity of drop counting gauges is provided by the volume of the drops dispensed by the nozzle and is in the order of 0.005 mm of rainfall, thus making this instrument suitable to measure light precipitation rates. However, this measuring principle assumes that the instrument generates droplets of constant volume at any given frequency, in a one-to-one relationship. This behavior has an upper limit, at which the water flux from the internal nozzle starts to be irregular and stepwise continuous. This leads to the failing of a one-to-one relationship between the measured drops frequency and rainfall intensity, leading to relevant inaccuracies. Furthermore, since the drop generating mechanism usually has a complex geometry, these instruments are highly susceptible to clogging in the absence of appropriate and frequent maintenance. Therefore, a stand-alone installation of drop counting gauges is discouraged.

4.2.3 Non-catching type instruments

Instruments based on non-catching (often contact-less) principles require no funnel, nor any container, to collect precipitation but rather sense individual hydrometeors while they cross the measuring area or impact on the sensor. Usually, they can provide additional information, further than precipitation accumulation or intensity, such as the particle size distribution and the hydrometeors fall velocity and for this reason they are known as disdrometers. The measurement of precipitation is generally based on optical, microwave, acoustic or impact principles.

Optical methods usually employ infrared or laser light sources to measure the light scatter due to the falling particles in a sample volume or the attenuation of the laser power detected while single particles cross the measuring section. Other instruments use imaging techniques in the visible portion of the light spectrum to recognise particles in the field-of-view of special cameras. Microwave disdrometers use small size radars, upwards oriented, to measure the signal spectrum backscattered by falling particles on a sampling volume near the ground. Acoustic or impact sensors measure either the vibration or the displacement induced by precipitation particles when impacting on a detector surface.

Due to the ability of disdrometers to provide additional information on precipitation with respect to traditional gauges and their low maintenance requirements, their employment in NWS networks is increasing in recent years. However, the last WMO field intercomparison of precipitation intensity gauges (Vuerich et al., 2009; Lanza and Vuerich, 2012) highlighted that disdrometers have lower performance when compared to traditional raingauges. The reason of this behaviour can be attributed to a lack of standardized calibration and recognised testing methods in the laboratory that are traceable to international standards.

4.3 Accuracy of tipping-bucket and weighing gauges

Rainfall intensity measurements are traditionally performed by means of TBRs and WGs, still representing the most popular and widespread types of automatic raingauges presently employed worldwide. In both cases, systematic biases due to the inherent characteristics of the counting device have a strong influence on the measurement of rainfall, with an increasing impact as the rain rate increases.

TBRs underestimate rainfall at high intensities because of the rainwater amount that is lost during the tipping movement of the bucket. Marsalek (1981) explains the mechanism that leads to the underestimation bias as follows: water losses are observed because the tipping movement starts at that instant in time when the bucket is completely filled, but the bucket tipping around its rotation axis requires a certain amount of time and during this time window the incoming rain is not measured. Marsalek (1981) estimated this measurement bias by testing three different TBRs in the laboratory, calculating the ratio between the measured rainfall intensity and the actual value, and found that the bias is larger than 10% for rain rates higher than 200 mm h^{-1}. This bias is known as the systematic mechanical error and can be adjusted by means of laboratory calibration.

Initially, in the static calibration, the volume of each compartment of the bucket is adjusted using a calibrated pipette to set the volume of water needed to trigger the tipping movement. This volume can be set equal to the nominal one by adjusting the position of the stop screws. The static calibration inevitably leads to an underestimation bias over the whole rainfall intensity range, that increases with the rainfall intensity.

In the single-point calibration the bucket volume is sometimes adjusted to a value smaller than the nominal one, to compensate the rainfall underestimation at the desired reference intensity. For instance, after performing the single-point calibration, Marsalek (1981) obtained measurement biases confined within 10%. Rainfall measurements based on the single-point calibration, therefore, exhibit overestimation at intensities lower than the reference one and underestimation at higher rain rates. Nevertheless, hydro-meteorological services and instrument manufacturers still often rely on single-point calibration.

To adjust rainfall measurements over the entire rainfall intensity (*RI*) range, correction curves obtained by means of dynamic calibration must be applied. The correction curve is specific of each single raingauge, depending on the mechanical design, the environmental conditions where the instrument operates and its state of wear. Therefore, dynamic calibration should be periodically repeated during the service life of any instrument. The dynamic calibration procedure is described in the

European Norm EN 17277:2019 (see next section). Dynamic calibration is based on the hypothesis that the *RI* is constant within a chosen period (e.g., a sampling time equal to one minute). During this period, the number of tips is counted, and the associated *RI* is derived by multiplying this number by the nominal volume of the bucket. This simplification does not properly represent some typical behaviour of the rainfall continuous phenomenon and generates the so-called sampling error. When the amount of rainfall in the sampling time is not sufficient to cause the system to tip, the volume of water that remains in the buckets is not measured. During consecutive events, both the presence of a certain amount of water previously stored in the bucket before the start of a new event and the amount of water that remains inside the bucket at the end of each event affect the measurement. During light precipitation events many isolated tips are recorded with remarkable overestimation of the actual precipitation rate at the corresponding time step and underestimation in the contiguous steps, as remarked e.g., by Molini et al. (2001), Wang et al. (2008), Habib et al. (2001).

Note that sampling errors mostly affect the calculation of rainfall rates, while the cumulated rainfall depth, over sufficiently long intervals, can be considered as reliable. Sampling errors can be reduced by measuring the inter-tip time instead of counting the number of tips within a constant sampling time. The inter-tip time was adopted in the works of Colli et al. (2013a), Stagnaro et al. (2016), which concluded that the knowledge of the inter-tip time is essential to improve the performance of TBRs.

WG measurements are also subjected to systematic biases associated with the filtering algorithm used to remove the signal noise from the acquisition of the measurement. This algorithm produces a systematic delay that is called step response. It varies from few seconds to a few minutes depending on the gauge design and model. This delay assumes a relevant role in case of high-resolution *RI* time series. The EN 17277:2019 standard describes the laboratory procedure needed to detect the time consant and defines it as the time employed by the instrument to measure 63.2% of the reference intensity value (see next section).

4.3.1 Literature works about calibration

In the work of Niemczynowicz (1986), three different TBRs were tested at about twenty different rainfall intensity values. First, to check the balance of the buckets, a constant water flow was supplied to the funnel of the tested gauge, simulating a constant rainfall intensity of about 60 mm h^{-1}. The volume of water running out of each compartment of the bucket was collected separately for about 20 tips and the stop screws were regulated so that both compartments contained approximately the same volume of water (equal to the nominal one) before triggering the tipping movement. Then, dynamic calibration was performed: the water running out from both compartments was collected for about 30 to 200 tips; its volume was measured with a burette, and the time required to obtain the same number of tips was recorded by a stopwatch. Finally, knowing the number of tips, the time and the total water volume, the exact value of rainfall intensity was compared to the tilting rate measured

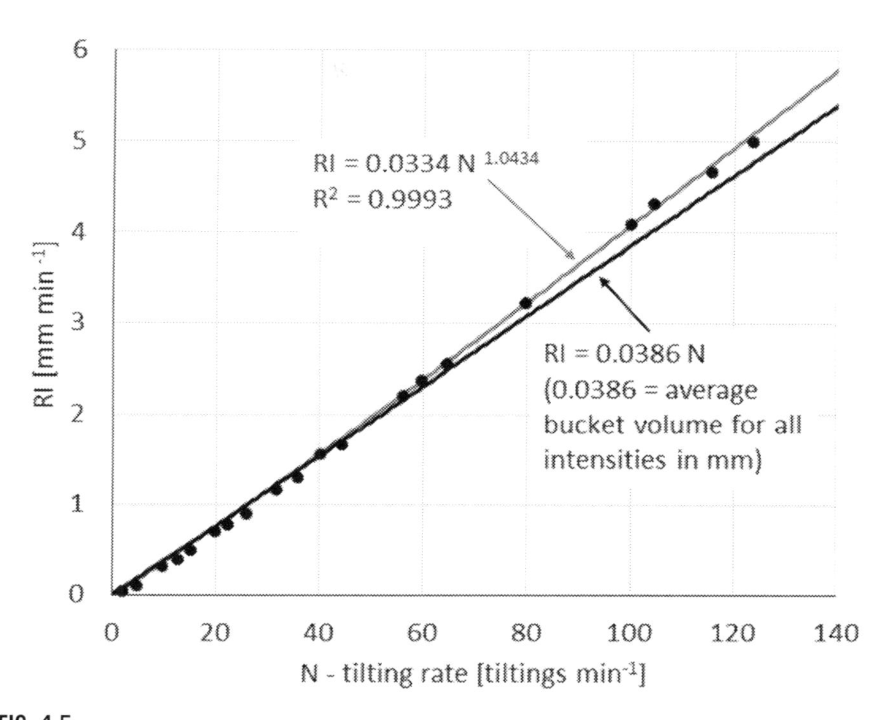

FIG. 4.5

Sample correction curve obtained after dynamic calibration. (Data from Niemczynowicz, 1986).

by the instrument. A sample correction curve is reported in Fig.4.5, where the actual *RI* values are plotted against the tilting rate, with a power law equation that fits the experimental data quite closely.

In the work of Costello and Williams (1991) TBRs were tested in the laboratory by recording the inter-tip time. Results were compared with intensities computed using a fixed sampling time (from 15 s to 1 h). The comparison revealed that intensity values measured using a fixed sampling time are significantly lower than those computed with the inter-tip method (Fig.4.6).

In the work of Colli et al. (2013a), measurements from co-located TBRs and drop counting gauges installed at the weather station of the Hong Kong International Airport were compared. The Drop Counter (DC) gauge measurements were assumed as the reference (actual) rainfall intensity for precipitation events up to 100 mm h^{-1}, recorded at a resolution of 10 s. The TBR was tested in the laboratory, obtaining the calibration curve of the instrument. The dynamic calibration procedure was based on the direct comparison between a generated constant flow rate and the measured rainfall intensity at various *RI* levels, covering the whole operational range of the instrument. Raw TBR data available from the field, at the resolution of one-minute, were corrected. To compare the gauge measurements and investigate the sampling issue, a "virtual" TBR rainfall intensity record was derived from the DC

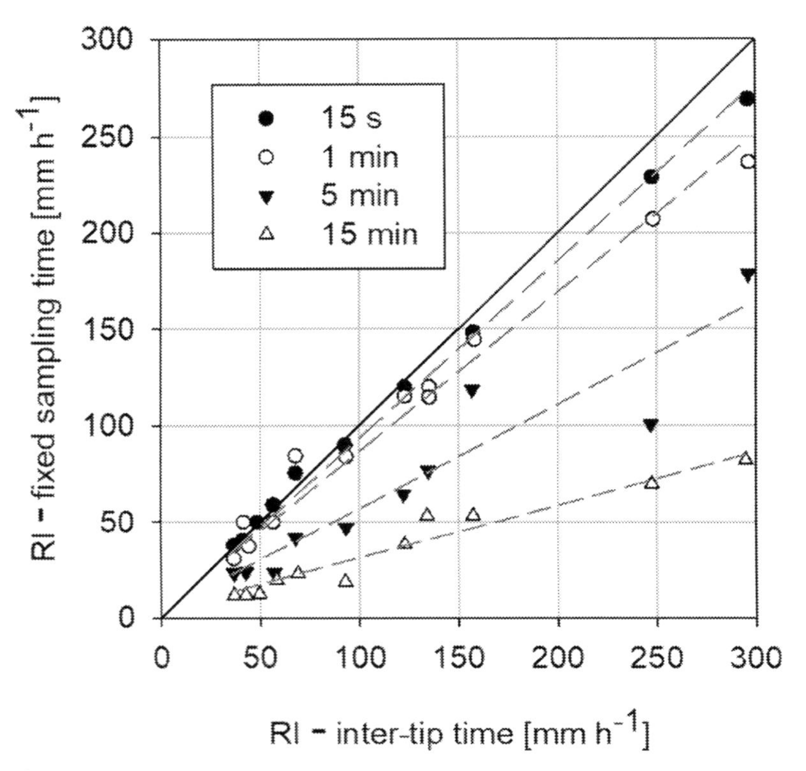

FIG. 4.6

RI measurements in mm h^{-1} (vertical axis) obtained by using different fixed sampling times vs those derived by recording the inter-tip time (horizontal axis in mm $^{h-1}$) at a one-minute resolution. (Data from Costello and Williams, 1991).

drop frequency series. The behaviour of the virtual TBR at a resolution of 0.5 mm was simulated by recording one tip every time the cumulated water volume observed by the DC reached the nominal capacity of the bucket (ideal behaviour) and reporting the calculation of the rainfall intensity at the one-minute time scale. In Fig. 4.7 (panel *a*), a sample series of rainfall intensity records from the DC at 10 s time resolution is depicted, with superimposed the simulated 0.5 mm tip counts (black bars). The distance between bars is equal to the inter-tip time, therefore, the nominal depth of water corresponding to the bucket volume was allotted to the entire inter-tip period rather than to one single minute as in common practice. In panel *b*, the same DC data aggregated at the one-minute time resolution are shown, used as the reference *RI* time series. The raw TBR measurements obtained by counting the number of tips in each minute are shown in panel *c*. By recording the inter-tip time, the same TBR measurements can be expressed as reported in panel *d*. By comparing panels *b*, *c*, and *d* the inter-tip method is confirmed to improve the interpretation of the measurement especially concerning the central portion of each event. Nevertheless, some marked differences between the reference data (panel *b*) and the *RI* measurements corrected

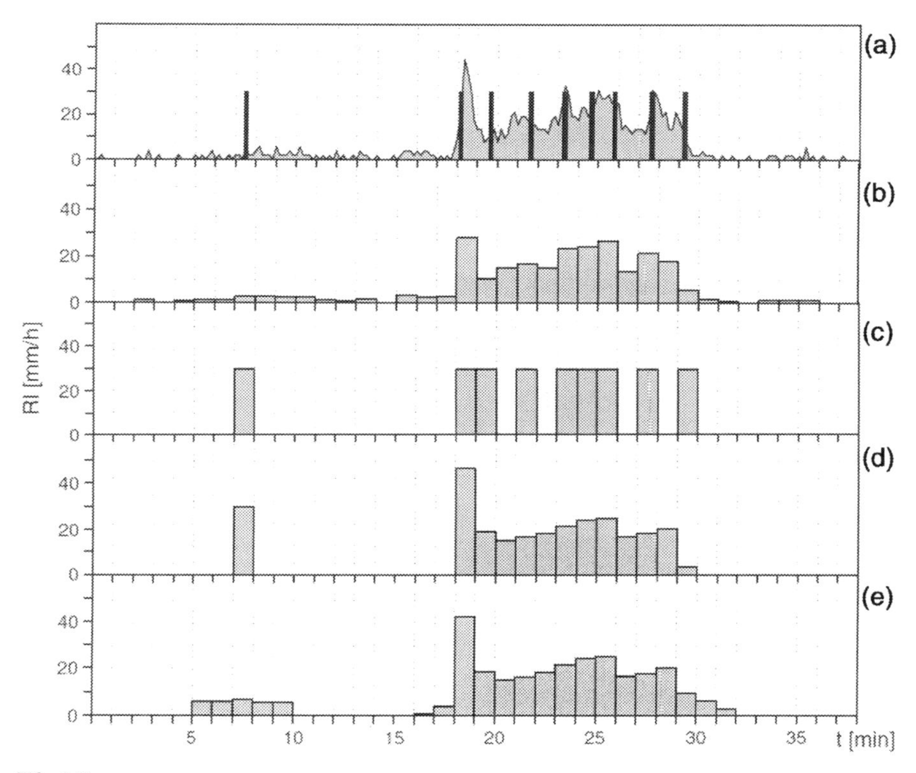

FIG. 4.7

Sample series of rainfall intensity records from (a) the DC at 10 s time resolution with superimposed the simulated 0.5 mm tip counts, (b) the reference intensity time series aggregated at the 1-min time scale, (c) the raw TBR rainfall intensity figures at the one-minute time scale obtained from tip counts only, (d) the corrected TBR rainfall intensity figures at the one-minute time scale obtained from the inter-tip time correction method, and (e) the corrected TBR rainfall intensity figures at the one-minute time scale obtained from the smoothing method applied on the inter-tip time corrected time series. (Source: Colli et al., 2013a).

with the inter-tip method (panel *d*) persist at the boundaries of the event and where isolated tips occur. Further progresses in the adjustment of the TBR recorded rain-rate series were obtained by running a smoothing algorithm to mimic the actual hyetograph profile observed when a rain event starts, finishes, or evolves through light intensities. The associated result is shown in panel *e*.

In the work of Colli et al. (2013b), a laboratory investigation was conducted to evaluate the accuracy and precision of a WG under unsteady reference *RI* conditions. Three different laboratory test conditions were applied: single and double step variations of the reference flow rate and a simulated real-world event. The measurements obtained from the WG were compared with those derived from a traditional TBR under the same laboratory conditions. The TBR measurements were corrected to account for systematic mechanical errors and comparison is also proposed after applying further

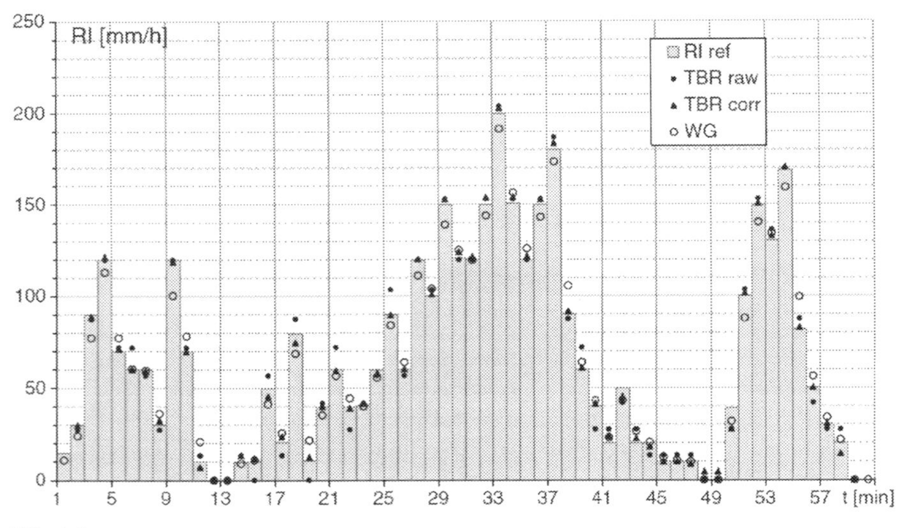

FIG. 4.8

Comparison between the *RI* values (mm h^{-1}) measured by the tested WG (white circles) and TBR (with and without correction for sampling errors, black circles and triangles, respectively) during a real-world event simulation. The reference intensity is also reported as a background histogram. (Source: Colli et al., 2013b).

algorithms to reduce the sampling errors. Results revealed that the performance of the investigated WG under unsteady conditions is comparable or even lower than what can be obtained from more traditional TBRs, even in case corrections for sampling errors are not applied. In Fig. 4.8, the temporal pattern of the simulated rainfall intensity (RI_{ref}), the TBR measurements evaluated by both considering the sole correction for dynamic calibration (TBR_{raw}) and after applying a correction algorithm for the sampling errors at low rain rates (TBR_{corr}), and the WG measurements are reported. This presentation highlights the effect of the response time of the WG on the measurements. Indeed, when the *RI* increases the WG underestimates the measurement while the opposite occurs when the *RI* decreases. TBR measurements, both corrected with dynamic calibration and corrected for the sampling errors, better agree with the reference.

4.4 Calibration procedures and standardization

Various calibration procedures for precipitation gauges have been developed, depending on the measurement principle employed by the instruments. For catching type gauges these procedures have been standardized by national and international standardisation bodies, while non-catching type gauges still suffer the lack of widely recognized standards.

Calibration is generally performed in a laboratory, where dedicated equipment is used to provide the reference quantities, while verification of the instrument calibration can be performed in the field, using appropriate devices.

4.4.1 Calibration procedures

Storage gauges are the simplest raingauges and only require calibration of the storage volume. This is performed by pouring a known volume of water into the gauge and checking the correspondent water level reached in the container. When automatic sensors are employed to record the level of collected water in the storage tank, proper calibration of the sensor (float, acoustic or pressure level probe) is needed.

For catching-type gauges, the recent EN 17277:2019 introduced the first international standard about instrument calibration and the associated testing procedures. Based on this norm, instruments can be classified according to their performance, measured by means of standardised laboratory calibration tests.

In case of TBRs, the norm first requires that accurate balancing of the two buckets is achieved. The required test consists in measuring the time of tipping of each bucket when operating under known and constant flow rates and evaluating the relative difference between the average tipping time of each bucket and the overall one. The relative difference must be less than ± 5% for each bucket to pass the test, and the coefficient of variation must be limited as well, so that 80% of the times of tipping are included within ± 5%.

To assess the performance of TBRs under varying rainfall rate, dynamic calibration procedure is required by the EN 17277:2019. Tests are performed under different steady state conditions of the reference equivalent rainfall intensity, and the relative deviation between the measured and the reference rainfall intensities is calculated.

WGs require calibration of the weighing sensor, usually performed by means of calibrated weights placed in the gauge container to simulate an equivalent volume of water. However, this calibration procedure can assess the precision of the weighing system in static conditions, while dealing with precipitation intensity measurements (that are highly variable in time) the dynamic response of the instrument must be evaluated. In the dynamic calibration procedure, a dedicated test is also prescribed for this type of gauges to assess their step response, expressed in terms of the instrument time constant, i.e., the amount of time that is required by the instrument to measure 63.2 % of the reference intensity value.

Traditional calibration procedures for drop counting raingauges consist in pouring a known volume of water into the gauge funnel and measuring the number of drops generated by the nozzle to obtain the drop volume. As described above, the recent study by Stagnaro et al. (2018) showed that the volume dispensed by the nozzle varies with the flow rate. Therefore, following the same procedures described for the dynamic calibration and measuring the water volume and the overall number of dispensed droplets provided to the gauge at different rainfall intensities, it is possible to obtain a calibration curve that relates the drop volume to the measured dripping frequency.

4.4.2 Laboratory dynamic calibration

The dynamic calibration procedure was first assumed as a standard test in the Italian norm UNI 11452:2012 and then included in the technical report CEN/TR

16469:2013. At present, the EN 17277:2019 requires that calibration is performed in a certified laboratory, equipped with a dedicated setup able to provide a known and constant flow rate for a sufficient period (at least 30 minutes) for each tested rainfall intensity within the operational range declared by the manufacturer. The reference flow rate is verified by weighing the water volume provided to the instrument and the corresponding duration of each test. The reference rainfall intensity values to be used in the tests are specified, as well as the required uncertainty for the flow generating system.

The calibration device must be capable to acquire the measures provided by the gauge under test at regular intervals. Each measurement is then compared with the reference one to calculate the relative percentage error as follows:

$$e_{rel}[\%] = \frac{RI_{meas} - RI_{ref}}{RI_{ref}} \times 100 \quad (4.1)$$

where, RI_{meas} is the liquid precipitation intensity measured by the gauge and RI_{ref} is the reference equivalent rainfall intensity provided by the calibration device.

The prescribed test report must contain, in tabular form, the average value and the 10° and 90° percentiles of the relative percentage error distribution, for each value of the tested reference intensity. A dynamic calibration curve based on the observed data can be determined either theoretically or in the form of a best-fit regression function (linear, second order polynomial or power law function).

It is also noted that, at some reference intensities, raingauges equipped with a funnel may suffer "storage" conditions when water accumulates inside the instrument's collector before it is conveyed to the counting device, and the occurrence of storage during the test must be visually detected and annotated.

4.4.3 In-field verification of catching-type instruments

In-field verification of the calibration of raingauges should be performed periodically during regular maintenance operations and between two laboratory calibration tests. The field verification aims to detect possible malfunctions of the gauge and any deterioration of the instrument performance. The procedure follows the same principles of laboratory calibration but involves a limited number of rainfall intensity values for the test and requires a lower volume of water, to reduce the overall test duration. The calibration device must be able to generate an equivalent constant water flow for any single test and to reproduce at least three different flow rates. A description of a suitable field calibration device is provided in WMO (2014).

The field calibration device developed at the University of Genova is made of a water tank equipped with a set of air intakes and water output nozzles. Different combinations of such intakes and nozzles allow to generate various flow rates. The device is equipped with an electronic system able to detect the emptying time of the water volume and the capacity of the instrument is calibrated in the laboratory, to serve as a reference. A sample image of the operation of this device in the field is shown in Fig. 4.9.

FIG. 4.9

Portable calibration device for field verification of catching-type raingauges in operational conditions.

The measurements of the gauge under test are then compared to the reference equivalent rainfall intensity provided by the field calibrator at the time scale provided by the gauge (the norm requires a one-minute time resolution), and the relative difference can be calculated. Results from different tests provide information on the status of the calibration and can be compared to the calibration curve obtained in the laboratory to assess the need to perform additional maintenance or re-calibration.

4.4.4 Rain gauge classification

The EN 17277:2019 provides a classification of catching-type rainfall intensity gauges based on the performance obtained from suitable tests carried out in certified laboratory. This classification is based on the instrument performance alone and does not relate to the measuring principle, nor the technical characteristics of the gauge.

Three performance classes are indicated in the norm according to the dynamic calibration and step response test results, as follows:

Class A: Class A rainfall intensity gauges shall have maximum deviations within ± 3% against the reference rainfall intensity at the temporal resolution of 1 min. Weighing raingauges shall also have a time constant less than or equal to 1 min.

Class B: Class B rainfall intensity gauges shall have maximum deviations within ± 5% against the reference rainfall intensity at the temporal resolution of 1 minute. Weighing raingauges shall also have a time constant less than or equal to 1 min.

Class C: Class C rainfall intensity gauges shall have maximum deviations within ± 10% against the reference rainfall intensity at the temporal resolution of 1 min. This also applies to weighing raingauges where the time constant is less than or equal to 1 min. If the time constant is greater than 1 min the maximum deviations shall be within ± 5%.

In case a raingauge has a maximum deviation greater than ± 10% in measuring the reference rainfall intensity at the temporal resolution of 1 min it cannot be classified according to this standard.

Furthermore, for TBRs it is required that the balancing test of the buckets is passed and consistency of all information provided by the gauge is ensured. Any inconsistency between the precipitation intensity output at one-minute resolution and other quantities provided by the instrument (e.g., the precipitation amount) must be declared.

The class attribution certificate must contain the average value and the 10th and 90th percentiles of the relative percentage error distribution ($e_{rel}[\%]$), for each value of the tested reference precipitation. These values are presented in the form of a table and can be shown in a graph using the non-parametric boxplot representation. The dynamic calibration curve, obtained theoretically or by fitting the relative errors of tested precipitation intensities, can be reported in the certificate to allow correcting the readings. Fig. 4.10 shows the performance of a TBR and a WG in terms of relative errors $e_{rel}[\%]$. The same instrument can be attributed different classes of performance over different measuring ranges.

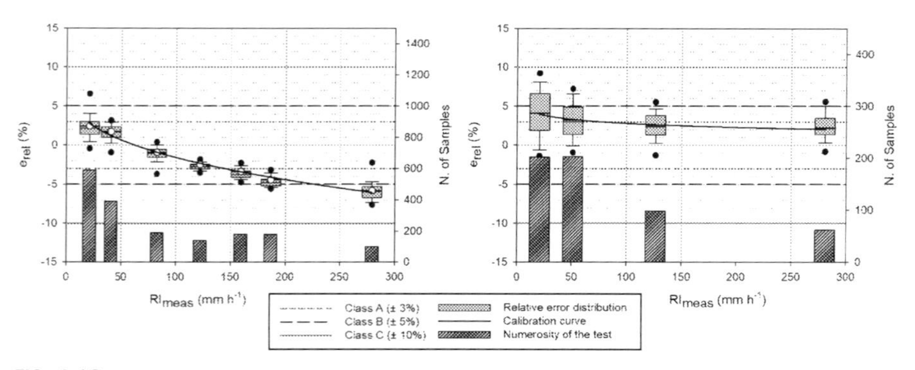

FIG. 4.10

Relative percentage errors at various equivalent reference intensities and calibration curves for a TBR (left) and a WG (right). Grey bars indicate the sample size according to the right-hand vertical axis.

4.5 The impact of wind on rainfall measurements

The influence of wind on precipitation measurements is due to the interaction between the gauge body and the airflow. This aerodynamic effect deviates the hydrometeors from their undisturbed trajectories (Nešpor and Sevruk, 1999; Folland, 1988), and is responsible for a significant reduction of the collection performance. For liquid precipitation, Sevruk (1982) reported that the typical magnitude of the wind-induced losses (undercatch) for the precipitation amount is 2–10%, while Pollock et al. (2018) reported an observed undercatch of about 10% to 23% from field experiments performed at a lowland and upland site, respectively. Nevertheless, the implementation of adjustments based on correction curves in operational conditions is still uncommon.

The wind-induced bias on precipitation measurements is addressed in the literature using both numerical simulation (Computational Fluid Dynamics – CFD – with particle tracking models) and experiments (field campaigns and wind tunnel tests). The main factors of influence are the gauge geometry, the wind speed, and the type of precipitation (liquid or solid) and its characteristics, including the crystal type, Particle Size Distribution (PSD) and precipitation intensity (Thériault et al., 2012; Colli et al., 2015).

Adjustment curves can be derived using data from experimental sites equipped with different precipitation gauges in operational conditions, and a reference one. The WMO recommends (WMO, 2014) using a gauge placed in a pit as the reference instrumental configuration for liquid precipitation, with the gauge orifice at ground level, sufficiently distant from the nearest edge of the pit to avoid in-splashing. A strong plastic or metal anti-splash grid with a central opening for the gauge must cover the pit, except for the central opening where the gauge orifice is located (construction details are provided by the EN 13798:2010). Because of the absence of wind-induced bias, pit gauges generally report more precipitation than any elevated gauge.

4.5.1 Field experiments

In field studies, the ratio between the precipitation measured by a gauge in operational conditions, h_{meas} (usually in [mm]) for a given wind speed, U_{ref} (m s^{-1}), and the reference one, h_{ref} (mm), is called the collection efficiency (CE):

$$CE = \frac{h_{meas}\left(U_{ref}\right)}{h_{ref}} \tag{4.2}$$

An example of adjustment curve derived from field measurements was obtained by Wolff et al. (2015). In that work measurements recorded at the Haukeliseter (Norway) experimental site during WMO SPICE (solid precipitation intercomparison experiment) from two WGs and a double fence intercomparison reference (DFIR) were used. The DFIR is the reference installation recommended by the WMO for solid precipitation (Goodison et al., 1998). In the work of Wolff et al. (2015) precipitation was classified in solid, mixed, or liquid, according to the air temperature thresholds reported by Sims and Liu (2015). For temperatures $T < -2°C$ the precipitation was assumed as snow, for $T > 2°C$ the precipitation was assumed as rain, while mixed precipitation was assumed in the range $-2°C \leq T \leq 2°C$. Adjustment curves as a function of the wind speed (U_{ref}) and air temperature (T) were formulated as follows, based on a three-year data set:

$$CE = \left[1-\tau_1-(\tau_2-\tau_1)\frac{e^{\left(\frac{T-T_\tau}{s_\tau}\right)}}{1+e^{\left(\frac{T-T_\tau}{s_\tau}\right)}}\right]e^{-\left(\frac{U_{ref}}{\theta}\right)^{\beta}}+\tau_1+(\tau_2-\tau_1)\frac{e^{\left(\frac{T-T_\tau}{s_\tau}\right)}}{1+e^{\left(\frac{T-T_\tau}{s_\tau}\right)}}+\sigma(T)\varepsilon \quad (4.3)$$

where β and θ are two fitting parameters, T_τ is the temperature threshold and defines the transition between the two limits above, while s_τ indicates the fuzziness between rain and snow, and $\sigma(T)$ is a parameter governing the variance of the measurement error.

4.5.2 Numerical simulation approaches

Adjustment curves derived from field experiments alone are characterised by their strict dependence on the site where the field test is geographically located, the associated precipitation and wind climatology, and the reliability and accuracy of the assumed reference gauge. A theoretically based approach, based on real world observations for proper validation of the algorithms and results, allows to achieve a complete coverage of various local climatological characteristics, precipitation microphysics, gauge shapes and wind conditions. This can be achieved by exploiting the potential of numerically solving the basic equations of fluid motion and of particle-fluid interactions, as in the works of Nešpor and Sevruk (1999), Colli et al. (2015), Cauteruccio and Lanza (2020), Cauteruccio et al. (2021c).

CFD simulation of the airflow fields around precipitation gauges when impacted by wind can be conducted by using different models solving, in particular, the Unsteady Reynolds Average Navier Stokes (URANS) or the Reynolds Average Navier Stokes (RANS) equations and Large Eddy Simulations (LES). Then particle trajectories are modelled by using a particle tracking model based on the particle motion equation. The motion of a particle in a fluid, such as air or water, is strictly related to the particle Reynolds number (Re_p) and the associated drag coefficient (C_D). The drag coefficient is a dimensionless quantity used to represent the resistance of an object in motion in a fluid and it is associated with the cross-sectional area of the object.

Nešpor and Sevruk (1999), conducted numerical simulations on three cylindrical gauges with different sizes and varying the shape of the collector rim. The airflow velocity (magnitude and directional components) was first calculated using a time average approach (RANS) based on the $k-\varepsilon$ turbulence closure model, where k is the *turbulence kinetic energy* and ε is the *energy dissipation per unit mass*. Then liquid particle trajectories were computed by using a Lagrangian Particle Tracking (LPT) model. A one-way coupled model, which neglects the interaction between particles and the effect of the particles on the air was employed. Spherical particles were separately simulated for each diameter and the collection efficiency was evaluated by computing the integral, over the particle size distribution, of the number of particles collected by the gauge with respect to the total precipitation.

This simulation scheme was adopted also by Thériault et al. (2012), Colli et al. (2015, 2016a, 2016b) for solid precipitation, by increasing the details of the computational mesh to better capture the airflow features. In the work of Thériault et al. (2012), different crystal types were modelled by using a power law parametrization of the terminal velocity, volume, density and cross section of the particles and a fixed value of C_D for each crystal type. In the work of Colli et al. (2016a), LES were conducted on a WG with chimney shape, both in unshielded and shielded configurations. In the work of Colli et al. (2016b) solid precipitation was modelled by defining two macro categories: wet and dry snow and results were compared with field measurements in terms of collection efficiency. The obtained *CE* curves for the two macro categories act as upper and lower thresholds of the wide spreading of experimental data obtained from the Haukeliseter (Norway) experimental site.

Colli et al. (2015) obtained a better fit of the collection efficiency curves with real-world data by calculating the snowflake trajectories while accounting for the dependence of the aerodynamic C_D on the local Reynolds number of solid particles, Re_p. The C_D of a falling particle is influenced by the instantaneous particle-to-air magnitude of velocity through the Re_p and at each time step the particle trajectory is obtained by updating the Re_p value and the associated C_D.

In the work of Cauteruccio et al. (2021b) the LPT model used by Colli et al. (2015) for solid precipitation was modified by introducing drag coefficient equations suitable for liquid precipitation. The drag coefficient equations were implemented for various ranges of the particle Reynolds number established a priori among those proposed in the literature by Folland (1988) and formulated starting from data published by Khvorostyanov and Curry (2005). This model was validated by means of a bespoken wind tunnel experimental setup designed within the activity of the Italian National project *PRIN 20154WX5NA* "Reconciling precipitation with runoff: the role of understated measurement biases in the modelling of hydrological processes" to release water drops in the incoming airflow and track their deformed trajectories while travelling close to the gauge collector.

4.5.3 Theoretical adjustment curves

The validated numerical approach based on CFD and LPT simulations allows to derive adjustment curves for the wind-induced bias by simulating different gauge shape, precipitation intensity and wind speed combinations.

In the work of Cauteruccio and Lanza (2020), CFD simulations already performed by Colli et al. (2018) for a cylindrical gauge were adopted, together with the LPT model validated by Cauteruccio et al. (2021b), to derive *CE* curves as a function of wind speed (U_{ref}) and parameterized with the *RI*. Easy to use adjustment curves were obtained as a function of wind speed and the rainfall intensity measured by the gauge (RI_{meas}).

The advantages of using *RI* instead of the air temperature (as shown above) were demonstrated by Colli et al. (2020). The authors analysed WMO-SPICE quality controlled 30-min accumulation data from the Marshall field-test site (CO, USA) and revealed that the wind-induced undercatch of precipitation gauges is best correlated with the measured precipitation intensity, rather than temperature (as widely used e.g., in Wolff et al., 2015; Kochendorfer et al. 2017a, 2017b), in addition to wind speed. This result was confirmed by the analysis of data from other field test sites, such as CARE (Canada) and Haukeliseter (Norway), and showed a consistent behaviour under different climatological conditions. The measured precipitation intensity has indeed the advantage of including information about the particle size distribution (Pruppacher and Klett, 2010).

In the work of Cauteruccio and Lanza (2020), the catch ratio, r (-), defined as the ratio between the number of particles captured by the gauge collector in disturbed airflow conditions, $n(d)$, and the maximum number of particles, $n_{max}(d)$, captured in undisturbed conditions was computed for drops assumed spherical with diameter $d = 0.25$, 0.5, and 0.75 mm and then from 1 to 8 mm, with bin size of 1 mm. After introducing a suitable *PSD*, indicating the number of particles $N(d)$ per unit volume of air and per unit size interval having a volume equal to the sphere of diameter d, the integral of the catch ratio over the range of diameters (from 0 to d_{max}) provides the numerical *CE* in the form:

$$CE\left(U_{ref}\right) = \frac{\int_0^{d_{max}} \rho_p \, V_p \, n(d) \, N(d) \, \mathrm{d}d}{\int_0^{d_{max}} \rho_p \, V_p \, n_{max}(d) \, N(d) \, \mathrm{d}d} \tag{4.4}$$

where ρp, V_p are the density and volume of particles with diameter d.

The *CE* curve was derived from the calculated catch ratios by assuming that the microphysical characteristics of precipitation are those obtained from disdrometer measurements in the Italian territory. The *PSD*s, provided by Caracciolo et al. (2008) for various *RI* classes, were fitted with the typical exponential function (Marshall and Palmer, 1948) (hereinafter *MP*), and the associated parameters N_0 (intercept) and Λ (slope) were adopted to calculate the *CE* values (here called raw *CE* values) per each *RI* class.

A three-step procedure was used to derive a suitable parameterization of the *CE* curves based on rainfall intensity. First, for each *RI* class, the *CE* curve was obtained as a function of wind speed (U_{ref}) from the raw values of the numerical simulation results by fitting a four-parameter sigmoidal function ($CE = f(U_{ref})$ – see Eq. (4.5)) and using the experimental N_0 and Λ parameters associated with each *RI* class.

$$CE\left(U_{ref}\right) = y_0 + \frac{a}{1 + e^{-\frac{\left(U_{ref} - x_0\right)}{b}}} \tag{4.5}$$

The dependency on *RI* of the obtained parameters N_0 and Λ, attributed to the mean value of each *RI* class, was then fitted with power law curves, and a new sigmoidal *CE* curve was obtained per each *RI* class from the fitted parameters. The definition of N_0 and Λ as a function of *RI* allows to obtain the *MP* parameters for any desired *RI* value, within the range investigated. After calculating the *CE* values for the associated *RI* and U_{ref}, the *CE* curves could be obtained again as a sigmoidal best-fit.

Finally, to obtain a simple formulation for the *CE* as a function of both *RI* and U_{ref} in the investigated ranges, the four parameters of the sigmoidal function (y_0, x_0,

a and b) were also fitted with power law (for a) and logarithmic (for b, x_0, y_0) curves as a function of RI. The best-fit curves for each sigmoidal parameter as a function of RI are listed below and shown in Fig. 4.11 with the associated correlation factors.

$$a = 0.2213RI^{-0.17} \tag{4.6a}$$

$$b = 0.1191\,ln(RI) - 4.1365 \tag{4.6b}$$

$$x_0 = 0.5222\,ln(RI) + 4.4164 \tag{4.6c}$$

$$y_0 = 0.0166\,ln(RI) + 0.8645 \tag{4.6d}$$

The derived numerical CE curve is a function of the actual rainfall intensity. However, in operational practice, the only knowledge available for rainfall intensity is that measured by the gauge, which is therefore affected by the wind-induced bias. The adjustment curve was further derived, for application purposes, as a direct relationship between the reference rainfall intensity (RI) and the measured one (RI_{meas}), so that the actual rainfall intensity (and the wind-induced bias) can be calculated starting from the measurement provided by the gauge, once the wind velocity is known.

Recalling that, in the field, the CE is defined as the ratio between the precipitation measured by the gauge (RI_{meas}) and the reference one (RI), it is possible to derive from Eq. (4.5), the RI_{meas} associated with each (U_{ref}, RI) couple within the investigated range. The results can be depicted in the (RI, RI_{meas}) plane for each wind speed, as shown in Fig. 4.12. The RI_{meas} values calculated from Eq. (4.5) for the mean values

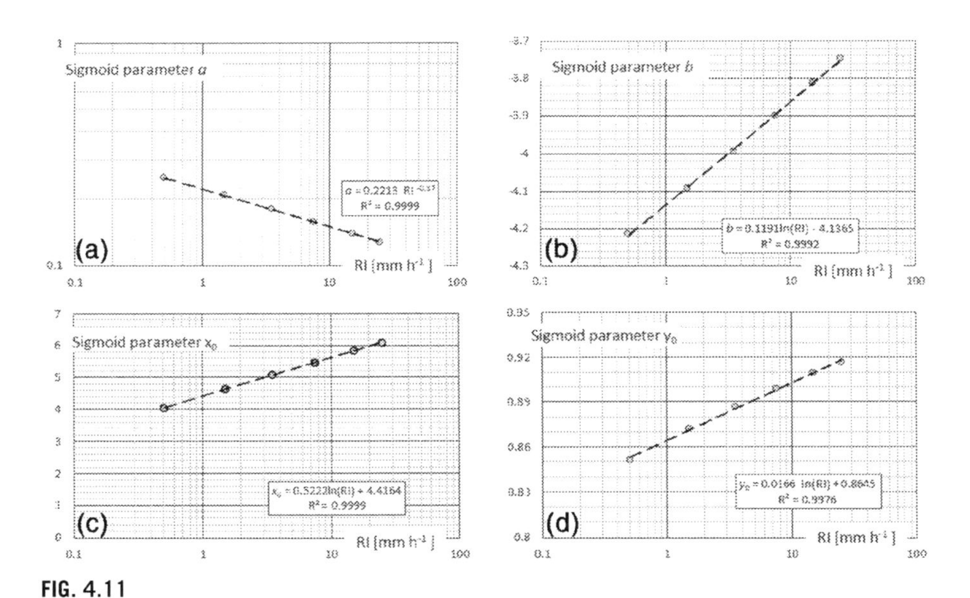

FIG. 4.11

Parameters of the sigmoidal curve associated with the mean value of each RI class (circles) and fitted by either power law or logarithmic curves (dashed lines). (a) parameter a, (b) parameter b, (c) parameter x_0, (d) parameter y_0. (Source: Cauteruccio and Lanza, 2020).

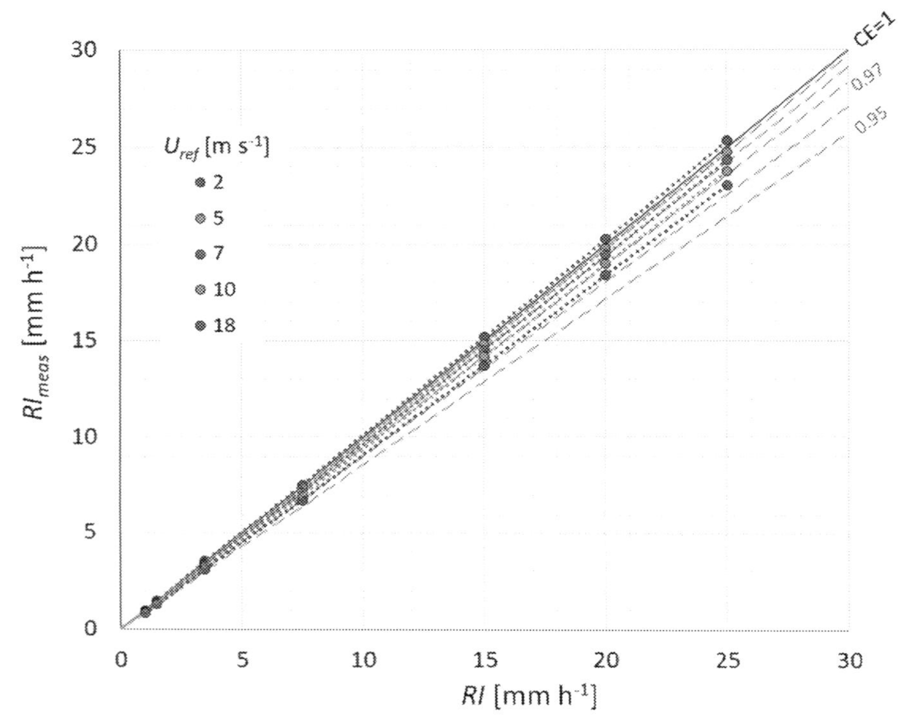

FIG. 4.12

Adjustment curves (dotted lines) for a cylindrical gauge and constant *CE* lines (dashed lines). Circles indicate the results of the performed numerical simulations and are colour-coded according to the wind speed (U_{ref}). (Source: Cauteruccio and Lanza, 2020).

of each *RI* class and the simulated wind speed are depicted with markers, while the associated adjustment curves were obtained as best-fit power law curves as follows:

$$RI_{meas} = \alpha\left(U_{ref}\right) RI^{\beta\left(U_{ref}\right)} \tag{4.7}$$

The coefficient α and exponent β of the adjustment curves for the simulated wind speed are summarized in Table 4.1 and their functional dependency on wind speed

Table 4.1 Parameters α and β of the power law best-fit curves at different wind speeds (U_{ref}). (Source: Cauteruccio and Lanza, 2020).

U_{ref} (m s^{-1})	2	5	7	10	18
α	1.0057	0.9668	0.9413	0.9099	0.8726
β	1.0023	1.0076	1.0106	1.0138	1.0175

is expressed in Eqs. 4.8a and 4.8b. The correlation factor is equal to 0.973 and 0.997 for α and β, respectively.

$$\alpha = 1.0621\, U_{ref}^{-0.066} \quad (4.8a)$$

$$\beta = 0.9971\, U_{ref}^{-0.007} \quad (4.8b)$$

In Fig. 4.12, the diagonal (continuous black line) indicates $CE = 1$, while the grey dashed lines correspond to CE values from 0.95 to 0.99. The adjustment curve for $U_{ref} = 2$ m s^{-1} is located above the diagonal, due to the overcatch provided by simulation results due to the balance between the drop size and the wind speed, while, the other curves, at higher wind speeds, are lower due to the wind-induced undercatch. Moreover, the adjustment curves are described by power law functions, with coefficients (see Table 4.1) that are increasingly different from one (the linear case when CE is constant) as the wind speed increases, reflecting the trend toward higher values of CE while the rainfall intensity increases. Although the deviation is small, it is possible to note in the graph that the adjustment curve at e.g., 18 m s^{-1} progressively diverges from the dashed line at $CE = 0.96$ with increasing rainfall intensity (toward the right-hand side of the graph), while they are practically superimposed at a low RI (left-hand side of the graph). This means that, at any given wind speed, the collection efficiency varies with the rainfall intensity, rather than lying on a curve (line) at CE = constant, as predicted by the existing experimental studies (Colli et al., 2020).

The obtained formulation for the adjustment curve is an easy-to-use tool that allows to correct (in real-time or a posteriori) rainfall intensity measurements obtained in windy conditions by employing catching-type precipitation gauges.

4.6 Rain gauge network design

The rainfall process is highly variable and erratic in both space and time, and the local measurements obtained at a single location are hardly representative of the overall magnitude and pattern of rainfall over wide areas. On the other hand, most hydrological processes respond to the spatial distribution of rainfall over the surface area of an either natural or artificial watershed, and their understanding and quantification largely rely on the knowledge of the spatial variability of rainfall. To characterize spatiotemporal patterns of the rainfall process over wide areas, measurements from sufficiently dense networks of raingauges are used.

Together with proper siting and traceability of measurements, the number of gauges used in a rainfall monitoring network is a primary indicator of its overall quality and accuracy. Many aspects contribute to determine the network density requirements, including the purpose of the measurement (multi-purpose networks are the norm) and the morphological characteristics of the region where the network is deployed. The type and accuracy of gauges used, and the temporal resolution of the measurement must be selected based on the expected application of the measured data (the most restrictive parameters apply in case of multi-purpose networks).

Fit-for-purpose environmental networks are presently invoked by some network managers and international organisations (see e.g., WMO, 2018) to make the selection of instruments and network design simpler and more focused on specific targets. However, it is well-known that environmental data from a network of raingauges prove to be useful for a variety of users and applications. Focusing the specifications of a monitoring network on any single purpose is a tremendous waste of money and resources for citizens and would result in the inevitable loss of important information. On the contrary, it is advisable that – at least those monitoring networks that are managed as public services by national or regional agencies – are designed to meet the requirements of a wide range of users. The focus should be posed, instead, on the accuracy and traceability of measurements (as described in the previous sections), so that the reliability of monitoring networks is enhanced in view of a renewed role of observations in a scientific era dominated by modelling, which generally undervalues the principals of precise and accurate measurements.

The World Meteorological Organization describes hydrological network design in their guide to hydrological practices (WMO, 2008) as an evolutionary process in which a minimum coverage (minimum network) is established early in the development of an area and the network is then upgraded periodically until an optimum network is attained. The initial network should be composed of the minimum number of stations which the collective experience of hydrological agencies of many countries has indicated to be necessary to initiate planning for the economic development of the water resources. The minimum densities of various types of hydrological stations (including precipitation) are recommended in WMO (2008) for different climatic and geographic zones, as reported in Table 4.2 in terms of the extension of the territory "covered" by each gauge in the network. These values must be interpreted as a general indication, based on the most typical applications, and are not necessarily satisfactory for either a specific hydrological application or the hydrological characteristics of a specific watershed.

Table 4.2 Recommended minimum densities of precipitation monitoring stations (area in km^2 per station). (Source: WMO, 2008).

	Precipitation	
Physiographic unit	**Non-recording**	**Recording**
Coastal	900	9,000
Mountains	250	2,500
Interior plains	575	5,750
Hilly/undulating	575	5,750
Small islands	25	250
Urban areas	-	10–20
Polar/ari	10,000	100,000

Once the minimum network is operational, regionalized hydrological relationships, interpreted information and models can be formulated for estimating general hydrological characteristics, including rainfall and runoff at any location in the area. The basic network should be adjusted over time until regional hydrological relationships can be developed for ungauged areas that provide the appropriate level of information (generally resulting in increasing the densities of hydrological stations). Appropriate network-design techniques are identified by WMO (2008) as based on simple hydrological characteristics, regression relationships, or more complex network analysis using generalized least squares methods.

A typical approach for raingauge networks is to describe the rainfall process in terms of its correlation structure in time and space and to estimate the variance of the sample mean areal rainfall (Rodriguez-Iturbe and Mejia, 1974). By expressing the variance as a function of correlation in time and space, and the network parameters, the influence of the network design scheme on the variance of the estimated values is derived. These methods require that a choice be made about the accuracy in the estimate of the areal rainfall mean that is deemed satisfactory. However, many other methods have been developed, based on various statistical tools, to assess the optimality of the network design.

Yeh et al. (2011) synthesise the broad range of methods proposed in the literature for optimizing the density and location of raingauge stations in a monitoring network, by including economically efficient techniques (Duckstein et al. 1974), clustering techniques (Burn and Goulter 1991), least square methods and entropy (Markus et al. 2003), covariance and variance analyses (Shih 1982; Rodriguez-Iturbe and Mejia 1974), kriging (Hughes and Lettenmaier, 1981; Kassin and Kottegoda 1991; Abtew et al. 1995; Pardo-Igu´zquiza, 1998; St-Hilaire et al., 2003; Zimmerman, 2006; Cheng et al., 2008), information entropy (Harmancioglu, 1987; Awumah and Goulter, 1990; Krstanovic and Singh, 1992; Al-zahrani and Husain, 1998; Ozkul et al., 2000; Kawachi et al., 2001; Masoumi and Kerachian, 2008; Mogheir et al., 2006; Yoo et al., 2008; Mogheir et al., 2009), geographic information systems (GIS) with fuzzy logic and hydrological simulation models (Strobl et al. 2006), and combined kriging and information entropy (Chen et al., 2008).

Shahidi and Abedini (2018) note that there is a close interaction between the process time scale and the type of objective function used in those methods. Entropy-based approaches are considered as more customized to short time scale suitable for flood forecasting, while long time scale is more suitable for water budget studies and consistent with variance-based approaches. The criterion used to assess the optimality of network design is specific of the methods employed, and in variance-based methods some measure of accuracy is used to minimise the variance of the residuals (i.e., to maximize the network information content). Geostatistical interpolation methods, such as kriging, require numerous iterations between the optimizer and the objective function, and take a considerable time to find the numerical value of the objective function due to numerous matrix inversions during the design process.

The reader is further referred to the comprehensive review by Mishra and Coulibaly (2009), where an historical overview of hydrometric network design (including raingauge networks) is provided, with a discussion on new developments and challenges in the design of optimal hydrometric networks. The authors provide an extensive survey of methodological developments and the issue of uncertainty in hydrometric network design, together with the evolution in data collection techniques and technology, with comments on the future of hydrometric network design.

4.7 Concluding remarks

In-situ rainfall measurements experienced little conceptual innovation for many decades now. Very traditional instruments, based on the tipping-bucket or weighing measuring principles, are still extensively used worldwide to provide the present knowledge of the magnitude and variability of the forcing input of the land phase of the hydrological cycle. Limitations of such measurements are well known and recent research efforts at the national and international scale highlighted the need to account for such drawbacks in the modelling of hydrological processes, the calibration of areal estimates from remote sensors and the assessment of trends in the statistical analysis of extreme events.

However, accurate calibration of the measurement instruments, optimal raw data interpretation algorithms, and suitable corrections for the most relevant instrumental and environmental biases are still too rarely employed by NMSs, hydrological network managers and even experienced users. This implies that the state of the art in rainfall monitoring is presently based on measurements performed with a much lower degree of accuracy that is technically feasible.

This chapter aims to convey the awareness of the importance of high-resolution and high-quality rainfall measurements in hydrology, to support both physically based and stochastic modeling of hydrological processes. Such an awareness is expected to lead decision makers to invest in the standardization of the instrument calibration and correction procedures, to remove measurement biases, and to ensure metrological traceability so that even the residual measurement uncertainty can be suitably quantified. This would allow users to achieve fully informed exploitation of the information content of rainfall datasets in all relevant applications.

References

Abtew, W., Obeysekera, J., Shih, G., 1995. Spatial variation of daily rainfall and network design. Trans. Am. Soc. Agric. Eng. 38 (3), 843–845.

Al-Zahrani, M., Husain, T., 1998. An algorithm for designing a precipitation network in the south-western region of Saudi Arabia. J. Hydrol. 205 (3–4), 205–216.

Awumah, K., Goulter, I., 1990. Assessment of reliability in water distribution networks using entropy based measures. Stoch. Hydrol. Hydraul. 4 (4), 309–320.

Biemans, H., Hutjes, R.W.A., Kabat, P., Strengers, B., Gerten, D., Rost, S., 2009. Effects of precipitation uncertainty on discharge calculations for main river basins. J. Hydrometeorol. 10 (4), 1011–1025.

Burn, D.H., Goulter, I.C., 1991. An approach to the rationalization of streamflow data collection networks. J. Hydrol. 122 (1–4), 71–91.

Caracciolo, C., Porcù, F., Prodi, F., 2008. Precipitation classification at mid-latitudes in terms of drop size distribution parameter. Adv. Geosci. 16, 11–17.

Cauteruccio, A., Lanza, L.G., 2020. Parameterization of the collection efficiency of catching-type rain gauges based on rainfall intensity. Water 12 (12), 3431.

Cauteruccio, A., Chinchella, E., Stagnaro, M., Lanza, L.G., 2021c. Snow particle collection efficiency and adjustment curves for the Hotplate© precipitation gauge. J. Hydrometeorology 22 (4), 941–954. https://doi.org/10.1175/JHM-D-20-0149.1.

Cauteruccio, A., Colli, M., Stagnaro, M., Lanza, L.G., Vuerich, E., 2021a. In-situ precipitation measurements. In: Foken, T (Ed.), Springer Handbook of Atmospheric Measurements. Springer Nature, Switzerland.

Cauteruccio, A., Brambilla, E., Stagnaro, M., Lanza, L.G., Rocchi, D., 2021b. Wind tunnel validation of a particle tracking model to evaluate the wind-induced bias of precipitation measurements. Water Resour. Res. https://doi.org/10.1002/essoar.10504288.1.

CEN/TR 16469:2013, 2013. Hydrometry - Measurement of the Rainfall Intensity (Liquid Precipitation): Requirements, Calibration Methods and Field Measurements. European Committee for Standardization, Bruxelles, Belgium.

Chen, Y.C., Chiang, W., Yeh, H.C., 2008. Rainfall network design using kriging and entropy. Hydrol. Process. 22 (3), 340–346.

Cheng, K.S., Lin, Y.C., Kiou, J.J., 2008. Rain-gauge network evaluation and augmentation using geostatistics. Hydrol. Process. 22 (14), 2554–2564.

Colli, M., Lanza, L.G., Chan, P.W., 2013a. Co-located tipping-bucket and optical drop counter RI measurements and a simulated correction algorithm. Atmos. Res. 119, 3–12.

Colli, M., Lanza, L.G., La Barbera, P., 2013b. Performance of a weighing rain gauge under laboratory simulated time-varying reference rainfall rates. Atmos. Res. 131, 3–12.

Colli, M., Lanza, L.G., Rasmussen, R., Thériault, J.M., Baker, B.C., Kochendorfer, J., 2015. An improved trajectory model to evaluate the collection performance of snow gauges. J. Appl. Meteorol. Climatol. 54 (8), 1826–1836.

Colli, M., Lanza, L.G., Rasmussen, R., Thériault, J.M., 2016a. The collection efficiency of shielded and unshielded precipitation gauges. Part I: CFD airflow modelling. J. Hydrometeor. 17 (1), 231–243.

Colli, M., Lanza, L.G., Rasmussen, R., Thériault, J.M., 2016b. The collection efficiency of unshielded precipitation gauges. Part II: modeling particle trajectories. J. Hydrometeor. 17 (1), 245–255.

Colli, M., Pollock, M., Stagnaro, M., Lanza, L.G., Dutton, M., O'Connell, P.E., 2018. A computational fluid-dynamics assessment of the improved performance of aerodynamic raingauges. Water Resour. Res. 54 (2), 779–796.

Colli, M., Stagnaro, M., Theriault, J.M., Lanza, L.G., Rasmussen, R., 2020. Adjustments for wind-induced undercatch in snowfall measurements based on precipitation intensity. J. of Hydrometeor. 21 (5), 1039–1050.

Costello, T.A., Williams, H.J., 1991. Short duration rainfall intensity measured using calibrated time-of-tip data from a tipping bucket raingage. Agric. For. Meteorol. 57 (1–3), 147–155.

Duckstein, L., Davis, D.R., Bogardi, I., 1974. Applications of decision theory to hydraulic engineering, Symposium of ASCE hydraulic division specialty conference. ASCE.

EN 13798:2010, 2010. Hydrometry – Specifications for a Reference Raingauge Pit. European Committee for Standardization CEN/TC 318 – Hydrometry, Bruxelles, Belgium.

EN 17277:2019, 2019. Hydrometry - Measurement requirements and classification of rainfall intensity measuring instruments. European Committee for Standardization, Bruxelles, Belgium.

Folland, C.K., 1988. Numerical models of the raingauge exposure problem, field experiments and an improved collector design. Q. J. R. Meteorol. Soc. 114, 1485–1516.

Goodison, B.E., Louie, P.Y.T., Yang, D., 1998. WMO Solid Precipitation Measurement Intercomparison, Geneva, Switzerland. IOM Rep. No. 67 and WMO/TD No. 872, 318.

Habib, E., Krajewski, W.F., Kruger, A., 2001. Sampling errors of tipping-bucket rain gauge measurements. J. Hydrol. Eng. 6 (2), 159–166.

Harmancioglu, N., Yevjevich, V., 1987. Transfer of hydrologic information among river points. J. Hydrol. 91 (1–2), 103–118.

Hughes, J.P., Lettenmaier, D.P., 1981. Data requirements for kriging: estimation and network design. Water Resour. Res. 17 (6), 1641–1650.

Kassin, A.H.M., Kottegoda, N.T., 1991. Rainfall network design through comparative kriging methods. Hydrol. Sci. 36 (3), 223–240.

Kawachi, T., Maruyama, T., Singh, V.P., 2001. Rainfall entropy for delineation of water resources zones in Japan. J. Hydrol. 246 (1–4), 36–44.

Khvorostyanov, V.I., Curry, J.A., 2005. Fall velocities of hydrometeors in the atmosphere: refinements to a continuous analytical power law. J. Atmos. Sci. 62 (12), 4343–4357.

Kochendorfer, J., Rasmussen, R., Wolff, M., Baker, B., Hall, M.E., Meyers, T., Landolt, S., Jachcik, A., Isaksen, K., Brækkan, R., Leeper, R., 2017a. The quantification and correction of wind-induced precipitation measurement errors. Hydrol. Earth Syst. Sci. 21 (4), 1973–1989.

Kochendorfer, J., Nitu, R., Wolff, M., Mekis, E., Rasmussen, R., Baker, B., Earle, M.E., Reverdin, A., Wong, K., Smith, C.D., Yang, D., Roulet, Y.-A., Buisan, S., Laine, T., Lee, G., Aceituno, J.L.C., Alastrué, J., Isaksen, K., Meyers, T., Brækkan, R., Landolt, S., Jachcik, A., Poikonen, A., 2017b. Analysis of single-alter-shielded and unshielded measurements of mixed and solid precipitation from WMOSPICE. Hydrol. Earth Syst. Sci. 21 (7), 3525–3542.

Krstanovic, P.F., Singh, V.-P., 1992. Evaluation of rainfall network using entropy II: application. Water Resour. Manag. 6 (4), 295–314.

La Barbera, P., Lanza, L.G., Stagi, L., 2002. Tipping bucket mechanical errors and their influence on rainfall statistics and extremes. Water Sci. Technol. 45 (2), 1–9.

Lanza, L.G., Stagi, L., 2008. Certified accuracy of rainfall data as a standard requirement in scientific investigations. Adv. Geosci. 16, 43–48.

Lanza, L.G., Vuerich, E., 2012. Non-parametric analysis of one-minute rain intensity measurements from the WMO Field Intercomparison. Atmos. Res. 103, 52–59.

Markus Momcilo, H., Knapp, V., Tasker, G.D., 2003. Entropy and generalized least square methods in assessment of the regional value of streamgauges. J. Hydrol. 283 (1–4), 107–121.

Marsalek, J., 1981. Calibration of the tipping bucket raingage. J. Hydrol. 53 (3–4), 343–354.

Marshall, J.S., Palmer, W.M.K., 1948. The distribution of raindrops with size. J. Meteorol. 5, 165–166.

Masoumi, F., Kerachian, R., 2008. Assessment of the groundwater salinity monitoring network of the Tehran region: application of the discrete entropy theory. Water Sci. Technol. 58 (4), 765–771.

Mishra, A.K., Coulibaly, P., 2009. Developments in hydrometric network design: a review. Rev. Geophys. 47 (2), RG2001.

Mogheir, Y., Singh, V.P., de Lima, J.L.M.P., 2006. Spatial assessment and redesign of a groundwater quality monitoring network quality using entropy theory, Gaza Strip, Palestine. Hydrogeol. J. 14 (5), 700–712.

Mogheir, Y., de Lima, J.L.M.P., Singh, V.P., 2009. Entropy and multiobjective based approach for groundwater quality monitoring network assessment and redesign. Water Resour. Managem. 23 (8), 1603–1620.

Molini, A., La Barbera, P., Lanza, L.G., Stagi, L., 2001. Rainfall intermittency and the sampling error of tipping-bucket rain gauges. Phys. Chem. Earth C Sol. Terr. Planet. Sci. 26 (10–12), 737–742.

Molini, A., Lanza, L.G., La Barbera, P., 2005. The impact of tipping-bucket rain gauge measurement errors on design rainfall for urban scale applications. Hydrol. Processes 19 (5), 1073–1088.

Morbidelli, R., Corradini, C., Saltalippi, C., Flammini, A., 2022. Time Resolution of Rain Gauge Data and its Hydrological Role. In: Morbidelli, R. (Ed.), Rainfall. Modeling, Measurement and Applications. Elsevier, Amsterdam, pp. 171–216. doi:10.1016/C2019-0-04937-0.

Nešpor, V., Sevruk, B., 1999. Estimation of wind-induced error of rainfall gauge measurements using a numerical simulation. J. Atmos. Ocean. Technol. 16 (4), 450–464.

Niemczynowicz, J., 1986. The dynamic calibration of tipping-bucket raingauges. Nord. Hydrol. 17 (3), 203–214.

Ozkul, D.S., Harmancioglu, N.B., Singh, V.P., 2000. Entropy-based assessment of water quality monitoring networks. J. Hydrol. Eng. 5 (1), 90–100.

Pardo-Igu, zquiza, 1998. E: optimal selection of number and location of rainfall gauges for areal rainfall estimation using geostatistics and simulated annealing. J. Hydrol. 210 (1–4), 206–220.

Pollock, M.D., O'Donnell, G., Quinn, P., Dutton, M., Black, A., Wilkinson, M.E., Colli, M., Stagnaro, M., Lanza, L.G., Lewis, E., Kilsby, C.G., O'Connell, P.E., 2018. Quantifying and mitigating wind-induced undercatch in rainfall measurements. Water Resour. Res. 54 (6), 3863–3875.

Pruppacher, H.R., Klett, J.D., 2010. Microphysics of Clouds and Precipitation: Atmospheric and Oceanographic Sciences Library. Springer, Dordrecht, Netherland, p. 954.

Rodriguez-Iturbe, I., Mejia, J.M., 1974. The design of rainfall networks in time and space. Water Resour. Res. 10 (4), 713–728.

Sevruk, B., 1982. Methods of Correction for Systematic Error in Point Precipitation Measurement for Operational Use. World Meteorological Organization, Geneva, Switzerland.

Shahidi, M., Abedini, M.J., 2018. Optimal selection of number and location of rain gauge stations for areal estimation of annual rainfall using a procedure based on inverse distance weighting estimator. Paddy Water Environ 16, 617–629.

Shih, S.F., 1982. Rainfall variation analysis and optimization of gauging systems. Water Resour. Res. 18 (4), 1269–1277.

Sims, E.M., Liu, G., 2015. A parameterization of the probability of snow–rain transition. J. Hydrometeor. 16 (4), 1466–1477.

Stagnaro, M., Colli, M., Lanza, L.G., Chan, P.W., 2016. Performance of post-processing algorithms for rainfall intensity using measurements from tipping-bucket rain gauges. Atmos. Meas. Tech. 9 (12), 5699–5706.

Stagnaro, M., Cauteruccio, A., Colli, M., Lanza, L.G., Chan, P.W., 2018. Laboratory assessment of two catching type drop-counting rain gauges. Geophysical Research Abstracts, EGU General Assembly 20 12407.

St-Hilaire, A., Ouarda, T.B.M.J., Lachance, M., Bobe´e, B., Gaudet, J., Gignac, C., 2003. Assessment of the impact of meteorological network density on the estimation of basin precipitation and runoff: a case study. Hydrol. Process. 17 (18), 3561–3580.

Strangeways, I., 2010. A history of rain gauges. Weather 65 (5), 133–138.

Strobl, R.O., Robillard, P.D., Shannon, P.D., Day, R.L., McDonnell, A.J., 2006. A water quality monitoring network design methodology for the selection of critical sampling points: Part I. Environ. Monit. Assess. 112 (1), 137–158.

Thériault, J.M., Rasmussen, R., Ikeda, K., Landolt, S., 2012. Dependence of snow gauge collection efficiency on snowflake characteristics. J. Appl. Meteorol. Climatol. 51 (4), 745–762.

UNI 11452:2012: hydrometry - measurement of rainfall intensity (liquid precipitation) - metrological requirements and test methods for catching type gauges, UNI Ente Italiano di Normazione, 2012.

Vuerich E., Monesi C., Lanza L.G., Stagi L., Lanzinger E., 2009. WMO field intercomparison of rainfall intensity gauges, World Meteorological Organisation Instruments and Observing Methods, Report 99.

Wang, J., Fisher, B.L., Wolff, D.B., 2008. Estimating rain rates from tipping bucket rain gauge measurements. J. Atmos. Oceanic Technol. 25 (1), 43–55.

WMO, 2008. World Meteorological Organization: Guide to Hydrological Practices, Volume I: Hydrology – From Measurement to Hydrological Information. Sixth edition Geneva, Switzerland, p. 296.

WMO World Meteorological Organization: Guide to Meteorological Instruments and Methods of Observation: (CIMO guide No. 8), 2014 edition, updated in 2017.

WMO, 2018. World Meteorological Organization: Towards fit-for-purpose environmental measurements, In: WMO (Ed.), Proc. WMO/CIMO Technical Conference on Meteorological and Environmental Instruments and Methods of Observation (CIMO TECO-2018), WMO, Geneva, Switzerland.

Wolff, M.A., Isaksen, K., Petersen-Øverleir, A., Ødemark, K., Reitan, T., Brækkan, R., 2015. Derivation of a new continuous adjustment function for correcting wind-induced loss of solid precipitation: results of a Norwegian field study. Hydrol. Earth Syst. Sci. 19 (2), 951–967.

Yeh, H.C., Chen, Y.C., Wei, C., Chen, R.H., 2011. Entropy and kriging approach to rainfall network design. Paddy Water Environ 9 (3), 343–355.

Yoo, C., Jung, K., Lee, J., 2008. Evaluation of rain gauge network using entropy theory: comparison of mixed and continuous distribution function applications. J Hydrol. Eng. 13 (4), 226–235.

Zimmerman, D.L., 2006. Optimal network design for spatial prediction, covariance parameter estimation, and empirical prediction. Environmetrics 17 (6), 635–652.

CHAPTER

Rainfall estimation by weather radar

5

Marco Borga[a], Francesco Marra[b], Marco Gabella[c]

[a]*Department of Land Environment Agriculture and Forestry, University of Padova, Padova, Italy*
[b]*Institute of Atmospheric Sciences and Climate, National Research Council (CNR-ISAC), Bologna, Italy*
[c]*Meteoswiss, Locarno Monti, Switzerland*

5.1 Introduction

Precipitation is a crucial element of the hydrological cycle and its quantitative measurement has been traditionally gathered by raingauges (Michaelides et al., 2009). A basic problem with raingauges, however, is the fact that they are point measurements. This means that their limited spatial representativeness can only be increased indirectly through temporal accumulation (Uijlenhoet, 1999; Villarini et al., 2008). Even then, the spatial representativeness of raingauges remains unclear as it will depend on the dynamics of the rainfall process. Moreover, accumulation of raingauge measurements reduces their ability to capture the small-scale spatiotemporal structure of rainfall (Sideris et al., 2014; Lengfeld et al., 2020).

Weather radar may help overcome the fundamental limitation of rainfall information from raingauges. Weather radars can provide complete spatial and temporal coverage of an area from one single measurement site and as such they are ideally suited for quantitative, fine resolution rainfall measurement over a wide range of spatial and temporal scales.

RADAR is the acronym for "RAdio Detection And Ranging." The term "radio" refers to the use of electromagnetic waves with wavelengths in the so-called radio wave portion of the spectrum, which covers a wide range from 10^4 km to 1 cm. According to Battan (1973), radar can be defined as "the art of detecting by means of radio echoes the presence of objects, determining their direction and range, recognizing their character and employing the data thus obtained." The principle of radar remote sensing is based on the transmission of a coded radio signal, the reception of a backscattered signal from the volume of interest and on inferring the properties of the objects contained in that volume by comparing the transmitted and received signals. In the case of radar meteorology, the objects in the scattering volume are in principle hydrometeors (precipitation particles), although occasionally the ground

RAINFALL. Modeling, Measurement and Applications. DOI: https://doi.org/10.1016/B978-0-12-822544-8.00016-0

surface or other targets such as insects and birds may be detected as well. Hydrometeors can be raindrops, but snowflakes and ice crystals as well. The main interest in this chapter lies obviously in the raindrops.

Radar technology was intensively developed for military use in the period before and during World War II. During the war, radar operators noticed echoes on radar screens caused by weather phenomena. After the war, scientists studied how to use radars for detecting precipitation. The quantitative use of radar measurement motivated an important body of works starting in the late 1940s with the finding of a relationship between the radar observables (radar reflectivity) and the rain intensity. The physical understanding of radar measurement progressed well during the following decades. However, the limited possibilities of signal processing and archiving at that time, made this knowledge almost impossible to apply to hydrological and meteorological applications. During the following three decades the main question explored by hydrologists was the complementarity between rain gauge networks and radar (see for instance Wilson and Brandes, 1979; Collier, 1996; Creutin et al., 1987; Krajewski, 1987). In spite of the quality of the theoretical frameworks used, the answers were somewhat discouraging. The adjustment techniques were only able to reduce the bias between gauge and radar measurements, but large discrepancies remained on individual measurements. Advances on the radar-based quantitative rainfall estimation came with the development of new radar systems with volumetric scanning and the digitisation of the signal processing. Zawadski (1984) summarized this new direction as follows: "The accuracy of radar estimates at ground will only be improved by addressing the various sources of error in a painstaking and a meticulous manner. The combination of hardware and software made available by the technology of today permits a complexity of radar data processing which should be helpful in reducing the errors discussed here" (the paper established the relative importance of the different sources of errors).

Villarini and Krajewski (2010) provide a comprehensive review of different sources of uncertainties in single polarization radar. The inclusion of Doppler capabilities into the single polarization radars, that is, the possibility to estimate the radial velocity of the raindrops by quantifying changes in wavelength due to the Doppler effect, has helped to reduce some of the above uncertainties (Chong et al., 2000; Tabary et al., 2001; Yamamoto et al., 2011). Further advances in radar technology have led to the development of polarimetric Doppler radar in the last two decades (Michaelides et al., 2009; Islam and Rico-Ramirez, 2014), which has the capability of transmitting and receiving in both the horizontal (H) and vertical (V) polarization states, and additional uncertainties can be reduced through the use of this technology.

Weather radar networks are now covering a majority of the densely populated areas of the world (Fig. 5.1), and several of these networks include polarimetric radar, thus providing additional information on the target precipitation particles. For most of the weather services, digitized radar archives are also available since year 2000, although longer records are indeed available (Saltikoff et al., 2019). Radar-based rainfall estimates provide fundamental information to understand the role of space-time rainfall variability on flood response (Sangati et al, 2009; Zoccatelli

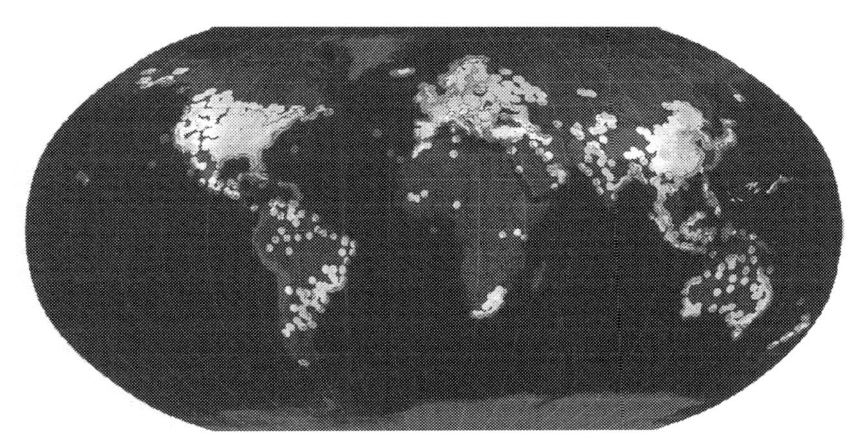

FIG. 5.1 A map of weather radar coverage in the world (from Saltikoff et al., 2019).

et al., 2010, 2011; Nikolopoulos et al., 2014; Amengual et al., 2021). Radar data are now used for flood and flash flood forecasting (Borga et al., 2000, 2006, 2014; Hossain et al., 2004), water resources management (Delrieu et al., 2009), landscape modelling for the simulation and prediction of geomorphic changes in landscape (Marra et al., 2014, 2017), risk management (Panziera et al., 2016; Marra et al., 2019), and climatology (Overeem et al., 2009; Marra and Morin, 2015; Saltikoff et al., 2019). Because of its ability in capturing information on precipitation particles, polarimetric weather radars represent unique tools for validating satellite precipitation estimates and to support corresponding algorithm developments (Wang and Carey, 2005; Wen et al., 2011).

This chapter provides an outline of the principles of precipitation estimation by means of weather radar, with a coverage of the main techniques for weather radar observation and of the methods used to generate rainfall products starting from weather radar observables.

5.2 Rainfall microstructure

Our understanding of the way in which electromagnetic signals transmitted by weather radars interact with rain and consequently our ability to use these instruments to accurately estimate rainfall intensities in space and time depends fundamentally on our knowledge and understanding of the structure of rainfall at the raindrop-scale.

A fundamental property of rainfall in this respect is the so-called drop size distribution (DSD, in the following). It represents the mean number of raindrops per unit of raindrop diameter interval and per unit volume of air. Thus, the notion of DSD is a mixture of two different concepts, namely that of the spatial distribution of raindrops in a volume of air (which controls the raindrop concentration) and that of the probability distribution of their sizes. A device called a disdrometer is usually employed to measure the DSDs. For their description, associated measurement errors and data processing, see Cao and Zhang (2009), Montopoli et al. (2008), and Tokay et al. (2002, 2005).

The reflectivity factor Z is the most important parameter for radar interpretation. The factor derives from the Rayleigh scattering model and is defined as the sixth moment of the DSD. It can be written as:

$$Z = \int_0^\infty N(D) D^6 dD \tag{5.1}$$

where N(D) is the DSD with diameters between D and D + dD in a unit volume. Hence, although Z is called the *radar* reflectivity factor, it is a purely meteorological quantity that is independent of any radar property. The radar reflectivity may span a wide range of orders of magnitude and is generally reported in units of decibels (dB).

If the effects of wind (notably up and downdraughts), turbulence and raindrop interaction are neglected, the (stationary) rain rate R is related to the DSD according to:

$$R = 6 \times 10^{-4} \int_0^\infty N(D) v(D) D^3 dD \tag{5.2}$$

where v(D) represents the functional relationship between the raindrop terminal fall speed in still air v and the equivalent spherical raindrop diameter D.

The conversion of the radar reflectivity factor Z, which is observed thanks to radar measurements, to rain rate R is a crucial step in radar rainfall estimation. On the basis of measurements of rain DSD s at the ground and an assumption about the v(D)-relationship, regression analysis permits to derive Z–R relationships. There exists overwhelming empirical evidence (e.g. Battan, 1973) that such relationships generally follow power laws as follows:

$$Z = aR^b \tag{5.3}$$

where *a* and *b* are coefficients that may vary from one location to the next and from one season to the next. Battan's (1973) classical treatise on radar meteorology quotes a list of 69 such empirical power law *Z–R* relationships derived for different climatic settings in various parts of the world. Uijlenhoet (2001) derived a mean power law relationship based on the 69 empirical *Z-R* relationships and found that this relationship is surprisingly close to the most widely used *Z–R* relationship,

$$Z = 200R^{1.6} \tag{5.4}$$

(Marshall et al., 1955). This may be an explanation for the success of Eq. (5.4) for many different types of rainfall in many parts of the world. Using the classical exponential rain DSD introduced by Marshall and Palmer (1948) as an example, Uijlenhoet (2001) demonstrated how the definitions of the radar reflectivity factor *Z* and the rain rate *R* in terms of that rain DSD naturally lead to the ubiquitous power law *Z–R* relationships.

Another fundamental property of rainfall is the raindrop shape. Raindrops are elliptically shaped with the major axis in the horizontal plane when falling freely in the atmosphere. The oblateness of the drop is related to the drop size. Drop sizes smaller than 1 mm are spherical, but the shape becomes more oblate as the size increases. The ratio between the maximum vertical and horizontal axis, termed the axis ratio, is used to exemplify the drop shape as a function of its equivolumetric diameter. The representation of the axis ratio with respect to the equivolumetric diameter plays a central role in the definition of polarimetric radar signatures.

5.3 Radar principles

Fig. 5.2 shows a typical radar and radar site. As most radar antennas, the directional antenna (2–8.5 m) is inside the radome (which aims to protect the antenna from wind and rain) on top of the tower, which is of the order of 10–30 m or more in height. The tower is used to elevate the antenna above local obstructions, such as trees and buildings.

Electromagnetic waves at fixed preferred frequencies are transmitted from the directional antenna into the atmosphere in a rapid succession of short pulses. The pulse length and range processing determine the range resolution of the radar data. A parabolic reflector in the antenna system concentrates the electromagnetic energy in a conical-shaped beam that is highly directional. The width of the beam increases with range, for example, a nominal ° beam spreads to 0.9, 1.7, and 3.5 km at ranges of 50, 100, and 200 km, respectively. For a pulse radar, the short bursts of electromagnetic energy are absorbed and scattered by the illuminated meteorological and nonmeteorological targets. Some of the scattered energy is reflected back to the

FIG. 5.2 The Monte Grande weather radar in Veneto, Italy. The figure shows the tower and the radome.

radar antenna and receiver. Since the electromagnetic wave travels at the speed of light (that is, 2.99×10^8 m s^{-1}), the range of the target can be determined by measuring the time between the transmission of the pulse and its return. Between the transmissions of successive pulses, the receiver listens for any return of the wave. The return signal from the target is commonly referred to as the radar echo. Volumetric observations of the atmosphere are normally made by scanning the antenna at a fixed elevation angle and subsequently incrementing the elevation angle in steps at each revolution.

The radar equation (see below) relates the power returned from the target to the radar characteristics. The power returned provides an estimate of the amount of precipitation in the resolution volume. This estimate depends on the assumption of the type of precipitation particles (raindrops, ice hydrometeors, etc.) and their size distribution in the resolution volume.

The power measurements are determined by the total power backscattered by the target within a volume being sampled at any one instant in time. This volume is called the pulse volume or sample volume. The pulse volume dimensions (which determine the resolution of the radar) are dependent on the radar pulse length in space (h) and the antenna beam widths in the vertical (φ) and the horizontal (θ). The beam width, and therefore the pulse volume, increases with range.

The location of the pulse volume in space is determined by the position of the antenna in azimuth, the elevation angle, the range to the target and also by the non-linear propagation path of the radar beam away from the radar. Electromagnetic waves propagate in straight lines in a homogeneous medium. However, the atmosphere is vertically stratified and the rays change direction depending on the changes in the refractive index (which is a function of air temperature and moisture). The amount of bending of electromagnetic waves can be predicted by using the vertical profile of temperature, moisture and pressure (Bean and Dutton, 1966). Under normal atmospheric conditions, the waves travel in a curve bending slightly earthward. When the waves encounter precipitation and clouds, part of the energy is absorbed and part is scattered in all directions, including back to the radar site.

Ground-based weather radars work mainly in three frequency bands: S, C, and X. Choice of the band for a particular location is made on the basis of a trade-off between the measuring range of reflectivity depending on the amount of signal attenuation and the cost of the radar. S-band (2.7–2.9 GHz) is well suited for detecting heavy rain at very long ranges (up to 300 km), as it is least affected by attenuation. However, this is also the most expensive. X-band (9.3–9.5 GHz) weather radars are more sensitive to hydrometeor than S- or C-band weather radars when measuring up to a range of 50 km. Attenuation of the signal by rain is however strongest in the case of X-band radars (compared to S- and C-band radars) and strongly limits quantitative precipitation estimation. On the other hand, X-band weather radars are the least expensive. C-band (5.6–5.65 GHz) represents a compromise between range and reliability of reflectivity measurements and cost.

5.4 The radar equation

The working principle of precipitation radar is the emission of short pulses of microwave radiation to the target precipitation particles. The energies scattered back from the target particles are then received as reflected power. Subsequently, the received power is related to the physical characteristics of the precipitation particles. Under the assumption that the precipitation particles are small compared to the radar wavelength, implying Rayleigh scattering approximation, the meteorological radar equation is given as (Probert-Jones, 1962):

$$P_r = \frac{\pi^3 P_t g^2 \theta \varphi h |K|^2 \sum D_i^6}{1024 ln2\, \lambda^2 r^2} \tag{5.5}$$

where P_r is received power, P_t is transmitted power, g is antenna gain, θ is horizontal beamwidth, φ is vertical beamwidth, h is pulse length, K is dielectric factor, D_i is scatterer diameter, λ is radar wavelength, and r is distance between sample and radar antenna. For summation over a unit volume, the equation can be expressed as:

$$P_r = \frac{C|K|^2 Z}{r^2} \tag{5.6}$$

where C is the radar constant that includes all the fixed radar characteristics (antenna gain, pulse width, wavelength, beam width, and pulse duration), and Z is the radar reflectivity factor.

These equations are based on the assumption that all scatterers are Rayleigh ($D << \lambda$), and the accurate value of Z will be derived, subject to knowing the appropriate value of dielectric factor K. However, at higher frequencies the Rayleigh approximation may not be correct since the targets could be Mie scatterers ($D \geq \lambda$). Particularly, at X-band frequencies and beyond, non-Rayleigh (Mie) scattering is more likely. In those cases, the Z factor can be adjusted to 'equivalent radar reflectivity factor', Z_e (Fritz and Chandrasekar, 2012; Islam et al., 2012). In conventional single polarization radar, there is only one polarization plane, producing $Z \sim Z_H$; but in polarimetric radar (or dual polarization radar), two orthogonal polarization beams are produced, at horizontal and vertical polarization planes (Cao et al., 2012). As a result, the radar return power at horizontal polarization can be compared with the radar return power at vertical polarization. The backscattering cross-sections of raindrops are introduced in both horizontal (H) and vertical (V) polarization planes, thus producing Z_H and Z_V, respectively. Consequently, additional information on precipitation particle properties can be gained and several other polarimetric radar signatures can be calculated. Islam and Rico-Ramirez (2014) provides an overview of polarimetric radar measurements, emphasizing the intrinsic signatures and their association to precipitation particle shapes, sizes and distributions.

The radar equation is developed based on several assumptions. Whenever these assumptions are not satisfied, the reflectivity may be considered in error. For example, if the target is not uniform or completely filled or is mixed, the equation is not appropriate.

Also, if the parameters in the equation, such as antenna gain, waveguide loss or pulse length, are incorrect then the radar constant will be in error and this will result in systematic biases in the conversion from power to reflectivity. In the following, various sources of error are discussed with respect to qualitative and quantitative applications.

5.5 Errors in radar rainfall estimation and their correction

Radar rainfall estimates are indirect but are also spatially dense and provide continuous monitoring of precipitation. The main drawback is that large observation errors may occur (Zawadzki, 1984; Villarini and Krajewski, 2010), which can be attributed to numerous meteorological, topographic and technical factors, such as the following: (1) variability of the Z-R relationship, (2) range-related errors due to the vertical variability of reflectivity and incomplete and/or nonuniform beam filling (Germann and Joss, 2002, 2004), (3) attenuation by intervening precipitation, (4) ground clutter, (5) beam blockage (Creutin et al.,1987), and (6) miscalibration of the electronic components (Seo et al., 1990). A review of the different sources of uncertainty can be found in numerous papers, such as, Krajewski and Smith, 2002, Michelson et al. (2005), Villarini and Krajewski (2010). An outline of the above-mentioned error sources and of the corresponding correction procedures is given below.

5.5.1 Variability of the Z-R relationship

Reflectivity and rainfall rate both depend on the DSD, which depends on the rainfall process and varies geographically. This implies that also the Z-R relationship varies geographically, with rainfall intensity, from storm to storm, and even within the same storm (e.g., Steiner and Smith, 2000; Uijlenhoet, 2001; Uijlenhoet et al., 2003; Smith et al., 2009). This variability unavoidably affects the rainfall estimates by radar in all of those studies and applications which use a fixed Z-R relationship (e.g., Brandes et al., 1999; Villarini and Krajewski, 2010).

Even though the most commonly used DSD is the exponential Marshall-Palmer distribution (Marshall and Palmer, 1948; Marshall et al., 1955), which corresponds to $a = 200$ and $b = 1.6$ in the Z–R relationship, the literature reports a large number of different Z-R relationships for different climatic conditions in different parts of the world (Battan et al., 1973). It is found that the multiplicative parameter a is smaller for stratiform rain and increases for increasing convective activity, while the exponent b behaves in the opposite manner, with smaller values for convective rain and larger for stratiform systems. This implies a larger sensitivity of the radar reflectivity to the rain rate for convective precipitation.

The variability in the Z-R relation may not be exclusively due to variability of the DSD. For example, in the radar equation the drops are assumed to be uniformly distributed in the sample volume. However, this assumption may not be fulfilled, and it is therefore important to evaluate the small-scale variability of the radar reflectivity: Miriovsky et al. (2004) used four disdrometers in a 1 km^2 area and found large variability.

Nevertheless, even if extensively studied throughout the years, there is still debate concerning the overall impact of the variability of the DSD, and therefore the Z-R relationship, on the precision of radar-rainfall estimates. Some earlier studies showed that the variability of the Z-R relation could represent a relatively small factor affecting the precision of the radar-rainfall estimates (Zawadzki, 1984; Smith, 1990; Joss and Waldvogel, 1990). Zawadzki (1984) wrote that "...the variability of the drop-size distribution is a relatively minor factor affecting the precision of radar estimates of rain rate" and that "...the variability of drop-size spectra introduces one of many errors and not the most severe at that." Joss and Waldvogel (1990) concur that this issue is frequently overemphasized. However, other studies showed how DSD variability explains a significant part of the errors in rainfall estimation when a single Z-R relation is used (Lee and Zawadzki, 2006) and that "the variety of physical processes leading to the variability of DSDs is a key limitation in radar rainfall estimation" (Lee and Zawadzki, 2005).

5.5.2 Ground clutter

In hilly and montainous terrain, the unwanted back-scattered signals generated from physical objects in the natural and built environment like topographic relief and buildings are termed ground clutter. Some ground clutter contamination is always present since the sides of the main lobe and the (lower) side lobes interact with the nearby terrain: an appropriate and careful design of antenna radiation pattern has a large impact on the intensity of ground clutter contamination. For given antenna costs, C-band wavelengths offer a narrower beam than S-band, and thus give better spatial resolution. A narrow beam and short pulse-length help to restrict the area of clutter contamination from mountain returns (see, e.g., Sec. 2.2.1 of Germann and Joss, 2004).

Clutter map have been used with the purpose of removing ground clutter. The patterns of radar echoes in non-precipitating conditions are used to generate "representative" a clutter map that is subtracted from the radar pattern collected in precipitating conditions. The problem with this "static" technique is that the pattern of ground clutter widely fluctuates in time. Such changes are primarily due to changes in propagation conditions; a prime example is anomalous propagation echoes which appear when an anomalous stratification of the atmospheric refractivity causes the radar beam to propagate with different curvatures and typically last several hours and then disappear. Micro changes (fluctuations and scintillations) of the refractive index of the troposphere cause short-term fluctuations in the pattern of ground echoes, which confound the use of clutter maps. That is why static maps tend to survive only at the end of smart, adaptive algorithms, as a conclusive test for very few not-yet classified high range resolution (short pulsewidth) polar radar bins. For instance, a sophisticated and efficient way to combine all the information available concerning radar returns (including statistical, dynamic tests on all the pulses that are average to derive the overall echo in the bin) is the decision tree algorithm, proposed by Germann and Joss (2004). Another example of decision-tree classifier for the separation of ground clutter from weather targets is the one presented by Steiner and Smith (2002). Alternatively, effective way for the identification of ground clutter is based on the analysis of the received radar signal in the spectral domain (e.g., the milestone paper by Passarelli et al., 1981). Once ground-clutter echoes have been identified, they can be modelled, simulated and analyzed in detail (e.g., Delrieu et al., 1995; Gabella and Perona, 1998; Hubbert et al., 2009). Thanks to the increase in knowledge and in computational power for modeling and statistical analysis, a novel point of view regarding ground clutter has emerged: it is no longer considered just a disturbance. Rather, it is used for monitoring radar hardware. For instance, Silberstein et al. (2008) have proposed a relative calibration adjustment (RCA) technique based on the probability distribution of an area containing hundreds of thousands clutter echoes close to a S-band radar operating in a tropical oceanic site, in order to monitor the stability of the reflectivity. Later on, Wolff et al. (2015) have shown that the RCA technique is capable of monitoring the radar reflectivity of other S-band radars, given proper generation of a suitable areal ground clutter map. The RCA technique operates with low range resolution radar bins (typically, 1° by 1° by 500 m) and indistinctly uses all the ground clutter values that are above a given high reflectivity threshold (e.g., 55 dBz) for more than a given per-centage of time. It does not perform a statistical analysis of individual radar bins nor considers other important spectral and polarimetric information. That is why it has been later proposed to analyze in detail spectral and polarimetric in-formation associated to an individual "Bright Scatterer" (BS). In order to be "bright", a point target with deterministic backscattering properties should be present at a near range and be hit by the antenna beam axis. For the Monte Lema radar (1625 m altitude) in Southern Switzerland, the first identified BS is the 90 m tall metallic tower on Cimetta (1633 m altitude, 18 km range). Gabella (2018) has shown that the peculiar Doppler and polarimetric signatures of the BS distinctively emerge in one single, high-range-resolution radar bin.

5.5.3 Beam blocking

Depending on the radar installation, the radar beam may be partly or completely occluded by the topography or obstacles located between the radar and the target. This results in underestimates of reflectivity and, hence, of rainfall rate. Visibility maps are used to correct for this error source. The visibility map indicates, as a function of range, elevation and azimuth angle, the percentage of the pulse volume that is visible from the radar antenna. Using a digital elevation model and a model for radar beam geometry and propagation, Pellarin et al. (2002) developed a model to compute the power loss due to partial or complete beam blockage by terrain. Germann et al. (2006) provides a detailed description of a beam blocking correction algorithm developed for use over the Swiss weather radar network. A similar methodology was used by Krajewski et al. (2006) to assess beam blockage severity. In the case of partial beam blockage, it is possible to apply an occultation correction, which increases the power received by the radar by a percentage that depends on the energy lost by the blockage based on the visibility maps. However, this approach has several limitations (e.g., Young et al., 1999); only standard propagation is considered, diffraction is neglected (geometric optics approach), side lobes in the whole solid angle cannot be modeled. Andrieu et al. (1997) discuss the issue related to beam blockage correction and show the relative error in radar-rainfall rate as a function of blocked beam (they decided not to apply any correction for beam blockage values larger than 60%). A pragmatic alternative is the one represented by empirical visibility maps, which can easily be obtained using long-term statistics of precipitation (Joss and Lee 1995). Empirical visibility maps have successfully been applied in several fields: Panziera et al. (2018) have identified and excluded from the hourly rainfall climatology some "bad" pixels (poor visibility or residual clutter contamination) (Beusch et al., 2018). Gugerli et al. (2020) have validated radar precipitation estimates over seven high-altitude Swiss glaciers and found that the regional variability of the underestimation factor could be mainly attributed to the empirical visibility of the radar network.

5.5.4 Range-related errors and the vertical profile of radar reflectivity

Due to the radar sampling geometry, radar rainfall estimation is necessarily subject to biases that are range related. Since the beam elevation angle is typically set to relatively high positive values to avoid beam occlusion and ground effects, the altitude of the radar sampling volume strongly depends on the radar range. This is even more important in mountainous terrain, where the volume visible by a radar is reduced because of ground clutter and elevated horizon. This often inhibits a direct view on precipitation close to the ground. When using radar measurements from aloft to estimate precipitation rates at ground level, the measurements must be corrected for the vertical change of the radar echo that is, the vertical profile of reflectivity (VPR), which usually describes the ratio between the reflectivity at a certain height and the reflectivity on the ground. The vertical variability of reflectivity is due to several effects including the growth (or evaporation) of hydrometeors, vertical air motion, change of hydrometeors' phase (ice to water in the bright band), and precipitation enhancement due to orography, among other effects (e.g., Joss and Waldvogel, 1990). For more stratiform precipitation systems, bright band enhancement is the most prominent source of range-dependent bias. The bright band is caused by the presence of a layer of enhanced reflectivity, which corresponds to the melting layer region, where snowflakes melt to form raindrops. The melting layer reflectivity enhancement takes place over a relatively thin cloud layer of about 300 to 500 meters. Precipitation overestimates up to a factor of 10 may result due to bright band effect (Smith, 1986). This rainfall overestimation occurs mainly at close and intermediate ranges (10 to 60 km), where most radars can resolve the melting layer structure. At further ranges bright band compensates for the reduced reflectivity aloft and at even longer ranges underestimation due to cloud overshooting occurs. Range-related errors mirror the vertical structure of the reflectivity field as it is perceived by the radar at various ranges.

Various correction procedures for range-related biases and vertical profile of radar reflectivity have been developed in the past three decades. The existing correction methods can be subdivided into three broad types (Zhang and Qi, 2010) depending on how the VPR is computed: climatological VPR, space-time averaged VPR, and modelled VPRs. Climatological VPRs are based on radar data averaged over long time periods (days, seasons, years) and over a certain spatial area (radar volume or well-visible regions) (Joss and Pittini, 1991; Joss and Lee, 1995). The advantage of this type of VPRs is that, once calculated, they are computationally inexpensive, based on actual radar data and thus always available. However, the climatological VPR assumes both spatial and temporal homogeneity, while in reality important variations may occur, for example, depending on whether precipitation is of stratiform or convective type.

Compared to the climatological VPRs, spatiotemporally averaged VPRs can better capture the temporal variations in reflectivity since these are based on a few volume scans only and regularly updated. They therefore also remain computationally inexpensive, and, among the few countries who correct for VPR in the operational processing, several are using some version of spatiotemporally averaged profiles (Joss and Lee, 1995; Germann and Joss, 2002). A number of authors used an inverse solution method or optimal linear estimation (e.g., Andrieu and Creutin, 1995; Andrieu et al., 1997; Borga et al., 1997; Vignal et al., 2001).

An alternative method to better account for spatial variability of profiles is to use a VPR model to obtain a profile at each location (e.g. Kitchen et al., 1994; Kirstetter et al., 2013). These can be determined using a set of physically based parameters in order to remain computationally inexpensive. The UK Met Office, for example, uses parameterizations for the melting layer (numerical weather prediction [NWP] model freezing level height), orographic growth (Hill, 1983) and the cloud top height (satellite infrared imagery) (Harrison et al., 2000).

Van den Heuvel et al. (2020), building on the availability of polarimetric data and a hydrometeor classification algorithm (Besic et al., 2016), explored the potential of machine learning methods to study the vertical structure of precipitation in Switzerland and to compute a more localized vertical profile correction.

5.5.5 Attenuation by intervening precipitation and due to a wet radome

In radar rainfall estimation, a common problem is signal attenuation, which is the reduction in power in electromagnetic radiation when passing through a medium of any density. Attenuation can be caused by gases in the atmosphere, clouds, rainfall, snow, and hail (e.g., Rinehart, 2004). We focus on rainfall attenuation since it is significant and most frequently encountered. Overall, this source of error almost exclusively affects radars with shorter wavelengths such as C-band and X-band radars (e.g., Anagnostou et al., 2006; Uijlenhoet and Berne, 2008). However, it could be greater than expected even for S-band radars.

Attenuation increases cumulatively with range with significant intervening rain; echoes from rain close to the radar are thus not attenuated as much as those at greater distance. Excessively attenuated radar echoes will be apparent as sectors or streaks of noticeably reduced returns in individual radar images, while less extreme attenuation (several decibels) or that which varies slowly in azimuth may be difficult to notice. However, these effects will appear as diminishing accumulations with increasing range so that any quantitative use of such data requires correction.

The conventional approach to attenuation correction involves the use of a power-law relationship between the reflectivity factor Z and the specific attenuation, k, in a similar manner to the reflectivity-rainfall relationship:

$$Z = \alpha k^{\beta} \tag{5.7}$$

The parameters α and β are dependent on the operating wavelength, the DSD and the temperature of the raindrops. This equation requires the unattenuated reflectivity not the measured reflectivity, which is affected by attenuation. The correction for attenuation then requires a cumulative procedure in range. In the correction algorithm, it is apparent that a consistent overestimate of reflectivity rapidly escalates, as the initial overestimates of reflectivity lead to overestimation of the specific attenuation which then causes subsequent reflectivities to be overestimated by an even greater extent, and so on. This form of attenuation correction is highly dependent on accurate calibration, yet without further information, it remains the only unapproximated correction option for single polarization measurements.

The numerical instabilities that affect the classical correction techniques may be avoided by using the mountain reference technique, based on the procedure reported in Bouilloud et al. (2009). The technique exploits variations in ground clutter return to estimate the total attenuation along the path (path integrated attenuation), assuming the reflectivity-attenuation relationship to be known (Marra et al., 2014).

Attenuation can also be caused by radome wetting. Most radar antennas are protected from wind and rain by a radome, usually made of fiber glass. The radome is engineered to cause little loss in the radiated energy. However, under intense rainfall, the surface of the radome can become coated with a thin film of water or ice, resulting in a strong azimuth-dependent attenuation. Combined with precipitation attenuation and at short wavelengths, the radar echoes may be heavily attenuated.

5.5.6 Electrical stability of radar system

The most crucial acitivities for a good radar system performance (including its accuracy and stability) are monitoring, maintenance and calibration. Monitoring describes the procedures to monitor the state, functionality and data quality of a radar system. Depending on the hardware and software of the radar, various parameters of the system can be monitored automatically and checked by a radar expert, ideally daily, or immediately in case of an alarm. Often during monitoring, decisions on intermediate maintenance or calibration are made. The purpose of the calibration is to provide radar measurements with a high predefined accuracy. In some cases, e.g. after hardware replacement due to maintenance, it can be necessary to repeat the calibration procedure. With the benefit of today's modern radar technology (e.g., low noise amplifiers (LNAs), fast and accurate Analog-to-Digital Converters) and with careful and regular calibration, it is possible to achieve high system stability. In this way, intrinsic uncertainties associated with the radar system itself are smaller than the uncertainties associated with the intrinsic variability of reflectivity of the radar target. For quantitative radar applications, high stability and accurate calibration are mandatory. Monitoring the stability of only the receiver chain or transmitter chain (one-way) is simpler than monitoring the stability of the entire radar system (two-way). To monitor the stability of the receiver chain, a reference power signal (instead of the received power coming from the antenna), is injected into the LNA input of the receiver and exactly that value (±a given uncertainty) is used for linking the given analogue-to-digital-unit value at the output of the digital receiver to the reference power value. No measurement is made of the power backscattered by a given object at a given distance. It is known that a given power on a logarithmic scale (dBm) corresponds to a given log-transformed analogue-to-digital-unit. In the case of an antenna-mounted receiver, an effective solution uses a noise source as the reference signal, taking advantage of its high temperature stability (Vollbracht et al., 2014). Monitoring the entire system's stability requires the assessment of losses (receive and transmit chains including waveguide, rotary joint, couplers, cables, radome, etc.), antenna gain and the accuracy of the antenna pointing angle. Assuring the stability of the entire system requires the calibration of the radar system against some known reference target (e.g. a metal sphere, a corner reflector with certified radar cross section) at various distances from the sensor itself. However, passive scatterers, such as large spheres or corner reflectors, are difficult to deal with, especially in heavy-cluttered mountainous terrain. There are two ways to overcome this difficulty: (a) total system stability (two-way) is occasionally (Gabella et al., 2013) or continuously (Kumagai et al., 1995) checked using active calibrators; (b) the problem is split into two simpler, complementary parts: (i) an external receiver is used as a one-way passive calibrator for checking the transmit chain (e.g. Reimann, 2013); (ii) the sun is used for calibrating (Gabella et al., 2016) and checking (Holleman et al., 2010) the receive chain. Results from the latter method were derived using single polariazation data acquired in 2008 (Holleman et al. 2010) during a period of quiet solar flux activity. More recently, it has been shown this method is also practicable during more active solar periods with both polarizations (Huuskonen et al., 2016; Gabella et al., 2015). The use of the sun is optimal in terms of cost/benefit. Solar monitoring can be carried out continuously.

Based on data from the Macaion weather radar concerning the storm event occurred on 2009 in the upper Adige river basin, Fig. 5.3 shows the effects of specific error sources, such as beam blockage, vertical profile of radar reflectivity, attenuation and wet radome attenuation, and their corrections as in Marra et al. (2014). The storm consisted of two successive convective bands, one of which directly affected the radar location causing strong attenuations due to wet radome and intervening heavy precipitation. These, together with beam overshooting at long distances, caused severe underestimation of the total storm rainfall. Radial underestimatin patterns typical of beam blockage are also clearly visible.

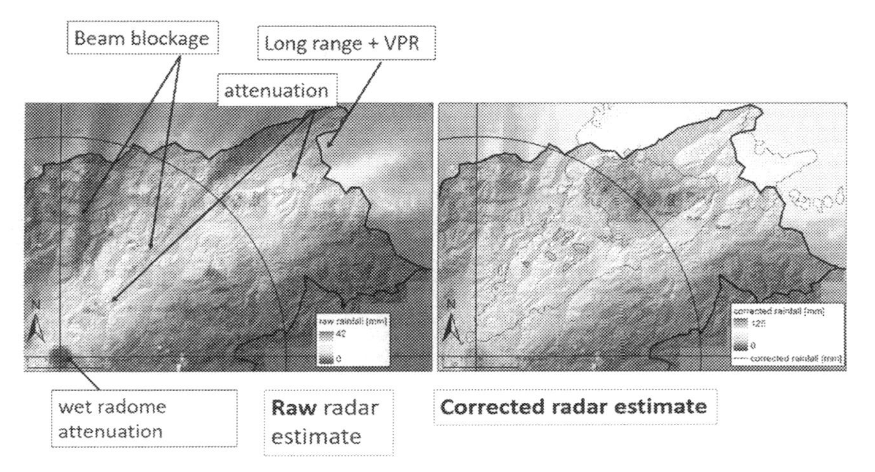

FIG. 5.3 Comparison of raw and corrected radar rainfall estimates, showing the effects of specific error sources, such as beam blockage, vertical profile of radar reflectivity, attenuation and wet radome attenuation, and the effects of their corrections.

5.6 Radar – raingauge merging and assessment of radar rainfall uncertainty

The idea of exploiting the complementarity between rain gauge networks and radar measurements in order to improve the accuracy of rainfall maps attracted the attention of hydrologists as soon as digitized radar observations were available. Various procedures, ranging from deterministic to statistical methods, were proposed for the so-called radar-raingauge merging (Wilson and Brands, 1979; Creutin et al., 1987; Seo et al., 1990; Gabella et al., 2000; Seo and Breidenbach, 2002; Gabella and Notarpietro, 2004; Gabella, 2004; Morin and Gabella, 2007; Cole and Moore, 2009; Velasco-Forero et al., 2009; Zappa et al., 2010; Sideris et al., 2014; Wolfensberger et al., 2021). The methods used for the purpose can be categorized as follows (Sideris et al., 2014): (1) computation of a constant multiplicative calibration factor (deterministic), (2) deterministic interpolation of the gauge to radar ratio, (3) statistical approaches based on multivariate analysis, (4) radar-raingauge probability distribution analysis, (5) geostatistical estimators, (6) Bayesian methods, and (7) machine learning methods.

The rain-gauge representativeness error is especially important in the context of combining radar and rain-gauge data since raingauges and weather radars are recording precipitation at different spatial scales. While the former is essentially a point scale measurement, the latter refers to a volume integral scale. In fact, in most cases, the ratio between observation scales (for standard C-band radar and a standard rain-gauge) is in the order of 10^7. Many studies have investigated the issue. Kitchen and Blackall (1992) showed that at an hourly scale in England the root-mean square error due to the spatiotemporal differences in sampling mode is as large as 150% of the mean rainfall.

Ciach and Krajewski (1999) introduced the error-variance separation method in order to assess the uncertainty associated with the area–point resolution difference (an approach that was also used and discussed in several other articles, e.g., Villarini et al., 2008; Seo and Krajewski, 2010). They estimated that, at hourly scales, the assumption of true areal rainfall estimation using raingauges accounts for approximately half of the overall discrepancies between radar and raingauges. In Florida, Habib and Krajewski (2002) estimated that the spatial sampling error accounts for approximately 40%–80% of the overall radar-gauge disagreement. Given the impact of the rain-gauge representativeness error, it is a common practice to employ aggregations over at least 1 h in order to mitigate the effect (Zawadzki, 1975; Sideris et al., 2014), but longer times might be required depending on the spatiotemporal scales of variability of precipitation.

Although radar errors have been widely studied and techniques have been developed to correct most of them, and rain gauge-radar merging methods are available to adjust radar-based estimates, residual errors are still intrinsic in radar rainfall estimates. An estimation of uncertainty of radar rainfall estimates and an assessment of uncertainty propagation in modelling applications is important to quantify the relative importance of the uncertainty associated to radar rainfall input in the overall modelling uncertainty.

A suitable tool for this purpose is the generation of radar rainfall ensembles (Germann et al., 2009; Cecinati et al., 2017). An ensemble is the representation of the rainfall field and its uncertainty through a collection of possible alternative rainfall fields, produced according to the observed errors, their spatial characteristics, and their probability distribution. The errors are derived from a comparison between radar rainfall estimates and rain gauge measurements, which is exploited to characterize the mean and covariance matrix of residual errors in radar precipitation estimates. For example, Germann et al. (2009) coupled the radar ensemble with a rainfall runoff model for a 2800 km^2 area in the southern Alps demonstrating the applicability of the method for a steep mountainous context.

5.7 Dual polarization

A major advent in the last two decades has been the implementation of dual polarization and Doppler techniques in an operational mode for several weather radar networks (Zrnic and Ryzhkov, 1990; Ryzhkov et al., 2005; Gourley et al., 2007). In addition to to the reflectivity Z, the differential reflectivity ZDR, the copolar correlation coefficient ρ_{HV}, the differential phase shift between the horizontal and vertical polarizations and other dual-polarization parameters can be measured for each radar sampling volume. Z_{DR} is defined as the log-transformed ratio between the copolar linear reflectivity measured using hori-zontal (zh) and vertical (zv) polarizations; it is expressed in dB and a value of 0 dB means that zh = zv. The differential reflectivity was introduced by Seliga and Bringi (1976) for a better estimate of rainfall, since it contributes to reduce the uncertainty associated with rain drop size distributions. Indeed, the information associated with ZDR is remarkable; however, the issue of a proper calibration remains a challenge for successful quantitative precipitation estimation (Ryzhkov et al., 2005). The copolar correlation coefficient represents the correlation between the copolar horizontal, HH, and vertical, VV. For a detailed description of the rather complex nature of this measurable, please refer to the electronic supplement (e06.1, ppt slides) accompanying the book by Fabry (2015). Here, we just remind that, being the module of the complex correlation coefficient between two orthogonal components (represented by two complex numbers) of the backscattered electromagnetic field within the radar sampling volume, it ranges between 0 (no correlation between the two polarizations) and 1 (perfect correlation). Another polarimetric quantity measured by the dual-polarization radar is the differential phase shift, Ψdp, between the phase of the copolar signal at horizontal and vertical polarization, respectively. Classification algorithms are used to distinguish several classes of liquid and frozen hydrometeors (Zrnic and Ryzhkov, 1990) prior to the application of specific rain rate estimators using various combinations of the radar measurables (Ryzhkov et al., 2005a). Polarimetry also offers some interesting possibilities such as self-calibration of reflectivity (Gorgucci et al., 1999), attenuation correction (Gorgucci and Baldini, 2007; Testud et al., 2000) and DSD retrieval (Bringi et al., 2003; Moisseev and Chandrasekar, 2007).

An exhaustive and clear description of the great potential of polarimetry is given in the monograph by Ryzhkov and Zrnic, 2019, which is rich of information regarding weather radar polarimetry and its applications. Regarding polarimetric radar rainfall estimators, the interested reader may refer to Chapter 10, Sections 3, 4, 5, 6 and 7 of the above-cited monograph (and corresponding literature at the end of the Chapter). A recent example using a C-band radar network in the mountainous terrain of Switzerland is given by Wolfensberger et al. (2021). Despite quantitative and quality improvement thanks to polarimetry, there is evidence that the density of existing operational radar networks is insufficient for a proper hydrological coverage at the national/continental scale if one considered a hundred of km range as a maximum practical limit for optimal spatial resolution. Two complementary solutions are currently being investigated:

The X-band frequency offers attractive design characteristics as compared to the S-band and C-band frequencies (reduced transmitted power and antenna size; reduced sensitivity to ground clutter) that are counterbalanced by a dramatic sensitivity to rain attenuation Anagnostou et al., 2009, 2010. In the recent years, a breakthrough has been achieved in radar QPE at X-band through the use of polarimetry and range-profiling algorithms (Testud et al., 2000). Several industrial (e.g., http://www.casa.umass.edu/) and research projects have been initiated to further develop this concept which has certainly a great future for radar QPE in rugged topography and/or in urban areas.

Although much less established, microwave links rainfall measurement also offers interesting potential with the deployment of wireless communication networks all over the world (Leijnse et al., 2007). A major advantage is related to the fact that such measurements are made close to the ground. First studies indicate that effects of rainfall variability are limited if the link length and frequency are chosen correctly (Berne and Uijlenhoet, 2007; Leijnse et al., 2007). For such settings, the main error sources are related to the wetting of antenna covers and to the resolution of power measurements.

5.8 Conclusions

The key central question that motivated the work by Berne and Krajewski (2013) ("Radar for hydrology: unfulfilled promise or unrecognized potential?") is still unanswered. Indeed, this chapter has shown that weather radar measurements have enormous potential in hydrological applications, though they are still not fully utilized. Among the reasons, the gap between the accuracy requirements for hydrological applications (hydrological flood modeling, for instance) and the uncertainties that still affect conventional radar rainfall estimates plays a key role.

The recourse to other sources of rainfall data (from raingauges, as an example) is needed to optimize the radar parameters and/or to control the radar QPE quality. Rather than a simple "radar calibration," such an operation should be seen as a multisensor merging requiring a detailed knowledge of the space-time structure of rainfall and of the instrumental/sampling errors associated with the various sensors used. The lack of absolute rainfall reference is often invoked as a limitation here; raingauge networks with adequate densities remain the only practical solution for obtaining such reference values for past data. The issue of representativeness and discrepancies in sampling scales is significant; time integration is a possible way to limit this influence. Other types of rainfall data such as disdrometers, microwave links, and satellite observations (in particular TRMM and GPM), microwave links and disdrometers could be progressively incorporated in such merging procedures.

References

Amengual, A., Borga, M., Ravazzani, G., Crema, S., 2021. The role of storm movement in controlling flash flood response: an analysis of the 28 September 2012 extreme event in Murcia, southeastern Spain. J. Hydrometeor. 22 (9), 2379–2392.

Anagnostou, M.N., Anagnostou, E.N., Vivekanandan, J., 2006. Correction for rain path specific and differential attenuation of X-band dual-polarization observations. IEEE Trans. Geosci. Remote Sens. 44 (9), 2470–2480.

Anagnostou, M.N., Kalogiros, J., Anagnostou, E.N., et al., 2009. Experimental results on rainfall estimation in complex terrain with a mobile X-band polarimetric weather radar. Atmos. Res. 94 (4), 579–595.

Anagnostou, M.N., Kalogiros, J., Anagnostou, E.N., et al., 2010. Performance evaluation of high-resolution rainfall estimation by X-band dual-polarization radar for flash flood applications in mountainous basins. J. Hydrol. 394 (1–2), 4–16.

Andrieu, H., Creutin, J.D., 1995. Identification of Vertical Profiles of Radar Reflectivity for Hydrological Applications Using an Inverse Method. Part I: Formulation. J. Appl. Meteorol. 34, 225–239.

Andrieu, H., Creutin, J.D., Delrieu, G., Faure, D., 1997. Use of a weather radar for the hydrology of a mountainous area. Part 1: Radar measurement interpretation. J. Hydrol. 193 (1–4), 1–25.

Battan, L.J., 1973. Radar Observation of the Atmosphere. University of Chicago Press, Chicago.

Bean, B.R., Dutton, E.J., 1966. Radio Meteorology. US Government Printing Office, Washington DC.

Berne, A., Delrieu, G., Creutin, J., Obled, C., 2004. Temporal and spatial resolution of rainfall easurements required for urban hydrology. J. Hydrol. 299, 166–179.

Berne, A., Uijlenhoet, R., 2007. Path-averaged rainfall estimation using microwave links: Uncertainty due to spatial rainfall variability. Geophys. Res. Lett. 2007, GL029409.

Berne, A., Krajewski, W.F., 2013. Radar for hydrology: unfulfilled promise or unrecognized potential? Adv. Water Res. 51, 357–366.

Besic, N., Figueras, J., Grazioli, J., Gabella, M., Germann, U., Berne, A., 2016. Hydrometeor classification through statistical clustering of polarimetric radar measurements: a semi-supervised approach. Atmos. Meas. Tech. 9, 4425–4445.

Beusch, L., Foresti, L., Gabella, M., Hamann, U., 2018. Satellite-Based Rainfall Retrieval: from Generalized Linear Models to Artificial Neural Networks. Remote Sensing 10.

Borga, M., Anagnostou, E.N., Krajewski, W., 1997. A simulation approach for validation of a bright band correction method. J. Appl. Meteorol. 36 (11), 1507–1518.

Borga, M., Anagnostou, E.N., Frank, F., 2000. On the use of real-time radar rainfall estimates for flood prediction in mountainous basins. J. Geophys. Res. 105 (D2), 2269–2280.

Borga, M., Esposti, S.D., Norbiato, D., 2006. Influence of errors in radar rainfall estimates on hydrological modelling prediction uncertainty. Water Resour. Res. 42, W08409.

Borga, M., Stoffel, M., Marchi, L., Marra, F., Jakob, M., 2014. Hydrogeomorphic response to extreme rainfall in headwater systems: flash floods and debris flows. J. Hydrol. 518, 194–205.

Brandes, E.A., Vivekanandan, J., Wilson, J.W., 1999. A comparison of radar reflectivity estimates of rainfall from collocated radars. J. Atmos. Ocean Technol. 16, 1264–1272.

Bringi, V.N., Chandrasekar, V., Hubbert, J., Gorgucci, E., Randeu, W.L., Schoenhuber, M., 2003. Raindrop Size Distribution in Different Climatic Regimes from Disdrometer and Dual-Polarized Radar Analysis. J. Atmospheric Sci. 60 (2), 354–365.

Cao, Q., Zhang, G.F., 2009. Errors in estimating raindrop size distribution parameters employing disdrometer and simulated raindrop spectra. J. Appl. Meteorol. Climatol. 48, 406–425.

Cao, Q., Yeary, M.B., Zhang, G.F., 2012. Efficient ways to learn weather radar polarimetry. IEEE Trans. Educ. 55, 58–68.

Cecinati, F., Rico-Ramirez, M.A., Heuvelink, G.B.M., Han, D., 2017. Representing radar rainfall uncertainty with ensembles based on a time-variant geostatistical error modelling approach. J. Hydrol. 548, 391–405.

Chong, M., Georgis, J.F., Bousquet, O., et al., 2000. Real-time wind synthesis from Doppler radar observations during the mesoscale alpine programme. Bull. Am. Meteorol. Soc. 81, 2953–2962.

Ciach, G.J., Krajewski, W.F., 1999. On the estimation of radar rainfall error variance. Adv. Water Resour. 22 (6), 585–595.

Cole, S.J., Moore, R.J., 2009. Distributed hydrological modelling using weather radar in gauged and ungauged basins. Adv. Water Res. 32 (7), 1107–1120.

Collier, C.G., 1996. Applications of Weather Radar Systems: A Guide to Uses of Radar Data in Meteorology and Hydrology, 2nd ed. John Wiley & Sons, Chichester, UK; New York, NY, USA.

Creutin, J.-.D., Delrieu, G., Lebel, T., 1987. Rain Measurement by Raingage-Radar Combination: A Geostatistical Approach. J. Atmos. Oceanic Technol. 5, 102–115.

Delrieu, G., Creutin, J.D., Andrieu, H., 1995. Simulation of radar mountain returns using a digitized terrain model. J. Atmos. Ocean. Technol. 12, 1039–1049.

Delrieu, G., Braud, I., Berne, A., Borga, M., Boudevillain, B., Fabry, F., Freer, J., Gaume, E., Nakakita, E., Seed, A., Tabary, P., Uijlenhoet, R., 2009. Weather radar and hydrology. Adv. Water Res. 32 (7), 969–974.

Fabry, F., 2015. Radar Meteorology: Principles and Practice, 1st ed., Cambridge University Press, Cambridge, United Kingdom, p. 256. ISBN 978-1-107-07046-2.

Fritz, J., Chandrasekar, V., 2012. Simulating radar observations of precipitation at higher frequencies from lower-frequency polarimetric measurements. J. Atmos. Oceanic Technol. 29, 1435–1454.

Gabella, M., Perona, G., 1998. Simulation of the orographic influence on weather radar using a -geometric-optics approach. J. Atmos. Oceanic Technol. 15, 1486–1495.

Gabella, M., Joss, J., Perona, G., 2000. Optimizing quantitative precipitation estimates using a noncoherent and a coherent radar operating on the same area. J. Geophys. Res. Atm. 105, 2237–2245.

Gabella, M., 2004. Improving operational measurement of precipitation using radar in mountainous terrain Part II: Verification and Applications. IEEE Geosci. Remote Sensing Lett. 1, 84–89.

Gabella, M., Notarpietro, R., 2004. Improving operational measurement of precipitation using radar in mountainous terrain. IEEE Geosci. Remote Sensing Lett. 1, 78–83.

Gabella, M., Sartori, M., Progin, O., Germann, U., 9–13 September, 2013. Acceptance tests and monitoring of the next generation polarimetric weather radar network in Switzerland, Proceedings of the IEEE International Conference Electromagnetics Advanced Applications. Torino, Italy, p. 4.

Gabella, M., Sartori, M., Boscacci, M., Germann, U., 2015. Vertical and Horizontal Polarization Observations of Slowly Varying Solar Emissions from Operational Swiss Weather Radars. Atmosphere 6, 50–59.

Gabella, M., Sartori, M., Boscacci, M., Germann, U., 2016. Calibration accuracy of the dual-polarization receivers of the C-band Swiss weather radar network. Atmosphere 7, 10.

Gabella, M., 2018. On the Use of Bright Scatterers for Monitoring Doppler, Dual-Polarization Weather Radars. Remote Sensing 10, 14.

Germann, U., Berenguer, M., Sempere-Torres, D., Zappa, M., 2009. REAL – Ensem-ble radar precipitation estimation for hydrology in a mountainous region. Q. J. R. Meteorol. Soc. 135 (639), 445–456.

Germann, U., Gianmario, G., Boscacci, M., Bolliger, M., 2006. Radar precipitation measurement in a mountainous region. Q. J. R. Meteorol. Soc. 132 (618A), 1669–1692.

Germann, U., Joss, J., 2002. Mesobeta profiles to extrapolate radar precipitation measurements above the Alps to the ground level. J. Appl. Meteorol. 41 (5), 542–557.

Germann, U., Joss, J., 2004. Operational measurement of precipitationin mountainous terrainWeather Radar: Principles and Advanced Applications, Series Physics of Earth and Space Environment Meischner P(ed). Springer-Verlag, Berlin, Germany, pp. 52–77.

Gorgucci, et al., 1999. A procedure to calibrate multiparameter weather radar using properties of the rain medium. IEEE Trans. Geosci. Remote Sens. 37 (1), 269–276.

Gorgucci, E., Baldini, L., 2007. Attenuation and differential attenuation correction of C-band radar observations using a fully self-consistent methodology. IEEE Geosci. Remote Sens. Lett. 4, 326–330.

Gourley, J.J., Tabary, P., Parent du Chatelet, J., 2007. A fuzzy logic algorithm for the separation of precipitating from nonprecipitating echoes using polarimetric radar observations. J. Atmos. Oceanic Technol. 24, 1439–1451.

Gugerli, R., Gabella, M., Huss, M., Salzmann, N., 2020. Can weather radars be used to estimate snow accumulation on alpine glaciers? - an evaluation based on glaciological surveys. J. Hydrometeorol. 21, 2943–2962.

Habib, E., Krajewski, W.F., 2002. Uncertainty analysis of the TRMM ground-validation radar-rainfall products: Application to the TEFLUN-B field campaign. J. Appl. Meteor. 41, 558–572.

Harrison, D.L., Driscoll, S.J., Kitchen, M., 2000. Improving precipitation estimates from weather radar using quality control and correction techniques. Meteorol. Appl. 7, 135–144.

Holleman, I., Huuskonen, A., Kurri, M., Beekhuis, H., 2010. Operational monitoring of weather radar receiving chain using the Sun. J. Atmos. Ocean. Technol. 27, 159–166.

Hossain, F., Anagnostou, E.N., Dinku, T., Borga, M., 2004. Hydrological model sensitivity to parameter and radar rainfall estimation uncertainty. Hydrol. Process. 18–17, 3277–3291.

Hubbert, J.C., Dixon, M., Ellis, S.M., Meymaris, G., 2009. Weather Radar Ground-clutter. Part I: Identification, Modeling, and Simulation. J. Atmos. Oceanic Technol. 26 (7), 1165–1185.

Huuskonen, A., Kurri, M., Holleman, I., 2016. Improved analysis of solar signals for differential reflectivity monitoring. Atmos. Measurement Techniques 9 (7), 3183–3192.

Islam, T., Rico-Ramirez, M.A., Han, D., et al., 2012. Performance evaluation of the TRMM precipitation estimation using ground-based radars from the GPM validation network. J. Atmos. Sol. Terr. Phys. 77, 194–208.

Islam, T., Rico-Ramirez, M.A., 2013. An overview of the remote sensing of precipitation with polarimetric radar. Prog. Phys. Geog. 38 (1), 55–78.

Joss, J., 1998. Operational Use of Radar for Precipitation Measurements in Switzerland. Vdf Hochschulverl. an der ETH Zürich, Zürich, Switherlands.

Joss, J., Lee, R., 1995. The application of radar-gage comparisons to operational precipitation profile corrections. J. Appl. Meteorol. 34, 2612–2630.

Joss, J., Waldvogel, A., 1990. Precipitation Measurements and Hydrology, a review. In: Atlas, D. (Ed.), Radar in Meteorology. Am. Met. Soc., Boston Mass, pp. 577–606.

Joss, J., Pittini, A., 1991. Real-time estimation of the vertical profile of radar reflectivity to improve the measurement of precipitation in an Alpine region. Meteorol. Atmos. Phys. 47 (1), 61–72.

Kirstetter, P.-E., Andrieu, H., Boudevillain, B., Delrieu, G.A., 2013. Physically based identification of vertical profiles of reflectivity from volume scan radar data. J. Appl. Meteorol. Climatol. 52 (7), 1645–1663.

Kitchen, M., Brown, R., Davies, A.G., 1994. Real-time correction of weather radar data for the effects of bright band, range and orographic growth in widespread precipitation. Quart. J. Roy. Meteorol. Soc. 120 (519), 1231–1254.

Kitchen, M., Blackall, R.M., 1992. Representativeness errors in comparisons between radar and gauge measurements of rainfall. J. Hydrol. 134, 13–33.

Krajewski, W.F., 1987. Cokriging Radar-Rainfall and Rain Gage Data. J. Geophys. Res. Atmos. 92, 9571–9580.

Krajewski, W.F., Smith, J.A., 2002. Radar hydrology: rainfall estimation. Adv. Water Resour. 25 (8–12), 1387–1394.

Krajewski, W.F., Ntelekos, A.A., Goska, R., 2006. A GIS-based methodology for the assessment of weather radar beam blockage in mountainous regions: Two examples from the US NEXRAD network. Comput. Geosci. 32 (3), 283–302.

Kumagai, H., Kozu, T., Satake, M., Hanado, H., Okamoto, K., 1995. Development of an active radar calibrator for the TRMM Precipitation Radar. IEEE Trans. Geosci. Remote Sens. 33, 1316–1318.

Lee, G., Zawadzki, I., 2005. Variability of drop size distributions: noise and noise filtering in disdrometric data. J. Appl. Meteorol. 44, 634–652.

Lee, G.W., Zawadzki, I., 2006. Radar calibration by gage, disdrometer, and polarimetry: theoretical limit caused by the variability of drop size distribution and application to fast scanning operational radar data. J. Hydrol. 328, 83–97.

Leijnse, H., Uijlenhoet, R., Stricker, J.N.M., 2007. Rainfall measurement using radio links from cellular communication networks. Water Resour. Res. 43 (3), W03201.

Lengfeld, K., Kirstetter, P.E., Fowler, H.J., Yu, J., Becker, A., Flaming, Z., Gourley, J., 2020. Use of radar data for characterizing extreme precipitation at fine scales and short durations. Environ. Res. Lett. 15, 085003.

Marra, F., Destro, E., Nikolopoulos, E.I., Zoccatelli, D., Creutin, J.D., Guzzetti, F., Borga, M., 2017. Impact of rainfall spatial aggregation on the identification of debris flow occurrence thresholds. Hydrol. Earth Syst. Sci. 21 (9), 4525–4532.

Marra, F., Morin, E., 2015. Use of radar QPE for the derivation of intensity–duration–frequency curves in a range of climatic regimes. J. Hydrol. 531, 427–440.

Marra, F., Nikolopoulos, E.I., Creutin, J.D., Borga, M., 2014. Radar rainfall estimation for the identification of debris-flow occurrence thresholds. J. Hydrol. 519 (Part B), 1607–1619.

Marra, F., Nikolopoulos, E.I., Anagnostou, E.N., Bárdossy, A., Morin, E., 2019. Precipitation frequency analysis from remotely sensed datasets: A focused review. J. Hydrol. 574, 699–705.

Marshall, J.S., Palmer, W.M., 1948. The distribution of raindrops with size. Journal of Meteorology 5, 165–166.

Marshall, J.S., Hitschfeld, W., Gunn, K.L.S., 1955. Advances in radar weather. Adv. Geophys. 2, 1–56.

Michaelides, S., Levizzani, V., Anagnostou, E.N., Bauer, P., Kasparis, T., Lane, J.E., 2009. Precipitation: measurement, remote sensing, climatology and modeling. Atm. Res. 94 (4), 512–533.

Michelson, D.B., Einfalt, T., Holleman, I., Gjertsen, U., Friedrich, K., Haase, G., Lindskog, M., Jurczyk, A., 2005. Weather Radar Data Quality in Europe – Quality Control and Characterization – Review. EUR 21955. Office for Official Publications of the European Communities, Luxembourg. ISBN 92-898-0018-6.

Miriovsky, B., Bradley, A.A., Eichinger, W.E., Krajewski, W.F., Kruger, A., Neslon, B.R., Creutin, J.-D., Lapetite, J.-M., Lee, G., Zawadzki, I., 2004. An experimental study of small-scale variability of radar reflectivity using disdrometer observations. J. Appl. Meteorol. 43 (1), 106–118.

Moisseev, D.N., Chandrasekar, V., 2007. Nonparametric estimation of raindrop size distributions from dual-polarization radar spectral observations. J. Atmos. Oceanic Technol. 24 (6), 1008–1018.

Montopoli, M., Marzano, F.S., Vulpiani, G., 2008. Analysis and synthesis of raindrop size distribution time series from disdrometer data. IEEE Trans. Geosci. Remote Sens. 46, 466–478.

Morin, E., Gabella, M., 2007. Radar-based quantitative precipitation estimation over Mediterranean and dry climate regimes. J. Geophys. Res. 112, D20108.

Nikolopoulos, E.I., Borga, M., Zoccatelli, D., Anagnostou, E.N., 2014. Catchment scale storm velocity: quantification, scale dependence and effect on flood response. Hydrol. Sci. J. 59 (7), 1363–1376.

Overeem, A., Buishand, T.A., Holleman, I., 2009. Extreme rainfall analysis and estimation of depth-duration-frequency curves using weather radar. Water Resour. Res. 45, W10424.

Panziera, L., Gabella, M., Zanini, S., Hering, A., Germann, U., Berne, A., 2016. A radar-based regional extreme rainfall analysis to derive the thresholds for a novel automatic alert system in Switzerland. Hydrol. Earth Syst. Sci. 20, 2317–2332.

Panziera, L., Gabella, M., Germann, U., Martius, O., 2018. A 12-year radar-based climatology of daily and sub-daily extreme precipitation over the Swiss Alps. Int. J. Climatol. 7–21.

Passarelli, R.E., Romanik, P., Geotis, S.G., Siggia, A.D., 1981. Ground-clutter rejection in the frequency domain (for radar meteor-ology applications), Preprints of the 20th Conf. on Radar Meteorology. Boston, MA, 1981, 295–300.

Pellarin, T., Delrieu, G., Saulnier, G.M., Andrieu, H., Vignal, B., Creutin, J.D., 2002. Hydrologic visibility of weather radars operating in mountainous regions: Case study for the Ardèche catchment (France). J. Hydrometeor. 3, 539–555.

Probert-Jones, J.R., 1962. The radar equation in meteorology. Q. J. R. Meteorolog. Soc. 88, 485–495.

Reimann, J., 2013. On fast, polarimetric non-reciprocal calibration and multi-polarization measurements on weather radars. PhD dissertation. DLR and Technische Universität Chemnitz, p. 161. ISRN DLR-FB-2013-36, ISSN 1434 8454.

Rinehart, R.E., 2004. Radar for Meteorologists. Univ. of North Dakota, p. 334.

Ryzhkov, A.V., Zrnic, D.S., 2019. Radar polarimetry for weather observations. Springer, p. 486.

Ryzhkov, A.V., Giangrande, S.E., Melnikov, V.M., Schuur, T.J., 2005a. Calibration issues of dual-polarization radar measurements. J. Atmos. Oceanic Technol. 22, 1138–1155.

Ryzhkov, A.V., Giangrande, S.E., Schuur, T.J., 2005b. Rainfall estimation with a polarimetric prototype of WSR-88D. J. Appl. Meteorol. 44 (4), 502–515.

Sangati, M., Borga, M., Rabuffetti, D., Bechini, R., 2009. Influence of rainfall and soil properties spatial aggregation on extreme flash flood response modelling: an evaluation based on the Sesia river basin, North Western Italy. Adv. Water Resour. 32 (7), 1090–1106.

Saltikoff, E., Haase, G., Delobbe, L., Gaussiat, N., Martet, M., Idziorek, D., Leijnse, H., Novák, P., Lukach, M., Stephan, K., 2019. OPERA the Radar Project. Atmosphere 10, 320.

Seliga, T., Bringi, V., 1976. Potential use of radar differential reflectivity measurements at orthogonal polarizations for measuring precipitation. J. Appl. Meteor. 15, 69–76.

Seo, D.-J., Krajewski, W.F., Bowles, D.S., 1990. Stochastic interpolation of rainfall data from rain gages and radar using cokriging. 1. Design of experiments. Water Resour. Res. 26 (3), 469–477.

Seo, B.C., Krajewski, W.F., 2010. Scale dependence of radar rainfall uncertainty: initial evaluation of NEXRAD's new super-resolution data for hydrologic applications. J Hydrometeorol 11 (5), 1191–1198.

Seo, D.-J., Breidenbach, J.P., 2002. Real-Time Correction of Spatially Nonuniform Bias in Radar Rainfall Data Using Rain Gauge Measurements. J. Hydrometeor. 3, 93–111.

Sideris, I.V., Gabella, M., Erdin, R., Germann, U., 2014. Real-time radar–rain-gauge merging using spatio-temporal co-kriging with external drift in the alpine terrain of Switzerland. Q. J. Royal. Met. Soc. 140 (680), 1097–1111.

Silberstein, D.S., Wolff, D.B., Marks, D.A., Atlas, D., Pippitt, J.L., 2008. Ground Clutter as a Monitor of Radar Stability at Kwajalein. RMI J. Atmos. Oceanic Technol. 25, 2037–2045.

Smith, C.J., 1986. The reduction of errors caused by bright bands in quantitative rainfall measurements made using radar. J Atmos Ocean Technol 3, 129–141.

Smith, P.L., 1990. Precipitation measurements and hydrology: panel report. Radar in Meteorology, Boston.

Smith, J.A., Hui, E., Steiner, M., Baeck, M.L., Krajewski, W.F., Ntelekos, A.A., 2009. Variability of rainfall rate and raindrop size distributions in heavy rain. Water Resour. Res. 45, 12.

Steiner, M., Smith, J.A., 2000. Reflectivity, rain rate, and kinetic energy flux relationships based on raindrop spectra. J. Appl. Meteorol. 39, 1923–1940.

Steiner, M., Smith, J.A., 2002. Use of three-dimensional reflectivity structure for automated detection and removal of non-precipitating echoes in radar data. J. Atmos. Oceanic Technol. 19, 673–686.

Tabary, P., Scialom, G., Germann, U., 2001. Real-time retrieval of the wind from aliased velocities measured by Doppler radars. J. Atmos. Oceanic Technol. 18, 875–882.

Testud, J., Le Bouar, E., Obligis, E., Ali-Mehenni, M., 2000. The rain profiling algorithm applied to polarimetric weather radar. J. Atmos. Oceanic Technol. 17, 322–356.

Tokay, A., Kruger, A., Krajewski, W.F., et al., 2002. Measurements of drop size distribution in the southwestern Amazon basin. J. Geophys. Res. – Atmos. 107 (D20), 8052.

Tokay, A., Bashor, P.G., Wolff, K.R., 2005. Error characteristics of rainfall measurements by collocated Joss-Waldvogel disdrometers. J. Atmos. Oceanic Technol. 22, 513–527.

Uijlenhoet, R., 1999. Parameterisation of rainfall microstructure for radar meteorology and hydrology. Doctoral dissertation. Wageningen University, The Netherlands, p. 279.

Uijlenhoet, R., 2001. Raindrop size distributions and radar reflectivity–rain rate relationships for radar hydrology. Hydrol. Earth Syst. Sci. 5 (4), 615–627.

Uijlenhoet, R., Steiner, M., Smith, J.A., 2003. Variability of raindrop size distributions in a squall line and implications for radar rainfall estimation. J. Hydrometeor. 4, 43–61.

Uijlenhoet, R., Berne, A., 2008. Stochastic simulation experiment to assess radar rainfall retrieval uncertainties associated with attenuation and its correction. Hydrol. Earth Syst. Sci. 12 (2), 587–601.

Van den Heuvel, F., Foresti, L., Gabella, M., Germann, U., Berne, A., 2020. Learning about the vertical structure of radar reflectivity using hydrometeor classes and neural networks in the Swiss Alps. Atmos. Meas. Tech. 13, 2481–2500.

Velasco-Forero, C.A., Sempere-Torres, D., Cassiraga, E.F., Gomez-Hernandez, J.J., 2008. Anon-parametric automatic blending methodology to estimate rainfall fields from rain gauge and radar data. Adv Water Resour 32 (7), 986–1002.

Vignal, B., Krajewski, W.F., 2001. Large-sample evaluation of two methods to correct range-dependent error for WSR-88D rainfall estimates. J Hydrometeorol 2 (5), 490–504.

Villarini, G., Mandapaka, P.V., Krajewski, W.F., Moore, R.J., 2008. Rainfall and sampling uncertainties: a rain gauge perspective. J. Geophys. Res. Atmos. 113 (11), 1–12.

Villarini, G., Krajewski, W.F., 2010. Review of the different sources of uncertainty in single polarization radar-based estimates of rainfall. Surv. Geophys. 31 (1), 107–129.

Vollbracht, D., Sartori, M., Gabella, M., 1–5 September, 2014. Absolute dual-polarization radar calibration: Temperature dependence and stability with focus on antenna-mounted receivers and noise source-generated reference signal, Proceedings of the 8th European Conference on Radar in Meteorology and Hydrology (ERAD2014). Garmisch-Partenkirchen, Germany, 91–102.

Wang, J.J., Carey, L.D., 2005. The development and structure of an oceanic squall-line system during the South China Sea Monsoon Experiment. Monthly Weather Review 133, 1544–1561.

Wen, Y.X., Hong, Y., Zhang, G.F., et al., 2011. Cross validation of spaceborne radar and ground polarimetric radar aided by polarimetric echo classification of hydrometeortypes. Journal of Applied Meteorology and Climatology 50, 1389–1402.

Wilson, J.W., Brandes, E.A., 1979. Radar Measurement of Rainfall—A Summary. Bull. Amer. Meteor. Soc. 60 (9), 1048–1060.

Wolfensberger, D., Gabella, M., Boscacci, M., Germann, U., Berne, A., 2021. RainForest: a random forest algorithm for quantitative precipitation estimation over Switzerland. Atmos. Meas. Tech. 14, 3169–3193.

Wolff, D.B., Marks, D.A., Petersen, W.A., 2015. General Application of the Relative Calibration Adjustment (RCA) Technique for Monitoring and Correcting Radar Reflectivity Calibration. J. Atmos. Oceanic Technol. 32 (3), 496–506.

Yamamoto, M.K., Mega, T., Ikeno, N., et al., 2011. Doppler velocity measurement of portable X-band weather radar equipped with magnetron transmitter and IF digital receiver. IEICE Trans. Commun. E94B, 1716–1724.

Young, C.B., Nelson, B.R., Bradley, A.A., Smith, J.A., Peters-Lidard, C.D., Kruger, A., Baeck, M.L., 1999. An evaluation of NEXRAD precipitation estimates in complex terrain. J. Geophys. Res.: Atmos. 104 (19), 691–703.

Zappa, M., Beven, K.J., Bruen, M., Cofiño, A.S., Kok, K., Martin, E., Nurmi, P., Orfila, B., Roulin, E., Schröter, K., et al., 2010. Propagation of Uncertainty from observing systems and NWP into hydrological models: COST-731 working group 2. Atmos. Sci. Lett. 11 (2), 83–91.

Zawadzki, I., 1975. On radar-raingage comparison. J. Appl. Meteor. 14, 1430–1436.

Zawadzki, I., 1984. Factors affecting the precision of radar measurements of rain, Proceedings of the 22nd Conference on Radar Meteorology. Boston, MA, USA, 251–256.

Zhang, J.A., Qi, Y.C., 2010. A real-time algorithm for the correction of brightband effects in radar-derived QPE. J. Hydrometeorol. 11, 1157–1171.

Zoccatelli, D., Borga, M., Zanon, F., Antonescu, B., Stancalie, G., 2010. Which rainfall spatial information for flash flood response modelling? A numerical investigation based on data from the Carpathian range, Romania. J. Hydrol. 394 (1–2), 148–161.

Zoccatelli, D., Borga, M., Viglione, A., Chirico, G.B., Blöschl, G., 2011. Spatial moments of catchment rainfall: rainfall spatial organisation, basin morphology, and flood response. Hydrology and Earth System Sciences 15, 3767–3783.

Zrnic, D.S., Ryzhkov, A.V., 1999. Polarimetry for weather surveillance radars. Bull. Am. Meteorol. Soc. 80, 389–406.

CHAPTER

Satellite rainfall estimation

Christopher Kidd[a,b], Vincenzo Levizzani[c]

[a]*Earth System Science Interdisciplinary Center, University of Maryland, MD, United States*

[b]*NASA Goddard Space Flight Center, Greenbelt, MD, United States*

[c]*Institute of Atmospheric Sciences and Climate, National Research Council (CNR-ISAC), Bologna, Italy*

6.1 Introduction

The measurement of precipitation across the Earth's surface is crucial for a range of scientific and societal applications (Kucera et al., 2013; Kirschbaum et al., 2017). For science, precipitation is a key component of the energy and water cycle through the evaporation of water, condensation and deposition as precipitation (Levizzani and Cattani, 2019). For society, monitoring and measuring precipitation is important for assessing groundwater storage, forecasting crop yields and combating water-borne diseases (Kirschbaum and Patel, 2016), and crucial for hydrological warning systems to mitigate the effects of, for example, flash floods. These applications necessitate measurements over a range of spatial (local to global) and temporal (instantaneous to climate) scales (Michaelides et al., 2009; Friedl, 2014). The challenge is to provide measurements of precipitation across the globe, with spatial and temporal resolutions and latencies that address the user requirements. Many scientific studies rely upon information on global precipitation, often contending with a range of values. For example, Herold et al. (2016) investigated how much rain falls over the land areas using both observational and model reanalysis products: they found that values range from about 5.5 mm day^{-1} (ERA40; Uppala et al., 2005) to 8.5 mm day^{-1} (HADEx2 - Donat et al., 2013; GHCNDEX-merged - Dittus et al., 2015). Most importantly, they showed that the differences in mean daily precipitation among the conventional observational datasets are as great as those among the models or reanalyses. These differences are crucial in studying the regional variability (Jurković and Pasarić, 2013) and long-term changes/trends in precipitation (New et al., 2001; Higgins and Kousky, 2013).

Conventional observations of precipitation from surface-based measurements, including rain gauges (Lanza et al., 2022) and weather radars (Borga et al., 2022), form the basis of many global precipitation products. However, the distribution of gauges over the Earth's surface is generally confined to land surface areas (Kidd et al., 2017) and the timely availability of data from these gauges is often problematic. The most comprehensive gauge-based precipitation data set is the Global Precipitation Climatology

RAINFALL. Modeling, Measurement and Applications. DOI: https://doi.org/10.1016/B978-0-12-822544-8.00005-6

Centre (GPCC) product available from 1901 to the present (Becker et al., 2013). The GPCC product incorporates gauge data from national networks as well as gauges that are included in other global or regional data sets, such as the Global Historical Climatology Network (GHCN; Menne et al., 2012) and the Asian-region APHRODITE data set (Yatagai et al., 2012). However, such global and regional gauge-based data sets must be generated with care to reduce errors and uncertainties in the final product.

The overall accuracy of the 'standard' accumulation gauge is discussed by Yang et al. (1998). Importantly, it should be noted that there is an element of random errors in any gauge measurements as noted by Ciach (2003) who, with a cluster of closely spaced gauges, noted that rain gauge errors are highly dependent upon on the local rainfall intensity and timescale. At the gauge scale, errors arise due to the deformation of the wind field around the gauge itself (Sevruk et al., 1991). Indeed, Chvíla et al. (2005) found wind-induced losses in thunderstorms while Duchon and Biddle (2010) noted under-catch using tipping bucket rain gauges in high intensity rain events and high winds. Some of these issues have been addressed through improving the design of gauges (e.g., Strangeways, 2004), although such design changes have generally yet to implemented and therefore challenges remain in providing reliable point rainfall measurements (Sieck et al., 2007), not least since different gauges lead to different responses (Saidi et al., 2014).

Beyond the individual gauge, the overall exposure of the gauge is also crucial: it should not be overshadowed yet should not be overly exposed (Sevruk and Kahlavova, 1994). At the region scale there are differences in gauge performance, especially over hilly or mountainous regions, which affect the precipitation-altitude relationships (Sevruk, 1997), even in regions of dense gauge networks (e.g., Dorninger et al., 2008). However, on a global scale the distribution of gauges is largely confined to land areas with dense population: consequently, much of the Earth's surface has poor or no gauge information (Kidd et al., 2017).

The development of weather radar has provided a significant source of precipitation information, particular for near real time monitoring (Wilson and Brandes, 1979; Heiss et al., 1990). Radars provide frequent and regular spatial information on precipitation at regional scales (Kitchen and Illingworth, 2011), although errors and uncertainties arise from variable reflectivity-rainrate relationships (Seo and Krajewski, 2010). While such radar observations may be used to generate longer term products (Fairman et al., 2015) careful processing is required to avoid artefacts in the final product. Despite the usefulness of radar-derived precipitation products, the distribution of weather radars typically matches that of the gauge networks, and consequently large areas of the Earth's surface are not covered by either gauge or radar measurements. There is therefore a need for precipitation measurements that are available across the whole of the Earth's surface: satellite observations have the potential to fulfil this need.

6.2 Satellites and sensors

Since the launch of the first meteorological satellite in 1960, many satellites have carried sensors capable of observing the Earth and its atmosphere. However, most of these sensors have not necessarily been specifically designed to measure precipitation.

Nevertheless, the precipitation community has been adept in developing schemes to retrieve precipitation from a variety of satellite observations. Not until the Tropical Rainfall Measuring Mission (TRMM; Kummerow et al., 1998; Simpson et al., 1988), launched in November 1997, and the Global Precipitation Measurement (GPM) mission (Hou et al., 2014) launched in February 2014, did dedicated satellite precipitation missions exist.

To maximise the availability of precipitation data a range of different satellite systems are utilised that can be split broadly into the Geostationary (GEO) visible (VIS) and/or infrared (IR) -based and Low Earth Orbit (LEO) passive microwave (PMW) or active microwave (AMW) -based systems (see Fig. 6.1). The operational GEO suite of sensors (Table 6.1) provides a consistent set of observations around the Equator with observations extending to about 60°N-60°S. A baseline for these observations is generally every 30 minutes at a nominal resolution of about 4 km (Janowiak et al., 2001), although new sensors can provide imagery every few seconds at 500 m resolution for limited regions and durations (Bessho et al., 2016). The more direct (passive) microwave measurements are available from about 10 LEO-based sensors: these provide less frequent observations (every few hours) with typically coarser spatial resolutions (5–15 km). These two complementary satellite systems are described in detail below. Observations made by the GEO and LEO based sensors are complementary and are often just used in conjunction with one another.

6.2.1 Satellite systems

The primary advantage of satellite observations is that they are capable of providing a global viewpoint, allowing regular observations of the Earth system including those related to clouds and precipitation. Observations used in the estimation of precipitation are typically derived from satellites in LEO or from GEO orbits: satellites in the former category orbit the Earth at about 850 km altitude or lower, while those in the latter group view the Earth from an altitude of about 36,000 km. Precipitation-capable LEO missions usually utilise sun-synchronous orbits that provide observations at the same local time each day. Current operational meteorological satellites in LEO include the European Organization for the Exploitation of Meteorological Satellites (EUMETSAT) Meteorological Operational satellite (MetOp) series (Klaes at al., 2007), the National Oceanic and Atmospheric Administration (NOAA) series (Goldberg et al., 2013), the Defense Meteorological Satellite Program (DMSP) series (Curtis and Adams, 1987) together with the Chinese FY-3 series (Yang, 2012) and the Russian Meteor-M series (Asmus et al., 2014). While these LEO satellites (Table 6.2) provide the foundation for the precipitation retrievals, the observations are limited to typically two per day per sensor, consequently multiple satellites are required to provide sufficient observations to capture the variability of precipitation. However, several LEO satellites occupy non-sun-synchronous orbits such as the TRMM, GPM and Megha-Tropiques missions. A useful feature of non-sun-synchronous orbits is that, over a period of time, they provide observations across different times of day, allowing the diurnal cycle to be sampled. Furthermore, their orbital path will intersect

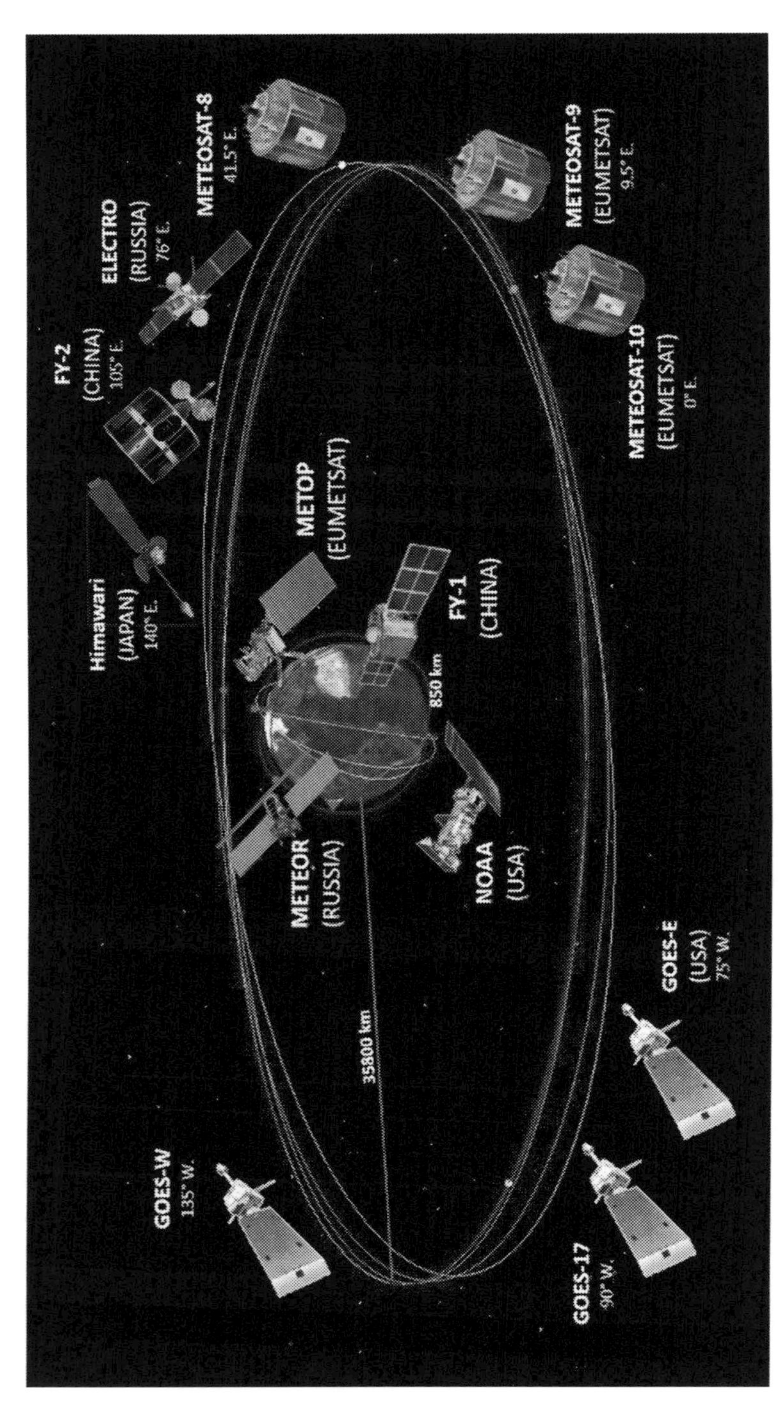

FIG. 6.1 The global observing system of geostationary (ca. 35800 km altitude) and low Earth orbiting satellites (ca. 850 km altitude).

Table 6.1 Geostationary satellites contributing the global observing system, allowing the capture of frequent, regular visible and infrared imagery.

Current geostationary meteorological satellites (35786 km altitude)					
Country	**Name**	**Acronym**	**launch**	**until**	**position**
US	Geostationary Operational Environmental Satellite (3rd generation) – GOES-East	GOES-16	19-Nov-16	≥2027	75.2°W
US	Geostationary Operational Environmental Satellite (3rd generation) – GOES-West	GOES-17	01-Mar-18	≥2029	137.5°W
Japan	Himawari (3rd generation)	Himawari-8	07-Oct-14	≥2029	140.68°E
Europe	Meteosat Second Generation (MSG) (Indian Ocean)	Meteosat-8	28-Aug-02	≥2022	41.5°E
Europe	Meteosat Second Generation (MSG)	Meteosat-9	21-Dec-05	≥2024	3.5°E
Europe	Meteosat Second Generation (MSG) (Rapid scan)	Meteosat-10	05-Jul-12	≥2028	9.5°E
Europe	Meteosat Second Generation (MSG) (Rapid scan)	Meteosat-11	15-Jun-15	≥2024	0.0°

those of the sun-synchronous satellites, thus allowing the cross-calibration of instantaneous satellite observations from different satellite sensors.

The GEO satellites, as the name suggests, remain (nearly) stationary with respect to their sub-satellite location located around the Earth's Equator. From their location their sensors are able to acquire frequent and regular images over a full (about 1/3) disc of the Earth (see Fig. 6.2). To provide (quasi-) global coverage a constellation of about five GEO satellites located around the Equator is needed. The current operational geostationary satellites are provided by the US NOAA Geostationary Operational Environmental Satellites (GOES-15/16/17; Goodman et al., 2020; Schmit et al., 2017), the European Meteosat Second Generation (MSG) Meteosat-8/9/10/11 (Schmetz et al., 2002) and the Japanese Himawari (Bessho et al., 2016), with additional geostationary satellites provided by the China Meteorological Administration (CMA) Feng-Yun (FY)-series, the Indian National Satellite System (INSAT)-3 series (Patel et al., 2018) and the Chinese FY-4 series. (Jiang and Liu, 2014).

6.2.2 Sensors and retrieval basis

The first meteorological imaging sensors captured relatively simple VIS and/or IR pictures of clouds allowing the delineation of clouds, and thus precipitation-bearing systems to be identified. Despite the indirect link of VIS/IR observations to precipitation (they only provide information of the cloud top properties) VIS/IR instruments have become the mainstay of satellite imagery for meteorology due to the frequency

Table 6.2 Characteristics of GPM international partner PMW radiometers.

Sensor	*SSMIS*	*AMSR2*	*TMI*	*GMI*	**MHS**	**SAPHIR**	**ATMS**
Satellite	*DMSP-F16, F17, F18, F19*	*GCOMW1*	*TRMM*	*GPM*	**NOAA18,19, MetOp-A, B**	**Megha-Tropiques**	**NPP**
Type	*Conical*	*Conical*	*Conical*	*Conical*	**Cross-track**	**Cross-track**	**Cross-track**
frequencies	*-* *-* *19.35VH* *22.235V* *37.0VH* *50.3-63.3VH* *91.65VH* *150H 183.31H -*	*6.925/7.3VH* *10.65VH* *18.70VH* *23.80VH* *36.5VH* *-* *89.0VH* *- - -*	*10.65VH* *18.70VH* *23.80VH* *36.5VH -* *89.0VH -* *- -*	*- 10.65VH* *18.70VH* *23.80V* *36.5VH* *-* *89.0VH* *165.6VH* *183.31V(2)*	- - - - - - 89V 157V 183.31H (2) 190.31V	- - - - - - - 183.31H(6) -	- - - 23.8 31.4 50-3-57.3 87-91 164-167 183.31(5) -
Sampling resolution (km)	*12.79 XT 12.59 AT*	*4.65 XT 4.28 AT*	*4.74 XT 13. AT*	*5.13 XT 13.19 AT*	16.87 XT 17.62 AT	6.73 XT 9.86 AT	16.06 XT 17.74 AT
Retrieval resolution (km)	*50 × 40 km*	*19 × 11 km*	*26 × 21 km*	*16 × 10 km*	15.88 × 15.88 km	10 × 10 km	16 × 16 km

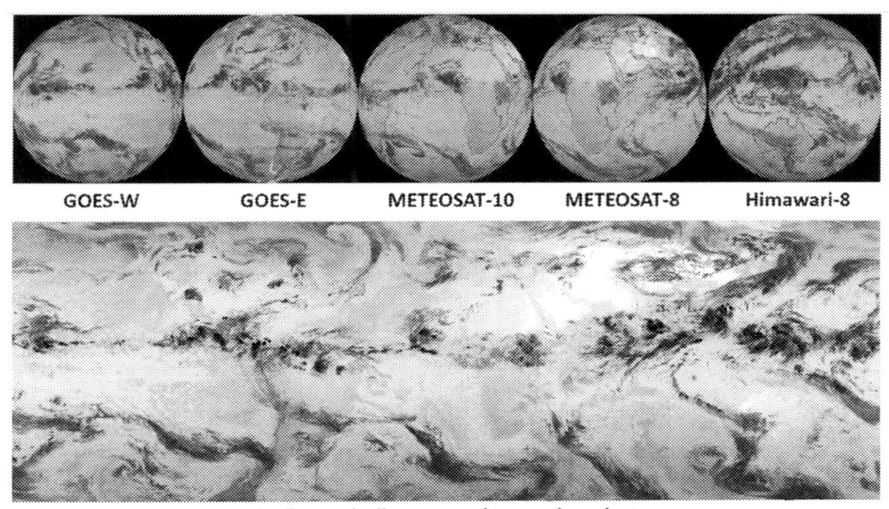

FIG. 6.2 Transformation of infrared imagery from the 5 operational geostationary satellite sensors to a quasi-global, 4 km product available every 30 min.

of observations (particularly from GEO) and the spatial resolution (from LEO) together with the low latency of the data to the users.

VIS techniques rely upon the reflectance of radiation from the cloud tops (therefore limited by daylight), and essentially record the presence/absence of clouds. Since IR observations measure the cloud top (or surface) temperature, techniques have exploited this noting that by setting a temperature threshold, areas of precipitation could be obtained reasonably well. By combining both the VIS/IR observations bi-spectral schemes were developed to identify the types of clouds and their association with precipitation (Behrangi et al., 2010). For example, clouds that had a high reflectance in the VIS ('bright') and low cloud top temperatures in the infrared ('cold') were more likely to be associated with heavy precipitation than dull, warm clouds. More recently, multi-spectral VIS/IR techniques have been developed exploiting data from the new generation of multi-spectral sensors, such as those on GEO satellites: these techniques provide information on cloud top characteristics including not only temperature (and hence height and pressure), but also information on the cloud particle sizes and phase (liquid vs solid) (King et al., 2012; Lensky and Rosenfeld, 1997; Platnick et al., 2003; Rosenfeld and Lensky, 1998).

While VIS/IR techniques offer the advantage of having frequent and regular observations at good resolution, their main drawback is that the relationship between the cloud top properties and the rain falling at the surface is generally poor, particularly at instantaneous timescales. More direct observations of precipitation are possible by measuring the upwelling radiation from the Earth's surface using microwave radiometers. At microwave frequencies the precipitation particles, rather than cloud particles, are the main factor affecting the radiation received by the satellite sensor. However, such sensors are currently only deployed on LEO satellites since a large antenna is required to obtain a sufficiently good spatial resolution. One consequence of the LEO-only sensors is that they only provide intermittent samples and therefore multiple satellites are required to provide adequate temporal sampling of precipitation (Huffman et al., 2016).

The frequencies that PMW sensors use fall into two main categories: 'imaging' which observe within atmospheric window channels and; 'sounding' which are at, or close to, the atmospheric absorption bands (and are typically less sensitive to the surface background) (Aonashi and Ferraro, 2020). While instruments are often described by the category they fall under (i.e., 'imager' or 'sounder') these sensors may include either or both imaging and sounding frequencies. Sensors are further categorised by their scan geometry: conically-scanning instruments, typically imagers, are usually preferred for precipitation retrievals due to consistent Earth incidence angle (EIA), polarization and resolution. Cross-track instrument, typically sounders, scan perpendicular to the satellite's track and consequently have a variable EIA that affects the spatial resolution and polarization of the observations.

The basis for measuring precipitation at microwave frequencies is based upon observing the effects of the precipitation itself. Over water surfaces, such as the ocean, the surface emissivity is low resulting in a small amount of radiation being received by the radiometer. Over such 'radiometrically cold' surfaces rain droplets increase the amount of radiation received by the radiometer, thus allowing a measure of precipitation to be made, albeit a measure along the atmospheric column being observed. Land surfaces have a higher and typically more variable emissivity ('radiometrically warm'): consequently, the additional radiation from the rain droplets is not discernible. However, by using higher frequencies, precipitation can scatter the upwelling radiation, and thus results in a decrease in the received signal. By using multi-frequency observations, a range of precipitation characteristics can be deduced.

The most direct satellite-based observations of precipitation come from AMW sensors (or radars). Similar to surface-based radar systems, the instruments emit a microwave pulse of energy that is then backscattered by any object it encounters, including precipitation: the backscattered radiation can then be converted into a precipitation measurement. Although this backscatter-to-precipitation relationship maybe ambiguous, one big advantage of these sensors is that they provide a vertical profile of the precipitation. However, precipitation radars have only been flown on the TRMM mission (Kummerow et al., 1998, 2000) with its Precipitation Radar (PR; Kozu et al., 2001), and more recently on the GPM mission (Hou et al., 2014) with the Dual-frequency Precipitation Radar (DPR; Kojima et al., 2012). The Cloud Profiling Radar (CPR) on the CloudSat (Stephens et al., 2002) mission has also provided valuable information on light rainfall and snowfall due to the sensitivity of the instrument. However, the small number of these radars and their narrow (or nadir-only) swath width result in poor temporal sampling.

6.3 Precipitation retrievals from satellite observations

Since the launch of TIROS-1, the first meteorological satellite with imaging capabilities (Anderson, 2010), a range of algorithms, techniques and schemes have been devised to generated precipitation measurements from satellite observations. Simple relationships between the satellite observations and surface precipitation may be used to generate an estimate of the precipitation based upon sound physical properties (e.g., Kidd et al., 1998; Kidd and Levizzani, 2011; Tapiador et al., 2012). As computational power increased, relationships based upon radiative transfer modelling were developed (Tripoli and Cotton, 1980; Cotton and Tripoli, 1986; Mugnai et al., 1990), together with multi-spectral, multi-sensor techniques to blend together satellite observations from different sensors, as well as those from surface observations (e.g., Adler et al., 1994; Huffman et al., 1997, 2007). Crucially, the development of techniques is often driven by the need to address user requirements, considering spatial and temporal scales and sampling, as well as the latency of the product. However, there is often a gulf between the expectations and requirements of the user community and the technical and feasibility of generating such precipitation products. Ultimately many schemes have to seek a compromise between providing a reasonable estimate of the surface precipitation to meet user requirements with the set of satellite observations available.

6.3.1 VIS/IR techniques

The range of techniques utilizing VIS/IR observations is now well established. VIS/IR retrieval schemes benefit from the frequent and consistent temporal sampling of observations that are available from GEO-based sensor, together with spatial resolutions (ca. 4 km) that are generally commensurate with many user applications. While VIS imagery has the best resolution, potential products are limited since they are limited to daytime use only. Consequently, the retrieval schemes have focused on IR imagery. IR-based techniques rely upon the notion that clouds with low cloud-top temperatures are taller/deeper and will therefore precipitate more. However, the relationship between cloud top temperature and the precipitation near to the Earth's surface is somewhat indirect and affected by variations in regional regimes and of the life cycle of the clouds themselves. Multi-spectral infrared channels may be used to improve the retrieval of cloud top characteristics, such as particle size and phase (King et al., 2012), but the indirectness of cloud-top properties to surface precipitation persists.

With the exception of a few AVHRR-based schemes (Ebert, 1996), IR-only satellite precipitation retrieval techniques exploit GEO-based observations due to their high temporal and spatial sampling capabilities. The benchmark IR-only technique is the GOES Precipitation Index (GPI; Arkin and Meisner, 1987) which assigns 3 mm h^{-1} to any cloud top temperature below 235 K. The GPI was developed to work at monthly/2.5° resolution scales, although it works surprisingly well at a range of finer temporal/spatial scales. A number of schemes have used the GPI as a basis, generating adjusted GPI schemes to account for regional and temporal variations in the cloud-top temperature to precipitation relationships (e.g., Xu et al., 1999; Kidd et al., 2003). Other IR-based precipitation products have been developed to distinguish between the intense precipitation of convective cores and lighter stratiform precipitation regions (e.g., Adler and Negri, 1988; Scofield and Kuligowski, 2003). More recently retrieval schemes have exploited neural networks to better identify and retrieve precipitation. The PERSIANN-CCS technique (Hong et al., 2004) uses cloud texture information (e.g., spatial variability, minimum cloud top temperature, etc.) to better define the surface precipitation. However, the delineation of precipitation in IR products remains problematic since cold thin clouds may be confused with cold deep clouds leading to 'false' precipitation, while warm shallow clouds which are precipitating may be missed. Overall, IR-based products perform better in tropical regimes where the cloud-top temperature to surface precipitation relationship is more robust (Kidd et al., 2019).

6.3.2 Passive microwave schemes

Observations made by PMW sensors are sensitive to liquid water and ice in the atmosphere. Low frequency channels (up to about 40 GHz) are generally more directly sensitive to the liquid hydrometeors, while the higher-frequency channels (>40 GHz) are more sensitive to solid (ice) particles, although this is dependent upon the concentrations of liquid/ice and the particle sizes. Over the oceans the increased emission from the liquid water droplets at the lower frequencies leads to increasing brightness temperatures (Tb) which can be used to retrieve precipitation. However, increasing ice scattering is present at 37 GHz in high intensity events which can reduce the Tbs, thus leading to potentially ambiguous retrievals. The high emissivity of land surfaces reduces the utilisation of low frequency channels for the detection of precipitation and the high frequency channels must be used. Scattering by precipitation-sized ice particles results in a reduction in the brightness temperatures. Thresholding approaches of the Tb depression at the various PMW frequencies have been developed (e.g., Laviola and Levizzani, 2011) and have been shown to be simple to implement and suitable for operational environments (Laviola et al., 2013). However, it should be noted that not all clouds have ice present and that the ice in the upper levels of a cloud might be a good representation of precipitation at the surface. Ideally, techniques use a combination of PMW frequencies to retrieve precipitation over different surface types (Kummerow, 2020).

Many precipitation retrieval techniques using these basic radiometric properties have achieved a degree of success (Kidd et al., 1998). However, providing an unambiguous retrieval of precipitation over different surfaces is often difficult and compounded by non-unique radiometric signature to hydrometeor profile/surface rainfall relationships. Computing the hydrometeor-radiative transfer relationships provided not only a basic understanding of the basic relationships, but also allowed the investigation of using these relationships to develop retrieval techniques. In particular, not only can the surface precipitation be retrieved, but also the vertical profile of the hydrometeors through the atmosphere. An example of such a physically-based precipitation retrieval scheme, is the Goddard Profiling (GPROF) scheme (Kummerow et al., 2015), which is designed to provide estimates of surface precipitation and vertical profiles constrained by model information (2-m temperature and total precipitable water) and ancillary data sets (surface types). The GPROF scheme is designed to work across a range of observations from PMW sensors to ensure consistent precipitation estimates. Meanwhile the Precipitation Retrieval and Profiling Scheme (PRPS; Kidd, 2016, 2018) was developed for cross-track sounding instruments (such as the Microwave Humidity Sounder, MHS, and the Sondeur Atmosphérique du Profil d'Humidité Intertropicale par Radiométrie (SAPHIR) using the GPM DPR measurements as the calibrator.

EUMETSAT's Satellite Application Facility on Support to Operational Hydrology and Water Management (H-SAF; Mugnai et al., 2013) generates a suite of precipitation products. A Cloud Dynamics and Radiation Database (CDRD; Casella et al., 2013) was developed by using a regional/mesoscale model, within a cloud resolving model (CRM), to produce a large set of precipitation simulations and their associated meteorological/microphysical vertical profiles (Casella et al., 2015). Similarly, a large database of cloud-resolving model simulations over Europe and Africa is used by the Passive microwave Neural network Precipitation Retrieval (PNPR; Sanò et al., 2016). Based on a single neural network, it is designed to retrieve surface precipitation from cross-track sounding instruments.

Further development of the radiative transfer models also allowed the generation of 1-D variational approaches to compute not only precipitation (Boukabara et al., 2011), but also a range of other geophysical parameters (e.g., Schröder et al., 2018). The Microwave Integrated Retrieval System (MiRS; Boukabara et al., 2011; Grassotti et al., 2020) is an iterative physical inversion system to extract profiles of temperature, moisture, liquid cloud, and hydrometeors, together with the surface emissivity and skin temperature. Such schemes perform well over the ocean where the surface background is relatively homogeneous. The Unified Microwave Ocean Retrieval Algorithm (UMORA; Hilburn and Wentz, 2008) retrieves sea surface temperature, surface wind speed, columnar water vapour, columnar cloud water, and surface rain rate, while the Hamburg Ocean Atmosphere Parameters and Fluxes from Satellite Data (HOAPS-3; Andersson et al., 2010) provides turbulent heat fluxes, evaporation, precipitation, freshwater flux and related atmospheric variables over the global ice-free ocean. As with all schemes, careful inter-sensor calibration is required to ensure a homogeneous time series.

Despite the advantage of the basis of the PMW techniques being more direct, the resolutions of PMW sensors are much poorer than those of IR sensors. The most recent PMW instruments have resolutions of about 5 km at best at the high-frequency channels (ca. 89 GHz), while at lower frequency channels are typically about 20 km or coarser. Furthermore, PMW instruments are only available on low-Earth-orbiting satellites and therefore a maximum of only two observations per day per satellite are possible: a number of satellites and sensors are needed to adequately resolve the variability of precipitating systems (Huffman et al., 2016).

6.3.3 Active microwave (radar) retrievals

The launch of the TRMM mission in 1997 heralded a new era in the measurement of precipitation from space through the inclusion of the TRMM Precipitation Radar (PR), to be followed by the GPM Dual-frequency Precipitation Radar (DPR; Iguchi, 2020) from 2014 onwards, providing vertical profiles of hydrometeors down to about 1 km above the Earth's surface. While the PR consisted of a single Ku-band radar, the DPR sensor has both of Ku- and Ka-band radars providing cross-track swaths of 245 km and 125 km (also 245 km after 21/05/2018; Iguchi, 2020), respectively, with a spatial resolution of ca. 5.2 km, and minimum detectable precipitation intensity of about 0.5 mm h^{-1}. The processing of the radar backscatter from the hydrometeors is constrained by several factors, such as stratiform/convective fraction, presence/absence of a bright-band, etc, as well as accounting for path attenuation through the use of the surface return (Iguchi et al., 2015). However, the retrieval capabilities from such radar systems depends on the Ku-Ka-band sensitivities to precipitation (Toyoshima et al., 2015).

The AMW sensors provide the most direct measure of precipitation. However, the utility of their products is limited by long revisit times (typically days), and north-south extent of the data (37°N-37°S for the PR and 68°N-68°S for the DPR). Despite this, such radar data is invaluable for the generation of databases used by PMW retrieval schemes (such as GPROF and PRPS) and for their calibration/validation (You et al., 2020). In addition, the capability of these sensors to generate detailed three-dimensional information allows a greater insight and understanding into precipitation systems to be made. Furthermore, the longevity of the PR, and now the DPR, allows climatological distributions of precipitation rates to be developed (Liu and Zipser, 2015).

6.3.4 Combined satellite retrievals

No single satellite precipitation product is capable of completely capturing the true surface precipitation, not least since different users have different requirements. Combining observations is seen as a way of harvesting information on precipitation that can be combined into a single, improved product, such as through the use of multiple observations to provide a better understanding between the observations and the surface precipitation (e.g., cloud delineation or constraining retrievals). A good example is the use of active (PR) and passive (TMI) observations gathered by TRMM to reduce the retrieval ambiguities by minimising the differences in the physical modelling of precipitation (Haddad et al., 1997). A similar scheme forms the basis of the GPM COmbined Radar-Radiometer algorithm (CORRA) that uses observations from both the DPR and GMI (Grecu et al., 2016; Grecu and Olson, 2020).

A main thrust of combining the information from various satellite observations has been to improve the temporal and spatial sampling of precipitation products, while ensuring that it is consistent and available frequently with low latency (observation to user). The PMW precipitation products, although more direct than the IR products, tend to be relatively infrequent and intermittent, while the IR products are less direct but more frequent and regular. Better precipitation products should then be possible through exploiting the strengths of each by combining both the IR and PMW measurements.

Several empirical schemes have been developed to exploit the directness of the PMW retrievals with the better temporal and spatial sampling of IR observations. Although simple IR techniques, such as the GPI product, are surprisingly good, particularly at relatively coarse temporal/spatial scales, many techniques have biases that vary by season and region. While it is possible to adjust the parameters within the GPI (such as the threshold and the rain rate), the key is to vary them spatially and through time. Examples of techniques that modify the IR thresholds and rain rates include those of Adler et al. (1993) and Xu et al. (1999), the latter forming the basis of the current PERSIANN suite of schemes (Hsu et al., 2020). Kidd et al. (2003) developed a combined PMW-IR algorithm that used regional/spatial histograms of PMW derived precipitation intensities to set the IR cloud top temperature thresholds for different rain intensities. The use of neural networks was explored by Tapiador et al. (2004a,b) to extract precipitation information from combined IR and PMW observations. Similarly, the Self-Calibrating Multivariate Precipitation Retrieval (SCaMPR) technique (Kuligowski, 2002, 2010) updates their IR-to-rainfall relationships based upon PMW observations.

The calibration of the IR observations by the PMW retrievals has shown a good degree of promise. However, such an approach has two main drawbacks. First, the spatial and temporal scale over which the calibration is undertaken is crucial for capturing the variations in the relationship yet providing sufficient information for a robust calibration (Kidd and Muller, 2010). Second, the distribution of the precipitation is determined by the cloud as observed by the IR and therefore may still not relate to the distribution of the precipitation at the surface. Since precipitation estimates derived from PMW observations are demonstrably better than those from the IR, particularly for instantaneous retrievals (Ebert, 1996), it is logical to use these PMW estimates and augment them with information from the IR observations. Schemes were therefore devised which generate motion vectors from the frequent/regular IR data to advect the PMW precipitation estimates (Joyce et al., 2004). These 'merged' schemes allow precipitation estimates to be produced at the temporal sampling of the IR data, but with the PMW precipitation estimates.

There are currently three main operational merged products available providing surface precipitation information at scales of about 10 km/30 minutes or better. These are the JAXA Global Satellite Mapping of Precipitation (GSMaP; Aonashi et al., 2009; Kubota et al., 2007, 2020; Ushio et al., 2009), the NOAA CPC Morphing Technique (CMORPH; Joyce et al., 2004), and the NASA Integrated Multi-satellitE Retrievals for GPM (IMERG; Huffman et al., 2014, 2020). Although these schemes differ, they are based around three similar algorithm stages: (i) individual PMW measurements are generated/acquired to estimate the surface precipitation; (ii) motion vectors are derived from IR imagery (or from models), to move the precipitation between the PMW estimates, and; (iii) if necessary, IR-derived estimates fill in gaps if no PMW estimate is available. Subtle differences arise between these schemes as a consequence of options taken at each of the three stages. For example, IR temperature change is used by GSMaP versus the IR-generated wind vectors of CMORPH; the morphing technique is used in different versions, with initial versions of CMORPH using just forward morphing, while later version of CMORPH and IMERG use both forward and backward morphing, with extended gaps between PMW estimates also being treated differently. Despite such schemes having the potential to provide a continuous record of precipitation over time, the complexities in the generation of such products should not be underestimated, primarily due to the reliance upon the original PMW retrievals, the accuracy of which varies between the sources of the data provided by the individual instruments. In addition, the mechanisms used to move/morph the precipitation between individual PMW retrievals and how one PMW retrieval is mapped (in time, space, and magnitude) onto the next PMW retrieval can lead to significant variations. Furthermore, these schemes do not necessarily cope well with situations such as orographic precipitation or short-lived precipitation events.

To meet the requirements of the user community many satellite precipitation products incorporate conventional surface data (e.g., gauges) to some extent: this of often necessary for final 'research-quality' to correct for any systematic bias in the satellite estimates. Indeed, the inclusion of surface data sets in long term precipitation records is essential to ensure consistency. Such an example is the TRMM Multisatellite Precipitation Analysis (TMPA). The TMPA combines precipitation estimates from multiple satellites, and gauge analyses (if available) to generate a precipitation product at 0.25° × 0.25° every 3 hours over 50°N-50°S. The near real time product is calibrated using TRMM precipitation products, while gauges are used for later versions. The TMPA is now superseded by the IMERG product that provides global precipitation products at 0.1° × 0.1° every 30 minutes. Although many techniques, such as TMPA and IMERG, combine multi-sourced datasets for global precipitation products, a number of studies have concentrated upon combining regional multi-sourced observations to generate improved precipitation products. For example, Tesfagiorgis and Mahani (2014) developed an approach to blend satellite, radar, and gauge products using a weighted successive correction method and Bayesian spatial model to fill gaps in the available ground-based radar precipitation data over the Colorado basin. More recently Jurczyk et al. (2020) describes the development of a regional precipitation estimate over Poland that combines gauge data, radar data and satellite data to generate 1-km, 10-minute products.

Some newer techniques have started to incorporate model information into their combined precipitation products, thus generating a satellite-gauge-model product. For example, the Multi-Source Weighted-Ensemble Precipitation (MSWEP; Beck et al., 2017a,b) scheme generates a combined 3-hourly product at 0.1° resolution using the combined products themselves. At the regional scale, Zhu et al. (2019) used a Bayesian model averaging to combine the TMPA and PERSIANN-CDR satellite-based products with the National Centers for Environment Prediction-Climate Forecast System Reanalysis (NCEP-CFSR) data to improve their streamflow simulations.

However, as noted above, all such combination schemes are only as good as their components: Awange et al. (2019) found, in a study over Australia and Africa, that the MSWEP product underestimated monsoon rainfall in northern Australia and had large uncertainties in eastern and southern Africa, missing some of the precipitation extremes.

6.3.5 Limitations to retrievals

The derivation of 'precipitation' from satellite observations encompasses rainfall, snowfall and hail, however, most retrieval schemes have concentrated upon rainfall since it is arguably the easiest of the three to measure (both from satellite and surface) to any degree of accuracy. As retrieval techniques have developed and improved in accuracy attention has shifted towards research into estimating snowfall and hail, not least due to the availability of data from the GPM mission aimed at improving mid- to high-latitude precipitation estimation. The measurement of snowfall, in particular, is crucial for applications such as weather monitoring and forecasting as well as hydrological applications to high-latitude/high-altitude basins: on the global scale it is deemed essential for closing the terrestrial water budget (Levizzani et al., 2011).

The challenges facing the retrieval of snowfall from satellite sensors relate to the signature of snowfall in the satellite observations. Over the Polar regions, clouds are often warmer than the surface background making the IR cold cloud-rainfall relationships obsolete. In the PMW the snowfall signature is often weak over ocean, while the signal over land resulting from the scattering by ice particles is often obscured by the emission of water vapour above particularly with low level/shallow precipitation (Liu and Seo, 2013). For AMW measurements the backscatter from snow particles is very small and often below the sensitivity of the radar. Furthermore, the rainfall-reflectivity relationships cannot be applied to snowfall, and if the precipitation layer is shallow (<1 km) it will not be detected by the satellite-based radars.

Consequently, much research is currently being conducted into improving the estimates and measurements of falling snow. The identification of snowfall is largely restricted to the higher PMW frequencies (89 GHz and above) where scattering processes associated with precipitation dominate (You et al., 2017): in this region it is important to distinguish between the ice hydrometeors and water droplets. Furthermore, identification of snowfall is hindered by an incomplete understanding of snow microphysics in mixed-phase clouds, as well as the complexity of the radiative properties of snowflakes and ice crystals. High-frequency PMW radiometers, such as the Advanced Microwave Sounding Unit-B (AMSU-B) and the MHS has enabled snowfall to be detectable, although a priori constraints are often needed. Nevertheless, physical models have been developed to identify snowfall over land (Skofronick-Jackson et al., 2004). To overcome the negation of the scattering effect by upper level water vapour, Liu and Seo (2013) applied a statistical approach to improve retrievals, at least for certain snowfall situations. More recently, Ebtehaj and Kummerow (2017) showed that a combination of low (10-19 GHz) and high frequency (89-166 GHz) observations allows snowfall to be detected from GMI observations. Meng et al. (2017)

retrieved snowfall using a 1D-Var approach, allowing multiple parameters to be assessed in the calculations, to optimise the final retrieval. Nevertheless, the combined availability of observations from the CPR on CloudSat and the GPM DPR has allowed significant advances both in terms of snowfall detection and retrieval (Casella et al., 2017; Panegrossi et al., 2017), but also through improved model simulations of snowflake shapes (Liu, 2008; Skofronick-Jackson et al., 2013). The CPR has also generated a global census of shallow cumuliform snow clouds (Kulie et al., 2016; Kulie and Milani, 2018), unobtainable from other satellite observations. A quantification of the causes of the differences among snow retrievals using DPR and CPR was recently provided by Skofronick-Jackson et al. (2019).

Although perhaps the least common form of precipitation, falling hail is of great importance not only for meteorology, but for agri-business where losses associated with hail damage can be significant. Studies by Cecil (2009, 2011) and Cecil and Blankenship (2012) have shown a strong relationship between the occurrence of hail and depressions in the TMI 37 GHz and 85 GHz Tbs, and showed that the satellite climatologies were consistent with those of surface observations. An AMSU-based hail detection algorithm developed by Ferraro et al. (2015) generated a 12-year climatology (2000-2011) of hail occurrence over the Continental US. The same climatology was used by Laviola et al. (2020) to design a probabilistic growth model of hailstones with the aim of a global application using the Tbs of the PMW sensors in orbit. Marra et al. (2017) demonstrated the capabilities of the GPM Core Observatory (CO) for intense hailstorm detection.

Precipitation retrievals from satellite over complex orography represent one more relevant source of error since retrieval schemes are not capable of capturing the orographic enhancement of precipitation through convection developing over mountain slopes. Several attempts to overcome this key problem have been recently made and most of them rely on models to describe orographic enhancement. The GSMaP product adopts some kind of real orographic correction (Yamamoto and Shige, 2015; Yamamoto et al., 2017).

In arid regions precipitation estimation schemes overestimate precipitation at the ground due to sub-cloud evaporation (Dinku et al., 2011) and this translates in a notable limitation in desert areas (Dinku et al., 2018).

6.4 Validation of satellite precipitation products

An essential part of the development of satellite precipitation retrieval schemes is the validation (or at least verification) of the products with surface reference data. This is both of importance to the algorithm developer to help ensure that the retrieval scheme is operating as expected, and to the product user as a measure of the accuracy of the precipitation product being developed (Kidd et al., 2018). Key to any validation exercise is the availability and robustness of the reference data to be used, generally derived from high quality gauge and radar networks. Ideally the validation is done over a range of scale, from the instantaneous through to monthly, and from footprint-scale through to 2.5° scale: this range of scales ensures the products are not scale dependent and therefore are useful to a range of applications.

Gauge data are the primary source of validation data. However, their time-integrated point measurements are not necessarily directly compatible with the satellite products, and consequently gauges tend to be used at relatively coarse temporal and spatial scales (e.g., daily 1° × 1°) to ensure sufficient gauges are available within the area of interest. In addition, the distribution of gauges across the Earth's surface is poor (Kidd et al., 2017), resulting in most of the gauge-based validation studies being confined to specific regions, and hence specific climatic regimes. The instantaneous, spatial measurements of surface-based radars provide data at scales that are commensurate with the satellite products and therefore are ideal for finer scale validation (i.e., footprint/instantaneous scales), particularly when high density/resolution gauge data is also available (Kirstetter et al., 2015).

A number of concerted precipitation intercomparison and validation projects have been carried out, albeit many years ago. These include the NASA Precipitation Intercomparison Projects (Adler et al., 2001) designed to evaluate the (then) new SSM/I PMW retrieval schemes. At roughly the same time GEWEX organised the Algorithm Intercomparison Program that compared a range of retrieval products, including those from VIS/IR observations, PMW and combined schemes (Ebert et al., 1996). The outcome of these comparisons showed that at the instantaneous scales the PMW schemes were certainly best, although at the monthly scale, the same PMW schemes performed no better than the IR schemes as a result of their intermittent sampling, and certainly less well than the merged precipitation products. However, it was also noted that it was difficult to show that any single precipitation estimate to be best (Adler et al., 2001). Moreover, it was noted that the performance of the products varied over the different regions and seasons.

More recently the International Precipitation Working Group (IPWG; Levizzani et al., 2018) has led a program on continuous assessment of precipitation over a number of validation regions (Ebert et al., 2007; Kidd et al., 2012), with updates provided by Kidd et al. (2020). These intercomparisons, aimed at providing developers and users of their products with an assessment of their performance, are carried out in near real time with daily, 0.25° × 0.25° resolution, using surface gauge and/or radar data. It is acknowledged that validation is required at a range of different scales: as a complement to the IPWG validation work, the GPM mission had undertaken Ground Validation (GV) work, providing a multi-tier approach (Houze et al., 2017; Kirstetter et al., 2020; Petersen et al., 2020). These dedicated GV campaigns are designed to collect detailed information from multiple sensors carried on multiple platforms, including satellites, aircraft and at the surface (Hou et al., 2014; Skofronick-Jackson et al., 2015). This strategy gathers a comprehensive set of observations of precipitation enabling a better understanding of not only how the precipitation is observed by the satellite sensors, but also the detailed characteristics of the precipitation itself (Jensen et al., 2016; Skofronick-Jackson et al., 2015; Houze et al., 2015, 2017). A summary of the GPM GV field campaigns can be found in Table 26.3 in Petersen et al. (2020).

The quality of surface-based radar data has improved through the deployment of new equipment and through better processing of the radar data. In addition, the availability of and access to such radar data enables near real time validation to be carried out (see Fig. 6.3). Long-term validation of instantaneous precipitation products from the GPM constellation was carried out by Kidd et al. (2018) using high quality radar and gauge data over Europe and the southeast US. Overall, there was good agreement between the satellite and surface products, typically with correlations of about 0.5 over both regions. The DPR-Ku radar product had the highest overall correlation and skill score, followed by the GMI product and then other sensors. However, time series of the statistical scores indicated that the results over Europe showed much more seasonal variability than those over the southeast US, suggesting that the meteorological regimes in the mid-latitudes still pose challenges. Large scale (global/monthly) validation analysis by Beck et al. (2017b) was carried out on several global precipitation products, including satellite-gauge and model reanalyses, finding they had varying degrees of correspondence to the surface gauge data over different regions.

One of the obstacles currently facing validation exercises is that the satellite precipitation products often contain the same gauge data that is used in the validation, consequently such products appear to work well. However, since the products and validation use similar gauges it becomes more of a comparison of how the gauge data was integrated into the product, rather than a validation of the product itself. Of course, any agreement between the precipitation product and gauge data is only valid for regions where the gauges exist, and is no measure of their performance in regions beyond the gauge coverage, particularly so over the oceans where little or no gauge data exists.

Finally, it is worth reiterating the conclusions of previous intercomparison/validation exercises, in that no single precipitation product stands out as being superior to any other: each product has its own subtleties and nuances, whether being better over some regions certain time scales or at particular resolutions than others.

6.5 Case study: hurricane Laura observed by the GPM Core Observatory

One of the main goals of TRMM was the imaging and analysis of storms across the Tropics, a role that has been inherited by the GPM mission. The GPM CO (Hou et al., 2014; Skofronick-Jackson et al., 2017; Kidd et al., 2020) carries both AMW and PMW sensors, the DPR and GMI respectively. The combination of the two sensors allows accurate observations (Draper et al., 2015; Wentz and Draper, 2016) and consistent retrievals to be made (Wilheit et al., 2015; Berg et al., 2016). The GMI (PMW) sensor provides observation across a swath of 885 km in 13 channels from 10 to 183 GHz, allowing the retrieval of surface precipitation. The DPR, although having a narrower swath of 245 km, provides information on the horizontal and vertical distribution of the precipitation, thus generating a 3D field of precipitation observations.

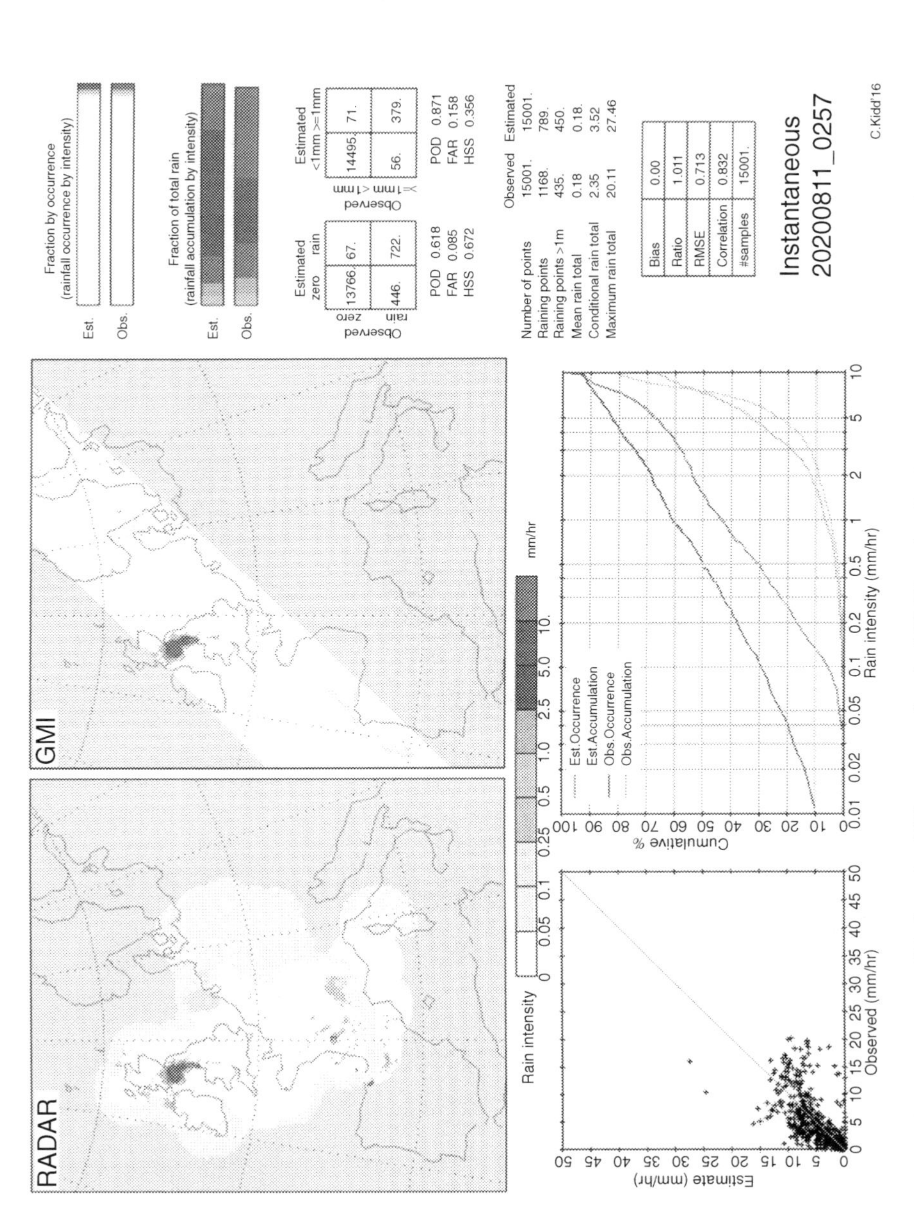

FIG. 6.3 Example of instantaneous validation of a GMI overpass for the GPROF retrieval scheme over western Europe using the Nimrod European surface radar network as ground validation. 11 August 2020 0257 UTC.

Since their launches, TRMM and GPM have observed hundreds of Tropical Cycles across the Tropics, one of the most recent ones being Hurricane Laura.

Tropical Cyclone Laura tracked across the Caribbean striking Haiti and Cuba before gathering strength over the Gulf of Mexico on 26th August 2020, where it rapidly intensified reaching hurricane strength with wind speeds of up to 241 km h^{-1}. On 27th August 2020 it made landfall along the Louisiana coast of the US, causing at least 59 deaths along its path and many billions of dollars in damages.

The GPM satellite operates in a low-inclination, low Earth orbit. While this orbit allows the satellite sensors to provide observations with relatively good resolution, this type of orbit does not permit the continual observation of any part of the Earth's surface, only intermittent observations when the satellite passes overhead. Therefore, capturing individual storms cannot be guaranteed. However, fortunately the orbit of the GPM allowed Hurricane Laura to be observed at 1311 UTC on 27th August 2020, shortly after it made landfall in Louisiana: the visualisation of this overpass is shown in Fig. 6.4. The surface (2D) precipitation is obtained using observations made by the GMI and processing through the GPROF retrieval scheme, while the 3D precipitation field is generated by the DPR observations. The intense precipitation at the centre of the hurricane can be seen (red and purple colours) together with the precipitation associated with the convective cores, as shown by the vertical profiles. The blue colours at the top of the vertical profiles show the frozen precipitation at the top of the hurricane due to these being about the freezing level. The full visualisation can be accessed at https://svs.gsfc.nasa.gov/4855.

6.6 Recent developments and future directions

An overview of the most recent developments and a glimpse of the future of satellite precipitation observations can be found in Kummerow et al. (2020). While it is very difficult to speculate on the future of satellite missions, recent developments and future challenges are based on the needs to better understand and predict precipitation changes in weather hydrology, and the climate system. From the weather perspective it is clear that the improvement of future precipitation missions with regard to observing cycle, resolution, timeliness and uncertainty level are deemed necessary and translate into improvements on high quality individual products as well as merged products. Hydrological needs vary from monitoring and predicting water availability in water-scarce areas (especially developing countries) to ensuring an efficient use of water where it is often used without adequate plans (most of the industrialized countries). Thus, accurate observations are the key element of hydrological predictions (e.g., Lettenmaier 2017). Finally, the climate system understanding poses grand challenges (https://www.wcrp-climate.org/grand-challenges/grand-challenges-overview, last accessed 21 Sep. 2020), including those that are directly related to observations of precipitation: "understanding and predicting the extreme weather", "clouds, circulation and climate sensitivity" and "water for the food baskets of the world".

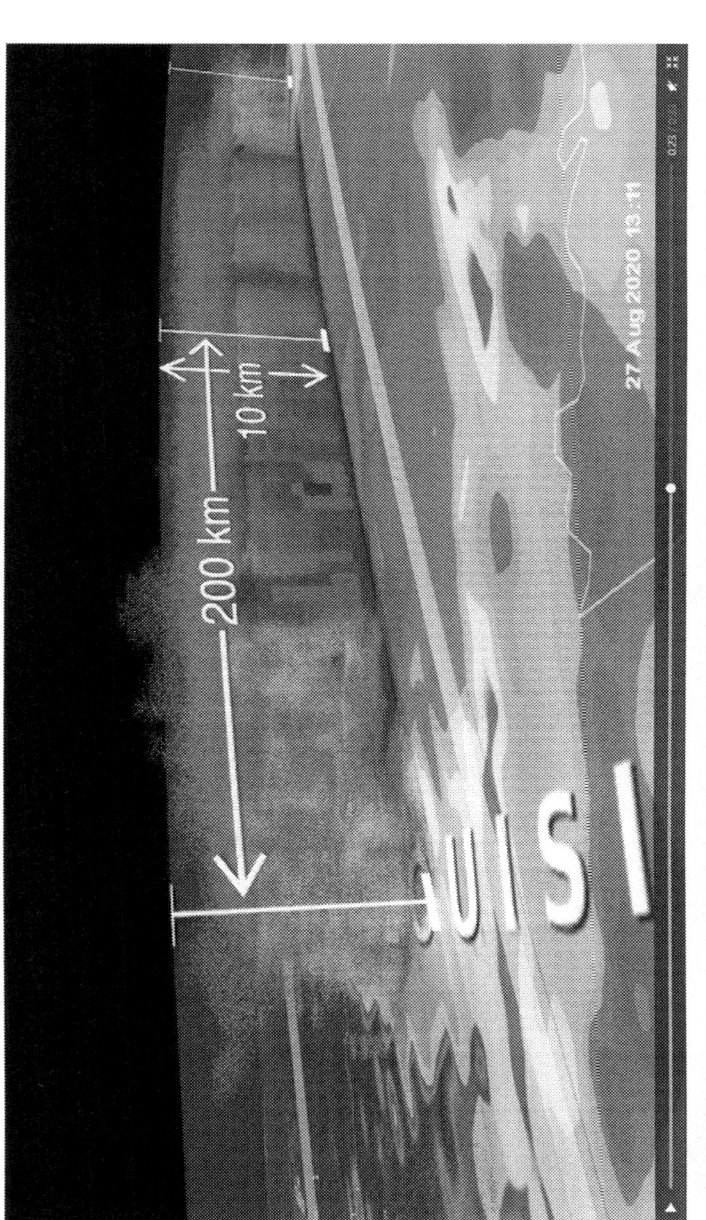

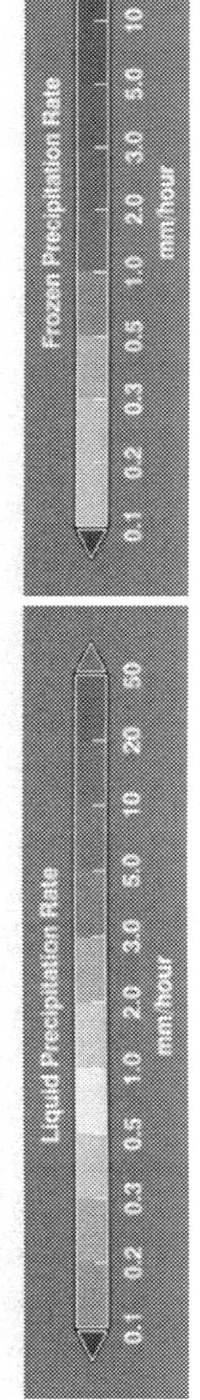

FIG. 6.4 Precipitation associated with Hurricane Laura shortly after landfall as measured by the GPM Core Observatory at 1311 UTC on 27 August 2020. The surface precipitation (foreground) is derived from the GMI while the 3D precipitation measurements are provided by the DPR sensor. The blue shading represents the ice above the freezing level.

(Still image from https://svs.gsfc.nasa.gov/4855)

Scientific/societal demands are broken down into specific science questions, such as: How does climate change affect precipitation, i.e. "is the total rainfall amount increasing?" or "is the extreme rainfall event increasing?" Another question is: "how climate change affects the global water cycle?" A deeper understanding of each element of the atmospheric water budget (e.g., water vapor, clouds and precipitation), hydrological processes (e.g., runoff, floods and droughts, and biogeoscience aspects such as vegetation and forest), must be combined into a unified view. This is the most important challenge that requires thinking the present and future missions having in mind not only "conventional" sensors and satellites, but also new observing systems that gather information on variables related with precipitation and neglected so far (e.g., evapotranspiration, river discharge, water storage, snow water equivalent).

On the technological side it is foreseen that precipitation radars will play a greater role. They will likely have any combination of higher sensitivity, wider swath, higher spatial resolution and reduced surface clutter extent than the current systems to meet the evolving requirements described above. The need for identifying more particle types and cloud processes also points to the desirability of simultaneous multi-frequency, polarization and Doppler velocity measurement (Kummerow et al., 2020). High sensitivity is essential for understanding the cloud-precipitation processes and observation of snowfall at high latitudes. Although the introduction of cloud radars is indispensable to detect drizzle and light precipitation, it is important to overlap the sensitivity ranges of radars to obtain more independent pieces of information, which are needed to better constrain largely underconstrained retrievals of cloud and precipitation properties. For very small drops, this may even require higher frequencies than are used today.

A number of radiometer advances have occurred or are occurring. Chief among these is the miniaturization of radiometer receivers that prompts for the launch of constellations of small satellites. Coupled with generally available CubeSats, this has led to a number of spaceborne PMW radiometers to be flown recently as well as manifested for the near future. Such small satellites can be launched and separated to form constellations that can increase sampling significantly at a reasonable cost. Yet, the requirement that the satellite remains small also points to small antenna system and thus limited spatial resolution. This key technological advance is likely to answer the need for maintaining PMW sensor constellations in orbit to serve the production of global precipitation data for the years to come.

6.7 Conclusion

Stephens and Kummerow (2007) articulated the need for a better definition of the atmospheric state and the vertical structure of clouds and precipitation to improve the information extracted from satellite observations. This has been the overarching reason for combining multiple frequency AMW and PMW measurements that offers some hope for constraining the atmospheric states needed to derive unique rain and snow rate signals. Radar improvements and the launch of constellations of high-quality radiometers for an adequate space-time sampling are likely to provide the observations needed for a substantial step forward in the quality of the retrieval algorithms. Without this step the available products will continue to show physical limitations that hinder their use in a number of applications.

One key aspect that underlies all aspects of climate and climate change is the need for long-term precipitation records that allow to extract changes in precipitation due to climate change. Accurate long-term (more than 30 years) data are necessary to reduce the effect of long-term variation of the Earth (El Niño Southern Oscillation-ENSO, solar cycle, Arctic oscillation, etc.). Given the time series that are available today, a significant requirement is thus to maintain and extend this time series of high-quality precipitation observations.

References

Adler, R.F., Negri, A.J, 1988. A satellite infrared technique to estimate tropical convective and stratiform rainfall. J. Appl. Meteor. 27 (1), 30–51. https://doi.org/10.1175/1520-0450(1988)027<0030:ASITTE>2.0.CO;2.

Adler, R., Huffman, G.J., Keehn, P.R, 1994. Global rain estimates from microwave adjusted geosynchronous IR data. Remote Sens. Rev. 11, 125–152. https://doi.org/10.1080/02757259409532262.

Adler, R.F., Negri, A.J., Keehn, P.R., Hakkarinen, I.M., 1993. Estimation of monthly rainfall over Japan and surrounding waters from a combination of low-orbit microwave and geosynchronous IR data. J. Appl. Meteor. 32, 335–356. https://doi.org/10.1175/1520-0450(1993)032%3C0335:EOMROJ%3E2.0.CO;2.

Adler, R.F., Kidd, C., Petty, G., Morissey, M., Goodman, H.M., 2001. Intercomparison of global precipitation products: the third precipitation intercomparison project (PIP–3). Bull. Amer. Meteor. Soc. 82, 1377–1396. https://doi.org/10.1175/1520-0477(2001)082%3C1377:IOGPPT%3E2.3.CO;2.

Anderson, G.D, 2010. The first weather satellite picture. Weather 65 (4), 87. https://doi.org/10.1002/wea.550.

Andersson, A., Fennig, K., Klepp, C., Bakan, S., Grassl, H., Schulz, J, 2010. The hamburg ocean atmosphere parameters and fluxes from satellite data—HOAPS-3. Earth Syst. Sci. Data 2, 215–234. https://doi.org/10.5194/essd-2-215-2010.

Aonashi, K., Ferraro, R.R, 2020. Microwave sensors, imagers and sounders. In: Levizzani, V., Kidd, C., Kirschbaum, D.B., Kummerow, C.D., Nakamura, K., Turk, F.J. (Eds.), Satellite Precipitation Measurement. Springer Nature, Cham, pp. 63–81. https://doi.org/10.1007/978-3-030-24568-9_4.

Aonashi, K., Awaka, J., Hirose, M., Kozu, T., Kubota, T., Liu, G., Shige, S., Kida, S., Seto, S., Takahashi, N., Takayabu, Y.N, 2009. GSMaP passive microwave precipitation retrieval algorithm: algorithm description and validation. J. Meteor. Soc. Japan 87A, 119–136. https://doi.org/10.2151/jmsj.87A.119.

Arkin, P.A., Meisner, B.N, 1987. The relationship between large scale convective rainfall and cold cloud over the western hemisphere during 1982-84. Mon. Wea. Rev. 115, 51–74. https://doi.org/10.1175/1520-0493.

Asmus, V.V., Zagrebaev, V.A., Makridenko, A., Milekhin, O.E., Solov'ev, V.I., Uspenskii, A.B., Frolov, A.V., Khailov, M.N., 2014. Meteorological satellites based on meteor-M polar orbiting platform. Russ. Meteor. Hydrol. 39, 787–794. https://doi.org/10.3103/S1068373914120012.

Awange, J.L., Hu, K.X., Khaki, M, 2019. The newly merged satellite remotely sensed, gauge and reanalysis-based multi-source weighted-ensemble precipitation: evaluation over Australia and Africa (1981–2016). Sci. Total Environ. 670, 448–465. https://doi.org/10.1016/j.scitotenv.2019.03.148.

Beck, H., van Dijk, A., Levizzani, V., Schellekens, J., Miralles, D., Martens, A.D.R.B, 2017a. MSWEP: 3-hourly 0.25° global gridded precipitation (1979–2015) by merging gauge, satellite, and reanalysis data. Hydrol. Earth Syst. Sci. 21, 589–615. https://doi.org/10.5194/hess-21589-2017 a.

Beck, H., Vergopolan, N., Pan, M., Levizzani, V., van Dijk, A., Weedon, G., Brocca, L., Pappenberger, F., Huffman, G.J., Wood, E, 2017b. Global-scale evaluation of 22 precipitation datasets using gauge observations and hydrological modelling. Hydrol. Earth Syst. Sci. 21, 6201–6217. https://doi.org/10.5194/hess-21-6201-2017 b.

Becker, A., Finger, P., Meyer-Christoffer, A., Rudolf, B., Schamm, K., Schneider, U., Ziese, M, 2013. A description of the global land-surface precipitation data products of the Global Precipitation Climatology Centre with sample applications including centennial (trend) analysis from 1901–present. Earth Syst. Sci. Data 5, 71–99. https://doi.org/10.5194/essd-5-71-2013.

Behrangi A., Hsu K., Imam B., Sorooshian S. 2010. Daytime precipitation estimation using bispectral cloud classification system. J. Appl. Meteor. Climatol. 49,1015–1031. https://doi.org/10.1175/2009JAMC2291.1.

Berg, W., Bilanow, S., Chen, R.Y., Datta, S., Draper, D., Ebrahimi, H., Farrar, S., Linwood Jones, W., Kroodsma, R., McKague, D., Payne, V., Wang, J., Wilheit, T., Yang, J.X, 2016. Intercalibration of the GPM microwave radiometer constellation. J. Atmos. Oceanic Technol. 33, 2639–2654. https://doi.org/10.1175/JTECH-D-16-0100.1.

Bessho, K., Date, K., Hayashi, M., Ikeda, A., Imai, T., Inoue, H., Kumagai, Y., Miyakawa, T., Murata, H., Ohno, T., Okuyama, A., Oyama, R., Sasaki, Y., Shimazu, Y., Shimoji, K., Sumida, Y., Suzuki, M., Taniguchi, H., Tsuchiyama, H., Ueasawa, D., Yokota, H., Yoshida, R, 2016. An introduction to Himawari-8/9—Japan's new-generation geostationary meteorological satellites. J. Meteor. Soc. Japan 94, 151–183. https://doi.org/10.2151/jmsj.2016-009.

Borga, M., Marra, F., Gabella, M., 2022. Rainfall Estimation by Weather Radar. In: Morbidelli, R. (Ed.), Rainfall. Modeling, Measurement and Applications. Elsevier, Amsterdam, pp. 109–134. doi:10.1016/C2019-0-04937-0.

Boukabara, S.-A., Garrett, K., Chen, W., Iturbide-Sanchez, F., Grassotti, C., Kongoli, C., Chen, R., Liu, Q., Yan, B., Weng, F., Ferraro, R., Kleespies, T.J, 2011. MiRS: an all-weather 1DVAR satellite data assimilation and retrieval system. IEEE Trans. Geosci. Remote Sens. 49, 3249–3272. https://doi.org/10.1109/TGRS.2011.2158438.

Casella, D., Panegrossi, G., Sanò, P., Milani, L., Petracca, M., Dietrich, S, 2015. A novel algorithm for detection of precipitation in tropical regions using PMW radiometers. Atmos. Meas. Tech. 8, 1217–1232. https://doi.org/10.5194/amt-8-1217-2015.

Casella, D., Panegrossi, G., Sanò, P., Marra, A.C., Dietrich, S., Johnson, B.T., Kulie, M.S, 2017. Evaluation of the GPM-DPR snowfall detection capability: comparison with CloudSat-CPR. Atmos. Res. 197, 64–75. https://doi.org/10.1016/j.atmosres.2017.06.018.

Casella, D., Panegrossi, G., Sanò, P., Dietrich, S., Mugnai, A., Smith, E.A., Tripoli, G.J., Formenton, M., Di Paola, F., Leung, W.-Y.H., Mehta, A.V, 2013. Transitioning from CRD to CDRD in Bayesian retrieval of rainfall from satellite passive microwave measurements: Part 2. Overcoming database profile selection ambiguity by consideration of meteorological control on microphysics. IEEE Trans. Geosci. Remote Sens. 51, 4650–4671. https://doi.org/10.1109/TGRS.2013.2258161.

Cecil, D, 2009. Passive microwave brightness temperatures as proxies for hailstorms. J. Appl. Meteor. Climatol. 48, 1281–1286. https://doi.org/10.1175/2009JAMC2125.1.

Cecil, D, 2011. Relating passive 37-GHz scattering to radar profiles in strong convection. J. Appl. Meteor. Climatol. 50, 233–240. https://doi.org/10.1175/2010JAMC2506.1.

Cecil, D., Blankenship, C.B, 2012. Toward a global climatology of severe hailstorms as estimated by satellite passive microwave imagers. J. Climate 25, 687–703. https://doi.org/10.1175/JCLI-D-11-00130.1.

Chvíla, B., Sevruk, B., Ondrás, M, 2005. The wind-induced loss of thunderstorm precipitation measurements. Atmos. Res. 77 (1–4), 29–38. https://doi.org/10.1016/j.atmosres.2004.11.032.

Ciach, G.J, 2003. Local random errors in tipping-bucket rain gauge measurements. J. Atmos. Oceanic Technol. 20, 752–759. https://doi.org/10.1175/1520-0426.

Cotton, W.R., Tripoli, G.J, 1986. Numerical-simulation of the effects of varying ice crystal nucleation rates and aggregation processes on orographic snowfall. J. Clim. Appl. Meteor. 25 (11), 1658–1680. https://doi.org/10.1175/1520-0450.

Curtis, J.A., Adams, L.J, 1987. Defense meteorological satellite program. IEEE Aerospace Electronic Syst. Mag. 2, 1317. https://doi.org/10.1109/MAES.1987.5005348.

Dinku, T., Ceccato, P., Connor, S.J, 2011. Challenges of satellite rainfall estimation over mountainous and arid parts of east Africa. Int. J. Remote Sens. 32, 5965–5979. https://doi.org/10.1080/01431161.2010.499381.

Dinku, T., Funk, C., Peterson, P., Maidment, R., Tadesse, T., Gadain, H., Ceccato, P, 2018. Validation of the CHIRPS satellite rainfall estimates over eastern Africa. Quart. J. Roy. Meteor. Soc. 144 (S1), 293–312. https://doi.org/10.1002/qj.3244.

Dittus, A.J., Karoly, D.J., Lewis, S.C., Alexander, L.V, 2015. A multi-region assessment of observed changes in the areal extent of temperature and precipitation extremes. J. Climate 28 (23), 9206–9220. https://doi.org/10.1175/JCLI-D-14-00753.1.

Donat, M.G., Alexander, L.V., Yang, H., Durre, I., Vose, R., Dunn, R.J.H., Willett, K.M., Aguilar, E., Brunet, M., Caesar, J., Hewitson, B., Jack, C., Klein-Tank, A.M.G., Kruger, A.C., Marengo, J., Peterson, T.C., Renom, M., Oria Rojas, C., Rusticucci, M., Salinger, J., Elrayah, A.S., Sekele, S.S., Srivastava, A.K., Trewin, B., Villarroel, C., Vincent, L.A., Zhai, P., Zhang, X., Kitching, S, 2013. Updated analyses of temperature and precipitation extreme indices since the beginning of the twentieth century: The HadEX2 dataset. J. Geophys. Res.-Atmos. 118, 2098–2118. https://doi.org/10.1002/jgrd.50150.

Dorninger, M., Schneider, S., Steinacker, R, 2008. On the interpolation of precipitation data over complex terrain. Meteor. Atmos. Phys. 101, 175–189. https://doi.org/10.1007/s00703-008-0287-6.

Draper, D.W., Newell, D.A., Wentz, F.J., Krimchansky, S., Skofronick-Jackson, G, 2015. The Global Precipitation Measurement (GPM) Microwave Imager (GMI): instrument overview and early on-orbit performance. IEEE J. Sel. Topics Geosci. Remote Sens. 8, 3452–3462. https://doi.org/10.1109/JSTARS.2015.2403303.

Duchon, C.E., Biddle, C.J, 2010. Undercatch of tipping-bucket gauges in high rain rate events. Adv. Geosci. 25, 11–15. https://doi.org/10.5194/adgeo-25-11-2010.

Ebert, E.E, 1996. Results of the 3rd algorithm intercomparison project (AIP-3) of the Global Precipitation Climatology Project (GPCP). *BMRC Research Report* No. 55. Bureau of Meteorology, Melbourne, Australia.

Ebert, E.E., Janowiak, J.E., Kidd, C, 2007. Comparison of near real time precipitation estimates from satellite observations and numerical models. Bull. Amer. Meteor. Soc. 88, 47–64. https://doi.org/10.1175/BAMS-88-1-47.

Ebert, E.E., Manton, M.J., Arkin, P.A., Allam, R.J., Holpin, G.E., Gruber, A, 1996. Results from the GPCP Algorithm Intercomparison Programme. Bull. Amer. Meteor. Soc. 77, 2875–2888. https://doi.org/10.1175/1520-0477.

Ebtehaj, A.M., Kummerow, C.D, 2017. Microwave retrievals of terrestrial precipitation over snow-covered surfaces: a lesson from the GPM satellite. Geophys. Res. Lett. 44, 6154–6162. https://doi.org/10.1002/2017GL073451.

Fairman Jr, J.G., Schultz, D.M., Kirshbaum, D.J., Gray, S.L., Barrett, A.I, 2015. A radar-based rainfall climatology of Great Britain and Ireland. Weather 70, 153–158. https://doi.org/10.1002/wea.2486.

Ferraro, R.R., Beauchamp, J., Cecil, D., Heymsfield, G., 2015. A prototype hail detection algorithm and hail climatology developed with the advanced microwave sounding unit (AMSU). Atmos. Res. 163, 24–35. https://doi.org/10.1016/j.atmosres.2014.08.010.

Friedl, L, 2014. GEO Task US-09-01a: Critical Earth observation priorities; Precipitation data characteristics and user types, 60. https://sbageotask.larc.nasa.gov/Weather_US0901a-FINAL.pdf. (Accessed 13 May 2021).

Goldberg, M.D., Kilcoyne, H., Cikanek, H., Mehta, A, 2013. Joint polar satellite system: the United States next generation civilian polar-orbiting environmental satellite system. J. Geophys. Res.-Atmos. 118, 13463–13475. https://doi.org/10.1002/2013JD020389.

Goodman, S.J., Schmit, T.J., Daniels, J., Redmon, R.J. (Eds.), 2020. The GOES-R Series: A New Generation of Geostationary Environmental Satellites. Elsevier, Amsterdam, p. 283.

Grassotti, C., Liu, S., Liu, Q., Boukabara, S.-A., Garrett, K., Iturbide-Sanchez, F., Honeyager, R, 2020. Precipitation estimation from the Microwave Integrated Retrieval System (MiRS). In: Levizzani, V., Kidd, C., Kirschbaum, D.B., Kummerow, C.D., Nakamura, K., Turk, F.J. (Eds.), Satellite Precipitation Measurement. Springer Nature, Cham, pp. 153–168. https://doi.org/10.1007/978-3-030-24568-9_9.

Grecu, M., Olson, W.S, 2020. Precipitation retrievals from satellite combined radar and radiometer observations. In: Levizzani, V., Kidd, C., Kirschbaum, D.B., Kummerow, C.D., Nakamura, K., Turk, F.J. (Eds.), Satellite Precipitation Measurement. Springer Nature, Cham, pp. 231–248. https://doi.org/10.1007/978-3-030-24568-9_14.

Grecu, M., Olson, W.S., Munchak, S.J., Ringerud, S., Liao, L., Haddad, Z.S., Kelley, B.L., McLaughlin, S.F, 2016. The GPM combined algorithm. J. Atmos. Oceanic Technol. 33, 2225–2245. https://doi.org/10.1175/JTECH-D-16-0019.1.

Haddad, Z.S., Smith, E.A., Kummerow, C.D., Iguchi, T., Farrar, M.R., Durden, S.L., Alves, M., Olson, W.S, 1997. The TRMM "Day-1" radar/radiometer combined rain-profiling algorithm. J. Meteor. Soc. Japan 75, 799–809. https://doi.org/10.2151/jmsj1965.75.4_799.

Heiss, W., McGrew, D., Sirmans, D, 1990. NEXRAD: next generation weather radar (WSR-88D). Microwave J 33 (1), 79–98.

Herold, N., Alexander, L.V., Donat, M.G., Contractor, S., Becker, A, 2016. How much does it rain over land? Geophys. Res. Lett. 43, 341–348. https://doi.org/10.1002/2015GL066615.

Higgins, R.W., Kousky, V.E, 2013. Changes in observed daily precipitation over the United States between 1950–79 and 1980–2009. J. Hydrometeor. 14, 105–121. https://doi.org/10.1175/JHM-D-12-062.1.

Hilburn, K.A., Wentz, F.J, 2008. Intercalibrated passive microwave rain products from the Unified Microwave Ocean Retrieval Algorithm (UMORA). J. Appl. Meteor. Climatol. 47, 778–794. https://doi.org/10.1175/2007JAMC1635.1.

Hong, Y., Hsu, K.L., Sorooshian, S., Gao, X, 2004. Precipitation estimation from remotely sensed imagery using an artificial neural network cloud classification system. J. Appl. Meteor. 43, 1834–1853. https://doi.org/10.1175/JAM2173.1.

Hou, A.Y., Kakar, R.K., Neeck, S.A., Azarbarzin, A., Kummerow, C.D., Kojima, M., Oki, R., Nakamura, K., Iguchi, T, 2014. The global precipitation measurement mission. Bull. Amer. Meteor. Soc. 95, 701–722. https://doi.org/10.1175/BAMS-D-13-00164.1.

Houze, R.A., Rasmussen, K.L., Zuluaga, M.D., Brodzik, S.R, 2015. The variable nature of convection in the tropics and subtropics: A legacy of 16 years of the tropical rainfall measuring mission satellite. Rev. Geophys. 53, 994–1021. https://doi.org/10.1002/2015RG000488.

Houze, R.A., McMurdie, L.A., Petersen, W.A., Schwaller, M.R., Baccus, W., Lundquist, J.D., Mass, C.F., Nijssen, B., Rutledge, S.A., Hudak, D.R., Tanelli, S., Mace, G.G., Poellot, M.R., Lettenmaier, D.P., Zagrodnik, J.P., Rowe, A.K., DeHart, J.C., Madaus, L.E., Barnes, H.C., Chandrasekar, V., 2017. The Olympic Mountains Experiment (OLYMPEX). Bull. Amer. Meteor. Soc. 98, 2167–2188. https://doi.org/10.1175/BAMS-D-16-0182.1.

Hsu, K.-L., Karbalee, N., Braithwaite, D., 2020: Improving PERSIANN-CCS using passive microwave rainfall estimation. In: Levizzani V., Kidd. C., Kirschbaum D.B., Kummerow C.D., Nakamura K., Turk F.J., Satellite Precipitation Measurement, Eds., Springer Nature, Cham, Advances in Global Change Research, 67, 375–391. https://doi.org/10.1007/978-3-030-24568-9_21.

Huffman, G.J., Levizzani, V., Ferraro, R.R., Turk, F.J., Kidd, C, 2016. Requirements for a robust precipitation constellation, 14th Special Meeting on Microwave Radiometry and Remote Sensing of the Environment (MICRORAD). Espoo, 37–41. https://doi.org/10.1109/MICRORAD.2016.7530500.

Huffman G.J., Bolvin D.T., Braithwaite D., Hsu K., Joyce R., Xie P. GPM Integrated Multi-Satellite Retrievals for GPM (IMERG) Algorithm Theoretical Basis Document (ATBD), p 30 http://pmm.nasa.gov/sites/default/files/document_files/IMERG_ATBD_V4.4.pdf, 2014.

Huffman, G.J., Bolvin, D.T., Nelkin, E.J., Wolff, D.B., Adler, R.F., Gu, G., Hong, Y., Bowman, K.P., Stocker, E.F, 2007. The TRMM Multisatellite Precipitation Analysis (TMPA): quasi-global, multiyear, combined-sensor precipitation estimates at fine scales. J. Hydrometeor. 8, 38–55. https://doi.org/10.1175/JHM560.1.

Huffman, G.J., Adler, R.F., Arkin, P., Chang, A., Ferraro, R., Gruber, A., Janowiak, J., McNab, A., Rudolf, B., Schneider, U, 1997. The global precipitation climatology project (GPCP) combined precipitation dataset. Bull. Amer. Meteor. Soc. 78, 5–20. https://doi.org/10.1175/1520-0477(1997)078%3C0005:TGPCPG%3E2.0.CO;2.

Huffman, G.J., Bolvin, D.T., Braithwaite, D., Hsu, K.-L., Joyce, R.J., Kidd, C., Nelkin, E.J., Sorooshian, S., Stocker, E.F., Tan, J., Wolff, D.B., Xie, P., 2020. Integrated multi-satellite retrievals for the Global Precipitation Measurement (GPM) Mission (IMERG). In: Levizzani, V., Kidd, C., Kirschbaum, D.B., Kummerow, C.D., Nakamura, K., Turk, F.J. (Eds.), Satellite Precipitation Measurement. Springer Nature, Cham, pp. 343–353. https://doi.org/10.1007/978-3-030-24568-9_19 2020.

Iguchi, T., 2020. Dual-frequency precipitation radar (DPR) on the Global Precipitation Measurement (GPM) mission's Core Observatory. In: Levizzani, V., Kidd, C., Kirschbaum, D.B., Kummerow, C.D., Nakamura, K., Turk, F.J. (Eds.), Satellite Precipitation Measurement. Springer Nature, Cham, pp. 183–192. https://doi.org/10.1007/978-3-030-24568-9_11 2020.

Iguchi, T., Seto, S., Meneghini, R., Yoshida, N., Awaka, J., Le, M., Chandrasekar, V., Kubota, T, 2015. GPM/DPR Level-2. Algorithm Theoretical Basis Doc. NASA/GSFC, Greenbelt, MD, p. 68. http://pps.gsfc.nasa.gov/Documents/ATBD_DPR_2015_whole_a.pdf. (Accessed 13 May 2021).

Janowiak, J.E., Joyce, R.J., Yarosh, Y, 2001. A real-time global half-hourly pixel-resolution infrared dataset and its applications. Bull. Amer. Meteor. Soc. 82, 205–217. https://doi.org/10.1175/1520-0477(2001)082,0205:ARTGHH.2.3.CO;2.

Jensen, M.P., Petersen, W.A., Bansemer, A., Bharadwaj, N., Carey, L.D., Cecil, D.J., Collis, S.M., Del Genio, A., Dolan, B., Gerlach, J., Giangrande, S.E., Heymsfield, A., Heymsfield, G., Kollias, P., Lang, T.J., Nesbitt, S.W., Neumann, A., Poellot, M., Rutledge, S.A., Schwaller, M., Tokay, A., Williams, C.R., Wolff, D.B., Xie, S., Zipser, E.J, 2016. The midlatitude continental convective clouds experiment (MC3E). Bull. Amer. Meteor. Soc. 97, 1667–1686. https://doi.org/10.1175/BAMS-D-14-00228.1.

Jiang, G., Liu, R, 2014. Retrieval of sea and land surface temperature from SVISSR/FY-2C/D/E measurements. IEEE Trans. Geosci. Remote Sens. 52 (10), 6132–6140. https://doi.org/10.1109/TGRS.2013.2295260.

Joyce, R.J., Janowiak, J.E., Arkin, P.A., Xie, P, 2004. CMORPH: A method that produces global precipitation estimates from passive microwave and infrared data at high spatial and temporal resolution. J. Hydrometeor. 5, 487–503. https://doi.org/10.1175/1525-7541(2004)005<0487:CAMTPG>2.0.CO;2.

Jurczyk, A., Szturc, J., Otop, I., Osrodka, K., Struzik, P, 2020. Quality-based combination of multi-source precipitation data. Remote Sens 12 (11), 1709. https://doi.org/10.3390/rs12111709.

Jurković, S.R., Pasarić, Z, 2013. Spatial variability of annual precipitation using globally gridded data sets from 1951 to 2000. Int. J. Climatol. 33, 690–698. https://doi.org/10.1002/joc.3462.

Kidd, C, 2018. GPROF CNES/ISRO Megha-Tropiques SAPHIR radiometer Precipitation Retrieval and Profiling Scheme, Level 2A precipitation product. NASA/GSFC, Greenbelt, MD, USA, https://arthurhou.pps.eosdis.nasa.gov/Documents/20180203_SAPHIR-ATBD.pdf. (Accessed 13 May 2021).

Kidd, C., Levizzani, V, 2011. Status of satellite precipitation retrievals. Hydrol. Earth Syst. Sci. 15, 1109–1116. https://doi.org/10.5194/hess-15-1109-2011.

Kidd, C., Levizzani, V., 2019. Quantitative precipitation estimation from satellite observations. In: Maggioni, V., Massari, C. (Eds.), Extreme Hydroclimatic Events and Multivariate Hazards in a Changing Climate. Elsevier, Amsterdam. https://doi.org/10.1016/B978-0-12-814899-0.00001-8.

Kidd, C., Muller, C., 2010. The combined passive microwave-infrared (PMIR) algorithm. In: Hossain, F., Gebremichael, M. (Eds.), Satellite Rainfall Applications for Surface Hydrology. Dordrecht, pp. 69–83. https://doi.org/10.1007/978-90-481-2915-7_5.

Kidd, C., Kniveton, D., Barrett, E.C, 1998. Advantage and disadvantages of statistical/empirical satellite estimation of rainfall. J. Atmos. Sci. 55, 1576–1582. https://doi.org/10.1175/1520-0469(1998)055,1576:TAADOS.2.0.CO;2.

Kidd, C., Kniveton, D.R., Todd, M.C., Bellerby, T.J, 2003. Satellite rainfall estimation using a combined passive microwave and infrared algorithm. J. Hydrometeor. 4, 1088–1104. https://doi.org/10.1175/1525-7541(2003)004,1088:SREUCP.2.0.CO;2.

Kidd, C., Tan, J., Kirstetter, P.-E., Petersen, W.A, 2018. Validation of the version 05 level 2 precipitation products from the GPM core observatory and constellation satellite sensors. Quart. J. Roy. Meteor. Soc. 144 (S1), 313–328. https://doi.org/10.1002/qj.3175.

Kidd, C., Bauer, P., Turk, F.J., Huffman, G.J., Joyce, R., Hsu, K.-L., Braithwaite, D, 2012. Inter-comparison of high-resolution precipitation products over northwest Europe. J. Hydrometeor 13, 67–83. https://doi.org/10.1175/JHM-D-11-042.

Kidd, C., Matsui, T., Chern, J., Mohr, K., Kummerow, C., Randel, D, 2016. Global precipitation estimates from cross-track passive microwave observations using a physically based retrieval scheme. J. Hydrometeor. 17, 383–400. https://doi.org/10.1175/JHM-D-15-0051.1.

Kidd, C., Huffman, G., Becker, A., Skofronick-Jackson, G., Kirschbaum, D., Joe, P., Muller, C, 2018. So, how much of the Earth's surface is covered by rain gauges? Bull. Amer. Meteor. Soc. 98, 69–78. https://doi.org/10.1175/BAMS-D-14-00283.1.

Kidd, C., Shige, S., Vila, D., Tarnavsky, E., Yamamoto, M.K., Maggioni, V., Maseko, B, 2020. The IPWG satellite precipitation validation effort. In: Levizzani, V., Kidd, C., Kirschbaum, D.B., Kummerow, C.D., Nakamura, K., Turk, F.J. (Eds.), Satellite Precipitation Measurement. Springer Nature, Cham, pp. 453–470. https://doi.org/10.1007/978-3-030-35798-6_1 2020.

Kidd, C., Takayabu, Y.N., Skofronick-Jackson, G.M., Huffman, G.J., Braun, S.A., Kubota, T., Turk, F.J, 2020. The Global Precipitation Measurement (GPM) Mission. In: Levizzani, V., Kidd, C., Kirschbaum, D.B., Kummerow, C.D., Nakamura, K., Turk, F.J. (Eds.), Satellite Precipitation Measurement. Springer Nature, Cham, pp. 3–23. https://doi.org/10.1007/978-3-030-24568-9_1 2020.

King, M.D., Kaufman, Y.J., Menzel, W.P., Tanré, D, 2012. Remote sensing of cloud, aerosol, and water vapor properties from the moderate resolution imaging spectrometer (MODIS). IEEE Trans. Geosci. Remote Sens. 30 (1), 2–27. https://doi.org/10.1109/36.124212.

Kirschbaum, D.B., Patel, K, 2016. Precipitation data key to food security and public health. EOS. https://eos.org/meeting-reports/precipitation-data-key-to-food-security-and-public-health.

Kirschbaum, D.B., Huffman, G.J., Adler, R.F., Braun, S., Garrett, K., Jones, E., McNally, A., Skofronick-Jackson, G., Stocker, E., Wu, H., Aitchik, B.F, 2017. NASA's remotely-sensed precipitation: a reservoir for applications users. Bull. Amer. Meteor. Soc. 98, 1169–1194. https://doi.org/10.1175/BAMS-D-15-00296.1.

Kirstetter, P.-E., Petersen, W.A., Kummerow, C.D., Wolff, D.B, 2020. Integrated multi-satellite evaluation for the Global Precipitation Measurement: impact of precipitation types on spaceborne precipitation estimation. In: Levizzani, V., Kidd, C., Kirschbaum, D.B., Kummerow, C.D., Nakamura, K., Turk, F.J. (Eds.), Satellite Precipitation Measurement. Springer Nature, Cham, pp. 583–608. https://doi.org/10.1007/978-3-030-35798-6_7.

Kirstetter, P.-E., Hong, Y., Gourley, J.J., Schwaller, M., Petersen, W., Cao, Q, 2015. Impact of sub-pixel rainfall variability on spaceborne precipitation estimation: evaluating the TRMM 2A25 product. Quart. J. Roy. Meteor. Soc. 141, 953–966. https://doi.org/10.1002/qj.2416.

Kitchen, M., Illingworth, A, 2011. From observations to forecasts - Part 13: the UK weather radar network – past, present and future. Weather 66, 291–297. https://doi.org/10.1002/wea.861.

Klaes, K.D., Cohen, M., Buhler, Y., Schlüssel, P., Munro, R., Luntama, J.-P., von Engeln, A., Ó Clérigh, E., Bonekamp, H., Ackermann, J., Schmetz, J, 2007. An introduction to the EUMETSAT polar system. Bull. Amer. Meteor. Soc. 88, 1085–1096. https://doi.org/10.1175/BAMS-88-7-1085.

Kojima, M., Miura, T., Furukawa, K., Hyakusoku, Y., Ishikiri, T., Kai, H., Iguchi, T., Hanado, H., Nakagawa, K, 2012. Dual-frequency precipitation radar (DPR) development on the Global Precipitation Measurement (GPM) core observatory, Proc. SPIE 8528. Earth

Observing Missions and Sensors: Development, Implementation, and Characterization II, 85281A. https://doi.org/10.1117/12.976823.

Kozu, T., Kawanishi, T., Kuroiwa, H., Kojima, M., Oikawa, K., Kumagai, H., Okamoto, K., Okumura, M., Nakatsuka, H., Nishikawa, K, 2001. Development of precipitation radar onboard the tropical rainfall measuring mission satellite. IEEE Trans. Geosci. Remote Sens. 39, 102–116. https://doi.org/10.1109/36.898669.

Kubota, T., Shige, S., Hashizume, H., Aonashi, K., Takahashi, N., Seto, S., Hirose, M., Takayabu, Y.N., Ushio, T., Nakagawa, K., Iwanami, K., Kachi, M., Okamoto, K, 2007. Global precipitation map using satellite-borne microwave radiometers by the GSMaP project: production and validation. IEEE Trans. Geosci. Remote Sens. 45, 2259–2275. https://doi.org/10.1109/TGRS.2007.895337.

Kubota, T., Aonashi, K., Ushio, T., Shige, S., Takayabu, Y.N., Kachi, M., Arai, Y., Tashima, T., Masaki, T., Kawamoto, N., Mega, T., Yamamoto, M.K., Hamada, A., Yamaji, M., Liu, G., Oki, R, 2020. Global Satellite Mapping of Precipitation (GSMaP) products in the GPM era. In: Levizzani, V., Kidd, C., Kirschbaum, D.B., Kummerow, C.D., Nakamura, K., Turk, F.J. (Eds.), Satellite Precipitation Measurement. Springer Nature, Cham, pp. 355–373. https://doi.org/10.1007/978-3-030-24568-9_20.

Kucera, P.A., Ebert, E.E., Turk, F.J., Levizzani, V., Kirschbaum, D., Tapiador, F.J., Loew, A., Borsche, M, 2013. Precipitation from space: advancing Earth system science. Bull. Amer. Meteor. Soc. 94, 365–375. https://doi.org/10.1175/BAMS-D-11-00171.1.

Kulie, M.S., Milani, L, 2018. Seasonal variability of shallow cumuliform snowfall: a CloudSat perspective. Quart. J. Roy. Meteor. Soc. 144 (S1), 329–343. https://doi.org/10.1002/qj.3222.

Kulie, M.S., Milani, L., Wood, N.B., Tushaus, S.A., Bennartz, R., L'Ecuyer, T.S, 2016. A shallow cumuliform snowfall census using spaceborne radar. J. Hydrometeor. 17, 1261–1279. https://doi.org/10.1175/JHM-D-15-0123.1.

Kuligowski, R.J, 2002. A self-calibrating real-time GOES rainfall algorithm for short-term rainfall estimates. J. Hydrometeor. 3, 112–130. https://doi.org/10.1175/1525-7541(2002)003%3C0112:ASCRTG%3E2.0.CO;2.

Kuligowski, R.J, 2010. The self-calibrating multivariate precipitation retrieval (SCaMPR) for high-resolution, low-latency satellite-based rainfall estimates. In: Gebremichael, M., Hossain, F. (Eds.), Satellite Rainfall Applications for Surface Hydrology. Springer, Cham, pp. 39–48. https://doi.org/10.1007/978-90-481-2915-7-3.

Kummerow, C.D, 2020. Introduction to passive microwave retrieval methods. In: Levizzani, V., Kidd, C., Kirschbaum, D.B., Kummerow, C.D., Nakamura, K, Turk, F.J. (Eds.), Satellite Precipitation Measurement. Springer Nature, Cham, pp. 123–140. https://doi.org/10.1007/978-3-030-24568-9_7.

Kummerow, C.D., Barnes, W., Kozu, T., Shiue, J., Simpson, J, 1998. The Tropical Rainfall Measuring Mission (TRMM) sensor package. J. Atmos. Oceanic Technol. 15, 809–817. https://doi.org/10.1175/1520-0426(1998)015%3C0809:TTRMMT%3E2.0.CO;2.

Kummerow, C.D., Randel, D.L., Kulie, M., Wang, N.-Y., Ferraro, R., Munchak, S.J., Petkovic, V, 2015. The evolution of the Goddard PROFiling algorithm to a fully parametric scheme. J. Atmos. Oceanic Technol. 32, 2265–2280. https://doi.org/10.1175/JTECH-D-15-0039.1.

Kummerow, C.D., Tanelli, S., Takahashi, N., Furukawa, K., Klein, M., Levizzani, V, 2020. Plans for future missions. In: Levizzani, V., Kidd, C., Kirschbaum, D.B., Nakamura, K., Turk, F.J. (Eds.), Satellite Precipitation Measurement. Springer Nature, Cham, pp. 99–119. https://doi.org/10.1007/978-3-030-24568-9_6.

Kummerow, C., Simpson, J., Thiele, O., Barnes, W., Chang, A.T.C., Stocker, E., Adler, R.F., Hou, A., Kakar, R., Wentz, F., Ashcroft, P., Kozu, T., Hong, Y., Okamoto, K., Iguchi, T., Kuroiwa, H., Im, E., Haddad, Z., Huffman, G., Ferrier, B., Olson, W.S., Zipser, E., Smith, E.A., Wilheit, T.T., North, G., Krishnamurti, T., Nakamura, K, 2000. The status of the Tropical Rainfall Measuring Mission (TRMM) after two years in orbit. J. Appl. Meteor. 39, 1965–1982. https://doi.org/10.1175/1520-0450(2001)040%3C1965:TSOTTR%3E2.0.CO;2.

Lanza, L.G., Cauteruccio, A., Stagnaro, M., 2022. Rain Gauge Measurements. In: Morbidelli, R. (Ed.), Rainfall. Modeling, Measurement and Applications. Elsevier, Amsterdam, pp. 77–108. doi:10.1016/C2019-0-04937-0.

Laviola, S., Levizzani, V, 2011. The 183-WSL fast rainrate retrieval algorithm. Part I: retrieval design. Atmos. Res. 99, 443–461. https://doi.org/10.1016/j.atmosres.2010.11.013.

Laviola, S., Levizzani, V., Cattani, E., Kidd, C, 2013. The 183-WSL fast rainrate retrieval algorithm. Part II: Validation using ground radar measurements. Atmos. Res. 134, 77–86. https://doi.org/10.1016/j.atmosres.2013.07.013.

Laviola, S., Levizzani, V., Ferraro, R.R., Beauchamp, J, 2020. Hailstorm detection by satellite microwave radiometers. Remote Sens 12, 621. https://doi.org/10.3390/rs12040621.

Lensky, I.M., Rosenfeld, D, 1997. Estimation of precipitation area and rain intensity based on the microphysical properties retrieved from NOAA AVHRR data. J. Appl. Meteor. 36, 234–242. https://doi.org/10.1175/1520-0450(1997)036%3C0234:EOPAAR%3E2.0.CO;2.

Lettenmaier, D.P, 2017. Observational breakthroughs lead the way to improved hydrological predictions. Water Resour. Res. 53, 2591–2597. https://doi.org/10.1002/2017WR020896.

Levizzani, V., Cattani, E, 2019. Satellite remote sensing of precipitation and the terrestrial water cycle in a changing climate. Remote Sens 11, 2301. https://doi.org/10.3390/rs11192301.

Levizzani, V., Laviola, S., Cattani, E, 2011. Detection and measurement of snowfall from space. Remote Sens 3 (1), 145–166. https://doi.org/10.3390/rs3010145.

Levizzani, V., Kidd, C., Aonashi, K., Bennartz, R., Ferraro, R.R., Huffman, G.J., Roca, R., Turk, F.J., Wang, N.-Y, 2018. The activities of the International Precipitation Working Group. Quart. J. Roy. Meteor. Soc. 144 (S1), 3–15. https://doi.org/10.1002/qj.3214.

Liu, C., Zipser, E.J, 2015. The global distribution of largest, deepest, and most intense precipitation systems. Geophys. Res. Lett. 42, 3591–3595. https://doi.org/10.1002/2015GL063776.

Liu, G, 2008. Deriving snow cloud characteristics from CloudSat observations. J. Geophys. Res.Atmos. 113, D8. https://doi.org/10.1029/2007JD009766.

Liu, G., Seo, E.-K, 2013. Detecting snowfall over land by satellite high-frequency microwave observations: the lack of scattering signature and a statistical approach. J. Geophys. Res.-Atmos. 118, 1376–1387. https://doi.org/10.1002/jgrd.50172.

Marra, A.C., Porcù, F., Baldini, L., Petracca, M., Casella, D., Dietrich, S., Mugnai, A., Sanò, P., Vulpiani, G., Panegrossi, G, 2017. Observational analysis of an exceptionally intense hailstorm over the Mediterranean area: Role of the GPM core observatory. Atmos. Res. 192, 72–90. https://doi.org/10.1016/j.atmosres.2017.03.019.

Meng, H., Dong, J., Ferraro, R., Yan, B.H., Zhao, L.M., Kongoli, C., Wang, N.-Y., Zavodsky, B, 2017. A 1DVAR-based snowfall rate retrieval algorithm for passive microwave radiometers. J. Geophys. Res.-Atmos. 122, 6520–6540. https://doi.org/10.1002/2016JD026325.

Menne, M.J., Durre, I., Vose, R.S., Gleason, B.E., Houston, T.G, 2012. An overview of the Global Historical Climatology Network-Daily database. J. Atmos. Oceanic Technol. 29, 897–910. https://doi.org/10.1175/JTECH-D-11-00103.1.

Michaelides, S., Levizzani, V., Anagnostou, E.N., Bauer, P., Kasparis, T., Lane, J.E, 2009. Precipitation: measurement, remote sensing, climatology and modeling. Atmos. Res. 94, 512–533. https://doi.org/10.1016/j.atmosres.2009.08.017.

Mugnai, A., Cooper, H.J., Smith, E.A., Tripoli, G.J, 1990. Simulation of microwave brightness temperatures of an evolving hailstorm at SSM/I frequencies. Bull. Amer. Meteor. Soc. 71 (1), 2–13. https://doi.org/10.1175/1520-0477(1990)071%3C0002:SOMBTO%3E2.0.CO;2.

Mugnai, A., Casella, D., Cattani, E., Dietrich, S., Laviola, S., Levizzani, V., Panegrossi, G., Petracca, M., Sanò, P., Di Paola, F., Biron, D., De Leonibus, L., Melfi, D., Rosci, P., Vocino, A., Zauli, F., Pagliara, P., Puca, S., Rinollo, A., Milani, L., Porcù, F., Gattari, F, 2013. Precipitation products from the hydrology SAF. Nat. Hazards Earth Syst. Sci. 13, 1959–1981. https://doi.org/10.5194/nhess-13-1959-2013.

New, M., Todd, M., Hulme, M., Jones, P, 2001. Precipitation measurements and trends in the twentieth century. Int. J. Climatol. 21, 1889–1922. https://doi.org/10.1002/joc.680.

Panegrossi, G., Rysman, J.-F., Casella, D., Marra, A.C., Sanò, P., Kulie, M.S, 2017. CloudSat-based assessment of GPM Microwave Imager snowfall observation capabilities. Remote Sens 12, 1263. https://doi.org/10.3390/rs9121263.

Patel, P.N., Babu, K.N., Prajapati, R.P., Sitapara, V., Mathur, A.K, 2018. Day-1 INSAT-3DR vicarious calibration using reflectance-based approach over Great Rann of Kutch. J Indian Soc. Remote Sens. 46, 885–894. https://doi.org/10.1007/s12524-017-0729-z.

Petersen, W.A., Kirstetter, P.-E., Wang, J., Wolff, D.B., Tokay, A, 2020. The GPM Ground Validation Program. In: Levizzani, V., Kidd, C., Kirschbaum, D.B., Kummerow, C.D., Nakamura, K., Turk, F.J. (Eds.), Satellite Precipitation Measurement. Springer Nature, Cham, pp. 471–502. https://doi.org/10.1007/978-3-030-35798-6_2.

Platnick, S., King, M.D., Ackerman, S.A., Menzel, W.P., Baum, B.A., Riedi, J.C., Frey, R.A, 2003. The MODIS cloud products: algorithms and examples from Terra. IEEE Trans. Geosci. Remote Sens. 41 (2), 459–473. https://doi.org/10.1109/TGRS.2002.808301.

Rosenfeld, D., Lensky, I.M, 1998. Satellite-based insights into precipitation formation processes in continental and maritime convective clouds. Bull. Amer. Meteor. Soc. 79, 2457–2476. https://doi.org/10.1175/1520-0477(1998)079%3C2457:SBIIPF%3E2.0.CO;2.

Saidi, H., Ciampittiello, M., Dresti, C., Turconi, L, 2014. Extreme rainfall events: evaluation with different instruments and measurement reliability. Environ. Earth Sci. 72, 4607–4616. https://doi.org/10.1007/s12665-014-3358-7.

Sanò, P., Panegrossi, G., Casella, D., Marra, A.C., Di Paola, F., Dietrich, S, 2016. The new Passive microwave Neural network Precipitation Retrieval (PNPR) algorithm for the cross-track scanning ATMS radiometer: Description and verification study over Europe and Africa using GPM and TRMM spaceborne radars. Atmos. Meas. Tech. 9, 5441–5460. https://doi.org/10.5194/amt-9-5441-2016.

Schmetz, J., Pili, P., Tjemkes, S., Just, D., Kerkmann, J., Rota, S., Ratier, A, 2002. An introduction to Meteosat Second Generation (MSG). Bull. Amer. Meteor. Soc. 83, 977–992. https://doi.org/10.1175/1520-0477(2002)083%3C0977:AITMSG%3E2.3.CO;2.

Schmit, T.J., Griffith, P., Gunshor, M.M., Daniels, J.M., Goodman, S.J., Lebair, W.J, 2017. A closer look at the ABI on the GOES-R series. Bull. Amer. Meteor. Soc. 98, 681–698. https://doi.org/10.1175/BAMS-D-15-00230.1.

Schröder, M., Lockhoff, M., Fell, F., Forsythe, J., Trent, T., Bennartz, R., Borbas, E., Bosilovich, M.G., Castelli, E., Hersbach, H., Kachi, M., Kobayashi, S., Kursinski, E.R., Loyola, D., Mears, C., Preusker, R., Rossow, W.B., Saha, S, 2018. The GEWEX Water Vapor Assessment archive of water vapour products from satellite observations and reanalyses. Earth Syst. Sci. Data 10 (2), 1093–1117. https://doi.org/10.5194/essd-10-1093-2018.

Scofield, R.A., Kuligowski, R.J, 2003. Status and outlook of operational satellite precipitation algorithms for extreme-precipitation events. Wea. Forecasting 18, 1037–1051. https://doi.org/10.1175/1520-0434(2003)018,1037:SAOOOS.2.0.CO;2.

Seo, B., Krajewski, W.F, 2010. Scale dependence of radar rainfall uncertainty: initial evaluation of NEXRAD's new super-resolution data for hydrologic applications. J. Hydrometeor. 11, 1191–1198. https://doi.org/10.1175/2010JHM1265.1.

Sevruk, B, 1997. Regional dependency of precipitation-altitude relationship in the Swiss Alps. Climatic Change 36, 355–369. https://doi.org/10.1023/A:1005302626066.

Sevruk, B., Zahlavova, L, 1994. Classification system of precipitation gauge site exposure: evaluation and application. Int. J. Climatol. 14, 681–689. https://doi.org/10.1002/joc.3370140607.

Sevruk, B., Hertig, J.A., Spiess, R, 1991. The effect of a precipitation gauge orifice rim on the wind field deformation as investigated in a wind tunnel. Atmos. Environ. 25 (7), 1173–1179. https://doi.org/10.1016/0960-1686(91)90228-Y.

Sieck, L.C., Burges, S.J., Steiner, M, 2007. Correction to "challenges in obtaining reliable measurements of point rainfall. Water Resour. Res. 43, W06701. https://doi.org/10.1029/2007WR005985.

Simpson, J.R., Adler, R.F., North, G.R, 1988. A proposed Tropical Rainfall Measuring Mission (TRMM) satellite. Bull. Amer. Meteor. Soc. 69, 278–295. https://doi.org/10.1175/1520-0477(1988)069,0278:APTRMM.2.0.CO;2.

Skofronick-Jackson, G., Johnson, B.T., Munchak, S.J, 2013. Detection thresholds of falling snow from satellite-borne active and passive sensors. IEEE Trans. Geosci. Remote Sens. 15, 4177–4189. https://doi.org/10.1109/TGRS.2012.2227763.

Skofronick-Jackson, G., Kim, M.-J., Weinman, J.A., Chang, D.-E, 2004. A physical model to determine snowfall over land by microwave radiometry. IEEE Trans. Geosci. Remote Sens. 42, 1047–1058. https://doi.org/10.1109/TGRS.2004.825585.

Skofronick-Jackson, G., Kulie, M.S., Milani, L., Munchak, S.J., Wood, N.B., Levizzani, V, 2019. Satellite estimation of falling snow: A Global Precipitation Measurement (GPM) Core Observatory perspective. J. Appl. Meteor. Climatol. 58, 1429–1448. https://doi.org/10.1175/JAMC-D-18-0124.1.

Skofronick-Jackson, G., Hudak, D., Petersen, W., Nesbitt, S.W., Chandrasekar, V., Durden, S., Gleicher, K.J., Huang, G.-J., Joe, P., Kollias, P., Reed, K.A., Schwaller, M.R., Stewart, R., Tanelli, S., Tokay, A., Wang, J.R., Wolde, M, 2015. Global Precipitation Measurement Cold Season Precipitation Experiment (GCPEx): For measurement sake let it snow. Bull. Amer. Meteor. Soc. 96, 1719–1741. https://doi.org/10.1175/BAMS-D-13-00262.1.

Skofronick-Jackson, G., Petersen, W.A., Berg, W., Kidd, C., Stocker, E.F., Kirschbaum, D.B., Kakar, R., Braun, S.A., Huffman, G.J., Iguchi, T., Kirstetter, P.E., Kummerow, C., Meneghini, R., Oki, R., Olson, W.S., Takayabu, Y.N., Furukawa, K., Wilheit, T, 2017. The Global Precipitation Measurement (GPM) mission for science and society. Bull. Amer. Meteor. Soc. 98, 1679–1695. https://doi.org/10.1175/BAMS-D-15-00306.1.

Stephens, G.L., Kummerow, C.D, 2007. The remote sensing of clouds and precipitation from space: a review. J. Atmos. Sci. 64, 3742–3765. https://doi.org/10.1175/2006JAS2375.1.

Stephens, G.L., Vane, D.G., Boain, R.J., Mace, G.G., Sassen, K., Wang, Z., Illingworth, A.J., O'Connor, E.J., Rossow, W.B., Durden, S.L., Miller, S.D., Austin, R.T., Benedetti, A., Mitrescu, C., 2002. The CloudSat science team: the CloudSat mission and the A-train: a new dimension of space-based observations of clouds and precipitation. Bull. Amer. Meteor. Soc. 83, 1771–1790. https://doi.org/10.1175/BAMS-83-12-1771.

Strangeways, I.C, 2004. Improving precipitation measurement. Int. J. Climatol. 24, 1443–1460. https://doi.org/10.1002/joc.1075.

Tapiador, F.J., Kidd, C., Levizzani, V., Marzano, F.S, 2004a. A maximum entropy approach to satellite Quantitative Precipitation Estimation (QPE). Int. J. Remote Sens. 25, 4629–4639. https://doi.org/10.1080/01431160410001710000 a.

Tapiador, F.J., Kidd, C., Levizzani, V., Marzano, F.S, 2004b. A neural networks-based fusion technique to estimate half-hourly rainfall estimates at 0.1° resolution from satellite passive microwave and infrared data. J. Appl. Meteor. 43, 576–594. https://doi.org/10.1175/1520-0450(2004)043%3C0576:ANNFTT%3E2.0.CO;2 b.

Tapiador, F.J., Turk, F.J., Petersen, W., Hou, A.Y., García-Ortega, E., Machado, L.A.T., Angelis, C.F., Salio, P., Kidd, C., Huffman, G.J., de Castro, M, 2012. Global precipitation measurement: methods, datasets and applications. Atmos. Res. 104-105, 70–97. https://doi.org/10.1016/j.atmosres.2011.10.021.

Toyoshima, K., Masunaga, H., Furuzawa, F.A, 2015. Early evaluation of Ku- and Ka-band sensitivities for the Global Precipitation Measurement (GPM) Dual-frequency Precipitation Radar (DPR). SOLA 11, 14–17. https://doi.org/10.2151/sola.2015-004.

Tripoli, G.J., Cotton, W.R, 1980. A numerical investigation of several factors contributing to the observed variable intensity of deep convection over south Florida. J. Appl. Meteor. 19 (9), 1037–1063. https://doi.org/10.1175/1520-0450(1980)019<1037:ANIOSF>2.0.CO;2.

Tesfagiorgis, K.B., Mahani, S.E, 2014. A multi-source precipitation estimation approach to fill gaps over a radar precipitation field: a case study in the Colorado River Basin. Hydrol. Process. 29, 29–42. https://doi.org/10.1002/hyp.10103.

Uppala, S.M., Kållberg, P.V., Simmons, A.J., Andrae, U., Da Costa Bechtold, V., Fiorino, M., Gibson, J.K., Haseler, J., Hernandez, A., Kelly, G.A., Li, X., Onogi, K., Saarinen, S., Sokka, N., Allan, R.P., Andersson, E., Arpe, K., Balmaseda, M.A., Beljaars, A.C.M., Van De Berg, L., Bidlot, J., Bormann, N., Caires, S., Chevallier, F., Dethof, A., Dragosavac, M., Fisher, M., Fuentes, M., Hagemann, S., Hólm, E., Hoskins, B.J., Isaksen, L., Janssen, P.A.E.M., Jenne, R., McNally, A.P., Mahfouf, J.-F., Morcrette, J.-J., Rayner, N.A., Saunders, R.W., Simon, P., Sterl, A., Trenberth, K.E., Untch, A., Vasiljevic, D., Viterbo, P., Woollen, J, 2005. The ERA-40 re-analysis. Quart. J. Roy. Meteor. Soc. 131 (612), 2961–3012. https://doi.org/10.1256/qj.04.176.

Ushio, T., Sasashige, K., Kubota, T., Shige, S., Okamoto, K., Aonashi, K., Inoue, T., Takahashi, N., Iguchi, T., Kachi, M., Oki, R., Morimoto, T., Kawasaki, Z.-I, 2009. A Kalman filter approach to the Global Satellite Mapping of Precipitation (GSMaP) from combined passive microwave and infrared radiometric data. J. Meteor. Soc. Japan 87A, 137–151. https://doi.org/10.2151/jmsj.87A.137.

Wentz, F.J., Draper, D, 2016. On-orbit absolute calibration of the Global Precipitation Measurement Microwave Imager. J. Atmos. Oceanic Technol. 33, 1393–1412. https://doi.org/10.1175/JTECH-D-15-0212.1.

Wilheit, T., Berg, W., Ebrahimi, H., Kroodsma, R., McKague, D., Payne, V., Wang, J, 2015. Intercalibrating the GPM constellation using the GPM Microwave Imager (GMI). IEEE Int. Geosci. Remote Sens. Symp. (IGARSS). https://doi.org/10.1109/IGARSS.2015.7326996.

Wilson, J., Brandes, E, 1979. Radar measurement of rainfall - a summary. Bull. Amer. Meteor. Soc. 60, 1048–1058. https://doi.org/10.1175/1520-0477(1979)060%3C1048:RMORS%3E2.0.CO;2.

Xu, L., Gao, X., Sorooshian, S., Arkin, P.A., Imam, B, 1999. A microwave infrared threshold technique to improve the GOES Precipitation Index. J. Appl. Meteor. 38, 569–579. https://doi.org/10.1175/1520-0450(1999)038,0569:AMITTT.2.0.CO;2.

Yamamoto, M.K., Shige, S, 2015. Implementation of an orographic/nonorographic rainfall classification scheme in the GSMaP algorithm for microwave radiometers. Atmos. Res. 163, 36–47. https://doi.org/10.1016/j.atmosres.2014.07.024.

Yamamoto, M.K., Shige, S., Yu, C.-K., Chen, L.-W, 2017. Further improvement of the heavy orographic rainfall retrievals in the GSMaP algorithm for microwave radiometers. J. Appl. Meteor. Climatol. 56, 2607–2619. https://doi.org/10.1175/JAMC-D-16-0332.1.

Yang, D., Goodison, B.E., Metcalfe, J.R., Golubev, V.S., Bates, R., Pangburn, T., Hanson, C.L, 1998. Accuracy of NWS 8" standard nonrecording precipitation gauge: Results and application of WMO intercomparison. J. Atmos. Oceanic Technol. 15, 54–68. https://doi.org/10.1175/1520-0426(1998)015<0054:AONSNP>2.0.CO;2.

Yang, J., Zhang, P., Lu, N., Yang, Z., Shi, J., Dong, C, 2012. Improvements on global meteorological observations from the current Fengyun 3 satellites and beyond. Int. J. Digital Earth 5 (3), 251–265. https://doi.org/10.1080/17538947.2012.658666.

Yatagai, A., Kamiguchi, K., Arakawa, O., Hamada, A., Yasutomi, N., Kitoh, A, 2012. Aphrodite: constructing a long-term daily gridded precipitation dataset for Asia based on a dense network of rain gauges. Bull. Amer. Meteor. Soc. 93, 1401–1415. https://doi.org/10.1175/BAMS-D-11-00122.1.

You, Y., Wang, N.-Y., Ferraro, R., Rudlosky, S, 2017. Quantifying the snowfall detection performance of the GPM Microwave Imager channels over land. J. Hydrometeor. 18, 729–751. https://doi.org/10.1175/JHM-D-16-0190.1.

You, Y., Petkovic, V., Tan, J., Kroodsma, R., Berg, W., Kidd, C., Peters-Lidard, C, 2020. Evaluation of V05 precipitation Estimates from GPM constellation radiometers using KuPR as the reference. J. Hydrometeor. 21, 705–728. https://doi.org/10.1175/JHM-D-19-0144.1.

Zhu, Q., Gao, X., Xu, Y.-P., Tian, Y, 2019. Merging multi-source precipitation products or merging their simulated hydrological flows to improve streamflow simulation. Hydrol. Sci. J. 64 (8), 910–920. https://doi.org/10.1080/02626667.2019.1612522.

CHAPTER

7 Time resolution of rain gauge data and its hydrological role

Renato Morbidelli, Corrado Corradini, Carla Saltalippi, Alessia Flammini
Department of Civil and Environmental Engineering, University of Perugia, Perugia, Italy

7.1 Introduction

As widely recognized, rainfall data are essential for mathematical modelling of extreme hydrologic events, including droughts (Linke et al., 2015) and floods (Wilhelm et al., 2019), as well as for estimating quantity and quality of surface and subsurface water (Melone et al., 2006; Melone et al., 2008).

Ground-based radars can provide estimation of phase, quantity, and elevation of generic hydrometeors in the atmosphere (Wilson and Brandes, 1979; Austin, 1987; Fread et al., 1995; Smith et al., 1996; Seo, 1998; Assouline, 2020; Borga et al. 2022). Satellites can provide by visible and infrared radiation images, but also platforms for radiometers to obtain quantity and phase of hydrometeors (Barrett and Beaumont, 1994; Sorooshian et al., 2000; Kuligowski, 2002; Turk and Miller, 2005; Joyce et al., 2011; Kucera et al., 2013; Hou et al., 2014; Skofronick-Jackson et al., 2017; Kidd and Levizzani, 2022). However, only rain gauges at the earth surface provide direct point measurements of rainfall.

As widely descripted in Lanza et al. (2022), direct rainfall observations can be automatically recorded or not: non-recording gauges generally consist of open receptacles with vertical sides, in which the depth of precipitation is determined by a graduated measuring cylinder through human observation, while recording gauges are devices that automatically record the rainfall depth at specific time intervals (census gauges). This last category may be of weighing type, float type, tipping bucket type, and also includes the newer disdrometers that can measure the size distribution and velocity of falling rain drops. A weighing type rain gauge continuously records the weight of the receiving container plus the accumulated rainfall by means of a spring mechanism or a system of balance weights. A float type rain gauge has a chamber containing a float that rises vertically as the water level in the chamber rises. A tipping bucket type rain gauge operates by means of a pair of buckets. The rainfall first fills one bucket, which overbalances, directing the flow of water into the second bucket. The flip-flop motion of the tipping buckets is transmitted to the recording device and provides a very detailed measure of the rainfall amount and intensity (see also Lanza et al., 2022).

When the rainfall was recorded through human observation, a manual transcription of the total amount accumulated, typically during the last 24 h, was carried

RAINFALL. Modeling, Measurement and Applications. DOI: https://doi.org/10.1016/B978-0-12-822544-8.00008-1

out. Instead, after the introduction of automatic recordings, initially over paper rolls (Deidda et al., 2007) and then on digital supports, rainfall information at higher time-resolution (or temporal aggregation), t_a, became possible. Therefore, rainfall data observations and available in the archives are characterized by different t_a, depending on the selected rain gauge type, adopted recording system, and in addition on specific interest of the data manager.

From this historical excursus it is evident that prior to the advent of digital dataloggers rainfall data were characterized by coarse time-resolution, with possible effects on analyses based on their use. For example, many authors evaluated the effect of coarse time-resolutions on the estimation of annual maximum rainfall depths, H_d, with assigned durations, d (Hershfield and Wilson, 1958; Hershfield, 1961; Weiss, 1964; Harihara and Tripathi, 1973; Van Montfort, 1990; Faiers et al., 1994; Van Montfort, 1997; Young and McEnroe, 2003; Yoo et al., 2015; Papalexiou et al., 2016; Morbidelli et al., 2017). All these studies showed that, for durations comparable with t_a, the actual values of H_d may be considerably underestimated up to 50%. Furthermore, long series of H_d values may include a significant percentage of elements derived from rainfall data with coarse t_a that could be underestimated together with a considerable percentage of values obtained from continuous data (typically recorded in the last two/three decades). This aspect, as well as the relocation of stations, use of different rain gauge types with time and change of surrounding near the equipment, could produce significant effects on many related analyses, including the evaluation of rainfall depth-duration-frequency curves (Morbidelli et al., 2017) and trend estimate of extreme rainfalls (Morbidelli et al., 2018a).

The problem of underestimated annual maximum rainfall depths could be solved for durations greater than 1 h by adopting one of the methodologies suggested by the scientific literature, while the same cannot be easily done for the analysis of heavy rainfalls characterized by sub-hourly durations. In fact, long H_d series for $d < 1$ h are rarely available for most geographical areas.

An approximate but realistic estimation of the rain gauges operative in the entire world is in the range 150,000 to 250,000 (Sevruk and Klemm, 1989; New et al., 2001; Strangeways, 2007). Since in each geographical area there are networks characterized by very different histories and managed with specific interests, the time-resolution of the available rainfall data can be quite different.

The main objective of this chapter is to address the aforementioned issues with the aim of improving the use of historical extreme rainfall data making the series commonly available homogeneous with respect to t_a. Particular emphasis is given to the correction of H_d series in order to avoid distortions in the analysis of climatic changes and in the design of hydraulic structures.

More specifically, first the history of the temporal resolution of rainfall data at the global scale is presented. Furthermore, the length of H_d series with given aggregation times required to estimate an average adjustment factor for reducing the original errors involved in each series element is introduced. Then, a methodology to obtain homogeneous series of H_d from data derived through different t_a is suggested. Two problems of major interest are also considered: the sensitivity of the rainfall depth-duration-frequency curves to the corrections of the H_d series and an investigation directed to establish whether the underestimated H_d values significantly affect trend analyses for determining if climate change is having an impact on intensities and frequencies of extreme events. Finally, a representative case study referred to Umbria region (central Italy) is examined.

7.2 Rainfall data types

In all regions of the world, available rainfall data are characterized by different time-resolutions, mainly on the basis of the specific objective of the network manager and also of the technologic progress of the adopted recording devices. Currently, most rainfall amounts are continuously recorded in digital data-loggers, allowing the adoption of very short aggregation times, even equal to 1 minute.

A few decades ago rainfall data were recorded only over paper rolls and archived typically with t_a = 30 minutes or 1 h (see Fig. 7.1). Finally, especially before the Second World War, most rainfall data were characterized by a daily resolution and were manually recorded each day at the same local time (see Fig. 7.2).

7.3 Rainfall data time-resolution in different geographic areas of the world

In a recent analysis Morbidelli et al. (2020) developed a database containing information on rainfall time-resolution at the global scale, providing for each rain gauge station the complete t_a history and including the geographical coordinates of the installation sites. Overall, 25,423 rain gauge histories from 32 study areas (see Fig. 7.3) were collected. Location and main characteristics of the available rainfall data are given in Table 7.1.

Even though the collected stations do not cover all the countries of the world they are sufficiently representative of the typology of available local rainfall data.

Table 7.1 shows that only in a few cases the series of rainfall data started in the 19th century (e.g. 1867 Colorado-US, 1881 in Nicosia-Cyprus), while most began in early 20th century (e.g. 1916 in Tuscany-central Italy, 1945 in Argentina). For each study area the main characteristics (total series length and adopted t_a interval) of the longest record are shown in Fig. 7.4. As it can be seen, in some cases the t_a history of stations operating for over 200 years has been reconstructed, however in most study areas the longest series characterized by known t_a history is of about 100 years. Furthermore, only in a few study areas the t_a history is available for stations recently installed.

In almost all study areas, particularly when the rain gauge networks are very dated, recordings started in manual mode (Table 7.2) with a coarse time resolution, normally equal to 1 day, but in some cases equal to 1 month or to 1 year. The oldest manual data recordings in the database (San Fernando station, Spain, since 1805) are characterized by t_a equal to several days.

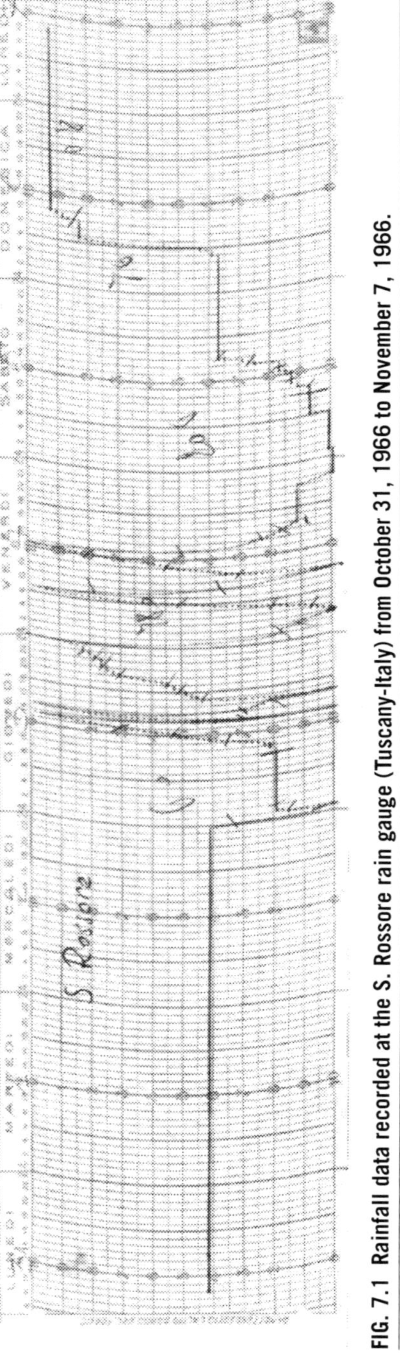

FIG. 7.1 Rainfall data recorded at the S. Rossore rain gauge (Tuscany-Italy) from October 31, 1966 to November 7, 1966.

(Source: Morbidelli et al., 2020).

(a)

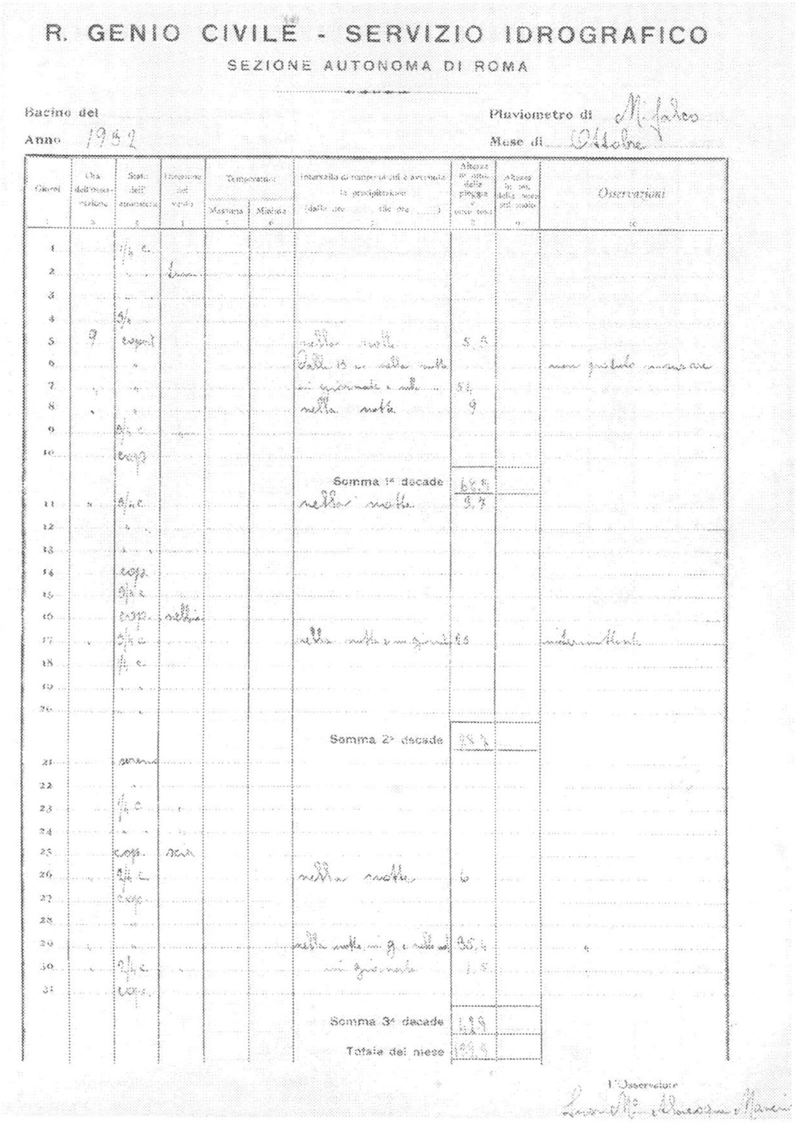

R. GENIO CIVILE - SERVIZIO IDROGRAFICO

SEZIONE AUTONOMA DI ROMA

Bacino del

Anno 1932

Pluviometro di M. falco

Mese di Ottobre

Somma 1ª decade

Somma 2ª decade

Somma 3ª decade

Totale del mese

L'Osservatore

FIG. 7.2 **(a) Manual recording of daily rainfall data during the month of October 1932 for Montefalco station (Umbria-central Italy), (b) Still considering the month of October 1932, transcription of daily rainfall recordings for some rain gauge stations in central Italy, including Montefalco.**

(Source: Morbidelli et al., 2020)

(b)

OSSERVAZIONI PLUVIOMETRICHE GIORNALIERE — **Anno 1932 - Mese di Ottobre**

Bacino o zona	STAZIONE	Tipo dell'apparecchio	Altezze di precipitazione in millimetri d'acqua osservate il giorno: 1	2	3	4	5	6	7	8	9	10	11	12	13	14	15	16	17	18	19	20	21	22	23	24	25	26	27	28	29	30	31	TOTALE MENSILE	Numero dei giorni con precip. super. o uguale ad 1 mm.
Zona II	Fossato di Vico	Pn.	—	—	—	—	—	**24.6**	9.4	14.2	—	—	—	—	—	—	—	11.0	6.4	—	—	—	—	—	—	—	—	—	8.9	—	5.7	—	9.0	89,2	8
id.	Gubbio	Pr.	—	—	0.4	**33.2**	26.7	2.2	9.7	1.1	6.0	—	7.0	—	0.4	1.4	4.4	—	—	—	—	—	—	—	—	—	1.5	20.2	0.2	0.2	8.8	12.0	—	135,4	13
id.	Padule	P.	—	—	—	3.0	—	8.0	—	—	10.0	—	**50.0**	—	—	—	11.0	—	—	—	—	—	—	—	—	—	—	25.0	—	1.0	—	—	—	108,0	7
id.	Pieve di Compresseto	Pn.	—	—	—	—	—	42.7	—	—	—	—	18.9	10.2	—	—	**60.3**	27.4	—	—	—	—	—	—	—	—	—	—	—	—	—	8.3	—	167,8	6
id.	Gualdo Tadino	Pr.	—	—	—	—	—	39.7	4.9	5.5	3.8	»	»	»	»	»	20.0	—	—	—	—	—	—	—	—	—	—	21.8	—	0.9	4.0	3.9	—	»	»
id.	Valfabbrica	P.	—	—	4.0	—	5.3	**14.2**	—	12.0	—	—	—	—	—	—	—	—	—	—	—	—	—	—	—	—	—	—	14.0	—	9.0	—	—	58,5	6
id.	Assisi	P.	—	—	—	—	—	12.0	—	—	—	—	—	—	—	—	0.5	—	**13.5**	—	—	—	—	—	—	—	3.0	10.0	—	—	10.6	—	—	49,6	5
id.	Pianello	P.	—	—	—	—	—	**26.3**	5.4	12.5	5.0	2.5	16.2	—	—	0.2	0.2	8.0	—	—	—	—	—	—	—	—	2.2	16.5	—	2.2	3.9	—	—	101,1	11
id.	Bastia	P.	—	—	0.2	—	—	**43.8**	4.1	10.1	1.1	0.2	6.2	0.3	—	1.0	—	5.1	3.0	—	—	—	—	—	—	—	3.0	8.8	—	6.8	2.8	1.8	—	92,3	12
id.	Annifo	Pn.	—	—	—	5.0	20.0	21.1	**72.1**	—	61.0	—	—	—	—	5.0	—	—	26 0	31.1	—	—	—	—	—	—	—	—	—	—	—	45.0	32.0	318,3	10
id.	Bagnara	Pn.	—	—	—	—	1.5	**41.6**	1.5	10.0	—	—	12.0	7.2	—	5.0	—	—	—	—	—	—	—	—	—	—	—	23.4	—	12.5	29.2	5.0	—	148,9	11
id.	Nocera Umbra	Pn.	—	—	—	3.0	**47.0**	**47.0**	10.0	11.4	—	3.0	16.0	10.0	—	1.0	3.0	20.0	7.0	—	—	1.0	—	—	—	—	—	13.0	21.0	3.0	4.0	—	5.2	225,6	18
id.	Valtopina	Pn.	»	»	»	»	»	»	»	»	»	»	12.9	—	—	3.1	—	16.9	—	10.7	—	—	—	—	—	—	10.4	—	—	—	5.4	—	—	»	»
id.	Rasiglia	Pn.	—	—	—			90.0			2.5	—	25.3	—	—	—	—	17.0	—	—	—	—	—	—	—	—	13.0	—	—	—	—	**90.0**	—	237,8	»
id.	Foligno	Pr.	—	—	—	—	0.6	**37.9**	7.8	6.9	0.3	—	—	0.4	—	0.4	0.6	8.7	11.6	—	—	—	—	—	—	—	0.2	7.5	—	1.8	22.2	2.2	—	109,1	9
id.	Bevagna	P.	—	—	—	—	—	9.3	—	7.1	—	—	**11.3**	4.5	—	—	—	10.2	9.5	—	—	—	—	—	—	—	—	—	—	5.4	7.1	—	5.6	70.0	9
id.	Mogliano	P.	—	—	27.0	15.3	—	—	7.4	—	—	—	**39.1**	—	5.2	—	—	—	—	—	—	—	—	—	—	—	19.4	—	8.0	—	—	—	11.3	132,7	8
id.	Spoleto	P.	»	»	»	»	»	»	»	»	»	»	»	»	»	»	»	»	»	»	»	»	»	»	»	»	»	»	»	»	»	»	»	»	»
id.	Spoleto (Catt. Amb. Agr.)	Pr.	»	»	»	»	»	»	»	»	»	»	»	»	»	»	»	»	»	»	»	»	»	»	»	»	»	»	»	»	»	»	»	»	»
id.	Montefalco	P.	—	—	—	—	5.3	—	**54.0**	9.0	—	—	3.7	—	—	—	—	—	25.0	—	—	—	—	—	—	—	—	6.0	—	—	35.4	1.5	—	139,9	8
id.	Trevi Umbro	P.	—	—	—	17.5	53.3	15.0	13.8	—	—	19.8	—	—	1.2	0.8	10.0	11.8	—	—	—	—	—	—	—	—	8.3	—	1.4	**70.0**	1.4	—	5.0	229,3	13
id.	Cannara	P.	—	—	—	—	**54.0**	14.6	—	—	—	4 6	11.2	—	—	1.2	—	26.8	—	—	—	—	—	—	—	—	6.4	—	—	33.0	2.3	—	4.9	159,0	10
Zona III	Casalina	P.	—	—	—	—	**29.0**	9.0	10.5	—	—	—	—	—	—	3.0	—	9.0	—	—	—	—	4.0	—	—	4.0	15.0	—	4.0	14.0	—	—	—	101,3	10
id.	Marcellano	P.	—	—	—	—	—	**50.8**	—	11.8	—	—	13.2	—	2.7	1.1	—	11.5	—	—	—	—	—	—	—	—	—	9.9	—	—	15.0	8.7	6.8	131,6	10
Zona IV	Panicale	P.	12.0	—	—	—	8.0	—	**29.0**	7.0	3.0	1.0	—	—	—	2.0	—	3.0	—	—	—	—	—	—	—	—	15.0	2.0	—	6.0	—	—	—	88,0	11
id.	S. Martino dei Colli	P.	—	—	0.2	—	—	**24.4**	1.5	8.2	3.4	0.5	3.2	6.7	—	1.8	—	4.4	—	—	—	—	—	—	—	—	3.7	23.4	—	—	3.5	—	—	78,9	10
id.	Corciano	P.	—	—	—	—	—	15.0	20.0	32.0	—	—	—	—	—	11.0	—	—	—	—	—	—	—	—	—	—	—	**47.0**	—	—	—	—	—	125,0	5
id.	Castiglione del Lago	P.	—	—	—	—	—	**29.0**	2.0	15.0	0.9	—	13.8	—	—	1.3	—	9.1	—	—	—	—	—	—	—	—	1.0	25.0	—	—	—	—	7.0	104,1	9
id.	Monte del Lago (1)	Pr.	—	—	—	—	—	**32.5**	1.8	—	15.0		—	11.0	—	—	—	1.0	—	—	—	—	—	—	—	—	—	—	—	—	4.0	—	—	65,0	6
id.	Isola Polvese	P.	—	—	—	—	—	**27.6**	3.7	17.6	0.6	—	5.6	0.4	—	1.9	—	—	—	—	—	—	—	—	—	—	—	18.0	—	—	4.1	5.3	—	84,8	8
id.	Passignano	P.	—	—	—	—	—	**25.8**	1.9	14.8	1.2	—	18.7	1.7	—	1.3	—	2.0	1.7	—	—	0.9	—	—	—	—	3.3	15.0	—	—	4.9	—	—	93,2	12
id.	S. Savino	P.	—	—	—	—	—	**31.0**	20.0	—	4.0	—	20.4	—	—	4.2	—	2.0	—	—	—	—	—	—	—	—	—	9.0	—	—	7.0	6.0	—	103,6	9
id.	Compignano	P.	—	—	—	—	—	**37.0**	2.5	9.0	—	—	9.0	—	—	—	—	9.5	1.0	—	—	—	—	—	—	—	—	15.0	—	—	3.5	—	—	86,5	8
id.	S. Venanzo	P.	—	4.2	7.3	—	—	—	—	15.8	**18.2**	7.5	13.0	—	—	—	—	—	6.4	10.5	8.3	—	—	—	—	—	—	—	—	—	0.2	—	—	91,4	9
Zona V	Fratta Todina	P.	—	—	—	—	—	29.0	8.0	17.0	—	—	4.0	—	—	—	—	—	**30.0**	—	—	—	—	—	—	—	2.0	17.0	—	—	16.0	1.0	—	124,0	9

(1) Osservazioni rilevate al pluviometro.

— 124 —

FIG. 7.2 *(Continued)*

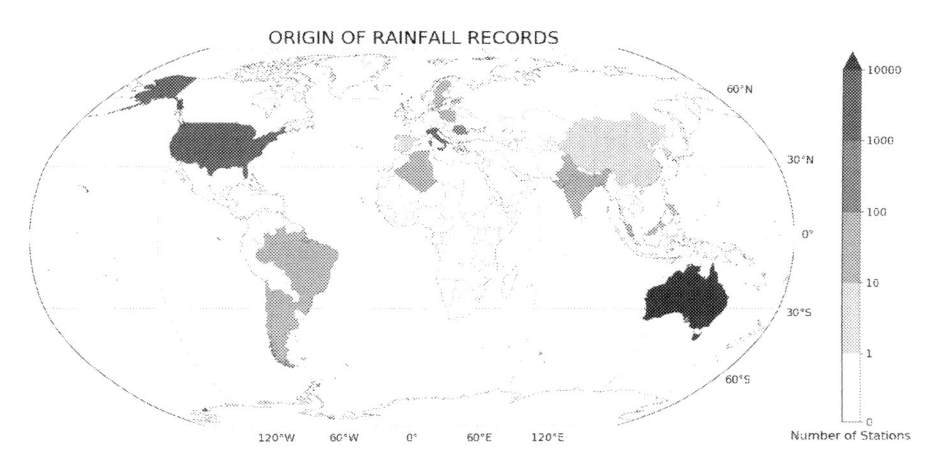

FIG. 7.3 Geographical position of the rain gauge stations considered by Morbidelli et al. (2020).
(Source: Morbidelli et al., 2020)

Table 7.1 Main characteristics of the available rainfall recordings for the rain gauge stations included in the database set up by Morbidelli et al. (2020).

Country (Area)	Rain gauges (number)	Record length min/ max (years)	Beginning of records (year)	Ending of records (year)	Time-resolution min/max (min)
Algeria (northern region)	30	9/41	1968	2010	1440
Argentina (Prov. Córdoba)	69	2/79	1941	2019	5/1440
Australia (whole country)	17,768	1/180	1805	2019	1/1440
Bangladesh (whole coun.)	35	19/72	1940	2019	180/1440
Brazil (eastern region)	2	35/54	1965	2019	1440
Brazil (northeast region)	18	3	2016	2018	10
Chile (El Rutal)	1	4	2011	2014	5
Chile (central region)	26	23/54	1959	2019	15/60
China (various areas)	7	5/11	2006	2017	10/30
Cyprus (central region)	7	54/139	1881	2019	10/518,400
Estonia (whole country)	51	3/133	1860	2019	10/1440
India (Tapi basin)	54	41/92	1930	2019	1/1440
Italy (Benevento)	2	49/135	1884	2019	10/43,200
Italy (Calabria region)	119	13/103	1916	2019	1/1440

(continued)

Table 7.1 Main characteristics of the available rainfall recordings for the rain gauge stations included in the database set up by Morbidelli et al. (2020). *Continued*

Country (Area)	Rain gauges (number)	Record length min/ max (years)	Beginning of records (year)	Ending of records (year)	Time-resolution min/max (min)
Italy (Sardinia region)	73	90/98	1921	2019	1/1440
Italy (Sicily region)	18	17/103	1916	2019	5/60
Italy (Tuscany region)	908	1/98	1916	2017	1/1440
Italy (Umbria region)	152	8/98	1915	2019	1/1440
Malaysia (Kuala Lumpur)	46	6/98	1879	2019	1/1440
Malta (whole country)	10	12/76	1922	2019	1/1440
Mongolia (western region)	2	49/57	1963	2019	1/720
Poland (whole country)	53	3/69	1951	2019	60/1440
Poland (Kujaw.-P. region)	10	1/159	1861	2019	5/43,200
Poland (Lubelskie region)	11	7/96	1922	2019	5/1440
Romania (whole country)	158	17/135	1885	2019	10/1440
South Korea (Seoul)	1	112	1907	2019	1/480
Spain (Andalusia region)	3	35/77	1942	2019	10/1440
Spain (Barcelona)	1	106	1914	2019	1/1440
Spain (Madrid)	1	100	1920	2019	10/1440
Spain (San Fernando)	1	184	1805	2019	1/>1440
Sweden (Uppsala region)	64	1/126	1893	2019	15/1440
USA (Colorado State)	5732	1/153	1867	2019	1/1440

(Source: Morbidelli et al., 2020)

Except for particular cases, mechanical recordings on paper rolls began in early 20th century. As an example, in the Morbidelli et al. (2020) database it can be found the existence of mechanical recordings carried out in the Alghero station (Italy-Sardina region) since 1927 and in the Campulung station (Romania) since 1949, in both cases with $t_a = 60$ minutes.

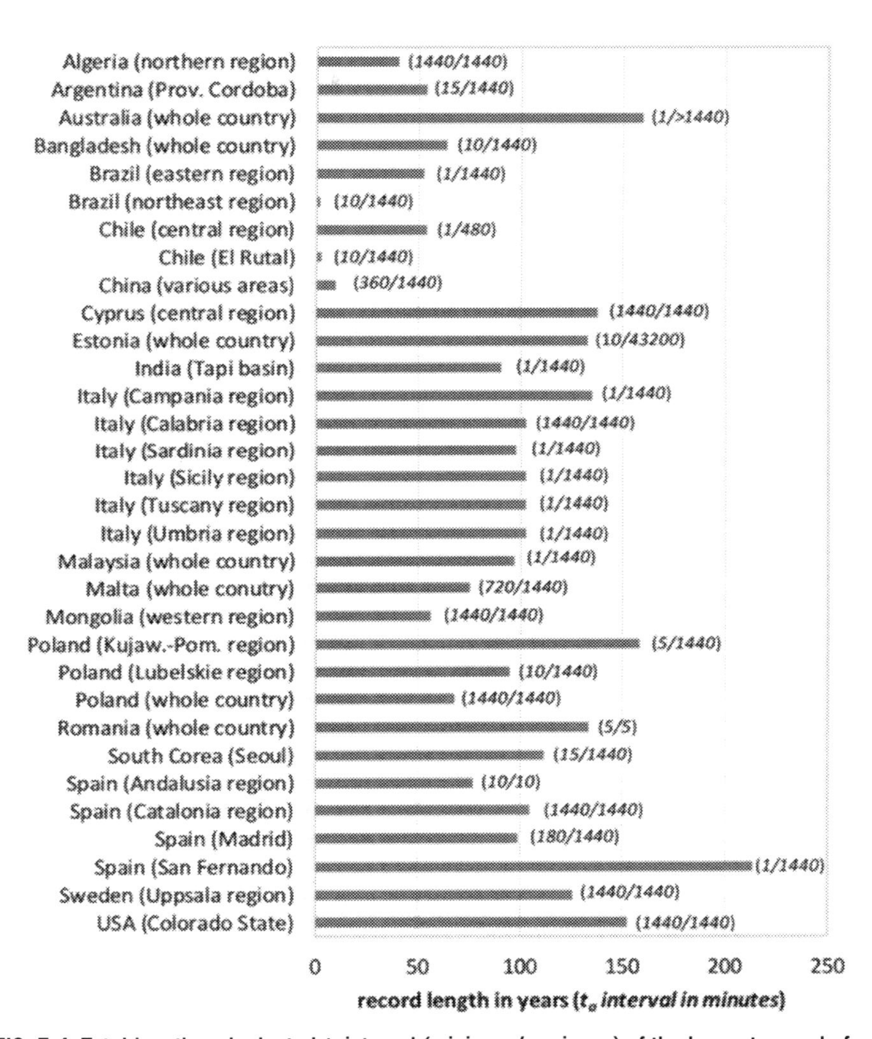

FIG. 7.4 Total length and adopted t_a interval (minimum/maximum) of the longest record of each study area considered in the database developed by Morbidelli et al. (2020).

(Source: Morbidelli et al., 2020)

Digital data logging began in the last decades of the 20th century with the consequence that analyses of the effects of climate change on short-duration (sub-hourly) heavy rainfalls could be unreliable in almost all geographic areas because of the too short length of rainfall series. Today the percentage of stations with data available at any time resolution (that is practically $t_a = 1$ minute) is very high. In the above cited database, examples of digital data characterized by $t_a = 1$ minute can be found in the Borgo S. Lorenzo station (Italy-Tuscany region) since 1991 and in the Valletta station (Malta) since 2006. Only in particular cases (e.g., Malaysia) long series of high resolution rainfall were taken out using automatic systems from strip carts of tipping bucket gauges (Deidda et al., 2007).

Table 7.2 Year of beginning of manual, mechanical and digital rainfall recordings for the study areas considered in Morbidelli et al. (2020).

Country (area)	Beginning of manual recording (year)	Beginning of mechanical recording (year)	Beginning of digitized recording (year)
Algeria (northern region)	1942	1967	-
Argentina (Prov. Córdoba)	1941	1941	1985
Australia (whole country)	1826	1920	1989
Bangladesh (whole coun.)	1867	1948	2003
Brazil (eastern region)	-	1965	-
Brazil (northeast region)	-	-	2016
Chile (El Rutal)	-	-	2011
Chile (central region)	-	1959	2012
China (various areas)	-	-	2006
Cyprus (central region)	1881	1911	2003
Estonia (whole country)	1860	-	2009
India (Tapi basin)	1925	1969	2012
Italy (Campania region)	1884	1921	2007
Italy (Calabria region)	1916	1916	1989
Italy (Sardinia region)	1921	1927	2007
Italy (Sicily region)	1832	1916	2002
Italy (Tuscany region)	1916	1928	1991
Italy (Umbria region)	1915	1928	1986
Malaysia (Kuala Lumpur)	-	1972	-
Malta (whole country)	1922	1957	2006
Mongolia (western region)	1963	-	2014
Poland (whole country)	1951	1963	2005
Poland (Kujaw.-P. region)	1861	1966	1997
Poland (Lubelskie region)	1922	-	1994
Romania (whole country)	1885	1898	2000
South Korea (Seoul)	1907	1915	2000
Spain (Andalusia region)	1942	-	1980
Spain (Catalonia region)	1885	1913	1988
Spain (Madrid)	-	1920	1997
Spain (San Fernando)	1805	-	1987
Sweden (Uppsala region)	1893	-	1986
USA (Colorado State)	1872	1948	1992

(Source: Morbidelli et al., 2020)

A wide heterogeneity of situations emerges, each conditioned by the specific politico-cultural history of the corresponding country. It is difficult to synthesize by individual figures and tables what happened in all the study areas considered in Morbidelli et al. (2020) as they sometimes contain and summarize the history of a single rain gauge, as in the case of the station installed in Madrid, while in other cases they involve a network with thousands of rain gauges, such as in the cases of Australia and Colorado (United States). However, Fig. 7.5 synthesizes the problem providing the percentage of rain gauges with specific t_a for all the stations except those located in Australia and Colorado (United States), because their huge number would make difficult the figure interpretation. Fig. 7.5 highlights that today, owing to the ease of continuous data recording, about 50% of the stations are working with $t_a = 1$ minute. The data recording with $t_a = 1440$ minutes will disappear within a short period.

An accurate analysis of the results in Morbidelli et al. (2020) also shows that most of the rain gauge stations changed the registration methods over the years. In many cases stations started working with daily manual recordings, then switched to mechanical recorders (t_a equal to 30 minutes or 1 h), more recently paired with digital data loggers capable of continuous recording. Both Table 7.2 and Fig. 7.5 show that these changes were not synchronized.

7.4 Effect of data time-resolution on the estimate of annual maximum rainfall depths

The errors in estimating extreme rainfalls associated with different values of d for a fixed t_a have been widely analyzed in the scientific literature. It is recognized that

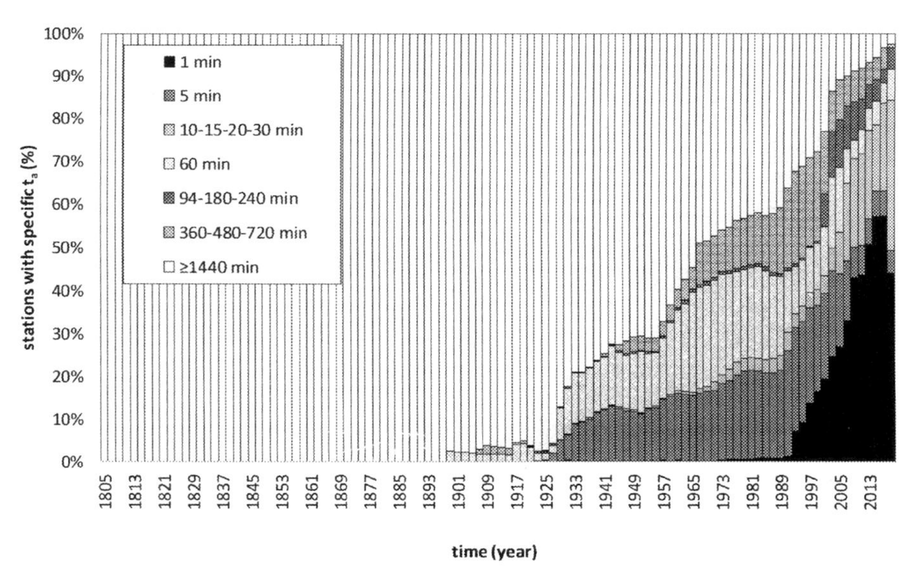

FIG. 7.5 Percentage of rain gauge stations with specific temporal aggregation, t_a, as a function of time.

All the stations included in the analysis by Morbidelli et al. (2020) except those located in Australia and Colorado (United States) are considered.

(Source: Morbidelli et al., 2020)

for d comparable with t_a the actual maximum rainfall depth may be underestimated (Hershfield and Wilson, 1958; Hershfield, 1961; Weiss, 1964; Harihara and Tripathi, 1973; Van Montfort, 1990; Faiers and al., 1994; Van Montfort, 1997; Young and McEnroe, 2003; Yoo et al., 2015; Papalexiou et al., 2016; Morbidelli et al., 2017). A few studies suggested effective methodologies for correcting the underestimate. Among these, Hershfield (1961) observed that for $d = t_a$ the results obtained from an analysis based on actual maxima were closely approximated through a frequency analysis of H_d with values multiplied by 1.13. Weiss (1964) used a probabilistic approach under the assumption of a uniform rainfall throughout the duration of interest and developed a relationship between the sampling ratio, t_a/d, and a sampling adjustment factor (SAF). The latter was defined as the average ratio of the real maximum rainfall depth for a given d to the maximum one deduced by a fixed recording interval. Young and McEnroe (2003) used high temporal resolution data from 15 rain gauges located in the Kansas City metropolitan area to derive a single empirical relationship between SAF and sampling ratio. This relation was found to provide adjustments consistent with other empirical studies (Miller et al., 1973). However, the length of the considered rainfall series (in the range 5.3–14.9 years, with average value of 9.6 years) was overly limited to draw a conclusion of general validity. Yoo et al. (2015) extended the probabilistic approach presented by Weiss (1964) considering several not uniform rainfall temporal distributions throughout the duration of interest that were found significantly related with the SAF. More recently, Morbidelli et al. (2017) defined a procedure to obtain quasi-homogeneous series of annual maximum rainfall depths that involves data derived through different temporal aggregations. They proposed a mathematical relation between average underestimation error and the ratio t_a/d to correct the H_d values. Overall, all these studies suggest that the SAF is dependent on both sampling ratio and d, with the latter involved because the shape of the rainfall temporal distribution is dependent on the rainfall duration.

7.4.1 Hyetograph shape and H_d underestimation

Following Burlando and Rosso (1996) and Boni et al. (2006), indicating by $x(t)$ the rainfall depth measured at time t at specific location, the accumulated rainfall recorded over a time interval d, $x_d(t)$, is given by:

$$x_d(t) = \int_{t}^{t+d} x(\xi)\,d\xi \tag{7.1}$$

and the annual maximum rainfall depth over a duration d can be expressed as:

$$H_d = max\left[x_d(t) : t_0 < t < t_0 - d + 1\ year\right] \tag{7.2}$$

where t_0 is the starting time of each year.

To determine H_d for a specific year, the knowledge of rainfall data characterized by any $t_a \le d$ is necessary. When $d = t_a$, independently of the rainfall pulse shape, the

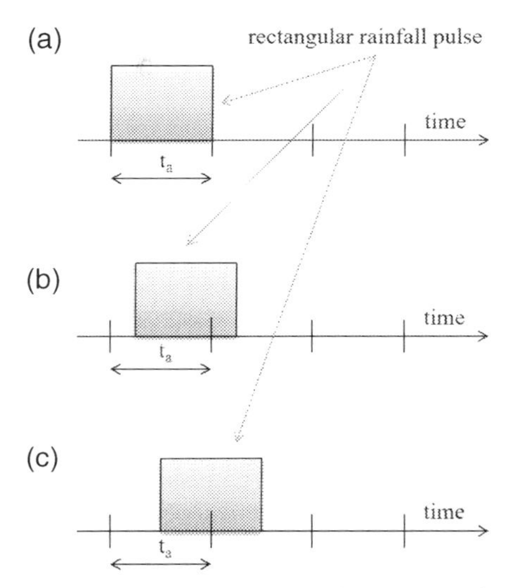

FIG. 7.6 Schematic representation of a rectangular rainfall pulse with duration $d = t_a$: (a) condition where a correct evaluation of H_d is possible, (b) condition with a generic underestimate of H_d, (c) condition with the maximum underestimate of H_d (equal to 50%).

(Source: Morbidelli et al., 2017)

H_d value is sometimes correctly estimated (Fig. 7.6a) but can also be underestimated (Fig. 7.6b–c) with errors up to 50%.

Despite the inability to correctly quantify the accuracy of a given H_d value, a representation of the average error for a time series containing a large number of elements can be made.

It is well-know that for each duration d, a long H_d series is affected by an average error depending on both t_a and the shape of the rainfall pulses. In the case of rectangular pulses, the average underestimate is equal to 25%, because each error assumes with the same probability of occurrence a value in the range 0–50%. This is consistent with the theoretical results by Yoo et al. (2015). However, it is widely recognized that the H_d values are determined by heavy rainfalls of erratic shape (Balme et al., 2006; Al-Rawas and Valeo, 2009; Coutinho et al., 2014). For example, Fig. 7.7 shows three hyetographs limited to the time intervals associated with the annual maximum rainfall depths for $d = 60$ minutes and continuously recorded by a rain gauge station located in central Italy. The hyetographs exhibit irregular shapes that for the sake of simplicity can be roughly considered of triangular type.

Under the assumption of a triangular rainfall pulse characterized by a duration d, the total rainfall depth, R_{pd}, is (Fig. 7.8a):

$$R_{pd} = \frac{dh}{2} \tag{7.3}$$

with h equal to the rainfall intensity peak.

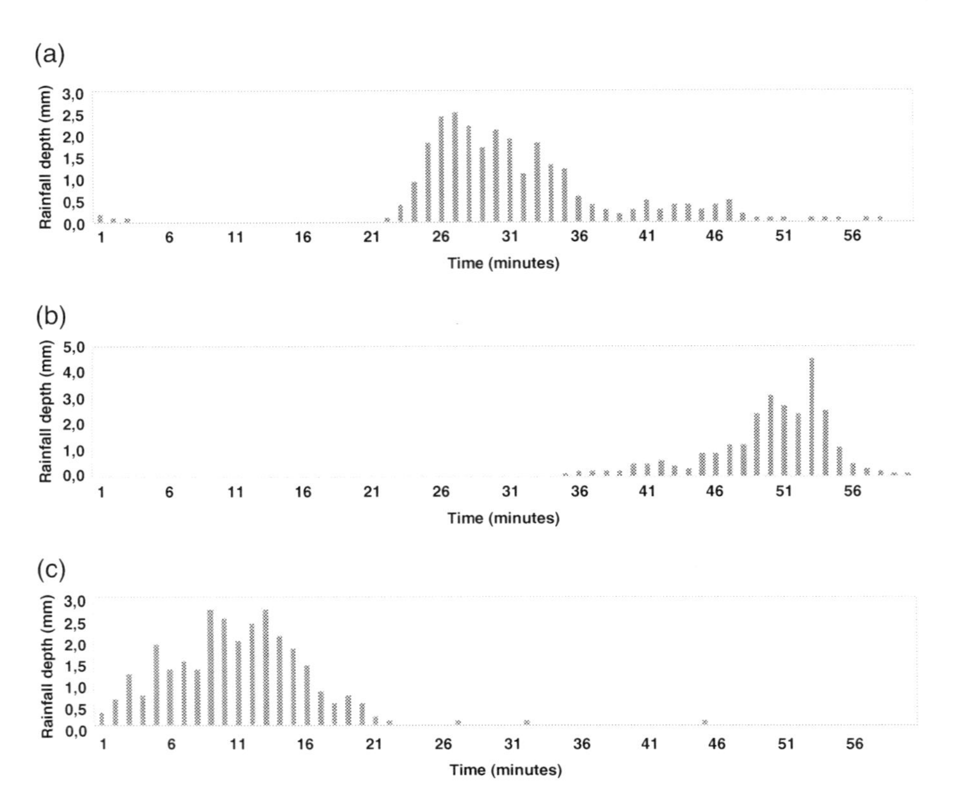

FIG. 7.7 Sample hyetographs continuously recorded at Cerbara station (Umbria Region, central Italy) and limited to the period involving annual maximum rainfall depths for duration equal to 60 minutes.

Moving windows starting on: (a) June 23, 1996 (9:10 a.m.), (b) July 25, 1997 (3:54 p.m.), (c) July 09, 2005 (6:03 p.m.).

When $t_a = d$, also with a triangular pulse the underestimate error of a single H_d is in the range 0-50%. Considering the possible pulse positions (Fig. 7.8b) the associated error is displayed in Fig. 7.8c. Its average value, E_a, obtained by integration through the pulse duration (see also Yoo et al., 2015) is given by:

$$E_a = \frac{1}{12} t_a h \tag{7.4}$$

This quantity may be expressed in terms of percentage of the rainfall pulse depth as:

$$E_{a\%} = 100 \frac{E_a}{R_{pd}} \tag{7.5}$$

For $t_a = d$, $E_{a\%}$ assumes the value 16.67% that is in agreement with the conclusions by Yoo et al. (2015).

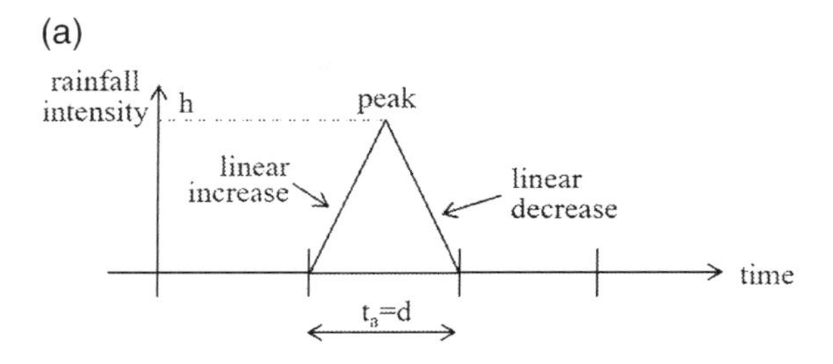

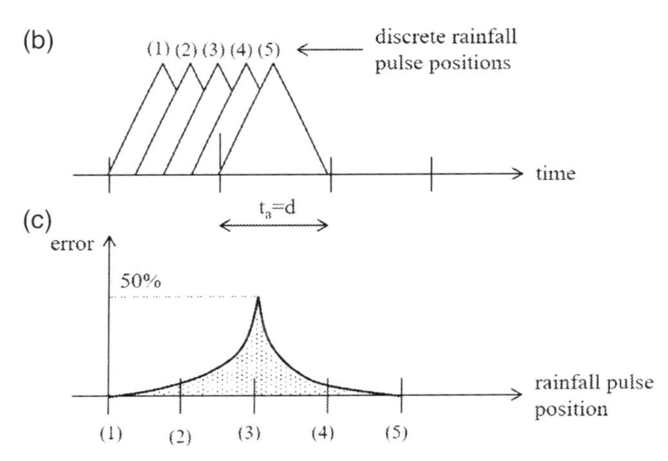

FIG. 7.8 Error in the annual maximum rainfall rate of duration *d*, H_d, referred to a triangular rainfall pulse is considered with *d* equal to the measurement aggregation time, t_a: (a) rainfall pulse details, (b) possible rainfall pulse positions, (c) errors associated with the above pulse positions.

(Source: Morbidelli et al., 2017)

However, an analysis of a considerable number of measured hyetographs performed for some rain gauge stations and using different d values highlights that in many cases before and after the peak the rainfall depth exhibits a steeper trend. Therefore, the actual value of $E_{a\%}$ should be less than 16.67%.

It can be deduced that, in principle, underestimation errors in determining the H_d values cannot be eliminated, independently of the adopted t_a. Moreover, the average error $E_{a\%}$ reduces when the ratio t_a/d decreases. For example, from Eqs. (7.3 and 7.5) it follows that:

$$E_{a\%}\left(d = nt_a\right) = \frac{1}{n} E_{a\%}\left(d = t_a\right) \quad n = 1, 2, \ldots \tag{7.6}$$

which implies that for t_a/d sufficiently small $E_{a\%}$ becomes negligible.

On the basis of the aforementioned analysis, if $d = t_a = 1$ minute for an extreme rainfall event of intensity equal to 300 mm/h the underestimate error becomes less than 1 mm. Further, considering that from a practical point of view the durations of interest for H_d are generally ≥5 minutes, rainfall data with $t_a = 1$ minute may be considered with negligible error as continuous data.

7.4.2 Development of average error relationships and correction of H_d series

If data with a coarse time-resolution are used, the underestimation error made in evaluating the annual maximum rainfall depth for a given d can be assumed as a random variable characterized by an exponential distribution with magnitude inversely correlated to H_d (Morbidelli et al., 2018b). In a H_d series the correction becomes significative only when a large number of underestimated elements is modified through the use of the average error. Morbidelli et al. (2017) deduced that a reliable estimate of the average error can be obtained only when the length of a series is approximately longer than 15–20 years, especially when $d \approx t_a$. Shorter series appear to have a trend of the average error rather irregular and unpredictable. The last result conflicts with the study by Young and McEnroe (2003) who examined data series with average length less than 10 years.

An overall analysis of the studies currently available suggests that:

1. For each specific year the single error is a random quantity with value in any case less than or equal to 50%
2. The average error depends on both t_a/d and d
3. The average errors can be approximately considered independent of the rain gauge location
4. The largest average error occurs for $d = t_a$ and from a theoretical point of view does not exceed 16.67%
5. For $d = nt_a$ the average error is less than or equal to $(1/n)16.67\%$
6. In any case an average error becomes reliable if its estimate is carried out using at least 15 to 20 years of observed rainfall data.

The aforementioned error characteristics account for the effect of temporal aggregation on H_d values, either for a specific year or for a long time series. On this basis it has been defined a methodology to improve the homogeneity of H_d series obtained from rainfall data associated to very different temporal aggregations.

Fig. 7.9 shows some significant relationships proposed by Weiss (1964), Young and McEnroe (2003), Morbidelli et al. (2017), to quantify the average underestimation error for different values of t_a/d. Weiss (1964) originally formulated a function that in terms of sampling adjustment factor can be expressed as:

$$E_{a\%} = 12{,}501\frac{t_a}{d} \quad [\%] \tag{7.7}$$

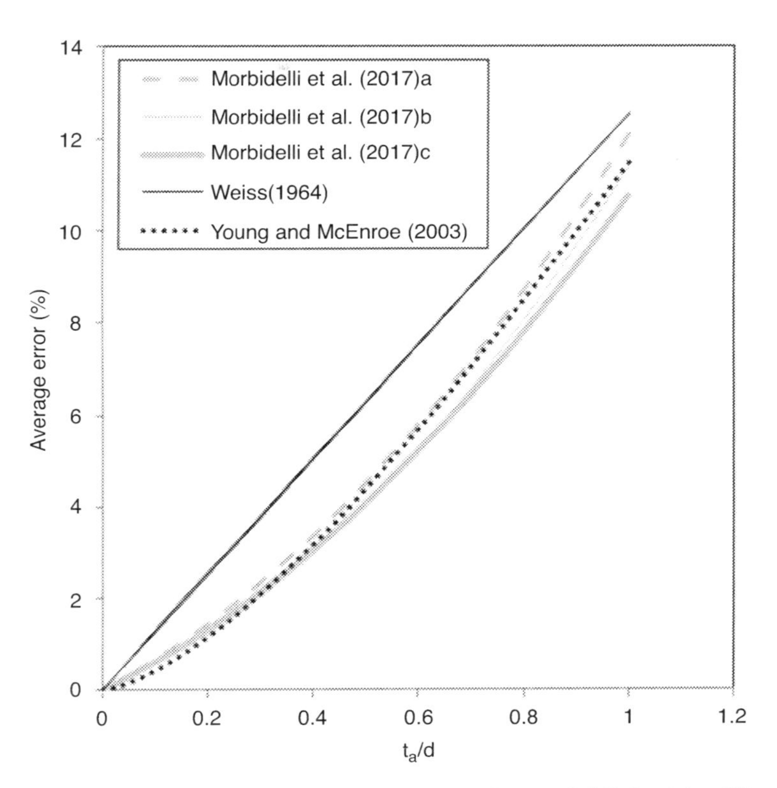

FIG. 7.9 Average underestimation error of the annual maximum rainfall depth for different values of the ratio between time-resolution, t_a, and duration, d, obtained by Eq. (7.7) (Weiss, 1964), Eq. (7.8) (Young and McEnroe, 2003), and Eq. (7.9) (Morbidelli et al., 2017).

In this last case, symbols "a", "b", and "c" stand for $d \leq 30$ minutes, d in the interval 30 to 180 min, and $d \geq 180$ minutes, respectively.

Weiss (1964) considered a probabilistic approach assuming a uniform rainfall rate through the accumulation period, while later Young and McEnroe (2003) used observed series of H_d to derive the following relation:

$$E_{a\%} = 100 - \frac{100}{1 + 0,13\left(\frac{t_a}{d}\right)^{1.5}} \quad [\%] \tag{7.8}$$

Finally, considering the link between d and the shape of the rainfall temporal distribution that influences the error magnitude (Yoo et al., 2015), Morbidelli et al. (2017) derived the following relations:

$$E_{a\%} = 6,14\left(\frac{t_a}{d}\right)^2 + 5,69\frac{t_a}{d} \quad [\%] d \leq 30 \text{ minutes} \tag{7.9a}$$

$$E_{a\%} = 6,7\left(\frac{t_a}{d}\right)^2 + 4,72\frac{t_a}{d} \quad [\%] 30\,\text{minutes} < d < 180\,\text{minutes} \tag{7.9b}$$

$$E_{a\%} = 5,2\left(\frac{t_a}{d}\right)^2 + 5,57\frac{t_a}{d} \quad [\%] d \geq 180\,\text{minutes} \tag{7.9c}$$

Given t_a/d and d, the correction to be applied to the H_d series obtained from data with coarse t_a can be quantified by Eqs. (7.9a–7.9c).

In principle the curves by Morbidelli et al. (2017) should be more appropriate because allow to take into account also the effects of the shape of the rainfall temporal distribution. As can be seen in Fig. 7.9, the Weiss (1964) equation, based on an ingenious theoretical intuition, provides larger average errors especially for intermediate values of the ratio t_a/d. However, the Young and McEnroe (2003) equation, although sometimes calibrated by using very short data sets, gives results comparable to those of Morbidelli et al. (2017).

7.5 Rainfall data time-resolution and its role in the hydrological applications

The marked heterogeneity in the t_a values, dependent on both the specific geographic area and the epoch, can influence subsequent analyses linked with H_d values, such as that concerning the assessment of the intensity-duration-frequency curves.

Morbidelli et al. (2017) showed that the use of long H_d series with underestimated values can lead to rainfall depth-duration-frequency curves with errors, slightly variable with both return period and rainfall duration, up to 10%. These errors, that affect the design of hydraulic structures, appreciably increase when the H_d series involve only values deduced through t_a much higher than 1 minute. In this context, possible effects of climate change on heavy rainfalls have to be also analyzed because they could have an important role in designing new structures as well as in restructuring the existing ones and this study could be distorted by the above errors in the H_d series.

Morbidelli et al. (2018a) showed that rainfall data with coarse time-resolution play a substantial role in the outcomes of very common statistical analyses (e.g. least-square linear trend, Mann-Kendall test, Spearman test, Sen's method) implemented to quantify the influence of climate change on intense rainfalls. The following major insights were drawn:

1. The underestimation errors due to coarse t_a produce significant effects on the least-squares linear trend analysis. The correction of the H_d values, performed by both deterministic and stochastic approaches, can change the sign of the trend from positive to negative, producing more evident effects for series characterized by a larger number of values with $t_a/d = 1$.
2. The non-parametric Mann-Kendall test (Mann, 1945; Kendall, 1975) and the Spearman rank correlation test (Khaliq et al., 2009), both with a significance

level of 0.05, are less sensitive than the least-squares linear trend to corrections of the H_d values.

3. The adoption of the innovative Sen method (Sen, 2012) produces different results if applied to uncorrected or corrected H_d values.
4. The hypothetical solution to consider only rainfall data with $t_a = 1$ minute is not feasible because in most geographic areas these data are only available for the last two/three decades (see also Morbidelli et al., 2020), while trend identification requires at least 60 years of data due to the effect of large-scale climate oscillations at multi-decadal time-scales (see also Willems, 2013).
5. The existence of anomalies in the H_d series due to different temporal aggregations of rainfall data cannot be detected by common homogeneity tests, e.g. the standard normal homogeneity test for a single break point (Alexandersson, 1986) or the Pettitt test (Pettitt, 1979). This is due to the fact that any consequent discontinuity, even though very evident, does not produce sufficiently large break points, at least for the annual maximum series.

7.6 Case study: Umbria region in central Italy

In order to better address the issues discussed in this chapter, the Umbria region (8456 km^2) is taken as a case study. This region is located in an inland zone of central Italy (Fig. 7.10) and is characterized by a complex orography along the eastern boundary, where the Apennine Mountains exceed 2000 m a.s.l. In the central and western areas orography is mainly of hilly type, with elevations ranging from 100 to 800 m a.s.l. A wide percentage of the study area is included in the basin of Tiber River that crosses the region from North to South-West receiving water from many tributaries, mainly located on the hydrographic left side.

On the basis of observations made by the rain gauge network shown in Fig. 7.10 and specified in Table 7.3, annual rainfall depth through the region ranges from 650 mm to 1450 mm with mean value of about 900 mm. Higher monthly rainfall values generally occur during the autumn-winter period with floods caused by widespread rainfall.

7.6.1 The history of the rainfall data time-resolution in the study area

In the Umbria region, as well as in many other Italian areas, the first available pluviometric recordings date back to the second decade of the last century. Specifically, the rain gauges of Cannara, Foligno and Perugia were installed in 1915, with data characterized by $t_a = 1440$ minutes. Fig. 7.11 shows the variation of the number of rain gauges with time, while Fig. 7.12 shows their spatial location in some particular years.

Initially, all the rain gauge stations (installed by the Italian National Hydrographic Service, INHS) were characterized by $t_a = 1440$ minutes. The first rain gauges with

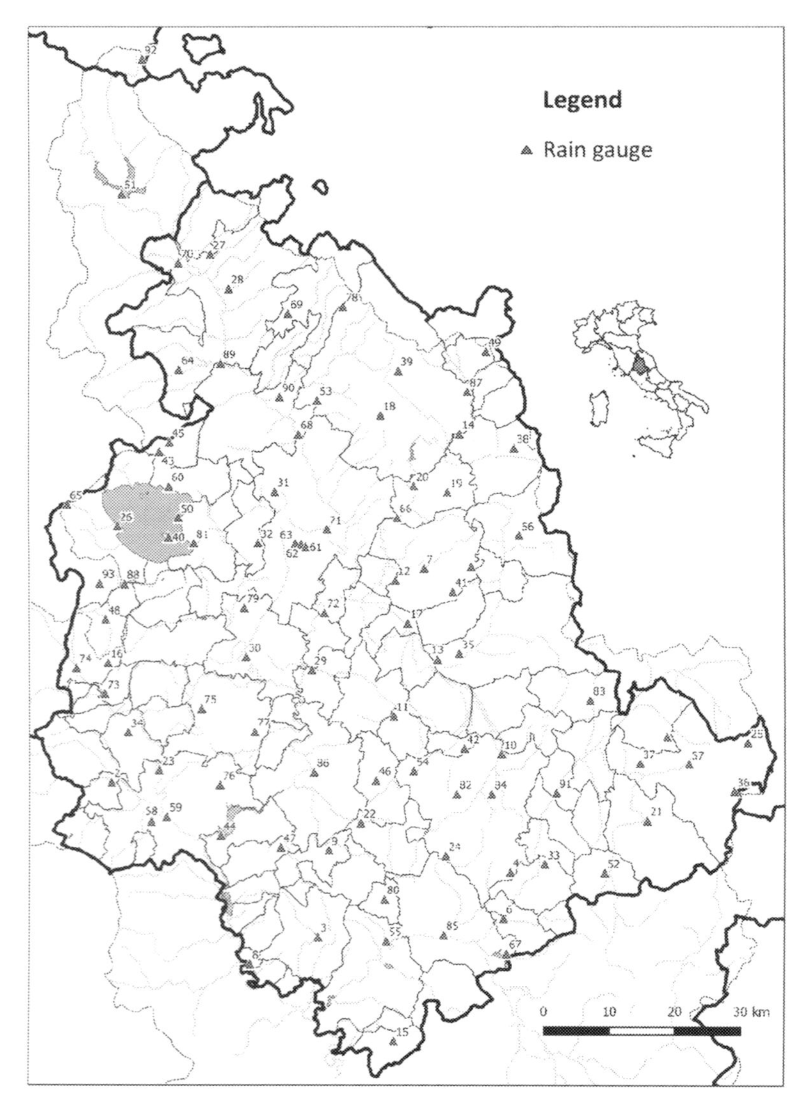

FIG. 7.10 General layout of the Umbria region (central Italy) showing the location of the rain gauges operated by the Regional Hydrographic Service (RHS), with identification numbers associated to Table 7.3.

recording on paper rolls were installed since 1927. The first stations equipped with digital data-logger (a group of 37 stations managed by the National Research Council) were set up in 1986, while the transition to digital of the INHS stations, that in the meantime became properties of the Regional Hydrographic Service (RHS), started since 1990 and was completed in 2011. Currently, all the rain gauge stations in the

Table 7.3 Main characteristics of the rain gauge stations operated in Umbria region (central Italy) by the regional hydrographic service (RHS). Identification (ID) numbers are associated to Fig. 7.10.

ID number	Rain gauge station	Altitude (m asl)	Available data period	ID number	Rain gauge station	Altitude (m asl)	Available data period
1	Abeto	946	1951–2014	48	Moiano	270	1994–2015
2	Allerona	153	1992–2015	49	Monte Cucco	1087	1996–2015
3	Amelia	321	1921–2015	50	Monte del Lago	260	1923–2015
4	Ancaiano	598	2008–2015	51	Montedoglio	393	2008–2015
5	Armenzano	708	1994–2015	52	Monteleone di S.	933	1953–2015
6	Arrone	221	1921–2015	53	Montelovesco	634	1921–2015
7	Assisi	408	1921–2001	54	Montemartano	625	2008–2015
8	Attigliano	64	1921–2015	55	Narni Scalo	109	1921–2015
9	Avigliano Umbro	482	2009–2015	56	Nocera Umbra	534	1921–2015
10	Azzano	235	1992–2015	57	Norcia	691	1921–2015
11	Bastardo	331	1992–2015	58	Orvieto	311	1921–2015
12	Bastia Umbra	203	1922–2015	59	Orvieto Scalo	109	1992–2015
13	Bevagna	212	1921–2015	60	Passignano sul T.	329	1921–2015
14	Branca	350	2009–2015	61	Perugia	440	1916–2010
15	Calvi dell'Umbria	305	1951–2015	62	Perugia S.G.	417	1992–2015
16	Campogrande	516	2009–2015	63	Perugia Sede	345	1995–2015
17	Cannara	191	1916–2015	64	Petrelle	342	1921–2015
18	Carestello	517	1998–2015	65	Petrignano del L.	304	1994–2015
19	CasaCastalda	730	1992–2015	66	Pianello	233	1922–2015
20	Casanuova	338	2009–2015	67	Piediluco	370	1921–2015
21	Cascia	604	1922–2015	68	Pierantonio	239	2005–2015
22	Casigliano	273	1992–2015	69	Pieve di Saddi	624	2008–2015
23	C. M. Bagni	158	2007–2015	70	Pistrino	292	2008–2015

(continued)

Table 7.3 Main characteristics of the rain gauge stations operated in Umbria region (central Italy) by the regional hydrographic service (RHS). Identification (ID) numbers are associated to Fig. 7.10. *Continued*

ID number	Rain gauge station	Altitude (m asl)	Available data period	ID number	Rain gauge station	Altitude (m asl)	Available data period
24	Castagnacupa	778	2009–2015	71	Ponte Felcino	205	1992–2015
25	Castelluccio di N.	1349	1921–2014	72	Ponte Nuovo di T.	174	1921–2015
26	Castiglione del L.	260	1921–2015	73	Ponte Santa Maria	240	1992–2015
27	Cerbara	310	1992–2015	74	Ponticelli	245	1992–2015
28	Città di Castello	304	1921–2015	75	Pornello	511	2009–2015
29	Collepepe	167	1999–2015	76	Prodo	431	1921–2015
30	Compignano	240	1921–2015	77	Ripalvella	453	1992–2015
31	Compresso	424	2008–2015	78	S. Benedetto V.	729	1992–2015
32	Corciano	306	1922–2015	79	S. Biagio della V.	257	1993–2015
33	Ferentillo	394	2008–2015	80	S. Gemini	299	1921–2015
34	Ficulle	440	1921–2015	81	S. Savino	260	1921–2015
35	Foligno	220	1916–2015	82	S. Silvestro	381	1992–2015
36	Forca Canapine	1652	1997–2015	83	Sellano	604	1951–2015
37	Forsivo	963	1992–2015	84	Spoleto	353	1921–2015
38	Gualdo Tadino	599	1921–2015	85	Terni	123	1921–2015
39	Gubbio	471	1921–2015	86	Todi	329	1921–2015
40	Isola Polvese	258	1928–2015	87	Torre dell'Olmo	554	2009–2015
41	La Bolsella	922	1994–2015	88	Tresa	267	2010–2015
42	La Bruna	240	2011–2015	89	Trestina	258	2007–2015
43	La Cima	791	1994–2015	90	Umbertide	305	1921–2015
44	Lago di Corbara	128	1963–2015	91	Vallo di Nera	307	1996–2015
45	Lisciano Niccone	300	1958–2008	92	Verghereto	1069	2002–2015
46	Massa Martana	328	1921–2015	93	Villastrada	350	1994–2009
47	Melezzole	548	2009–2015				

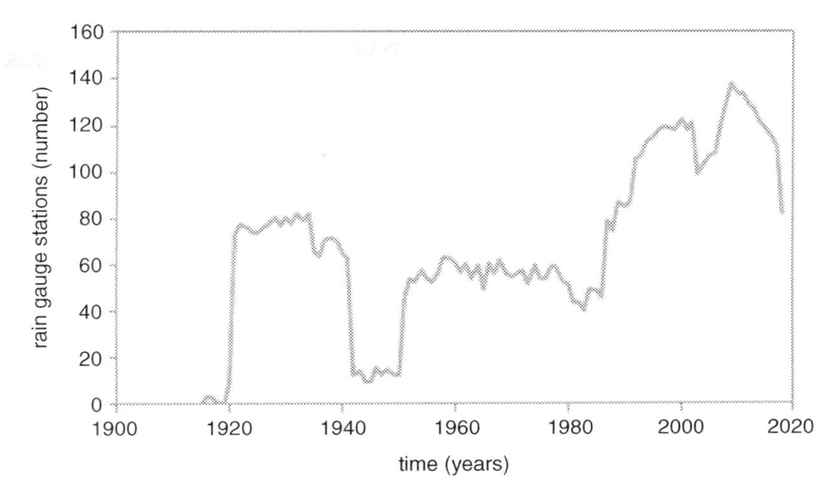

FIG. 7.11 Variation of the number of rain gauges with time in Umbria region, central Italy.
(Source: Morbidelli et al., 2020)

Umbria region are characterized by t_a = 1 minute, except 9 telemetering stations for which t_a = 5 minutes (Fig. 7.13).

Table 7.4 shows information on the t_a history for some representative stations of the Umbria region. Rain gauges are grouped as: 1) very old stations installed by the INHS that over the years have adopted all types of recording (initially manual with t_a = 1440 minutes, successively over paper rolls with t_a = 30 minutes, finally digital with t_a = 1 or 5 minutes); 2) stations installed by the INHS after the Second World War that have typically adopted only two different types of recording (initially manual, then digital); 3) stations installed by the RHS within the last three decades, all with t_a = 1 minute; 4) stations installed by the National Research Council since 1986, all with t_a = 1 minute.

7.6.2 The underestimation error of the annual maximum rainfall depth in the study region

For the analysis aimed at determining the underestimate of H_d in the study area, only the rain gauge stations characterized by continuous rainfall data for at least 20 years (12 out of 93) were considered.

Starting from the continuous rainfall data of these selected stations, aggregated data with the following t_a were obtained: 1 minute, henceforth denoted as "Observed"; 10, 15, 30, 60, 180, 360, 720 and 1440 minutes, henceforth denoted as "Generated". An example of this procedure is shown in Table 7.5 for rainfall data recorded at the Petrelle station.

For each selected station, considering some typical values of d (≤1440 minutes), with $d \geq t_a$, all H_d values may be easily determined by using both the "Observed" and "Generated" data.

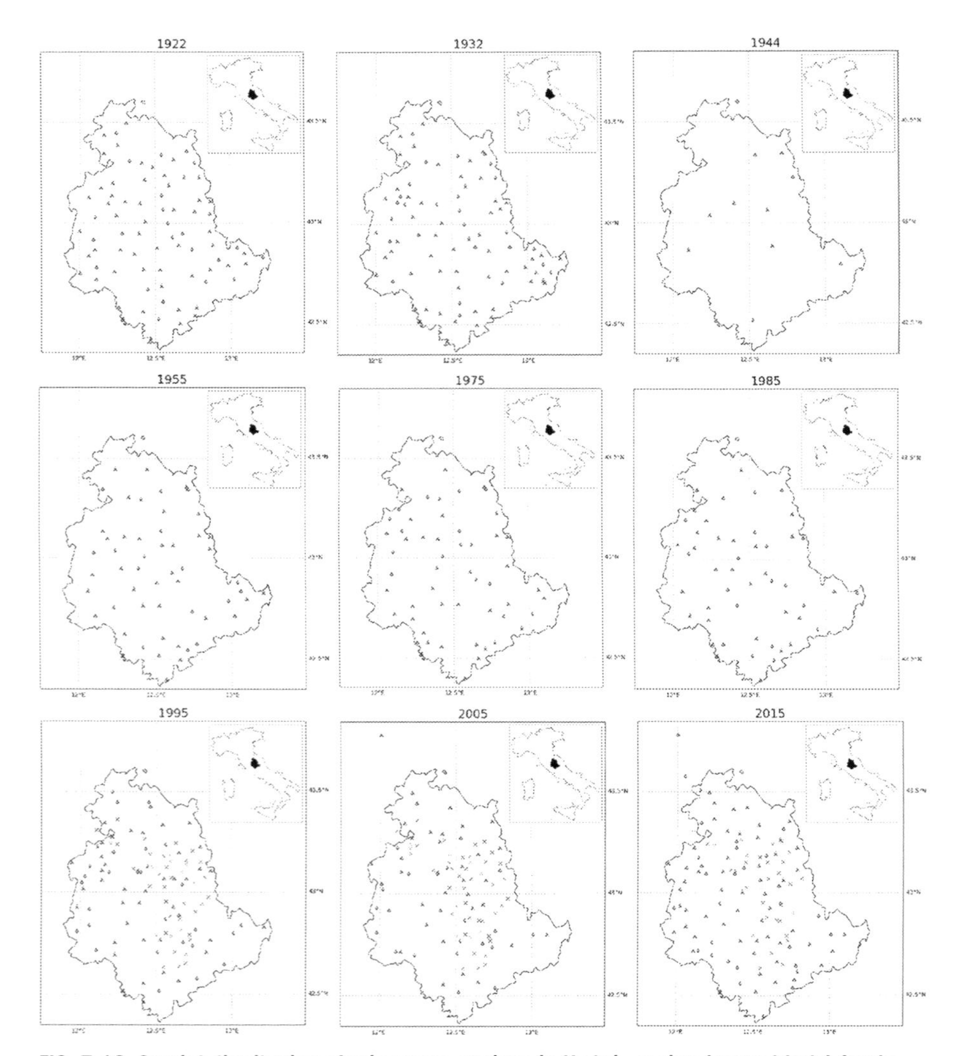

FIG. 7.12 Spatial distribution of rain gauge stations in Umbria region (central Italy) for the indicated years.

Symbol (▲) denotes rain gauge stations installed by the Italian National Hydrographic Service, symbol (x) rain gauges installed by the National Research Council.

Assuming each H_d value obtained from the "Observed" data as a benchmark, the H_d underestimate produced by the use of rainfall data with coarse t_a ("Generated") can be quantified. As representative cases, Tables 7.6 and 7.7 highlight the underestimation errors for the Bastia Umbria station considering temporal aggregations equal to 30 and 15 minutes, respectively. It can be seen that, for fixed t_a and d, errors randomly vary with years. For $t_a = d = 30$ minutes, the minimum underestimation error in Table 7.6 is practically negligible (0.32% in 1992), while the error increases to about 34% in 1997.

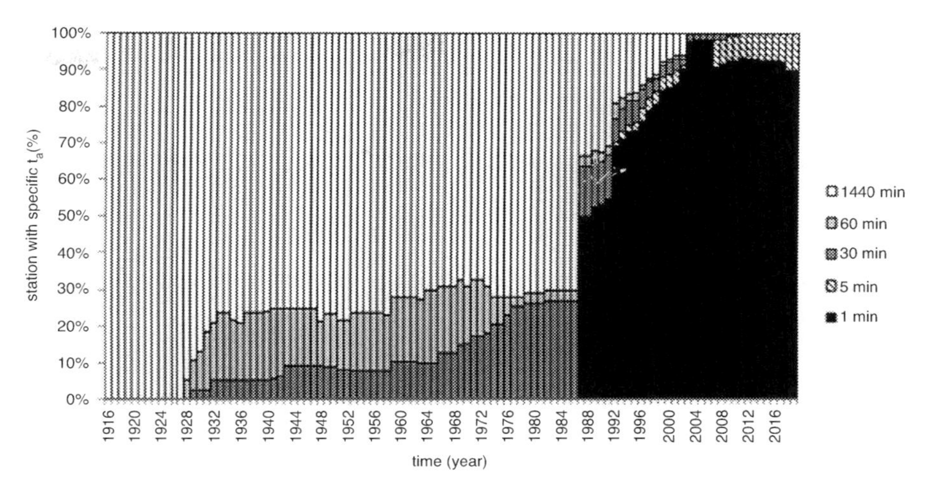

FIG. 7.13 Time evolution of the percentage of rain gauge stations in Umbria region (central Italy) with specific temporal aggregation, t_a.

(Source: Morbidelli et al., 2020)

Table 7.4 Different groups of representative rain gauge stations of the Umbria region (central Italy) with the time evolution of the adopted temporal aggregation, t_a.

Rain gauge station	From/to (year) t_a (min)	From/to (year) t_a (min)	From/to (year) t_a (min)	From/to (year) t_a (min)	From/to (year) t_a (min)
Very old stations installed by the Italian National Hydrographic Service					
Cannara	1915/1940 1440	1992/2020 1			
Foligno	1915/1927 1440	1928/1934 60	1938/1952 1440	1953/1973 60	1993/2020 1
Perugia	1915/1931 1440	1932/1996 30	2008/2010 1		
Todi	1921/1930 1440	1931/1942 60	1948/1958 1440	1959/1991 30	1992/2020 1
Stations installed by the Italian National Hydrographic Service after the Second World War					
Abeto	1951/1998 1440	2007/2020 1			
Calvi dell'Umbria	1951/2002 1440	2007/2020 5			
Lago di Corbara	1963/1992 1440	1993/2020 1			
Sellano	1951/2000 1440	2007/2020 5			

(continued)

Table 7.4 Different groups of representative rain gauge stations of the Umbria region (central Italy) with the time evolution of the adopted temporal aggregation, t_a. *Continued*

Rain gauge station	From/to (year) t_a (min)	From/to (year) t_a (min)	From/to (year) t_a (min)	From/to (year) t_a (min)	From/to (year) t_a (min)
Stations installed by the Regional Hydrographic Service					
Casa Castalda	1992/2020 1				
La Bruna	2011/2020 1				
Monte Cucco	1996/2020 1				
Ponte Felcino	1992/2020 1				
Stations installed by the National Research Council					
Cantinone	1986/2018 1				
Fosso Impiccati	2000/2018 1				
Monte Bibbico	1986/2018 1				
Valfabbrica	1986/2018 1				

(Source: Modified from Morbidelli et al., 2020)

In the considered period, on the average more significant errors occur when $t_a = d$, while they become less than 1% when $t_a/d<0.1$. A comparison of Tables 7.6 and 7.7 shows that the error magnitude, particularly in terms of average values, is mainly related to the ratio t_a/d. For example, values in the third column of Table 7.6 (where $d = 60$ minutes and $t_a/d = 0.5$) are comparable with those in the second column of Table 7.7 (where $d = 30$ minutes and $t_a/d = 0.5$). However, Table 7.8 shows that in some cases there are appreciable differences because the average underestimation errors depend on d, too. This justifies the choice of Morbidelli et al. (2017) who, for the correction to be applied to the H_d series, adopted a different equation for 3 different d-ranges (see Eq. [7.9]).

Furthermore, Fig. 7.14 indicates the existence of a limited dependence of the error magnitude on the rain gauge location.

In addition, Fig. 7.15 highlights the dependence of the average error on the data series length. The results of this analysis, performed using series with a length of at least 20 years, are synthesized through a few representative cases referred to $t_a/d = 1$ and $d >> t_a$, which determine extreme values of the average error. Fig. 7.15a–f indicate an irregular average error trend with increasing series dimension, independently of the ratio t_a/d. This is an expected result considering that H_d is a random variable.

Table 7.5 "Observed" and "Generated" rainfall data characterized by different temporal aggregations, t_a (in minutes), starting from January 1, 2006 at 0:00 a.m. at the Petrelle station (Umbria Region, central Italy).

"Observed" rainfall depth (mm)	"Generated" rainfall depth (mm)							
	t_a							
1	10	15	30	60	180	360	720	1440
0.0	0.6	0.6	0.7	2.2	12.8	20.5	23.8	32.1
0.1	0.1	0.1	1.5	4.8	7.7	3.3	8.3	2.6
0.1	0.0	0.6	1.3	5.8	2.7	7.1	0.1	5.5
0.1	0.2	0.9	3.5	6.7	0.6	1.2	2.5	0.0
0.1	0.5	0.4	3.9	0.8	5.9	0.1	5.5	0.0
0.1	0.8	0.9	1.9	0.2	1.2	0.0	0.0	0.0
0.1	0.2	1.2	5.7	0.2	1.1	0.0	0.0	0.0
0.0	0.5	2.3	1.0	1.9	0.1	2.5	0.0	0.0
0.0	0.6	2.2	0.6	0.6	0.1	5.2	0.0	0.0
0.0	0.6	1.7	0.2	0.2	0.0	0.3	0.0	0.0
0.0	1.3	0.8	0.1	0.1	0.0	0.0	0.0	0.0
0.0	1.6	1.1	0.1	0.3	0.0	0.0	0.0	0.0
0.0	1.7	2.6	0.1	0.9	0.0	0.0	0.0	0.0
0.0	1.4	3.1	0.1	1.1	0.0	0.0	0.0	0.0
0.0	0.8	0.4	1.8	3.9	0.0	0.0	0.0	0.0
0.0	0.5	0.6	0.1	0.2	2.5	0.0	0.0	0.1
0.0	0.7	0.5	0.3	0.4	4.7	0.0	0.0	0.2
0.1	0.7	0.1	0.3	0.6	0.5	0.0	0.0	0.9
0.0	1.2	0.1	0.1	0.6	0.3	0.0	0.0	0.0
0.0	2.3	0.1	0.1	0.3	0.0	0.0	0.0	0.0

(Source: Morbidelli et al., 2017)

However, it is possible to deduce the data series length required to obtain a reliable estimation of the average error. In most cases it should be approximately greater than 15 to 20 years, but for $d >> t_a$ (Fig. 7.15e) the average error magnitude is of minor importance even though shorter lengths are used.

The methodology proposed by Morbidelli et al. (2017) was widely validated, also with rainfall data from the following stations located in the Umbria region: Monte Cucco, Narni Scalo, Ponte Santa Maria and San Biagio della Valle. Each series of H_d values obtained with coarse t_a (10, 15, 30, ... minutes) was corrected by adding the quantity given by Eqs. (7.9a–7.9c) and then compared with the "Observed" series. The improvements obtained through the application of the developed methodology can be deduced from Fig. 7.16, where the average annual maximum rainfall depths, $\mu(H_d)$, for the "Uncorrected" and "Corrected" series are compared with the corresponding values derived from rainfall data with $t_a = 1$ min, $\mu(H_d)^{ta=1}$. As it can be seen, in all cases the residual average errors become of minor interest.

Table 7.6 Underestimation errors (in %) in the evaluation of the annual maximum rainfall depth considering rainfall data with temporal aggregation of 30 minutes and different durations, *d*, at the Bastia Umbra station (Umbria Region, central Italy).

	d (minutes)					
Year	**30**	**60**	**180**	**360**	**720**	**1440**
1992	0.32	0	0.41	0	0	0
1993	16.22	2.22	1.54	0.02	0.89	0.08
1994	0.76	1.26	0	0	0.03	0.03
1995	23.72	3.07	3.61	0.05	0	0
1996	9.70	8.85	0	1.48	0.81	0.02
1997	34.23	0.62	6.89	0.84	0	0
1998	30.32	5.28	2.54	0.99	0.29	0.02
1999	5.02	5.26	0	0	0	0
2000	10.95	4.90	0	0.08	0.21	0
2001	1.13	0	0.66	0	0	0
2002	7.51	12.79	1.67	1.17	0.53	0
2003	21.12	7.21	1.17	0.02	0	0
2004	11.27	4.95	0.17	0.24	0	0
2005	9.66	14.76	1.44	0.02	0.73	0
2006	14.09	7.70	0.51	4.19	0.01	0
2007	21.77	1.19	1.75	0.06	0.03	0.17
2008	2.21	4.14	4.94	0	0	0.09
2009	12.18	0.61	0.67	0	0	0
2010	5.16	0	0	0	2.03	0
2012	8.62	0	0	0	2.86	0.99
2013	1.55	1.14	0.18	0	0.18	0.13
2014	14.61	9.86	2.71	0.89	0	0.12
2015	3.95	21.74	1.60	0	0	0
Average	11.57	5.11	1.41	0.44	0.37	0.07

(Source: Morbidelli et al., 2017)

7.6.3 Influence of t_a on hydrological analyses in the study area

As a first hydrological analysis, the effect of the correction of H_d on the rainfall depth-duration-frequency curves was quantified. This issue is addressed through the description of both the adopted procedure and the results obtained for the representative rain gauge station of Gubbio (see also Fig. 7.10 and Table 7.3). For each duration, in addition to 24 values of H_d appropriately observed with $t_a = 1$ minute during the period with digital recording, 20 values obtained from data recorded earlier than 1992 with $t_a = 30$ minutes were used. This H_d series represents the "Uncorrected" one, while a series including the 24 values of H_d observed with $t_a = 1$ minute and the remaining 20 values modified by the methodology proposed

Table 7.7 Underestimation errors (in %) in the evaluation of the annual maximum rainfall depth considering rainfall data with temporal aggregation of 15 minutes and different durations, *d*, at the Bastia Umbra station (Umbria Region, Central Italy).

	d (minutes)					
Year	**30**	**60**	**180**	**360**	**720**	**1440**
1992	0.32	0	0.41	0	0	0
1993	4.27	2.22	0.09	0.02	0.89	0.03
1994	0.76	1.26	0	0	0.03	0.03
1995	0.99	3.07	3.61	0.05	0	0
1996	1.39	0.33	0	0	0.27	0.02
1997	6.64	0.62	0.33	0.84	0	0
1998	5.16	5.03	0.58	0.46	0.01	0
1999	5.02	5.26	0	0	0	0
2000	10.95	0.63	0	0.08	0.14	0
2001	1.13	0	0.13	0	0	0
2002	7.51	9.66	0	0.86	0.53	0
2003	0.41	1.03	1.17	0.02	0	0
2004	0.12	3.69	0.03	0.03	0	0
2005	9.66	1.56	1.44	0.02	0.73	0
2006	6.19	0	0.31	1.70	0.01	0
2007	5.44	1.19	0.45	0.06	0.03	0.17
2008	2.21	0.45	4.94	0	0	0.09
2009	12.18	0.61	0.07	0	0	0
2010	5.16	0	0	0	0	0
2012	8.62	0	0	0	1.15	0.44
2013	1.55	1.14	0	0	0.18	0.13
2014	1.37	0.35	1.81	0.89	0	0.12
2015	0	0	0.53	0	0	0
Average	4.22	1.66	0.69	0.22	0.17	0.04

(Source: Morbidelli et al., 2017)

by Morbidelli et al. (2017) is the "Corrected" series. The statistical analysis of each random variable H_d for different d was performed using the Generalized Extreme Value (Jenkinson, 1955; Coles, 2001) distribution function. Fig. 7.17 indicates that for durations up to 3 h the error expressed as a percentage of the annual maximum rainfall depth is slightly variable with both return period and duration and the use of the "Uncorrected" H_d series determines depth underestimations between 5% and 10%. Similar results were obtained for durations up to 24 h.

Then, considering that many H_d series involve a relevant number of possible underestimated values due to the availability of only coarse resolution rainfall data, an assessment of the influence of this aspect on trend analyses finalized to determine if climate change is having an impact on extreme events was carried out. Sixty H_d time

Table 7.8 Average underestimation errors (in %) in the evaluation of the annual maximum rainfall depth for some rainfall stations. Different values of duration, *d*, are considered. The symbol t_a denotes the aggregation time. In the last line the average values, representative of each duration, are shown.

Rain gauge station	d (minutes)					
	30	60	180	360	720	1440
	$t_a/d = 1$					
Bastardo	12.47	8.28	8.39	12.70	10.14	12.71
Bastia Umbra	11.57	13.30	13.95	12.70	11.74	7.50
Casa Castalda	12.81	8.72	10.00	15.26	12.05	11.13
Cerbara	13.18	10.61	10.93	10.49	10.47	11.11
Compignano	11.13	15.58	12.99	9.06	13.58	9.00
Forsivo	12.83	10.68	7.66	4.19	8.53	12.28
Gubbio	8.50	7.15	8.41	10.91	11.65	8.29
Montelovesco	14.25	13.98	9.00	6.70	9.14	10.63
Nocera Umbra	12.73	10.26	11.97	11.57	10.72	10.97
Petrelle	15.45	11.45	14.03	12.11	10.31	12.03
Ripalvella	10.74	12.77	13.11	14.03	11.26	10.87
San Silvestro	9.64	13.10	8.76	10.04	8.66	8.70
	12.11	11.32	10.77	10.81	10.69	10.43

(Source: Morbidelli et al., 2017)

series associated with durations 1 h, 3 h, 6 h, 12 h, 24 h and 48 h for 10 rain gauge stations with longest recordings (Amelia, Città di Castello, Foligno, Gualdo Tadino, Gubbio, Orvieto, Spoleto, Terni, Todi and Umbertide) were selected. An example of the available H_d series is shown in Table 7.9 for the rain gauge station of Città di Castello, where for $d = 1$ h there are 27 values (53% of the total) in the period 1921–1991, characterized by $t_a/d = 1$. Therefore, single underestimation errors up to 50% and average error equal to ~11% could be involved (Morbidelli et al., 2017). Each of the selected "Uncorrected" series was used to investigate the existence of possible trends induced by climate change. Specifically, the least-squares linear trend analysis, the non-parametrical Mann-Kendall test (Mann, 1945; Kendall, 1975), the Spearman rank correlation test (Khaliq et al., 2009), and the Sen's method (Sen, 2012) were considered. The same tests were also repeated on a "Corrected" version of the series in which the underestimation errors due to coarse temporal aggregation of historical rainfall data were minimized/eliminated by using Eqs. (7.9a–7.9c). In the deterministic approach adopted during the correction phase, an average correction was identically applied to all H_d values characterized by the same ratio t_a/d. To this aim, the average underestimation errors were determined. All corrections with average underestimation error less than 1% were neglected.

As a first test, a least-squares linear trend was fitted to the annual maxima for the 60 datasets. For the "Uncorrected" H_d series, the positive least-squares linear trends (28) slightly outnumbers the negative ones (27) (see values in bold in Table 7.10).

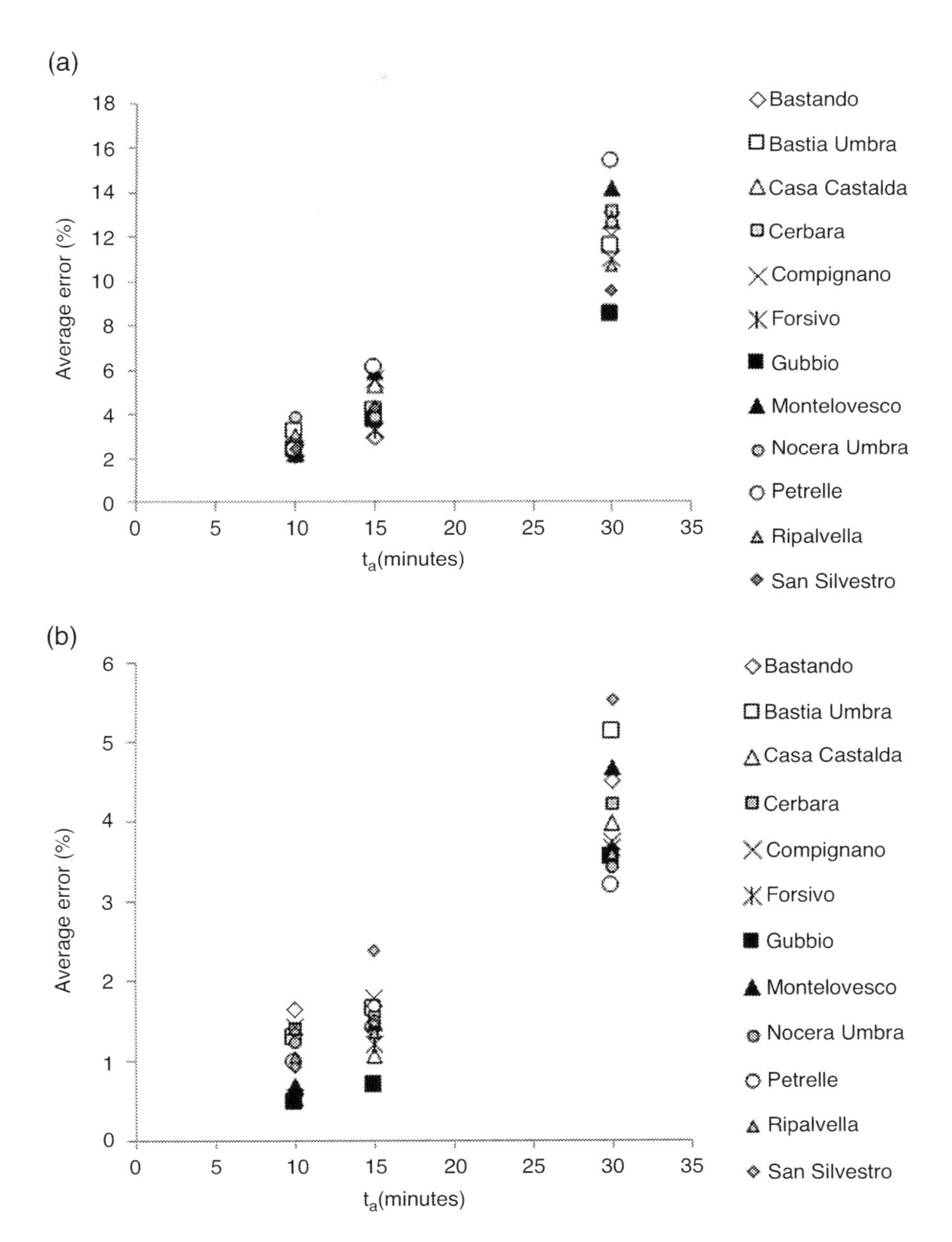

FIG. 7.14 Average underestimation error of the annual maximum rainfall depth as a function of the aggregation time, t_a, for two different durations: (a) $d = 30$ minutes, (b) $d = 60$ minutes.

Data for different stations are compared.

(Source: Morbidelli et al., 2017)

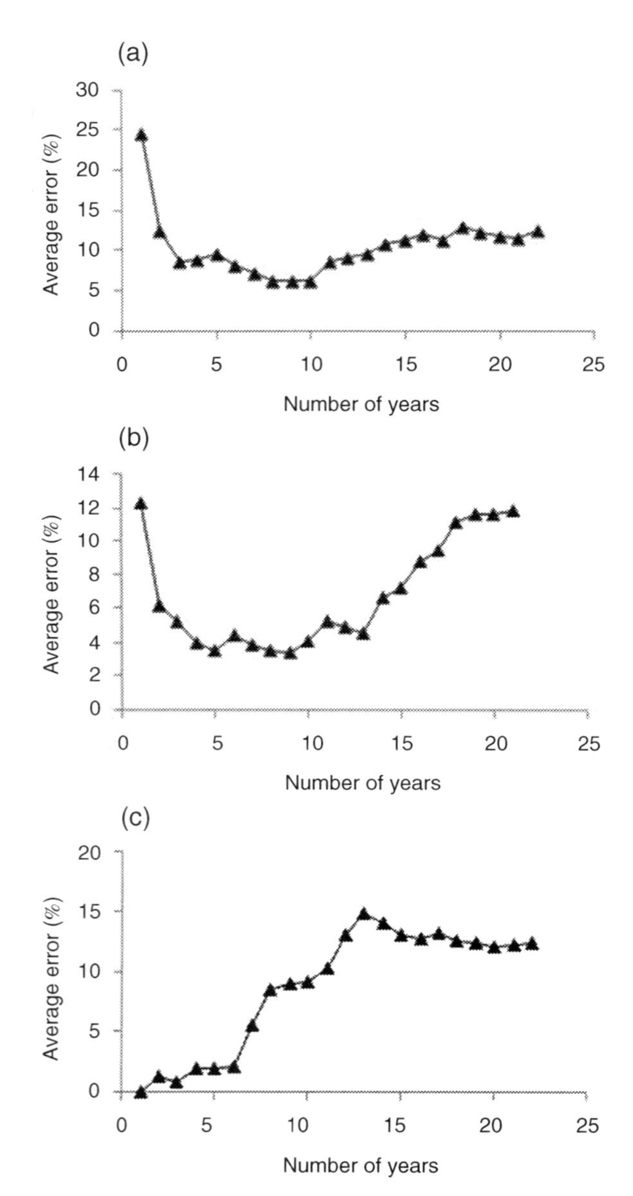

FIG. 7.15 Average errors in the evaluation of annual maximum rainfall depth, H_d, as a function of the number of years preceding the last H_d value for different combinations of aggregation time, t_a (in minutes), and duration, d (in minutes): (a) Bastardo station, $t_a = 30$, $d = 30$; (b) Cerbara station, $t_a = 30$, $d = 30$; (c) Ponte Santa Maria station, $t_a = 60$, $d = 60$; (d) Ripalvella station, $t_a = 60$, $d = 60$; (e) Nocera Umbra station, $t_a = 30$, $d = 180$; (f) Nocera Umbra station, $t_a = 30$, $d = 30$.

(Source: Morbidelli et al., 2017)

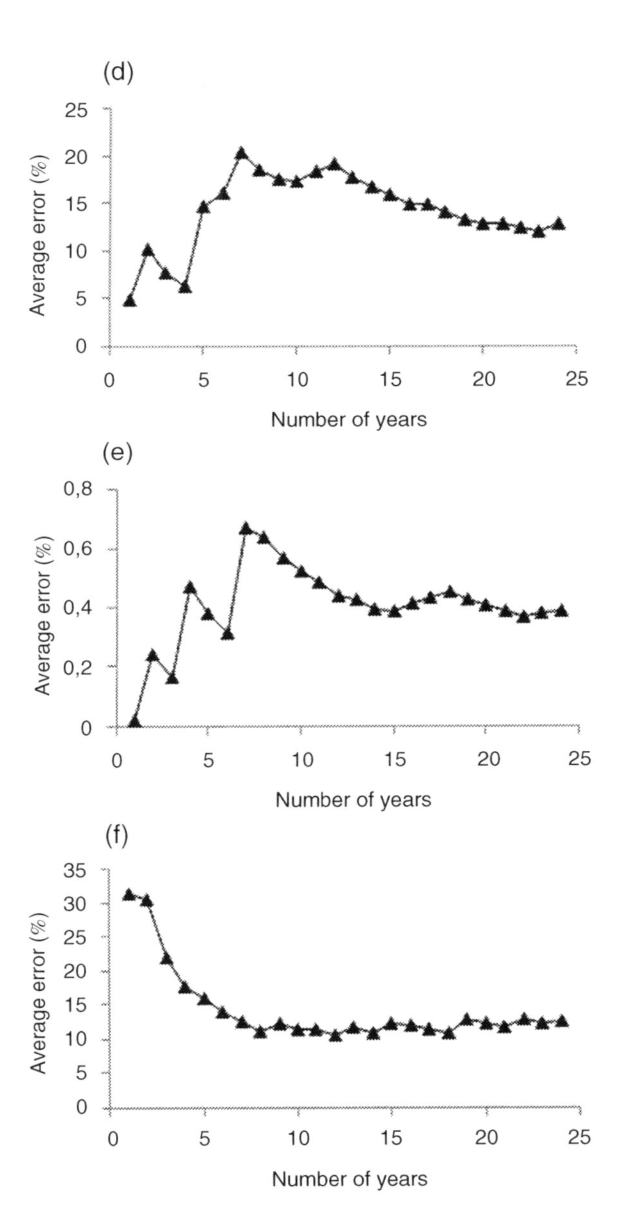

FIG. 7.15 (*Continued*)

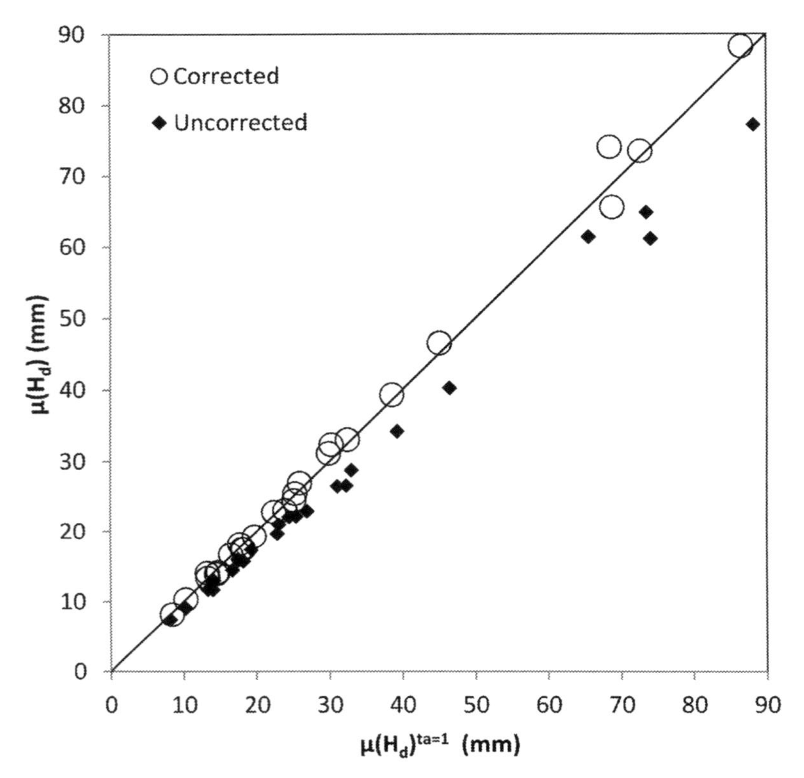

FIG. 7.16 Average annual maximum rainfall depths, $\mu(H_d)$, versus the corresponding values derived from rainfall data with $t_a = 1$ minute, $\mu(H_d)^{ta=1}$.

"Uncorrected" and "Corrected" stand for the values obtained before and after the application of the methodology proposed by Morbidelli et al. (2017), respectively. Rainfall data from Monte Cucco, Narni Scalo, Ponte Santa Maria, and San Biagio della Valle stations, located in the Umbria region. Only cases with duration equal to the aggregation time are shown.

(Source: Morbidelli et al., 2017)

Opposing results were obtained for the "Corrected" H_d series. In fact, after corrections, cases with negative least-square linear trends became 36, while the positive ones reduced to 22. Table 7.10 clearly shows that the corrections of H_d values produce significant effects on the least-square linear trends, particularly for series characterized by durations 1 h and 24 h in which the possibility to find values with $t_a/d = 1$ is particularly high.

Fig. 7.18 and 7.19 show a comparison between "Uncorrected" and "Corrected" annual maximum rainfall depths at Umbertide for durations 1 h and 24 h, respectively. It can be seen that the "Uncorrected" series show a positive trend (Figs. 7.18a and 7.19a) while the "Corrected" ones have a negative trend (Figs. 7.18b and 7.19b). The trends in Figs. 7.18 and 7.19 are not significant, but they are anyway important in terms of general trend evaluation.

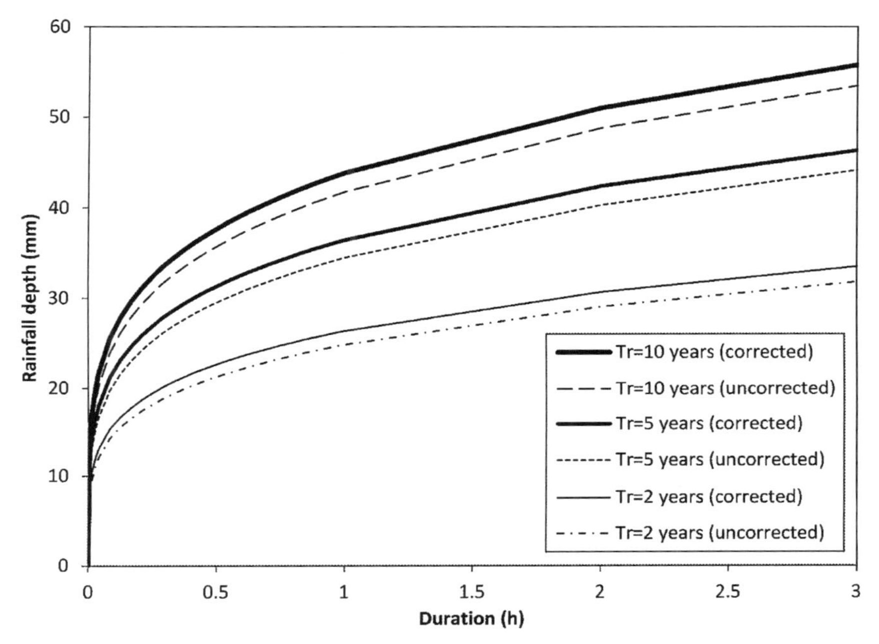

FIG. 7.17 Rainfall depth-duration-frequency curves for different return periods, T_r.

Comparison of curves obtained by "Uncorrected" H_d series and corresponding series "Corrected" by the Morbidelli et al. (2017) methodology. Sample rain gauge station of Gubbio in Umbria region.

(Source: Morbidelli et al., 2017)

The analysis of the 60 time series was also performed by the non-parametric Mann- Kendall (Mann, 1945; Kendall, 1975) and the Spearman rank correlation (Khaliq et al., 2009) tests, which are frequently used to detect trends in heavy rainfall (Fu et al., 2004; Houssos and Bartzokas, 2006; Ramos and Martinez-Casasnovas, 2006; Su et al., 2006; Fatichi and Caporali, 2009; Rashid et al., 2015). For the "Uncorrected" series, by using a significance level equal to 0.05, the percentage without relevant trends was independent of the adopted test. Specifically, for both tests this percentage was higher than 93%, while 2 series (Spoleto station with rainfall durations 24 h and 48 h) had a negative trend and 2 series (Gubbio station with rainfall durations 1 h and 24 h) a positive trend. After the correction, the Mann-Kendall and the Spearman rank correlation tests provided 3 cases with a negative trend and no positive case. In fact, also the Todi station with $d = 1$ h became characterized by a negative trend. It is interesting to note that in this last case the least-square linear trend was negative also before applying the correction by Eqs. (7.9a–7.9c), with slope of the linear regression equal to −0.05 mm/year. However, after the correction a pronounced negative trend was obtained (slope of the linear regression equal to −0.12 mm/year). Finally, it is evident that in the transition from "Uncorrected" to "Corrected" H_d series, the application of the Mann-Kendall and Spearman rank

Table 7.9 Rainfall data observed at Città di Castello: annual maximum rainfall rates (in mm) for different durations, d, and aggregation times, t_a.

	duration (h)							duration (h)					
year	1	3	6	12	24	48	year	1	3	6	12	24	48
1921					45.9	78.3	1965	28	49	70	108.2	111.2	132.6
1922	10				42.9	72.7	1966	33.5	37	46	62.4	67.4	69.2
1923					55.8	84.5	1967	29.6	39	39.2	39.4	39.6	42.4
1926					32.4	46.5	1968	27.8	39	57	73.6	74.2	84.4
1927					21.3	37.4	1969	26	37.4	49.8	58.6	61	65.3
1928					75	106.2	1970	18.4	21.8	23.8	33	45.8	58.4
1929	17.1	22.4	22.4	22.4	38.6	63.1	1971	14	25.8	30.2	43.2	48.2	58.2
1930	14	35	35	56.5	69	105	1972					36.4	57.6
1931					69	119.4	1973					59.2	59.6
1932	25.4	29.8	37	46	62.2	65.9	1982					68.2	70.6
1933	15	20.6		52.8	74	89	1989					56.6	86
1934	21						1990					31.2	61
1937					158.6	178.3	1991					51.2	61
1938	45.2	66.6	66.8	67	83	83	1992	35.5	61.3	67.7	72.6	74.7	83
1939	15.8	23.4	33	54.8	88	92.6	1993	36.7	42.5	43.6	44.3	44.8	71.4
1940		26.2			62.4	78.4	1994	34.8	35.4	36.5	50.4	65	83
1941					158	165.4	1995	24.1	26.7	32.4	44.1	62.8	63.4
1942	33.6	33.6	46.6	58.8	65.2	65.4	1996	21.1	30.4	52.7	73.7	78.9	106.4
1943	17.4	27.2	35.4	48.4	56.2	59.8	1997	18.4	30.9	35.9	55.9	80	93.2

	duration (h)							duration (h)					
1946	23.8	43	49.4	59.4	69	78	1998	21.4	29.6	31.2	36.1	54.7	58
1947	40.2	40.2	40.2	47.4	59	64.5	1999	27.3	28.6	34.4	40.1	43.7	66.3
1948	15.4	24.2	28	36	67.8	76.2	2000	22.1	33.2	37.7	51.9	74.8	93.9
1950	23	28.6	41.4	50	50.4	50.5	2001	23.2	25.8	45.9	49.9	50.5	59.6
1951	20				42.4	49.2	2002	20.7	26.8	29.2	38.8	51.5	60.5
1952					54	63.6	2003	23.1	28.6	30.5	32	43.5	70
1953					49.5	66	2004	27	27.1	27.2	30.2	48.7	60.5
1954					41	62.2	2005	26.8	35.1	39.1	50.2	77.7	89.3
1955	15	20	37	44	47.4	55.4	2006	22.1	33.5	46.4	65.4	83.1	86.8
1956					34.5	68.5	2007	18.7	24.3	29.8	31.1	39	47.3
1957					36	46.6	2008	23.6	24	29.4	42.4	62	76.6
1958	14.5				42.2	58.6	2009	25.1	25.5	25.9	26.5	38.1	53.7
1959					51.6	69.4	2010	28.1	31.1	51.3	70	73.4	83.5
1960					104	140	2011	21.4	29.5	37.3	38.7	40.4	50
1961					45.4	61.2	2012	31.2	32.4	35.7	65.3	119.6	163.4
1962	30.8	30.8	30.8	36	49	66.8	2013	25.2	31.8	46.1	68.8	93.5	106.4
1963	28.4	32.6	51	56.5	61	79.6	2014	22.5	23.7	32.6	45.3	47.1	69.1
1964	39.2	39.2	46.4	54.5	69.8	72.4	2015	19.6	24.8	28.6	48.1	67.3	70.6

t_a = 1 min

t_a = 1 h

t_a = 1 day

(Source: Morbidelli et al., 2018a)

Table 7.10 Slope (in mm/year) of the least-squares linear regressions of annual maximum rainfall depths series for the selected stations and for different rainfall durations. Values for the "Uncorrected" and "Corrected" series are reported with bold and normal characters, respectively.

Rain gauge station	Duration (h)					
	1	3	6	12	24	48
Amelia	**+0.01**	**-0.22**	**-0.22**	**-0.17**	**-0.05**	**-0.17**
	-0.05	-0.24	-0.24	-0.18	-0.13	-0.21
Città di Castello	**+0.04**	**-0.04**	**-0.06**	**-0.04**	**-0.01**	**-0.06**
	-0.01	-0.05	-0.07	-0.04	-0.12	-0.11
Foligno	**+0.05**	**-0.03**	**-0.04**	**0.00**	**+0.09**	**+0.07**
	+0.01	-0.04	-0.05	0.00	+0.04	+0.04
Gualdo Tadino	**+0.06**	**-0.02**	**-0.06**	**+0.14**	**+0.04**	**-0.05**
	+0.02	-0.03	-0.07	+0.14	-0.04	-0.09
Gubbio	**+0.11**	**+0.05**	**+0.10**	**+0.10**	**+0.19**	**+0.15**
	+0.07	+0.04	+0.09	+0.09	+0.11	+0.12
Orvieto	**-0.01**	**-0.11**	**+0.01**	**+0.04**	**+0.16**	**+0.14**
	-0.05	-0.12	+0.01	+0.03	+0.10	+0.10
Spoleto	**+0.06**	**+0.01**	**-0.08**	**-0.15**	**-0.25**	**-0.43**
	+0.03	-0.01	-0.08	-0.16	-0.33	-0.48
Terni	**+0.10**	**+0.19**	**+0.17**	**+0.10**	**0.00**	**-0.06**
	+0.06	+0.18	+0.17	+0.10	-0.07	-0.09
Todi	**-0.05**	**-0.19**	**-0.21**	**-0.24**	**-0.02**	**0.00**
	-0.12	-0.21	-0.22	-0.25	-0.11	-0.04
Umbertide	**+0.03**	**+0.07**	**0.00**	**0.00**	**+0.05**	**+0.06**
	-0.01	+0.06	-0.02	0.00	-0.02	+0.03

(Source: modified from Morbidelli et al., 2018a)

correlation tests produces different results particularly when $d = 1$ or 24 h for which the possibility to find H_d values with $t_a/d = 1$ is particularly high.

Sen's innovative method (Sen, 2012) was adopted in many analyses (Sonali and Kumar, 2013; Guclu, 2016; Oztopal and Sen, 2016; Guclu, 2018; Guclu et al., 2018), particularly for a comparison with classical methods. It is based on the realization of two sub-series (in ascending order) and offers the following advantages: 1) the chance to clearly visualize results on graph, and 2) the possibility to consider five trend types (monotonic increasing and decreasing, non-monotonic increasing and decreasing, and no trend conditions) instead of the usual three types (monotonic increasing and decreasing, and no trend condition). Similarly to the most classic methodologies, the adoption of Sen's method produced different results when applied to the "Uncorrected" and "Corrected" H_d series. As an example, Fig. 7.20 shows the application of the method to both the "Uncorrected" and "Corrected" H_d series observed at Todi considering $d = 1$ h. As it can be seen in Fig. 7.20a the "Uncorrected"

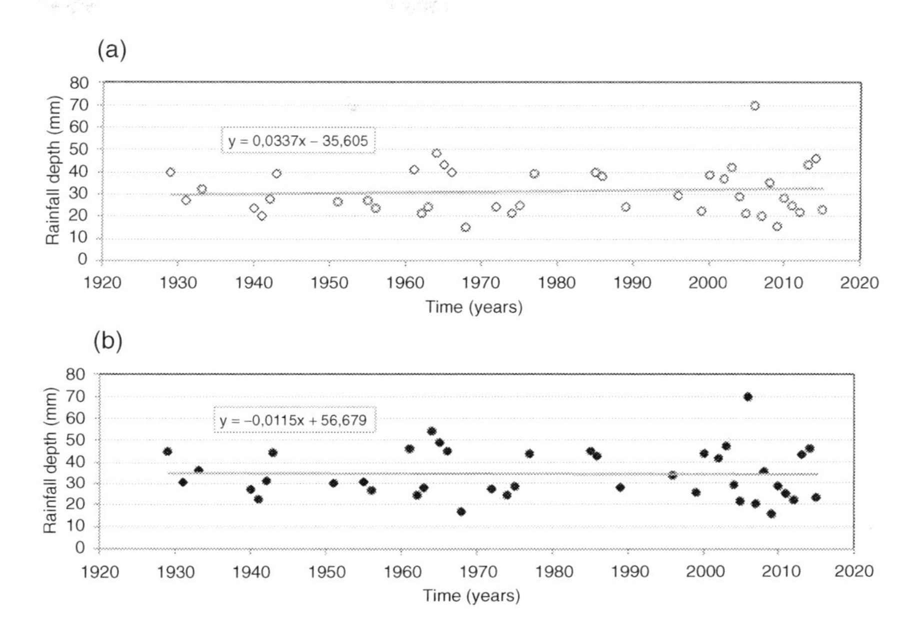

FIG. 7.18 Time sequence of annual maximum rainfall depths for duration 1 h with the corresponding linear trends: (a) "Uncorrected" series, (b) "Corrected" series.

Available data within the period 1929–2015 for the Umbertide station.

(Source: Morbidelli et al., 2018a)

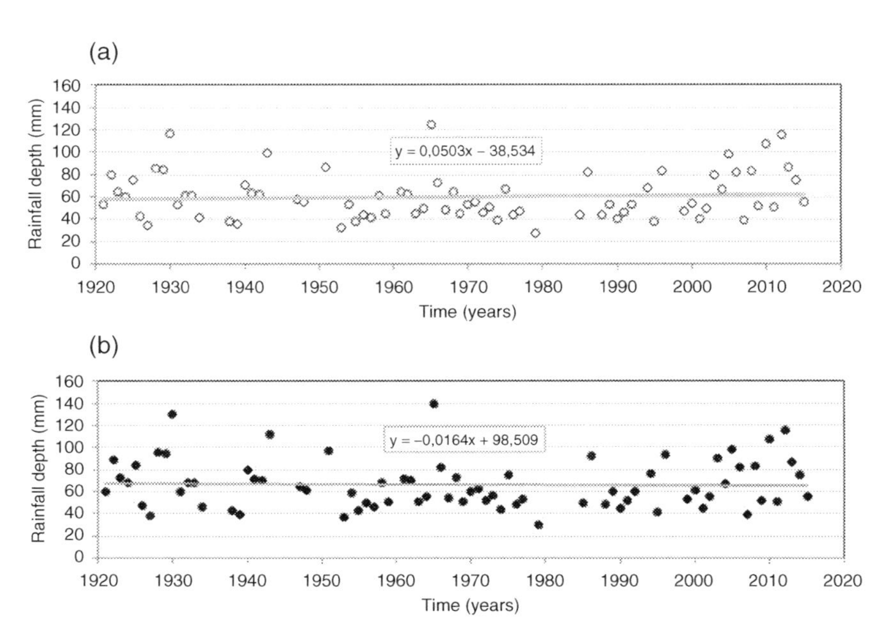

FIG. 7.19 Time sequence of annual maximum rainfall depths for duration 24 h with the corresponding linear trends: (a) "Uncorrected" series, (b) "Corrected" series.

Available data within the period 1921–2015 for the Umbertide station.

(Source: Morbidelli et al., 2018a)

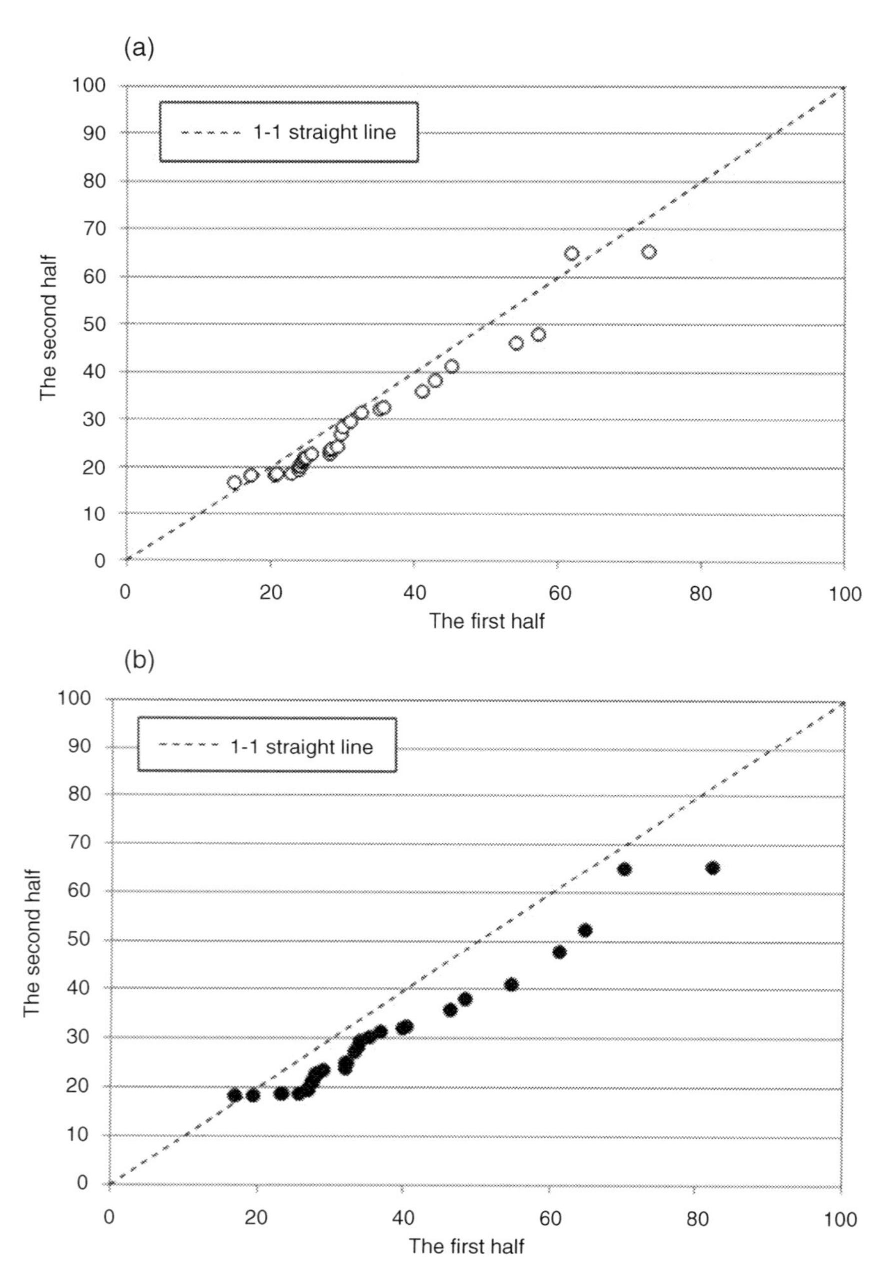

FIG. 7.20 Trend conditions according to Sen's method for the annual maximum rainfall depths of duration 1 h: (a) "Uncorrected" series, (b) "Corrected" series.

Available data within the period 1931–2015 for the Todi station. (Source: Morbidelli et al., 2018a) series shows a no trend condition because all values are close to 1:1 line, while in Fig. 7.20b the "Corrected" series is characterized by a clear monotonic decreasing trend (for further details on the results interpretation see Sen, 2012).

7.7 Conclusions

Historical rainfall data are available with different temporal aggregations, mainly within the range 1 minute-1 day, linked to the progress of recording systems through time. This chapter, in the first section, analyzes the time evolution of t_a on the basis of a large rainfall database of rain gauge networks operative in many geographic areas. Is has been pointed out that for the dated rain gauge networks, installed in the 19th century or during the first decades of the 20th century, recordings started in manual mode with coarse time-resolution, in the range 1 day-1 month, while mechanical recordings on paper rolls, with t_a typically in the range 30 minutes to 1 h, began in the first half of the 20th century. Digital data logger registrations started during the last two decades of the 20th century providing the possibility of any temporal aggregation, also equal to 1 minute; today the estimated percentage of this kind of stations in the world is about 50%. Most of older rain gauge networks have changed the registration methods during their operation period (in some cases more than one time) from manual to mechanical and finally to digital.

This chapter also demonstrates that, because of rainfall data available with coarse t_a, the use of "Uncorrected" H_d series for the determination of rainfall-depth-duration-frequency curves can lead to underestimates in the order of 10% and affect the most common methods used to evaluate trend signals in intense rainfall.

However, by using specific mathematical relations between the average underestimation error and the ratio t_a/d, each H_d value may be corrected obtaining H_d series appropriate for every analysis type.

References

Alexandersson, H, 1986. A homogeneity test applied to precipitation data. J. Climatol 6 (6), 661–675.

Al-Rawas, G.A., Valeo, C, 2009. Characteristics of rainstorm temporal distributions in arid mountainous and coastal regions. J. Hydrol. 376 (1-2), 318–326.

Austin, P.M., 1987. Relation between measured radar reflectivity and surface rainfall. Mon. Wea. Rev. 115 (5), 1053–1070.

Assouline, S, 2020. On the relationships between radar reflectivity and rainfall rate and kinetic energy resulting from a Weibull drop size distribution. Water Resour. Res. 56 (10), e2020WR028156.

Balme, M., Vischel, T., Lebel, T., Peugeot, C., Galle, S, 2006. Assessing the water balance in the Sahel: impact of small scale rainfall variability on runoff, part 1: rainfall variability analysis. J. Hydrol. 331 (1-2), 336–348.

Barrett, E.C., Beaumont, M.J, 1994. Satellite rainfall monitoring: an overview. Remote Sens. Rev. 11 (1-4), 23–48.

Boni, G., Parodi, A., Rudari, R, 2006. Extreme rainfall events: learning from raingauge time series. J. Hydrol. 327 (3-4), 304–314.

Borga, M., Marra, F., Gabella, M., 2022. Rainfall Estimation by Weather Radar. In: Morbidelli, R. (Ed.), Rainfall. Modeling, Measurement and Applications. Elsevier, Amsterdam, pp. 109–134. doi:10.1016/C2019-0-04937-0.

Burlando, P., Rosso, R, 1996. Scaling and multiscaling models of depth-duration-frequency curves for storm precipitation. J. Hydrol. 187 (1-2), 45–64.

Coles, S, 2001. An Introduction to Statistical Modelling of Extreme Value. Springer, London.

Coutinho, J.V., Almeida, C.D.N., Leal, A.M.F., Barbosa, L.R, 2014. Characterization of sub-daily rainfall properties in three rainfall gauges located in Northeast of Brazil, Evolving Water Resources Systems: Understanding, Predicting and Managing Water-Society Interactions. Proc of ICWRS2014. Bologna, Italy. IAHS Publ, 364, (June), pp. 345–350 .

Deidda, R., Mascaro, G., Piga, E., Querzoli, G, 2007. An automatic system for rainfall signal recognition from tipping bucket gage strip charts. J. Hydrol. 333 (2-4), 400–412.

Faiers, G.E., Grymes, J.M., Keim, B.D., Muller, R.A, 1994. A re-examination of extreme 24 hour rainfall in Louisiana, USA. Clim. Res. 4 (1), 25–31.

Fatichi, S., Caporali, E, 2009. A comprehensive analysis of changes in precipitation regime in Tuscany. Int. J. Climatol. 29 (13), 1883–1893.

Fread, D.L., Shedd, R.C., Smith, G.F., Farnsworth, R., Hoffeditz, C.N., Wenzel, L.A., Wiele, S.M., Smith, J.A., Day, G.N, 1995. Modernization in the National Weather Service River and Flood Program. Weather Forecast 10 (3), 477–484.

Fu, G., Chen, S., Liu, C., Shepard, D, 2004. Hydro-climatic trends of the Yellow River basin for the last 50 years. Clim. Change 65 (1–2), 149–178.

Guclu, Y.S, 2016. Comments on "comparison of Mann-Kendall and innovative trend method for water quality parameters of the Kizilirmac River, Turkey (Kisi and Ay, 2014)" and "an innovative method for trend analysis of monthly pan evaporations (Kisi, 2015)". J. Hydrol. 538, 878–882.

Guclu, Y.S, 2018. Alternative trend analysis: half time series methodology. Water Resour. Manag. 32 (7), 2489–2504.

Guclu, Y.S., Sisman, E., M.O, Yelegen, 2018. Climate change e frequency-intensity-duration (FID) curves for Florya station, Istambul. J. Flood Risk Manag. 11 (S1), S403–S418.

Harihara, P.S., Tripathi, N, 1973. Relationship of the clock-hour to 60-min and the observational day to 1440-min rainfall. Ind. J. Meteorol. Geophys. 24 (3), 279–282.

Hershfield, D.M, 1961. Rainfall frequency atlas of the United States for durations from 30 minutes to 24 hours and return periods from 1 to 100 yearsUS Weather Bureau Technical Paper N. 40. U.S. Dept. of Commerce, Washington, DC.

Hershfield D.M., Wilson W.T. Generalizing of rainfall-intensity-frequency data. 43: 499-506, 1958.

Hou, A.Y., Kakar, R.K., Neeck, S., Azarbarzin, A.A., Kummerow, C.D., Kojima, M., Oki, R., Nakamura, K., Iguchi, T, 2014. The global precipitation measurements mission. Bull. Amer. Meteor. Soc. 95 (5), 701–722.

Houssos, E.E., Bartzokas, A, 2006. Extreme precipitation events in NW Greece. Adv. Geosci. 7, 91–96.

Jenkinson, A.F, 1955. The frequency distribution of the annual maximum (or minimum) values of meteorological elements. Q J R Meteorol. Soc. 81 (348), 158–171.

Joyce, R.J., Xie, P., Janowiak, J.E, 2011. Kalman filter-based CMORPH. J. Hydrometeor. 12 (6), 1547–1563.

Kendall, M.G, 1975. Rank Correlation Methods. Griffin, London.

Khaliq, M.N., Ouarda, T.B.M.J., Gachon, P., Sushama, L., St-Hilaire, A, 2009. Identification of hydrological trends in the presence of serial and cross correlation: a review of selected methods and their application to annual flow regimes of Canadian rivers. J. Hydrol. 368 (1–4), 117–130.

Kidd, C., Levizzani, V., 2022. Satellite Rainfall Estimation. In: Morbidelli, R. (Ed.), Rainfall. Modeling, Measurement and Applications. Elsevier, Amsterdam, pp. 135–170. doi:10.1016/C2019-0-04937-0.

Kucera, P.A., Ebert, E.E., Turk, F.J., Levizzani, V., Kirschbaum, D., Tapiador, F.J., Loew, A., Borsche, M, 2013. Precipitation from space: advancing Earth system science. Bull. Amer. Meteor. Soc. 94 (3), 365–375.

Kuligowski, R.J, 2002. A self-calibrating real-time GOES rainfall algorithm for short-term rainfall estimates. J. Hydrometeor. 3 (2), 112–130.

Lanza, L.G., Cauteruccio, A., Stagnaro, M., 2022. Rain Gauge Measurements. In: Morbidelli, R. (Ed.), Rainfall. Modeling, Measurement and Applications. Elsevier, Amsterdam, pp. 77–108. doi:10.1016/C2019-0-04937-0.

Linke, A.M., O'Loughlin, J., McCabe, J.T., Tir, J., Witmer, F.D.W., 2015. Rainfall variability and violence in rural Kenya: investigating the effects of drought and the role of local institutions with survey data. Global Environ. Change 34 (1), 35–47.

Mann, H.B, 1945. Nonparametric tests against trend. Econometrica 13, 46–59.

Melone, F., Corradini, C., Morbidelli, R., Saltalippi, C, 2006. Laboratory experimental check of a conceptual model for infiltration under complex rainfall patterns. Hydrol. Proc 20 (3), 439–452.

Melone, F., Corradini, C., Morbidelli, R., Saltalippi, C., Flammini, A, 2008. Comparison of theoretical and experimental soil moisture profiles under complex rainfall patterns. J. Hydrol. Eng 13 (12), 1170–1176.

Miller, J.F., Frederick, R.H., Tracey, R.J, 1973. Precipitation-frequency atlas of the western United States. NOAA Atlas 2, National Weather Service, National Oceanic and Atmospheric Administration, U.S. Dept. of Commerce, Washington DC.

Morbidelli, R., García-Marín, A.P., Al Mamun, A., Atiqur, R.M., Ayuso-Muñoz, J.L., Bachir Taouti, M., Baranowski, P., Bellocchi, G., Sangüesa-Pool, C., Bennett, B., Oyunmunkh, B., Bonaccorso, B., Brocca, L., Caloiero, T., Caporali, E., Caracciolo, D., Casas-Castillo, M.C., Catalini, C.G., Chettih, M., ChowdhuryA.F.M.K., Chowdhury R., Corradini, C., Custò, J., Dari, J., Diodato, N., Doesken, N., Dumitrescu, A., Estévez, J., Flammini, A., Fowler, H.J., Freni, G., Fusto, F., García-Barrón, L., Manea, A., Goenster-Jordan, S., Hinson, S., Kanecka-Geszke, E., Kanti Kar, K., Kasperska-Wołowicz, W., Krabbi, M., Krzyszczak, J., Llabrés-Brustenga, A., Ledesma, J.L.J., Liu, T., Lompi, M., Marsico, L., Mascaro, G., Moramarco, T., Newman, N., Orzan, A., Pampaloni, M., Pizarro-Tapia, R., Puentes Torres, A., Rashid, M.M., Rodríguez-Solà, R., Sepulveda Manzor, M., Siwek, K., Sousa, A., Timbadiya, P.V., Filippos, T., Vilcea, M.G., Viterbo, F., Yoo, C., Zeri, M., Zittis, G., Saltalippi, C, 2020. The history of rainfall data time-resolution in different geographical areas of the world. J. Hydrol. 590, 125258.

Morbidelli, R., Saltalippi, C., Flammini, A., Cifrodelli, M., Picciafuoco, T., Corradini, C., Casas-Castillo, M.C., Fowler, H.J., Wilkinson, S.M, 2017. Effect of temporal aggregation on the estimate of annual maximum rainfall depths for the design of hydraulic infrastructure systems. J. Hydrol. 554, 710–720.

Morbidelli, R., Saltalippi, C., Flammini, A., Corradini, C., Wilkinson, S.M., Fowler, H.J., 2018a. Influence of temporal data aggregation on trend estimation for intense rainfall. Adv. Water Resour. 122, 304–316.

Morbidelli, R., Saltalippi, C., Flammini, A., Picciafuoco, T., Dari, J., Corradini, C, 2018b. Characteristics of the underestimation error of annual maximum rainfall depth due to coarse temporal aggregation. Atmosphere 9, 303.

New, M., Todd, M., Hulme, M., Jones, P.D, 2001. Precipitation measurements and trends in the twentieth century. Int. J. Climatol. 21 (15), 1899–1922.

Oztopal, A., Sen, Z, 2016. Innovative trend methodology applications to precipitation records in Turkey. Water Resour. Manag. 31 (3), 727–737.

Papalexiou, S.M., Dialynas, Y.G., Grimaldi, S, 2016. Hershfield factor revisited: correcting annual maximum precipitation. J. Hydrol. 524, 884–895.

Pettitt, A.N, 1979. A non-parametric approach to the change-point detection. Appl. Statist. 28 (2), 126–135.

Ramos, M.C., Martinez-Casasnovas, J.A, 2006. Trends in precipitation concentration and extremes in the Mediterranean Penedés-Anoia Region, NE Spain. Clim. Change 74 (4), 457–474.

Rashid, M.M., Beecham, S., Chowdhury, R.K, 2015. Assessment of trends in point rainfall using continuous wavelet transforms. Adv. Water Resour. 82, 1–15.

Sen, Z, 2012. Innovative trend analysis methodology. J. Hydrol. Eng. 17 (9), 1042–1046.

Seo, D.-J, 1998. Real-time estimation of rainfall fields using radar rainfall and rain gage data. J. Hydrol. 208 (1-2), 37–52.

Sevruk B., Klemm S. Catalogue of national standard precipitation gauges. Instruments and observing methods. Report No. 39, WMO/TD-No. 313, 1989.

Skofronick-Jackson, G., Petersen, W.A., Berg, W., Kidd, C., Stocker, E.F., Kirschbaum, D.B., Kakar, R., Braun, S.A., Huffman, G.J., Iguchi, T., Kirstetter, P.E., Kummerow, C., Meneghini, R., Oki, R., Olson, W.S., Takayabu, Y.N., Furukawa, K., Wilheit, T, 2017. The global precipitation measurement (GPM) mission for science and society. Bull. Amer. Meteor. Soc. 98 (8), 1679–1695.

Smith, J.A., Seo, D.-J., Baeck, M.L., Hudlow, M.D, 1996. An intercomparison study of NEXRAD precipitation estimates. Water Resour. Res. 32 (7), 2035–2045.

Sonali, P., Kumar, N.D, 2013. Review of trend detection methods and their application to detect temperature changes in India. J. Hydrol. 476, 212–227.

Sorooshian, S., Hsu, K.-L., Gao, X., Gupta, H.V., Imam, B., Braithwaite, D, 2000. Evaluation of PERSIANN system satellite-based estimates of tropical rainfall. Bull. Amer. Meteor. Soc. 81 (9), 2035–2046.

Strangeways, I, 2007. Precipitation: Theory, Measurement and Distribution. Cambridge University Press, Cambridge.

Su, B.D., Jiang, T., Jin, W.B, 2006. Recent trends in observed temperature and precipitation extremes in the Yangtze River Basin, China. Theor. Appl. Climatol. 83 (1), 139–151.

Turk, F.J., Miller, S.D, 2005. Toward improved characterization of remotely sensed precipitation regimes with MODIS/AMSR-E blended data techniques. Geosci. Remote Sens. 43 (5), 1059–1069.

Van Montfort, M.A.J, 1990. Sliding maxima. J. Hydrol. 118 (1–4), 77–85.

Van Montfort, M.A.J, 1997. Concomitants of the Hershfield factor. J. Hydrol. 194 (1–4), 357–365.

Weiss, L.L, 1964. Ratio of true to fixed-interval maximum rainfall. J. Hydraul. Div. 90 (1), 77–82.

Wilhelm, B., Ballesteros Canovas, J.A., Macdonald, N., Toonen, W., Baker, V., Barriendos, M., Benito, G., Brauer, A., Corella Aznar, J.P., Denniston, R., Glaser, R., Ionita, M., Kahle, M., Liu, T., Luetscher, M., Macklin, M., Mudelsee, M., Munoz, S., Schulte, L., St George, S., Stoffel, M., Wetter, O, 2019. Interpreting historical, botanical, and geological evidence to aid preparations for future floods. WIREs Water 6 (1), e1318.

Willems, P, 2013. Adjustment of extreme rainfall statistics accounting for multidecadal climate oscillations. J. Hydrol. 490, 126–133.

Wilson, J.W., Brandes, E.A, 1979. Radar measurement of rainfall. Bull. Amer. Meteor. Soc. 60 (9), 1048–1058.

Yoo, C., Park, M., Kim, H.J., Choi, J., Sin, J., Jun, C, 2015. Classification and evaluation of the documentary-recorded storm events in the Annals of the Choson Dynasty (1392–1910), Korea. J. Hydrol. 520, 387–396.

Young, C.B, McEnroe, B.M, 2003. Sampling adjustment factors for rainfall recorded at fixed time intervals. J. Hydrol. Eng. 8 (5), 294–296.

CHAPTER

Mean areal precipitation estimation: methods and issues

Ramesh S.V. Teegavarapu

Department of Civil, Environmental and Geomatics Engineering, Florida Atlantic University, Boca Raton, Florida, United States

8.1 Introduction

The mean areal precipitation (MAP) is an estimate of the average depth of precipitation over a region for a specific time interval. Accurate estimation of mean rainfall or mean areal precipitation (MAP) over a region or watershed is essential for hydrologic modeling efforts that rely on lumped simulation models. The MAP is also input to many single event-based models primarily used for the hydrologic design (Teegavarapu, 2013a; Teegavarapu, 2020b) and river forecasting applications (NWS, 2020). Observed precipitation from rain gauges (Lanza et al., 2022) and estimates from radar (Borga et al., 2022) or satellite-based (Kidd and Levizzani, 2022) sources can be used for the estimation of MAP. MAP estimation requires the availability of gap-free serially continuous error-free precipitation data from these sources. Datasets from these sources are seldom error-free and adequate quality assurance and quality control (QAQC) procedures are required for ensuring accurate precipitation datasets. Bias corrections using ground truth (i.e., ground-based observations from rain gauges) are employed to correct and validate radar and satellite-based precipitation estimates. Accuracy of MAP estimation depends on the characteristics (i.e., spatial orientation, density, number of observation sites) of the rain gauge monitoring network, topographic features of the region, temporal frequency of precipitation measurements, the spatial resolution of quantitative precipitation estimates (QPEs) from radar or satellite-based sources. There is also a need to re-evaluate the existing precipitation monitoring networks and design new or optimize the existing ones (Teegavarapu, 2018a) with considerations given to maximizing the information from these networks and any budgetary constraints.

In the past three decades the advances in spatial analysis and geoprocessing methods and tools, in addition to availability of powerful personal computing environments have benefited distributed hydrologic modeling approaches that use spatially varying precipitation data. Despite this widespread use of spatially distributed modeling approaches, design hydrology methods still rely on MAP estimates as inputs in the lumped models, and therefore estimation methods that can provide accurate MAP are of enormous interest. A summary of conceptually simple methods for MAP estimation was provided by Rainbird (1967). Comparative assessments of different methods

RAINFALL. Modeling, Measurement and Applications. DOI: https://doi.org/10.1016/B978-0-12-822544-8.00001-9

and recommendations for the best method for estimation of MAP are provided in several studies (Tabios and Salas, 1985; Singh and Chowdhury, 1986; Teegavarapu, 2016). Surveys of commonly used spatial interpolation methods for areal rainfall estimation provided by Ly et al. (2013). Teegavarapu (2016) discusses the advantages and limitations of different MAP estimation methods.

Discussion related to instruments used to measure precipitation, measurement errors, and uncertainty in the measurement process that are relevant to the MAP estimation process is beyond the scope of this chapter. Readers are advised to refer to works of Strangeways (2007), Linacre (1992), WMO (1983), Habib et al. (2010), Sene (2010), Teegavarapu (2013b), Lanza et al. (2022) for an exhaustive discussion on precipitation measurement devices, networks, radar-based precipitation estimation, and recent advances in hydrometeorology. A comprehensive review of radar-based rainfall estimation, analysis, and its use for hydrologic modeling was provided by Pathak and Teegavarapu (2018) and Borga et al. (2022). The existence of missing precipitation data is a major hindrance to MAP estimation and appropriate imputation approaches need to be used to obtain datasets without any gaps. Deterministic weighting and stochastic interpolation methods (Teegavarapu, 2018b; Yeggina et al. 2019) have been used for the creation of rainfall surfaces (fields) or gridded precipitation data, estimation of missing data, and bias corrections of radar data (Teegavarapu 2018c; Teegavarapu, 2013b). The methods and the associated references listed, and the features and limitations documented in this chapter can be useful for the selection of the best method for MAP estimation.

8.2 Mean areal precipitation

Mean areal precipitation (MAP), also referred to as average rainfall, equivalent uniform depth of rainfall over a region or area-based average rainfall, is required in many hydrologic modeling studies as the essential input to lumped hydrologic simulation models or rainfall-runoff models (Teegavarapu, 2016). The MAP estimate (θ_M) over a region with an area A and measured precipitation $\theta(x, y)$ at point (x, y) is given by Eq. 8.1 following Dingman (2015).

$$\theta_M = \frac{\iint_A \theta(x,y)\, dx\, dy}{A} \tag{8.1}$$

Available methods for MAP estimation range from conceptually simple methods with low calculation complexity to computationally intensive stochastic methods (Dingman, 2015; WMO, 1994; ASCE, 1996; Teegavarapu, 2016). The existing MAP estimation methods can be classified into two distinct categories as illustrated in Fig. 8.1 based on the utilization of rain gauge data and radar- or satellite-based data. The rain gauge-based MAP estimation methods can be further divided into two classifications (Fig. 8.1a) that use: (1) weighting methods that rely on the spatial arrangements of rain gauge sites in a region and (2) spatial interpolation-based methods to generate precipitation fields or surfaces using rain gauge data. These are referred to as point-based and point-to-areal interpolation

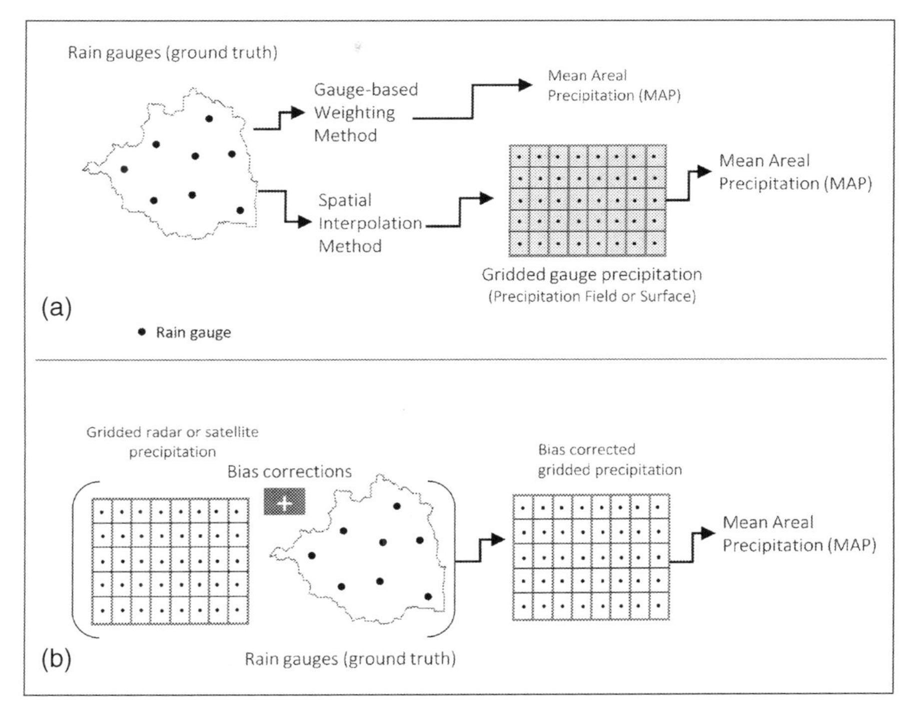

FIG. 8.1

Two major approaches for MAP estimation using rain gauge and radar or satellite datasets.

methods. The radar or satellite-based method relies on the generation of bias-corrected gridded precipitation data using ground truth (i.e., rain gauge measurements) to estimate MAP. Bias correction procedures also require spatial interpolation methods. Data from a numerical weather model can also be used as a first guess field to improve the precipitation surface generation using spatial interpolation methods. Absolute values or anomalies derived from long-term precipitation data (Yeggina et al., 2020; Wilmott and Robeson, 1995) can be used in spatial interpolation for the estimation of MAP. Several MAP estimation methods are discussed briefly in the next sections and then issues related to these methods are also elaborated.

8.3 Mean areal precipitation estimation: weighting methods

Weighting or direct weighted average methods use either distances or areas and precipitation values observed in a region to estimate MAP. These methods are computationally simpler compared to the stochastic spatial interpolation methods which will be discussed later in this chapter.

8.3.1 Arithmetic mean method

The arithmetic mean method (AMM) or unweighted mean method (UMM) (Singh and Birsoy, 1975) is a conceptually simple, objective, and naïve method for estimating mean precipitation depth over an area of interest (i.e., watershed or basin) using a mean of all available rain gauge-based observations (McCuen, 2004). The estimate given by Eq. 8.2 is derived for any temporal scale and is based on observations of precipitation at gauges that are within the area of interest. The MAP estimate (θ_M) using the mean method is given by Eq. 8.2 based on observations at nrg rain gauges with θ_j as observed precipitation value at rain gauge j.

$$\theta_M = \frac{\sum_{j=1}^{nrg} \theta_j}{nrg} \tag{8.2}$$

Gauges outside the region can be considered if there is a reason to believe that the observations recorded at these sites are representative of precipitation values observed at all other gauges within the region. The method is appropriate for regions with spatially uniform precipitation. AMM is not appropriate for regions with varying topographic and physiographic features that affect the precipitation patterns, intensities, and spatial orientation of storms. National Weather Service (NWS) (NWS, 2020) of the U.S. lists AMM as one of the six methods for MAP estimation. The estimate provided by this method may be biased in favor of those rain gauges that are most closely grouped (Singh and Birsoy, 1975). In some cases, a small number of rain gauges carefully selected in a region may provide an equally good estimate of MAP compared to that is obtained from many or all rain gauges.

8.3.2 Grouped area-aspect weighted mean method

The grouped area-aspect weighted mean method (GAAWMM) (Whitmore et al., 1961) is an area-weighted precipitation estimate obtained by using mean precipitation values from areas delineated inside a region of interest with similar altitude and aspect. Initially, the region is divided or delineated into primary altitudinal and secondary aspect zones (i.e., $z = 1, 2.. nz$). The area (A_z) of each zone z and mean precipitation (θ_z) for that zone is obtained and then an estimate of MAP (θ_M) for the region with nz areas is derived using Eq. 8.3.

$$\theta_M = \frac{\sum_{z=1}^{nz} A_z \theta_z}{\sum_{z=1}^{nz} A_z} \tag{8.3}$$

Topographical data is essential for this method and a digital elevation model (DEM) can be used for this purpose. Also, spatial analysis needs to be carried to obtain the aspect information.

8.3.3 Thiessen polygon-based weighting method

The Thiessen (Voronoi/Dirichlet/proximity) polygon method (Thiessen, 1911; Brassel and Reif, 1979; Boots, 1986, Teegavarapu and Chandramouli, 2006; Teegavarapu, 2016) as an objective and one of the most used methods for MAP estimation provides an area-weighted average of observations at all gauges within the region of interest (or a watershed). The areas required for the estimate are those associated with the polygons in which the rain gauges are located. The polygons are constructed by joining the point locations of gauges by straight lines to form the smallest triangles possible. This process refers to the generation of a triangular irregular network (TIN) which satisfies the Delauney criterion (Okabe et al. 2000). Then perpendicular bisectors are drawn that meet to form irregular polygons when bounded by the region of interest or watershed. If polygon areas extend beyond the boundary of the watershed, only portions of the areas within the region are considered for an estimate. The sizes and areas of the polygons and their clusters vary according to the spatial orientation of gauges in the monitoring network. Rain gauges outside the region may also be considered if the observations at these gauges are generally representative of precipitation patterns in the region. Also, any point inside a Thiessen polygon is closer to the point that is used to create the polygon than any other control point (i.e., rain gauge site) in the region. Therefore, it is safe to assume that any location inside the polygon will most likely have similar precipitation characteristics to those observed at a rain gauge located in that specific polygon or area. Thiessen polygons also referred to as Wigner-Seitz regions can be constructed manually using the procedure outlined or by using any spatial analysis software.

An illustration of Thiessen polygons created based on six rain gauges in a basin is shown in Fig. 8.2. The MAP (θ_M) is estimated with the help of weights (Eq. 8.4) and area-weighted calculation provided by Eq. 8.5. The variables β_j, θ_j, A_j and nrg represent the weight (Thiessen coefficient), precipitation amount, and polygon area associated with rain gauge j and the number of gauges respectively.

$$\beta_j = A_j \left\{ \sum_{j=1}^{nrg} A_j \right\}^{-1} \quad \forall j \tag{8.4}$$

$$\theta_M = \sum_{j=1}^{nrg} \beta_j \theta_j \tag{8.5}$$

Thiessen polygon method can also be used for the estimation of missing precipitation data (Guillermo et al., 1985). Also, the method as described does not include or benefit from physiographic or topographic information relevant to the region that influences the spatial variability of the precipitation. Modifications to this method have been made to incorporate basin topography information to improve MAP estimates. Mair and Fares (2011) used the Thiessen polygon method for MAP estimation in a mountainous region. Sen (1988) and Akin (1971) provided different weighting methods for estimation of MAP using polygons defined by rain gauge networks. To avoid re-creating the Thiessen polygon if a rain gauge is removed from the network

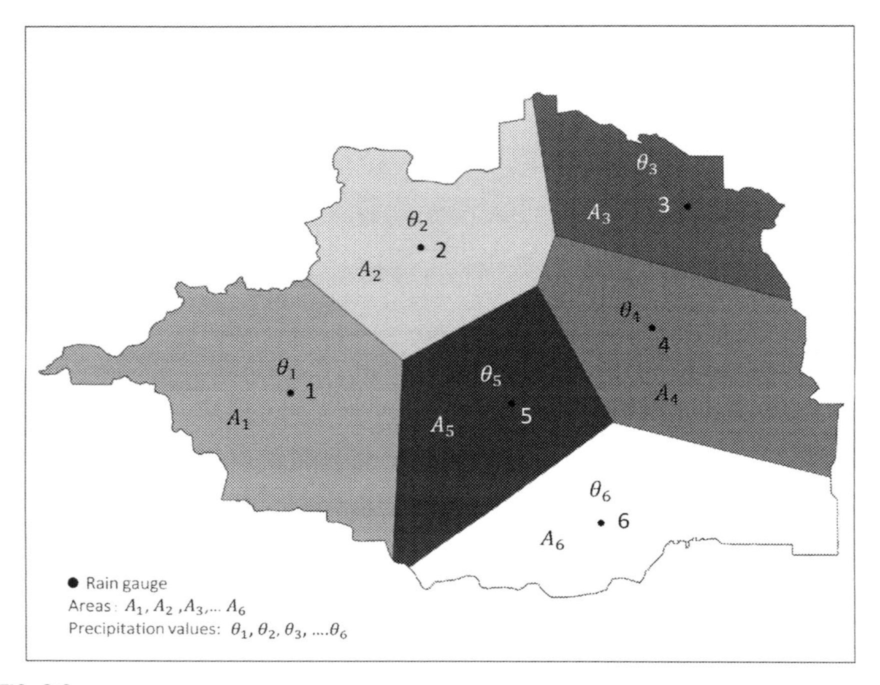

FIG. 8.2

Schematic illustration of the Thiessen polygon method with rain gauges and polygons.

due to missing data, it is recommended that the missing rain gauge value be first estimated. Sugawara (1992) discusses the limitations of Thiessen polygon-based weights and reports a method wherein weights are calculated based on observed and estimated discharges based on precipitation measurements. Abrupt regions created at the boundaries (Voronoi edges) of the Thiessen polygons may not be representative of the natural precipitation pattern in the region as they break the spatial contiguity of a naturally occuring process.

8.3.4 Isohyetal method

The isohyetal method or eyeball method (Linsley et al., 1949; France, 1985; Dingman, 2015; Shaw et al., 2011; Teegavarapu, 2016) uses isohyets (i.e., lines or contours of equal precipitation amounts) for MAP estimation. Isohyets are created using rain gauge measurements in a region by linear interpolation between any two points and adopting some form of smoothing or surface fitting methods. The accuracy and validity of the isohyets drawn using interpolated rainfall values need to be confirmed before they can be used in MAP estimation. The area-based weights are obtained using Eq. 8.6. MAP estimate (θ_M) as an area-weighted value is obtained using Eq. 8.7 based on average value (θ_i^{av}) of precipitation values associated with two isohyets

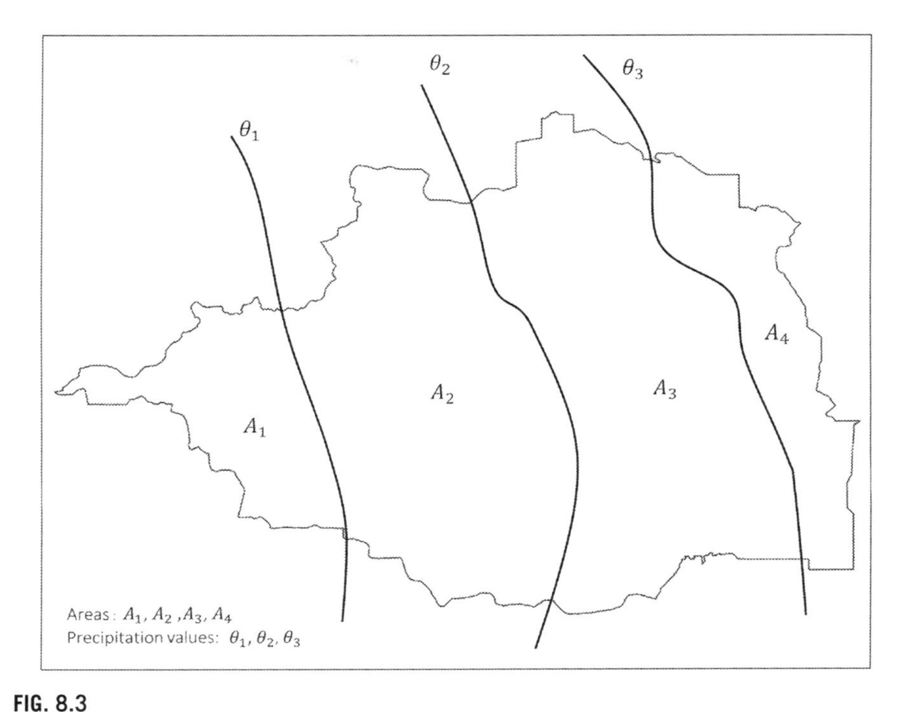

FIG. 8.3

Illustration of the isohyetal method for estimation of MAP.

enclosing an area as shown in Fig. 8.3. The variable *nis* refers to the total number of isohyets and β_j is the weight obtained using areas encompassed by isohyets. If an area (e.g., A_1) is bounded by basin or watershed boundary and an isohyet with precipitation value θ_1, then the average value (θ_1^{av}) equal to θ_1.

$$\beta_j = A_j \left\{ \sum_{j=1}^{nis} A_j \right\}^{-1} \quad \forall j \tag{8.6}$$

$$\theta_M = \sum_{i=1}^{nis} \beta_j \theta_i^{av} \tag{8.7}$$

The magnitudes of areas required for MAP estimation can be obtained by planimeter or using any spatial analysis software. The isohyets need to be developed by the analysts using topographic information and spatial variation of precipitation in the region. Also, isohyets need to be developed for each storm event for MAP estimation. It is generally assumed that precipitation between any two gauges varies linearly unless abrupt changes in topography indicate otherwise (Jain and Singh, 2005). This assumption of linear variation may not be always justified.

8.3.5 Centroidal distance method

The centroidal distance method (CDM) or proximate gauge method (Zeiger and Hubbart, 2017) uses information about distances from the centroid of the bounded region of interest (i.e., watershed or basin) to the rain gauges for estimation of MAP. Distances of the rain gauges from the centroid are calculated and the inverses of these distances ($d_{c,j}^{-f}$) raised to an exponent (f) are used as weights ($w_{c,j}$) (Eq. 8.8) in a distance weighted method along with precipitation measures at these gauges. An exponent (f) value of 2 is generally used for calculations. The variables $d_{c,j}$ and nrg refer to the distance of the gauge j to the centroid (c) of the region and the total number of rain gauges respectively. The MAP estimate (θ_M) is obtained using Eq. 8.9 and the CDM is illustrated in Fig. 8.4.

$$w_{c,j} = d_{c,j}^{-f} \left\{ \sum_{j=1}^{nrg} d_{c,j}^{-f} \right\}^{-1} \quad \forall j \tag{8.8}$$

$$\theta_M = \sum_{j=1}^{nrg} w_{c,j} \theta_j \tag{8.9}$$

Using the centroid of the region as the point of interest for the calculation of Euclidean distances and MAP seems arbitrary and the utility of CDM is debatable. The spatial variability of precipitation cannot be characterized for accurate MAP estimation using this approach.

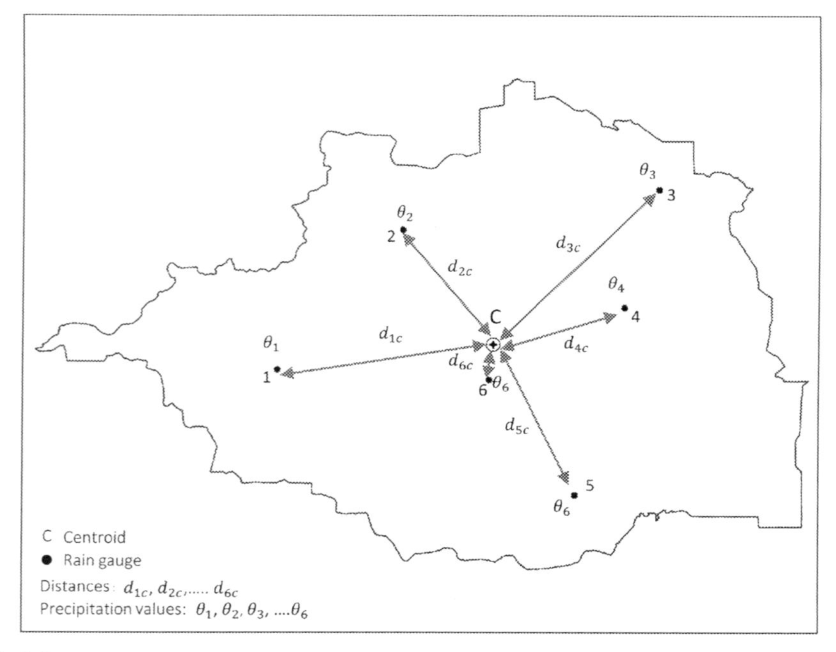

FIG. 8.4

Illustration of centroid distance calculations for estimation of MAP.

8.3.6 Line method

The line method (LM) (Goel and Aldabagh, 1979) uses the lengths of lines drawn between any two points with precipitation measurements in a weighted average estimation of MAP. The lines are drawn using all the gauges (nrg). The MAP estimate (θ_M) is obtained using Eq. 8.10 and the variable $dl_{i,j}$ is the length of the line from point i to point j (i.e., Euclidean distance between two rain gauges) as shown in Fig. 8.5. Rain gauges lying outside the region can be used for estimation of MAP with the line method.

$$\theta_M = \frac{\sum_{i=1}^{nrg} \sum_{j=1}^{nrg} dl_{i,j} \left(0.5 * \left[\theta_i + \theta_j\right]\right)}{\sum_{i=1}^{nrg} \sum_{j=1}^{nrg} dl_{i,j}} \tag{8.10}$$

8.3.7 Triangle method

The triangle method (TM) (Sumner 1988; Shuttleworth, 2012) involves the creation of triangles that are mostly equilateral using the rain gauges (or control points) as vertices. An illustration of the method using four rain gauges is shown in Fig. 8.6. The MAP is estimated using Eqs. 8.11 and 8.12. The area (A_n) covered by each triangle referred by index n is initially estimated. Total area (SAR) for all nt triangles

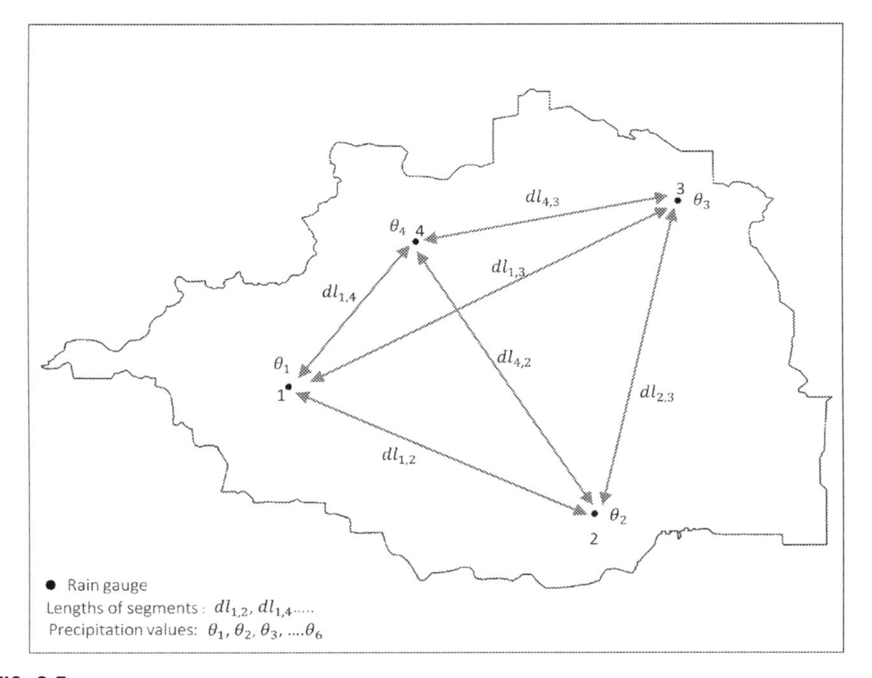

FIG. 8.5

Illustration of the line method for estimation of MAP.

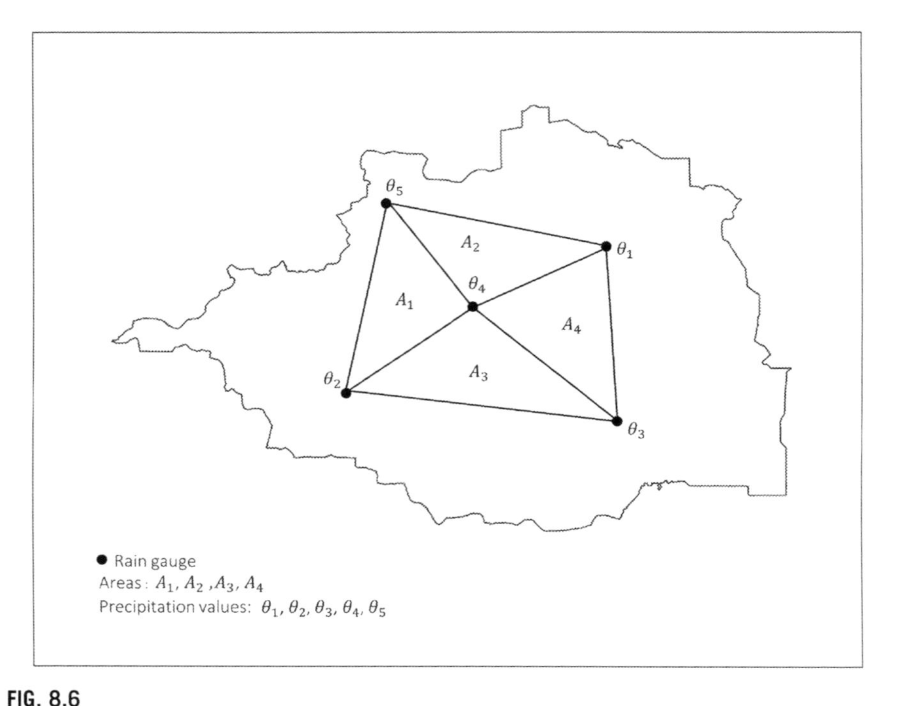

FIG. 8.6

Illustration of the triangle method for estimation of MAP.

is obtained using Eq. 8.11. The average precipitation (θ_n^{av}) for each triangle n is obtained using the average of values from three gauges forming the vertices of the triangle. For example, average precipation value θ_1^{av} is obtained using the mean of precipitation values (i.e., θ_2, θ_4 and θ_5) at three rain gauges as shown in Fig. 8.6. The MAP estimate (θ_M) is obtained using a weighted average (i.e., product of mean precipitation value and the area of each triangle over the total area from all the triangles) as given by Eq. 8.12. Gauges that are outside the region of interest can also be used in this method (Sumner, 1988).

$$SAR = \sum_{n=1}^{nt} A_n \tag{8.11}$$

$$\theta_M = \frac{\sum_{n=1}^{nt} A_n \theta_n^{av}}{SAR} \tag{8.12}$$

8.3.8 Two-axis method

The two-axis method (TAM) often referred to as Bethlahmy's two-axis method (Bethlahmy, 1976) is another MAP estimation method. The method is developed and

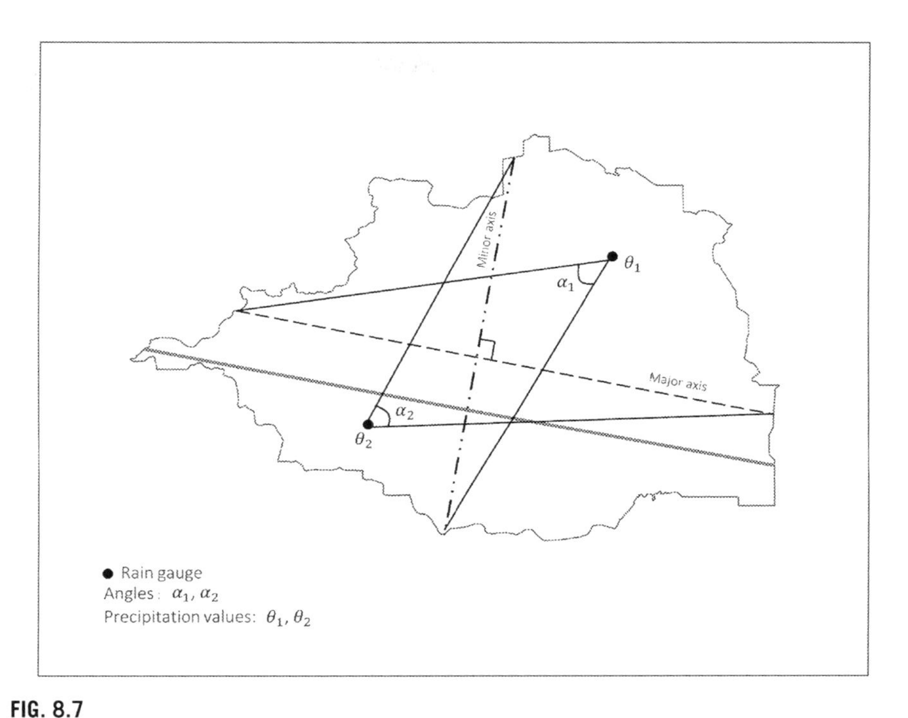

FIG. 8.7

Illustration of the two-axis method for estimation of MAP.

described in detail by Dingman (2015) and WMO (1994) and easier to use compared to Thiessen polygon and other weight-based methods. The two-axis method similar to line and triangle methods makes use of the spatial arrangement of rain gauges in a region and geometry to account for spatial variability of precipitation in the estimation of MAP. Estimation of MAP using TAM with two rain gauges is illustrated in Fig. 8.7. The MAP is obtained through a series of steps and they include: (1) using the region (or watershed boundary) the longest line that is possible is drawn; (2) a perpendicular bisector to this line referred to as minor axis and another bisector to this axis defined as the major axis is drawn; (3) lines from each gauge are drawn to join the farthest points on the minor and major axes; (4) the angles (e.g., α_1, α_2) that are formed by these lines are then estimated. The MAP estimate (θ_M) is obtained using Eqs. 8.13 and 8.14 using the sum of all angles (SA) defined based on nrg rain gauges and associated precipitation values (θ_i).

$$SA = \sum_{i=1}^{nrg} \alpha_i \tag{8.13}$$

$$\theta_M = \frac{\sum_{i=1}^{nrg} \alpha_i \theta_i}{SA} \tag{8.14}$$

8.3.9 Percent normal method

The percent normal method (PNM) (WMO, 1994) is applicable to regions where the physiography influences the spatial variability of rainfall at different temporal resolutions. Average precipitation for a storm event over a region is estimated by expressing the storm precipitation as a percentage of mean annual or mean seasonal precipitation and iso-percental maps are used for preparing isohyetal maps (WMO, 1994). While this method provides accurate estimates of MAP compared to others discussed in the previous sections, it is computationally intensive and requires historical precipitation records.

8.3.10 Gauge-weighted methods: issues

The weight-based methods discussed in the previous sections rely on the use of geometric features or measures (e.g., areas, Euclidean distances from observations to the central location of a region, triangle areas formed based on locations of the gauges, lengths of lines connecting gauges, angles formed by lines extended to major and minor axes drawn within the region) for estimation of MAP. While these methods are conceptually simple, accurate estimates of MAP are not possible without region-specific physiographic, climatic, and topographic information. These methods are objective and replicable and add to the variants of weighting methods using Euclidean distances or areas with limited applications in hydrologic modeling that need accurate estimates of MAP. The spatial partitioning of the region in some of the methods discussed earlier may be considered somewhat arbitrary. The actual variation of precipitation in a region is not considered for the development of these partitioning schemes.

8.4 Surface generation methods

The surface generation or fitting methods use point observations and spatial interpolation approaches for the creation of precipitation surfaces or fields. The surfaces generated are based on values estimated for a spatial element of a specific size or tessellation (i.e., grid size) or at a point. The first law of geography: "everything is related to everything else, but near things are more related than distance things" attributed to Tobler (Tobler, 1970) is the founding principle for any spatial interpolation method. The following sections discuss methods that use generated precipitation surfaces to obtain the estimates of MAP.

8.4.1 Inverse distance weighting method

The inverse distance (reciprocal-distance) weighting method (IDWM) is one of the most used exact, deterministic spatial interpolation methods for the generation of rainfall fields and estimation of MAP. This method follows Tobler's first law of geography in spirit. The reciprocal-distance method for estimation of precipitation value at location m (defined using the Cartesian coordinates: x, y) using the observed precipitation values at all sites. Euclidean distance ($d_{m,j}$) between the point of interest, m, and the gauge j is used to estimate the weight ($w_{m,j}$) is given by Eq. 8.15. The exponent attached to distance, f, is referred to as friction distance and it ranges typically between 1.0 to 4.0. The estimated value of precipitation ($\hat{\theta}(x,y)$) at a location (x, y) is obtained by Eq. 8.16. The MAP estimate (θ_M) is obtained using Eq. 8.17 using the average of all the precipitation values from ng points (or grid points) in the area (Ar). An example of the precipitation surface generated using the inverse distance method is shown in Fig. 8.8. Optimization methods can be used to obtain the best value of friction distance (f) using available precipitation data.

$$w_{m,j} = d_{m,j}^{-f} \left\{ \sum_{j=1}^{nrg} d_{m,j}^{-f} \right\}^{-1} \tag{8.15}$$

$$\hat{\theta}(x,y) = \sum_{j=1}^{nrg} w_{m,j}\theta_j \tag{8.16}$$

$$\theta_M = \frac{\sum_{x,y \in Ar} \hat{\theta}(x,y)}{ng} \tag{8.17}$$

The IDWM as an exact deterministic spatial interpolation method creates a smooth surface. The smoothness of the surface depends on the friction distance and the maximum and minimum values from the interpolated surface occur only at the sites with observations. Clustering of observation sites and outliers will influence the surface generated by the method. One of the issues associated with this method is the tent pole (or bull's eye) effect that generates isolated higher (peaks) or lower (troughs) values surrounding control points (i.e., rain gauge sites) with measured high or low values respectively. This artifact associated with interpolation can be addressed by using dense precipitation networks.

One of the variants of the inverse distance method proposed by Shepard (1968) can be used for spatial interpolation of precipitation values. Shepard's method considers not only Euclidean distance but also an angular distance that is specific to the spatial distribution of control points (i.e., rain gauges). Shepard's method also known as the angular distance weighting (ADW) method has been used in multiple studies for interpolation of monthly and daily climate data (Alexander et al., 2006; New et al., 2000). Conceptual improvements related to neighborhood selection for use in Shepard's method were proposed and evaluated by Yeggina et al. (2019).

8.4.2 Natural neighbor method

The natural neighbor method (NNM) also known as the area stealing method (Sibson, 1981), as an exact interpolator, uses Thiessen polygons to identify nearest neighbors useful for spatial interpolation. In this method, the natural neighbors to a point of interest are the closest Thiessen polygons. The execution of the NNM with multiple steps is illustrated in Fig. 8.9. Using the rain gauge sites in the region, Thiessen polygons are

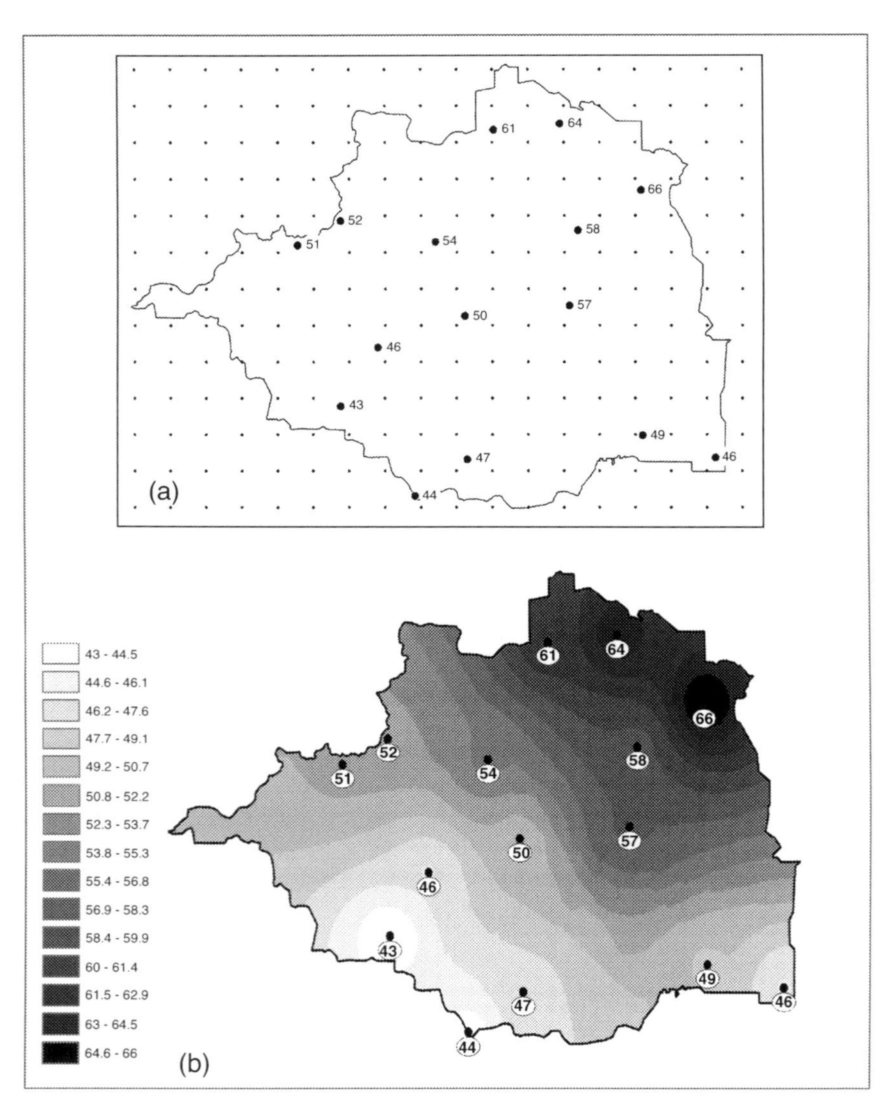

FIG. 8.8

Illustration of the precipitation surface generated for MAP estimation using inverse distance weighting method.

initially constructed (Fig. 8.9a). Another set of Thiessen polygons are created after a specific point (where the interpolated value is sought) (Fig. 8.9b) is added to the existing region along with a set of rain gauge locations. These polygons are overlaid on top of the previous set of polygons as shown in Fig. 8.9c. The variable ws_p is the weight assigned to a location (x, y) associated with the polygon p and is obtained by using polygon areas

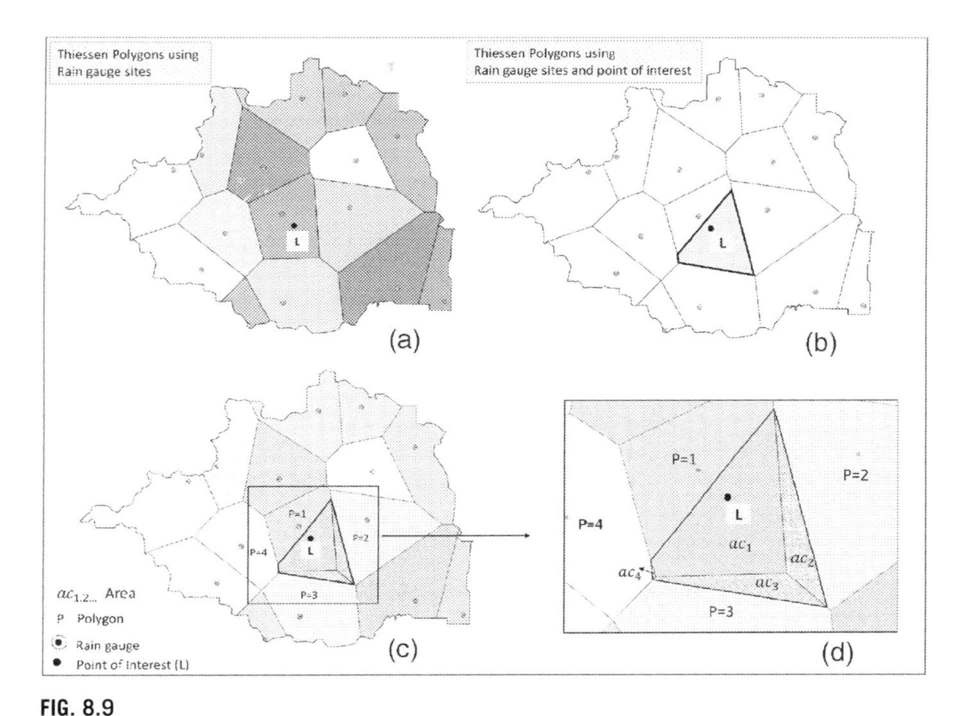

FIG. 8.9

Illustration of natural neighbor interpolation method.

(ac_p, $p = 1.. np$) as shown in Fig. 8.8d and calculated using Eq. 8.18. The weight (ws_p) is the percentage of overlap calculated based on the two Thiessen polygon sets. The estimate of precipitation value at a location L (x, y) is obtained using Eq. 8.19. The MAP estimate (θ_M) is obtained using Eq. 8.20 using ng points in a region or area (Ar).

$$ws_p = ac_p \left\{ \sum_{p=1}^{np} ac_p \right\}^{-1} \quad \forall p \tag{8.18}$$

$$\hat{\theta}(x, y) = \sum_{p=1}^{np} ws_p \theta_p \tag{8.19}$$

$$\theta_M = \frac{\sum_{x,y \in Ar} \hat{\theta}(x, y)}{ng} \tag{8.20}$$

The NNM produces a continuous precipitation surface that can be used for MAP estimation. The method is also conceptually superior to the inverse distance method as the Euclidean distance-based weights are now replaced by area-based weights. Also, there is no need to specify the number of neighbors, radius (i.e., constant Euclidean distance from the point of interest used in local interpolation), and arbitrary distance-based weights. Two points that are equidistant from a point of interest will be assigned equal weights in IDWM, whereas these two weights are not equal in the NNM. The weights are unit-less as they are in all weight-based methods. The NNM is computationally intensive compared to the inverse distance method as estimations of weights require area calculations.

8.4.3 Variance-dependent stochastic interpolation methods

Variance-dependent stochastic interpolation techniques based on geostatistical approaches (Isaaks and Srivastava, 1989) such as kriging (Krige, 1966) are also often used for estimation of MAP. Kriging and its variants require multiple steps: (1) the development of semi-variograms based on point data; (2) selection of authorized variogram models that can be fitted to semi-variograms; and (3) estimation of kriging weights. Kriging adopts a best linear unbiased estimator (BLUE) approach for the estimation of weights used in spatial interpolation. The approach minimizes the mean squared residual error (MSRE) of the residuals resulting from the interpolation surface-based estimates and observation values at rain gauges. Local and global variants of kriging based on the number of control points (i.e., sites with observations) can be developed. In kriging, the degree of spatial dependence generally is expressed using a semi-variogram. To construct semi-variogram observations at pairs of rain gauges are used as illustrated in Fig. 8.10. A typical

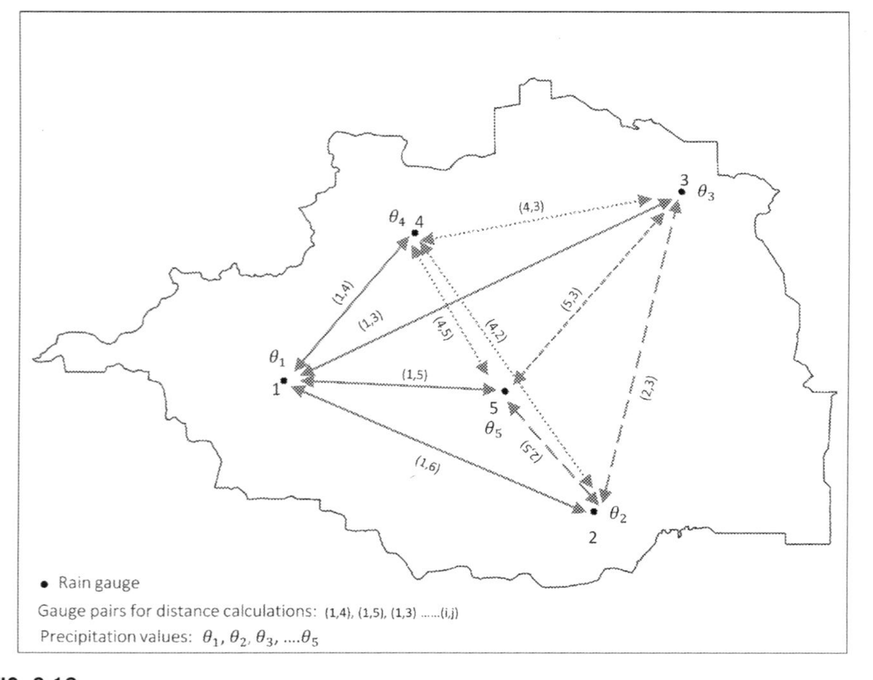

FIG. 8.10

Illustration of distance pairs for constructing semi-variogram.

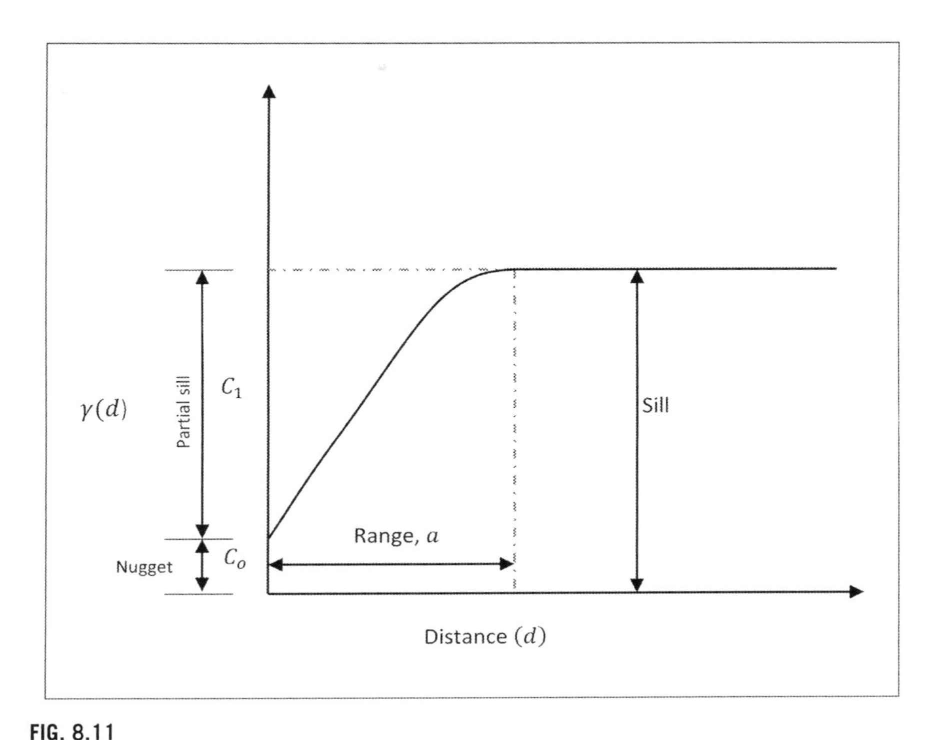

FIG. 8.11

Typical semi-variogram and different parameters.

semi-variogram and its different components such as nugget, partial sill, range, and sill are shown in Fig. 8.11.

The semi-variogram can be estimated using Eq. 8.21,

$$\gamma(d) = \frac{1}{2n(d)} \sum_{d_{ij}} \left(\theta_i - \theta_j\right)^2 \tag{8.21}$$

where $\gamma(d)$ is the semi-variance which is defined over observations θ_i and θ_j lagged successively by distance d. Surface interpolation using kriging depends on the semi-variogram model that is selected which must be fitted with a theoretical form that can be used to estimate the semi-variogram values at arbitrary separation distance values. Several semi-variogram models generally are tested before selecting a particular one. The four most widely used semi-variogram models (linear, spherical, exponential, Gaussian, and circular) are given by Eqs. 8.22, 8.23, 8.24, 8.25, and 8.26 respectively. The variables C_o, C_1, a and d refer to nugget, partial sill, range, and distance respectively, and are shown in Fig. 8.11. For all the semi-variogram models, if the lag distance (d) is greater than or equal to the range (a), then the semi-variance will be equal to the total value of nugget and partial sill as given by Eq. 8.27 for linear and spherical variogram models. The exponential and Gaussian functions approach the sill value asymptotically and 95% of the sill value is achieved at approximately lag distances of $3a$ and $\sqrt{3a}$ respectively.

$$\gamma(d)_1 = C_o + C_1\left[\frac{d}{a}\right] \qquad 0 \le d \le a \tag{8.22}$$

$$\gamma(d)_2 = C_o + C_1\left[\frac{1.5d}{a} - 0.5\left(\frac{d}{a}\right)^3\right] \qquad 0 < d \le a \tag{8.23}$$

$$\gamma(d)_3 = C_o + C_1\left[1 - \exp\left(-\frac{d}{a}\right)\right] \qquad d > 0 \tag{8.24}$$

$$\gamma(d)_4 = C_o + C_1\left[1 - \exp\left(-\frac{d^2}{a^2}\right)\right] \qquad d > 0 \tag{8.25}$$

$$\gamma(d)_5 = C_o + C_1\left[1 - \frac{2}{\pi}\cos^{-1}\left(\frac{d}{a}\right) + \frac{2d}{\pi a}\sqrt{1 - \frac{d^2}{a^2}}\right] \qquad 0 \le d \le a \tag{8.26}$$

$$\gamma(d)_k = C_o + C_1 \qquad k = 1, 2. \qquad d \ge a \tag{8.27}$$

Depending on the shape of the semi-variogram, several mathematical models are possible, including linear, spherical, circular, exponential, and Gaussian functional forms. Pentaspherical, cubic, Whittle's elementary correlation and Matérn function are a few other models. The summation of C_o and C_1 is referred to as sill, and the semi-variance at the range, a, equals the sill value. The values of C_o and C_1 are obtained using trial and error or an optimization formulation. Visual observation of the variogram cloud based on the available spatial data is carried out and the possible fit of the theoretical variogram model to the cloud is evaluated and confirmed for use. Weights in the kriging approach are based not only on the distance between the measured points and the prediction location (x, y) but also on the overall spatial arrangement among the measured points and their values. The weights mainly depend on the fitted model (i.e., the semi-variogram) to the measured points. The estimated value $(\hat{\theta}(x, y))$ at a location (x, y), is obtained by Eq. 8.28 using a total number of gauges nrg. The variable λ_j is the weight assigned to a location (x, y) associated with the observation site j.

$$\hat{\theta}(x, y) = \sum_{j=1}^{nrg} \theta_j \lambda_j \tag{8.28}$$

The weight λ_j is obtained from the fitted semi-variogram, and θ_j is the value of the precipitation at site j. The observed data are used twice, once to estimate the semi-variogram and then to interpolate the values. To avoid any systematic bias in the kriging estimates, a constraint on weights is enforced using Eq. 8.29.

$$\sum_{j=1}^{nrg} \lambda_j = 1 \tag{8.29}$$

The MAP estimate (θ_M) is obtained using Eq. 8.30 using all the estimates at all the locations (x, y) that belong to a region or area Ar.

$$\theta_M = \frac{\sum_{x,y \in Ar} \hat{\theta}(x,y)}{ng} \tag{8.30}$$

An example of an interpolated precipitation surface generated using ordinary kriging (OK) with first-order trend removal and exponential semi-variogram is shown in Fig. 8.12.

The variants of kriging (Karnieli, 1990) can be beneficial for the estimation of MAP. Co-kriging of radar and rain gauge data has been employed by Krajewski (1987) to estimate mean areal precipitation. Seo et al. (1990a, 1990b) described the use of co-kriging and indicator kriging for interpolating rainfall data. Chua and Bras (1982) report the use of kriging to estimate MAP for mountainous regions considering drift and spatial dependency of precipitation influenced by orographic effects. Goovaerts (2000) reports the use of simple kriging with varying local means, kriging with an external drift; and co-located kriging for incorporating a digital elevation model into the spatial prediction of rainfall. Phillips et al. (1992) evaluated the performance of three geostatistical methods for obtaining mean annual precipitation estimates on a regular grid of points in mountainous terrain. The methods evaluated include (1) kriging; (2) kriging elevation-detrended data; and

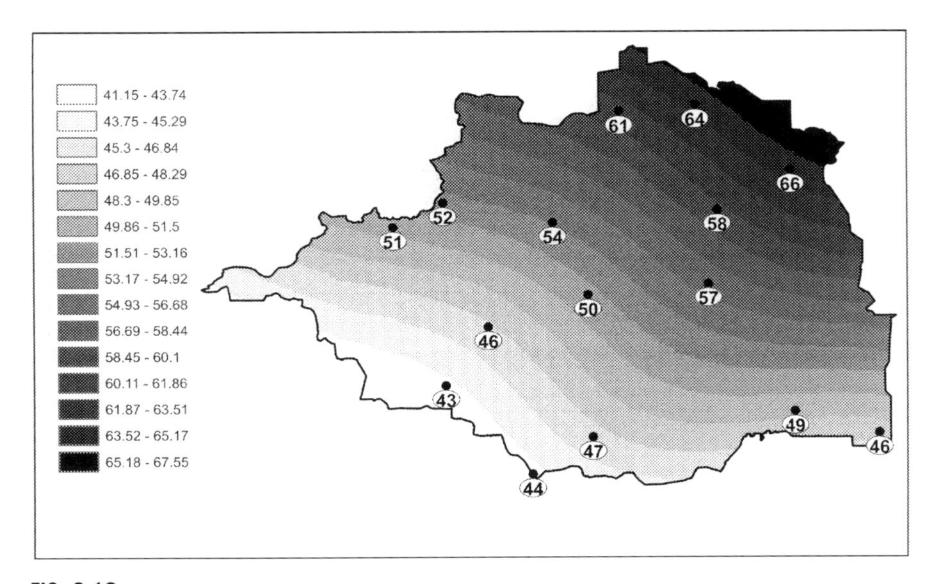

FIG. 8.12

Precipitation surface generated using ordinary kriging method.

(3) co-kriging with elevation as an auxiliary or ancillary variable. The last two methods provided better estimates compared to kriging. Rogelis and Werner (2013) investigated the use of ordinary kriging, universal kriging, and kriging with external drift with individual and pooled variograms for real-time rainfall field estimation in areas with complex topography. In their study, they found that interpolators using pooled variograms provided results compared to those obtained when the interpolators were applied to the storms individually. Mair and Fares (2011) used ordinary kriging (OK), and simple kriging with varying local means to estimate MAP in mountainous regions.

While kriging is one of the best stochastic inexact interpolation methods for rainfall surface generation, several drawbacks of this method should be noted. The limitations of kriging include difficulty in characterizing precipitation data using variograms based on sparse or missing precipitation data, selection of the appropriate semi-variogram model from an available set of theoretical (authorized) models, uncertainty associated with the assignment of the sill, and nugget parameters. In some instances, variants of kriging may result in negative precipitation estimates which would make MAP estimation invalid and inaccurate.

8.4.4 Trend surface and regression-based methods

Surface fitting methods can be used for the generation of precipitation surface using point measurements for estimation of MAP. Polynomial, Lagrange polynomial, and splines with or without tension, optimal interpolation via kriging, empirical orthogonal functions (EOF) can also be used for estimation of MAP as elaborated by Dingman (2015). Trend surface models (smoothing models) use polynomial functions of different degrees (orders) to fit the surfaces to observation points (i.e., rain gauges) in a region. Smooth and irregular surfaces may result depending on the nature of the polynomial or the degree of the polynomial adopted for the surface. First (linear), second (quadratic), and third (cubic) order polynomial trend surface models are described by the Eqs. 8.31, 8.32, and 8.33 respectively that can be used for estimation of precipitation value at any location defined by cartesian coordinates (x, y) in a region or area (Ar). The coefficients (a_o, a_1 .... a_9) in equations depending on the order of the model used are estimated based on precipitation observations at available locations using a least-squares method that minimizes the cumulative error based on estimates and observations. The total number of coefficients for an order of τ is equal to $(\tau + 1)(\tau + 2)/2$. The coefficients are obtained by minimizing the squared error (Eq. 8.34) based on observed ($\theta(x_i, y_i)$) and estimated ($\hat{\theta}(x_i, y_i)$) values of precipitation at gauges ($i = 1....nrg$). The MAP estimate (θ_M) is obtained by the average of precipitation values $\left(\hat{\theta}(x, y),\ x, y \in Ar\right)$ at all grid points using Eq. 8.35.

$$\hat{\theta}(x, y) = a_o + a_1 x + a_2 y \tag{8.31}$$

$$\hat{\theta}(x, y) = a_o + a_1 x + a_2 y + a_3 x^2 + a_4 xy + a_5 y^2 \tag{8.32}$$

$$\hat{\theta}(x,y) = a_o + a_1 x + a_2 y + a_3 x^2 + a_4 xy + a_5 y^2 + a_6 x^3 + a_7 x^2 y + a_8 xy^2 + a_9 y^3 \quad (8.33)$$

$$\sum_{i=1}^{nrg} \left(\theta(x_i, y_i) - \hat{\theta}(x_i, y_i)\right)^2 \quad (8.34)$$

$$\theta_M = \frac{\sum_{x,y \in Ar} \hat{\theta}(x,y)}{ng} \quad (8.35)$$

Examples of the surfaces generated by polynomial trend surface models based on precipitation observations available at different sites in the region are shown in Fig. 8.13. In the case of zero-order ($\tau = 0$), the surface field will have a constant value that is equal to the mean value of all precipitation measurements from the sites. The use of higher-order models may lead to overfitting of the trend surface thus reducing the model's ability to generalize the real variation of the precipitation in a region. In such cases, the precipitation surfaces generated may violate physical reality. The trend surface methods are inexact spatial interpolators and may be treated as regression-based methods if specific assumptions are satisfied and hence can be referred to as stochastic methods. Also, the order of the model can be selected based on testing the significance of coefficients using a conventional statistical hypothesis test (e.g., *F*-test).

Several other spatial interpolation methods such as natural neighbors, nearest neighbors, triangular irregular networks (TINs), trend surface models, and thin-plate splines (TPS) can be used for the estimation of MAP. An exhaustive review of several interpolation methods along with discussions about the assumptions needed, complexities encountered and computational requirements for accurate surface generation is provided by Burrough and McDonnell (1998) and Li and Heap (2011). The local and global polynomial approaches of different orders, spline with tension, and multiquadric methods have been used by Zeiger and Hubbart (2017). The application of a bi-cubic spline to estimate MAP was discussed by Shaw and Lynn (1972).

Trend surface analyses using linear, quadratic, and cubic functions for the estimation of MAP were reported by Mandeville and Rodda (1973). They conclude that trend surface methods provide better estimates of MAP compared to those from Thiessen polygon or isohyetal methods. Abtew et al. (1993) evaluated Thiessen polygon, inverse-distance, multi-quadratic, polynomial, optimal and point interpolation, ordinary and universal kriging for point and areal estimation of rainfall in South Florida. They conclude that multi-quadric, kriging, and optimal interpolation are the best three methods among those evaluated. The inverse-distance weighting method with optimal parameters (exponent and search radius) was used to estimation of spatial rainfall distribution by Chen and Liu (2012) and Noori et al. (2014). Mair and Fares (2011) used inverse distance weighting method (IDWM) and linear regression for the estimation of MAP. Lebel et al. (1987) used estimation error variance computed from a scaled climatological variogram model of the rainfall field as a criterion to evaluate three MAP estimation methods.

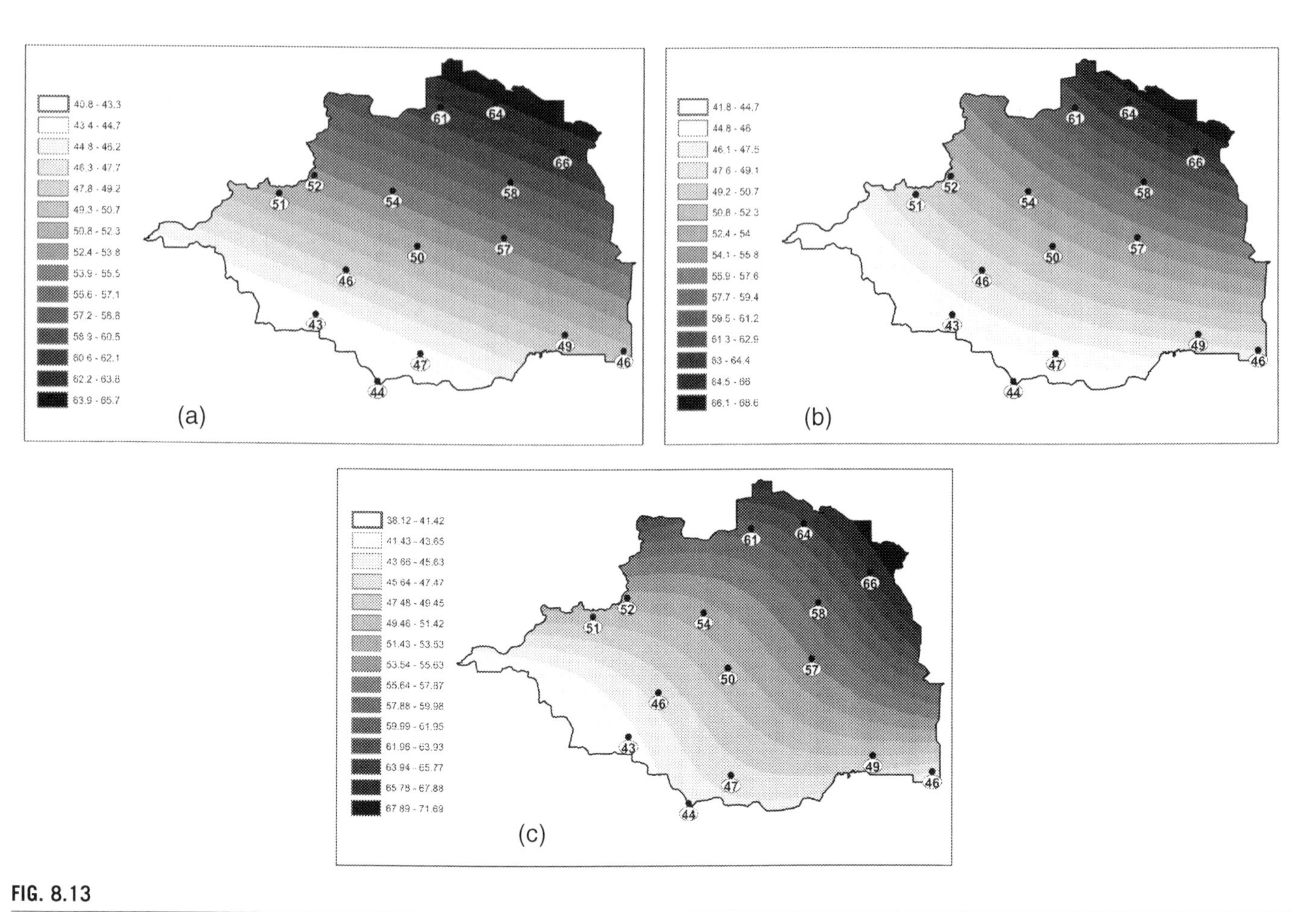

FIG. 8.13

Polynomial trend surface models of different orders for estimation of MAP: (a) location of rain gauges and precipitation values; trend surface models (b) first order; (c) second order; and (d) third order.

The hypsometric method is a surface fitting method and is particularly suitable for mountainous regions and requires four bi-variate functional relationships and they are (1) area-elevation; (2) station-precipitation; (3) station-elevation; and (4) area-precipitation. The procedures for the construction of curves, use of these curves, and estimation of MAP are available from WMO (1994).

Regression-based methods with auxiliary variables (i.e., time-invariant variables such as relief, elevation, slope, latitude, longitude, distance to the coast, land-sea gradient, proximity to large water bodies, and other characteristics) that are easy to measure can be used for the estimation of precipitation in a region. Parameter-elevation Regression on Independent Slopes Model (PRISM) developed by Daly et al. (1994; 2002) uses elevation, topography, and proximity to coastal area and distances as independent variables. Regions with high topographic variations can benefit from the use of PRISM data for the estimation of MAP. Descriptions and application of other methods suitable for desert environments and varying topography can be found in the works of Sen and Eljadid (2000) and Dawdy and Langbein (1960) respectively. AURELHY (Analysis Using the Relief for Hydrometeorology) is another method that uses topographic information for spatial interpolation of hydroclimatic variables including precipitation. The method combines principal components analysis (PCA), regression, and kriging.

8.4.5 Multiquadratic surface method

The multi-quadratic method (Shaw and Lynn, 1972; Shaw et al., 2011) uses a three-dimensional description of the rainfall surface. The coefficients of the mathematical function or equation characterizing rainfall surface are estimated using precipitation measurements at several points in space. The equation can be used to obtain the volume of rainfall by integration over the area of the catchment or basin and finally dividing the volume of rainfall by the catchment area. More details of this method can be found in the work by Shaw et al. (2011). Generation of authentic precipitation surface without any artifacts using point measurements in this method will lead to better characterization of spatial variability of rainfall and therefore improved estimations of MAP compared to those from Thiessen polygon and centroidal distance-based methods.

8.4.6 Spline interpolation methods

Thin-plate splines as exact spatial interpolators (Wang, 2015; Burrough and Donnell, 1998; Chang, 2019; Franke, 1982) can be used to create precipitation surfaces that can help estimate values at different locations in a region and finally the MAP. The equation for the regularized thin-spline surface (Wang, 2015) is given by Eq. 8.36.

$$\hat{\theta}(x,y) = \varepsilon_o + \varepsilon_1 x + \varepsilon_2 y + \sum_{j=1}^{nrg} k_j d_{m,j}^2 \log d_{m,j} \tag{8.36}$$

The variable $d_{m,j}$ is the distance from the point m (x, y) from the rain gauge j where the estimate is required, and ε_o, ε_1, ε_2, and k_j are the parameters to be estimated based on data from the total of *nrg* rain gauges. The parameters are estimated using three equations and are ultimately used for the generation of the surface. Wang (2015) indicates that thin-plate splines have problems in interpolation in data-poor areas and suggests that improved variants such as thin-plate splines with tension and regularized splines can be used to address some of these limitations. Chang (2019) and Franke (1982) describe four equations for the solution as a linear system of Eqs. (8.37–8.40) for obtaining values of coefficients (k_j, ε_o, ε_1, and ε_2). The distance $d_{m,j}$ is obtained using Eq. 8.37.

$$d_{m,j} = \left[\left(x - x_j \right)^2 + \left(y - y_j \right)^2 \right]^{1/2} \quad \forall j \tag{8.37}$$

$$\sum_{j=1}^{nrg} k_j = 0 \tag{8.38}$$

$$\sum_{j=1}^{nrg} k_j x_j = 0 \tag{8.39}$$

$$\sum_{j=1}^{nrg} k_j y_j = 0 \tag{8.40}$$

Two major conditions that are required to be fulfilled for spline interpolation are: (1) the surface needs to exactly pass through all the observation sites (i.e., rain gauges) and (2) the surface should have minimum curvature. Thin-plate splines have the same disadvantages as trend surface methods and are not suitable for multi-time period estimation of missing precipitation data if the thin-splines need to be fitted multiple times. Hutchinson (1995) and Ball and Luk (1998) reported success using splines for interpolation of precipitation data. Xia et al. (2001) reported the use of thin-splines, closest station, multiple linear regression techniques, and Shepard's method (Shepard, 1968) for estimation of daily climatological data. They indicated that the thin-splines method was the best among all the methods investigated. While thin-plate spline methods have several advantages over others, they tend to generate steep gradients in data-poor areas leading to compounded errors in the estimation process (Wang, 2015).

8.4.6.1 Thin-plate splines with tension

Thin-plate splines with tension (Mitas and Mitasova, 1993; Chang, 2019) can be used for the estimation of MAP. The splines with tension belong to the group of radial basis functions and are useful for the fitting surface to values that vary smoothly in a spatial domain. Radial basis functions are deterministic interpolators that are exact. Different radial basis functions are possible, and they are (1) thin plates, (2) thin-plate with splines, (3) regularized splines, (4) multiquadratic functions, and (5) inverse multi-quadratic splines. Thin-plate splines are also referred to as Laplacian smoothing splines. Thin-plate splines with a tension parameter allow for controlling the shapes of membranes (i.e., surfaces) passing through control points

(Franke,1985). The equation for estimated precipitation value ($\hat{\theta}(x, y)$) at a location m (x, y) is given by Eq. 8.41.

$$\hat{\theta}(x,y) = \vartheta(x,y) + \sum_{j=1}^{nrg} A_j \, R(d_{m.j}) \tag{8.41}$$

The variable $\vartheta(x, y)$ represents the trend function, A_j is a coefficient and $R(d_j)$ is the basis function. The basis function value is obtained by using Eq. 8.42.

$$R(d_j) = \frac{1}{2\pi\eta^2}\left[\ln\left(\frac{d_{m.j}\eta}{2}\right) + c + k_o\left(d_{m.j}\eta\right)\right] \forall j \tag{8.42}$$

The variables ϑ and A_j need to be estimated, $d_{m,j}$ is the distance between the rain gauge j and at any location m (x, y). The variable η is the tension (or weight) parameter, c is a constant (Euler's constant equal to 0.577215) and k_o is the modified Bessel function. When the tension parameter is set close to zero, the results from this method approximate to those from a thin-plate spline method.

The MAP estimate (θ_M) is obtained by the average of precipitation values $\left(\hat{\theta}(x,y),\, x, y \in Ar\right)$ at all grid points in a region (Ar) using Eq. 8.43.

$$\theta_M = \frac{\sum_{x,y \in Ar} \hat{\theta}(x,y)}{ng} \tag{8.43}$$

Spline interpolation methods with or without tension are ideally suited for smoothly varying processes and may not always be appropriate for regions with high spatial variability of precipitation. Rapid changes in the gradient of the surface may occur in the proximity of the observation sites (i.e., rain gauges).

8.4.7 Surface fitting methods: issues

Surface fitting methods are reliable and accurate if adequate point measurements are available to characterize the spatial variability of precipitation in a region. Clustering of rain gauges, networks with low rain gauge density, lack of long historical data for calibration and validation, issues with missing precipitation data are few factors that will severely influence the accuracy of these methods for estimation of MAP. The density and location (spatial placement in a region) of rain gauges will influence the accuracy of the MAP estimates using these methods especially in regions with high spatial variability of precipitation. The surface fitting methods are conceptually simple and computationally less intensive in comparison with other approaches such as kriging. Negative precipitation depths are possible when these methods are used, and these values can be set to zero values as part of the post interpolation corrections. The use of trend surface methods may result in overfitting of the trend surfaces by the selection of higher-order polynomials that may not represent the general variation of precipitation in the region. In the case of trend surface methods, precipitation data available at the gauges are used only for estimation of coefficients of the specific model (i.e., linear, quadratic, etc.) and not for estimation of values at all other sites other than the gauges. Variance deflation, over and underestimation of lower and higher-end extremes respectively, alteration of statistical distributions, and other characteristics of precipitation time series are inevitable consequences of spatial interpolation methods that are often used for the generation of gridded datasets (Teegavarapu et al., 2013; Goly and Teegavarapu, 2014). These major limitations of gridded data generated using spatial interpolation methods must be recognized.

8.5 Radar and satellite-based methods

Weather surveillance radar (WSR) (Raghavan, 2003; Sene, 2010; Pathak and Teegavarapu, 2018) and satellite-based precipitation estimates (Sene, 2010; Teegavarapu, 2013b) provide two alternative sources to rain gauge-based precipitation measurements for estimation of MAP. Radar-based precipitation estimates, even while becoming more reliable and accurate in the past two decades, require bias corrections before their use for hydrological modeling or any design purpose (Teegavarapu et al., 2017; Pathak and Teegavarapu, 2018). Radar- (Sene, 2010; Teegavarapu 2013b; Raghavan, 2003) and satellite-based (e.g. Tropical Rainfall Monitoring Mission (TRMM), (Teegavarapu, 2013b) and Global Precipitation Measurement (GPM) (Sene, 2010) precipitation data available at pre-specified spatial resolutions are now available for calculation of MAP.

Details of radar-based precipitation estimation and bias correction procedures are discussed exhaustively by (Teegavarapu, 2013b) and Sene (2010). Bias corrections may involve the use of spatially varying or constant corrections using inverse distance and negative exponential weighting methods. The corrections are obtained by spatially interpolating errors noted at control points (i.e., rain gauges) in the region and using them to adjust the quantitative precipitation estimates. The corrections can be additive or multiplicative based on these methods. In the case of multiplicative error corrections, the ratios of the gauge to radar-based precipitation values as mean of ratios or ratio of means are used. In the case of additive corrections, the error estimated using the differences in radar and rain gauge-based precipitation values are used. These corrections sometimes incorporate a factor that relates to rain gauge density in a region. While most of the bias corrections address systematic errors, random errors will still remain. Improvements in bias-corrected gridded precipitation products are possible using the information available from nearby rain gauges to a grid. Transformations of precipitation data from one grid size to another using spatial interpolation methods (Teegavarapu et al., 2012) are required in many situations to estimate MAP. Several areal grid-based weighting methods were evaluated for such transformations by Teegavarapu et al. (2012). Grimes et al. (1999) estimated MAP using an optimal merging of the estimates provided by satellite information and estimates obtained from rain gauges. Sokol (2003) reported the use of rain gauge and bias-corrected radar-based precipitation estimates for estimation of MAP.

Gridded precipitation data developed using rain gauge data (referred to MAPO) with the help of spatial interpolation methods and a radar/gauge multisensor product (referred to as MAPX) are used by the National Weather Service (NWS) of the U.S. (NWS, 2020) for estimation of MAP. The method is illustrated in Fig. 8.14. The estimate is an average value of all precipitation values from the grids with their centers within the basin or region of interest. The MAP estimate (θ_M) using the MAPX method that uses a radar-based gridded precipitation product is given by Eq. 8.44. The variables θ_i^r and ngc refer to the precipitation estimate and the total number of grids respectively.

$$\theta_M = \frac{\sum_{i=1}^{ngc} \theta_i^r}{ngc} \tag{8.44}$$

The reader is advised to refer to a comprehensive review of methods for estimation of MAP in mountainous and non-mountainous provided by Anderson (2002) and Jones (1983). A comparative analysis of MAP estimates from radar and rain gauges provided by (Johnson et al., 1999) indicated MAPX provides a competitive alternative to MAP estimated based on gauges. Cho et al. (2017) proposed and evaluated a polygon-based method referred to as the radar polygon method (RPM) for

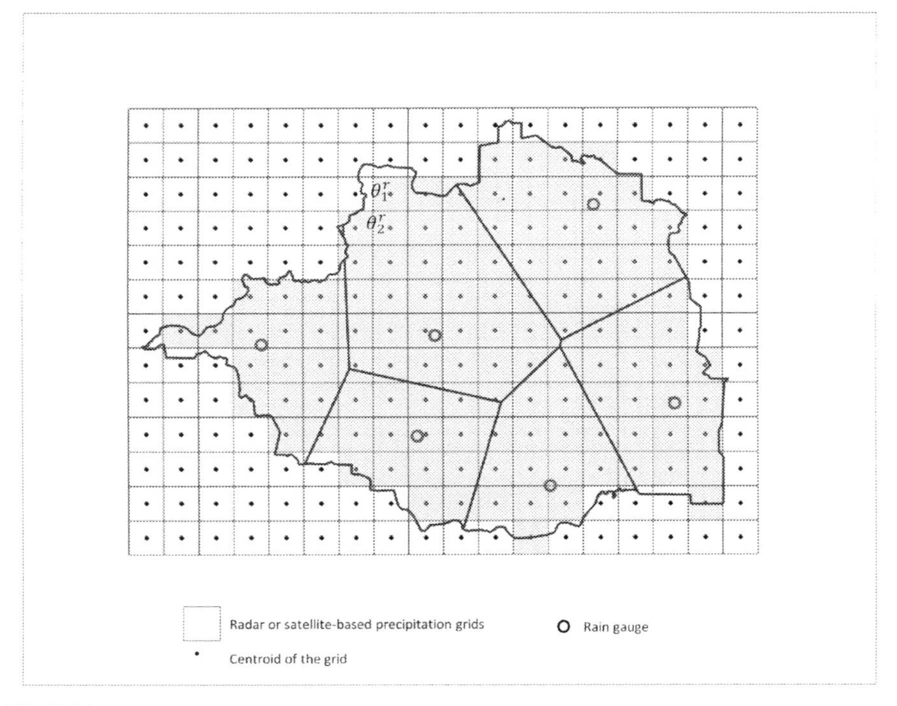

FIG. 8.14

Gridded MAP estimation using MAPX method.

areal rainfall estimation using radar rainfall data. Radar rainfall imageries are used along with gauge data to find regions that have similar characteristics or precipitation data at the gauges. This similarity assessment considers the competition among the rain gauges to obtain the radar grid (cell) location under its territory (Cho et al., 2017). Gridded radar data are evaluated using rainfall similarity (*RS*) (Cho et al. 2017) measure based on radar-estimate of precipitation (P_r) in a grid *j* and the rain gauge measured value (P_g) as given by Eq. 8.45.

$$\begin{aligned} RS_j &= 1 - \left[\frac{P_g - P_r}{P_g}\right]^2 \; if \left[\frac{P_g - P_r}{P_g}\right]^2 \le 1, \\ or\ RS_j &= 0,\ if \left[\frac{P_g - P_r}{P_g}\right]^2 > 1\ \forall j \end{aligned} \tag{8.45}$$

The *RS* values are obtained for each rain gauge and a rainfall similarity map for each rain gauge is created for different rain events. The rainfall similarity maps are superimposed to obtain a rainfall similarity score map that will help to define the territory (radar polygon) for each rain gauge. The similarity regions can be regarded as homogenous rain areas explored and used by Teegavarapu et al. (2013) in their study of precipitation extremes. An illustration of the RPM for estimation MAP is shown in Figs. 8.15a–b. Fig. 8.15a shows the radar polygon derived based on the rain gauges and Fig. 8.15b shows the geo-processed polygons to estimate MAP for the basin under consideration. Once the radar-based polygons are obtained the MAP value can be estimated using an area-weighted calculation similar to the one used in the Thiessen polygon method.

Radar or satellite-based QPEs can characterize the spatial variability of precipitation in a region better than a set of sparse point measurements (i.e., rain gauges) in a region. However, without bias corrections using reliable ground truth, the accuracy of MAP estimation based on these datasets is questionable. Radar or satellite-based precipitation estimates are the last resort datasets for MAP estimation in the case of ungauged basins.

8.6 Computationally intensive estimation methods

Methods using finite elements, artificial neural networks (ANNs), and objective analysis (OA) have been proposed and used for MAP. Hutchinson and Walley (1972) and Lal and Al-Mashidani (1978) proposed a finite element method (FEM) with altitude corrections and an approach using the product of distances enclosing or radiating from a rain gauge as weights for MAP estimation. The use of artificial neural networks (ANNs) for evaluation of the spatial distribution of precipitation was reported by Tsangaratos et al. (2014) and for MAP by Bryan and Adams (2002). Objective analysis (OA) proposed by Gandin (1965) can be used for the interpolation of precipitation data required for MAP. OA uses an optimal interpolation analysis

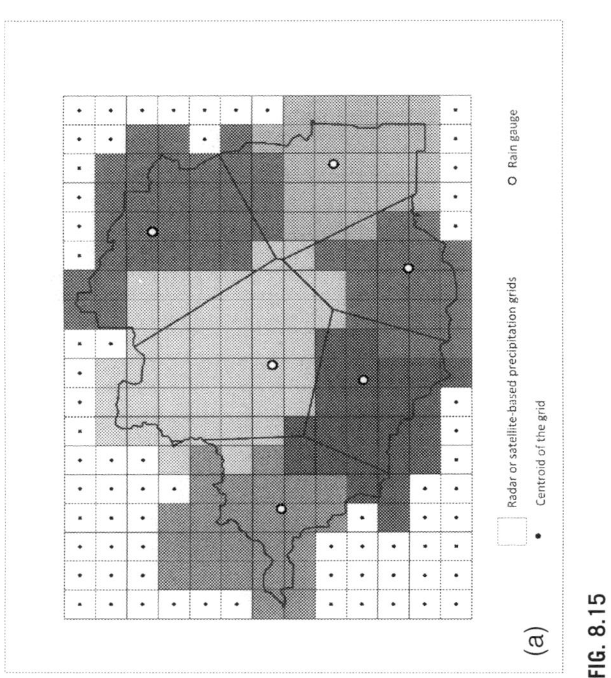

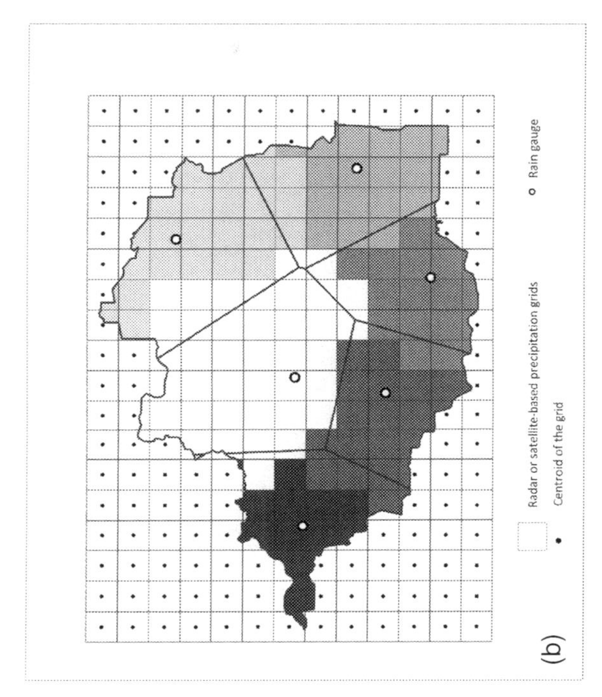

FIG. 8.15

Illustration of MAP estimation using radar polygon method.

scheme to obtain the best possible estimate of a meteorological field (e.g. precipitation in this context) at a regular network of grid points using observations at sites available in space by minimizing the mean square interpolation error for a large ensemble of analysis situations (Bergman, 1978). Conditional interpolation (CI) for the generation of gridded area-averaged daily precipitation values was proposed and evaluated by Hewitson and Crane (2005). The CI method uses a two-step procedure to identify the spatial pattern of wet/dry conditions and the magnitude of precipitation is estimated. The CI method attempts to overcome the limitations of conventional interpolation methods (e.g., weighted, surface fitting, and kriging methods) by addressing spatial representativity of stations as a function of the driving synoptic state and spatial discontinuity of precipitation fields (Hewitson and Crane, 2005). The ANNs, FEM, OA, and CI methods are computationally intensive and are data sensitive. MAP estimation using these methods requires extensive datasets and processing.

8.7 Mean areal precipitation estimation: issues

Several factors influence the accuracy of MAP methods (Teegavarapu, 2016) and they include: (1) rain gauge density and its characteristics (spatial arrangement and inter gauge distances and ability to characterize spatial and temporal variability of precipitation in a region), (2) topographic and physiographic variations of the region including the existence of separable homogeneous rain areas, (3) utility of rain gauge(s) that are outside of the domain of interest, (4) accuracy of rainfall measurements from rain gauges considering random, systematic and transcription errors, (5) availability of bias-corrected radar or satellite-based precipitation datasets at an appropriate spatial resolution (i.e., tessellation) considering the watershed size, (6) validated and quality assured and quality controlled (QAQCd) precipitation datasets that are error, anomaly and gap-free, and (7) spatial interpolation method used to generate rainfall fields or surfaces. All MAP estimation methods and their applications should consider and use basin-specific characteristics, physiographic features, and orographic effects, and merged data from different measurements and estimation sources of precipitation (i.e. multi-sensor estimates). Moulin et al. (2009) indicate one of the major sources of input uncertainty in hydrologic simulation models comes from the lack of representativeness of a discrete set of gauges of a network and from the necessity to interpolate the rain rates between these points. Issues with MAP estimates can be also noted when results from calibrated and validated hydrologic simulation models are evaluated.

8.7.1 Precipitation monitoring network adequacy

A network of recording or non-recording rain gauges is generally employed for the measurement of precipitation. Often these networks are designed based on the objectives of the planned hydrological project in a specific location aimed at observing spatial and temporal variations of precipitation patterns. The accuracy of the MAP estimate is highly dependent on the network gauge density as indicated by Masih et al. (2011). Recommendations exist for precipitation gauge network density (i.e., number of gauges in each area) in different climatic zones and topographic settings. These recommendations are difficult to implement sometimes due to budgetary limitations and field-level challenges in siting the rain gauges. Improved monitoring network designs based on geostatistical approaches (Teegavarapu 2018a) information-theoretic methods (Xing et al. 2013), entropy, and multi-objective-based techniques (Mogheir et al. 2013), and fuzzy theory and multiple criteria analysis methods (Chang and Lin, 2014) can help in the MAP estimation. Adequate monitoring network size and density are required for accurate estimation of MAP. Even though radar and satellite-based QPEs are available, ground-based rain gauge observations serving as ground truth are critical for improvement and bias corrections of gridded precipitation products derived from these sources.

8.7.2 Quality of precipitation measurements

The accuracy of the MAP estimate depends on the quality of precipitation data which also influences the results of hydrologic simulation models (Xu and Vandewiele, 1994). Rain gauge measurements of precipitation are prone to systematic and random errors. Systematic errors are easier to identify and correct compared to random errors. Errors in rain gauge measurements are generally introduced due to (1) inappropriate size of the collector, (2) evaporation loss, (3) in and out splash, (4) orientation, (5) placement, (6) observation errors, and (6) wind effects. A detailed discussion of errors and possible correction mechanisms are documented in many investigations by World Meteorological Organization (WMO, 1982, 1984, 1986) and Habib et al. (2010). WMO (1982) discusses methods used for adjusting raw precipitation data and WMO (1989a, 1989b) describes how the errors occur due to a variety of site-specific and instrument-related reasons. Precipitation amounts may be underestimated due to wind and the underestimations are related to the speed and the direction of the wind.

8.7.3 Missing precipitation data at rain gauges

Missing precipitation data at one or more rain gauges is a critical issue that needs to be addressed and resolved as MAP estimation methods require precipitation data at all sites. The existence of gaps in precipitation data at any site requires an interpolation method to fill the gaps first and then to estimate MAP using any one of the methods discussed in this chapter. Missing data can lead to the removal of one or more sites that are crucial for MAP estimation. Deterministic, time-series, trend surface generation, variance-dependent stochastic interpolation, and data-driven and data mining approaches can be used for the estimation of missing precipitation data. Many of these imputation approaches (Teegavarapu, 2016) fall under the category of spatial interpolation. Temporal interpolation methods (Pappas et al., 2014) can be used if the serial autocorrelation derived from precipitation time series is high for the first few temporal lags. Establishing linear and nonlinear relationships can help infill the data gaps in the series. In some cases, simple procedures such as last observation carry forward (LOCF) and baseline observation carry forward (BOCF) (Buuren, 2012) may also be appropriate. Spatial interpolation is the most used approach for the estimation of missing data at a site using observations from all remaining sites. Naïve deterministic methods include gauge and nearest neighbor mean (Paulhus and Kohler, 1952) and single best estimator (SBE) methods. SBE method uses the precipitation value from the closest neighbor site to the site with the missing data. The inverse distance weighting method uses reciprocal Euclidean distances and the normal ratio method (Paulhus and Kohler, 1952; McCuen, 2004) requires long-term historical data. The quadrant method (McCuen, 2004) divides the region into four regions (or quadrants) using the site of the interest at the intersections of two axes and uses the nearest neighbor in each quadrant based on Euclidean distance from the gauge with missing data for estimation of filling the gaps in data.

The Euclidean distance used in many weighting methods for missing data estimation (Teegavarapu and Chandrmouli, 2005) and MAP estimation (as in CDM) may not always serve as an appropriate surrogate that can quantify the inter-gauge relationships in all hydroclimatic and topographical settings. Recent studies have sought improved variants of weighting methods considering replacements for Euclidean distances. Statistical distance (Ahrens, 2006), Pearson correlation coefficient, modified distance based on the Thiessen polygons, the cumulative distance between geometric patterns derived from precipitation time series were proposed and evaluated by Teegavarapu and Chandramouli (2005). The use of angular distances in weighting methods that consider the spatial orientation of gauges (Shepard, 1968; Yeggina et al. 2019) led to improved estimates compared to those methods that use Euclidean distances. Probability space-based error measures and distribution similarity hypothesis test statistic metrics were evaluated by Teegavarapu, (2020b) as surrogates for distances in weighting methods to address this limitation. These replacements for distances have improved the accuracy of the imputation estimates.

In the case of spatial weighting methods, guidance is required for the selection of the local or global version of interpolation, the number of sites, weights, and identification of specific sites to obtain estimates of missing data. Improvements can be achieved in the accuracy of the estimation if all these factors are considered. Improved weighting methods and their variants were developed with help of optimal functional forms (Teegavarapu et al., 2009) association rules (Teegavarapu, 2009), proximity-based metrics, clustering, and nearest neighbor classification approach (Teegavarapu, 2014a), and probability space-based methods (Teegavarapu, 2020b). These works have led to variations in weighting functions, data-based real and binary value similarity measures as surrogates for distances or weights, optimal selection of sites, and post-interpolation corrections of estimates. Nonlinear programming models with binary variables were proposed and developed by Teegavarapu (2012) to improve inverse distance weighting methods and normal ratio methods by incorporating the objective site selection process. Data-driven techniques such as artificial neural networks (ANNs), correlation weighting approaches, and nearest-neighbor-based correction methods were developed and tested for infilling precipitation records by Teegavarapu et al. (2018). Stochastic spatial interpolation techniques that include various forms of kriging can be used for the estimation of missing data. Teegavarapu and Chandramouli (2005), Teegavarapu (2007) document the use of ordinary and universal function approximation-based kriging methods for estimation of missing data. Distance-based methods are not only used for estimation of MAP, but also for understanding spatial variability of precipitation extremes (Teegavarapu, 2013b). Radar-based precipitation estimates (Teegavarapu, 2018b) can also be used to infill the missing precipitation data provided the estimates are validated for accuracy and bias-corrected.

Universal function approximators such as artificial neural networks (ANNs) have been successfully used in the estimation of missing precipitation records (Kuligowski and Barros, 1998; Teegavarapu and Chandramouli, 2005). Data mining approaches using association rules are used to correct spatial interpolation estimates using a single best estimator (Teegavarapu, 2009). Clustering and nearest neighbor classification approaches are developed by Teegavarapu (2014a) to obtain improvements in estimates of missing data using groups of sites, partitioning of the data, and similarity metrics adopted from the field of numerical taxonomy. Recent work by Bardossy and Pegram (2014) evaluated the use of Copula-based methods for infilling precipitation records.

Filling of gaps in precipitation series using spatial or temporal interpolation methods may lead to changes in distributional characteristics, extreme precipitation indices, autocorrelation structure, rain or no-rain transitions in time (e.g. wet-to-dry) (Teegavarapu, 2014b), and trends assessed by filled datasets (Teegavarapu and Nayak, 2017). Also, filling of gaps may lead to under and over-estimation of higher and lower-end precipitation values respectively. These under and overestimations will impact the extreme precipitation indices generally evaluated in climate change and variability studies. Also, the adverse impacts of filling were found to be prominent when the filled data length is more than five percent of the total length of the time series. Statistical corrections of spatially interpolated estimates suggested by Teegavarapu (2014b) are useful to replicate the properties of filled data series to those of original precipitation series without any gaps. The adverse effects of filling and statistical corrections need to be evaluated before the filled data can be considered for hydrologic modeling and climate change and variability studies (Teegavarapu and Nayak, 2017).

8.7.4 Quality of radar and satellite-based precipitation estimates

Limitations of radar and satellite-based MAP estimation are generally related to coarser spatial resolution at which gridded precipitation products may not characterize the spatial precipitation variability in a region. While radar and satellite estimated precipitation data are ideally suited for MAP estimation, a ground-based reliable rain gauge network (ground truth) of sufficient density and with the spatially uniform placement of gauges in a region is critical as the data from the rain gauges are used for bias corrections of radar (Teegavarapu 2018c) and satellite data (Yeggina et al., 2020). Continuing enhancements in weather surveillance radar (WSR) technology (e.g., dual-polarization radar) in the recent decade have led to speculations about the possible disappearance of rain gauges in the future. Random errors and other limitations of radar remain although improvements in hydrometeor identification ability of weather surveillance radars (WSRs) have substantially improved over the past decade. With all the limitations of rain gauges aside, they are still essential for correcting and improving radar-based estimates and they are the only sources of precipitation measurements in regions with no weather radars.

8.7.5 Spatial interpolation approaches

All the MAP estimation methods other than gauge-based weighting methods discussed in this chapter use spatial interpolation. Deterministic interpolation approaches require fewer assumptions to be satisfied and are easy to develop and apply compared to stochastic methods. The former approaches do not provide information about uncertainty in estimates. The local or global variant of a specific spatial interpolation method may be beneficial depending on the spatial and temporal scale at which the spatial precipitation estimates are needed. Spatial interpolation methods may generate negative, unusually high, or low values and patterns that are not realistic and physically possible. The geostatistical approaches may provide negative values of precipitation whereas spline-based approaches may generate unrealistically smooth surfaces. Ordinary kriging requires evaluation of precipitation data for stationarity in space and a check for isotropic or anisotropic variation. Misspecification of the semivariogram model will impact the performance of kriging in generating reliable estimates of precipitation. Spatially interpolated precipitation estimates can be corrected using nearest neighbors as suggested by Yeggina et al. (2020).

8.8 Evaluation of mean areal precipitation estimation methods

The MAP estimates from different methods discussed in the chapter need to be evaluated before they can be used for hydrologic modeling or design. The accuracy and quality of MAP estimates from these methods can be evaluated using two approaches. In the first approach, the estimated precipitation values are compared with observed values from rain gauges (i.e. ground truth). The second approach compares the hydrologic simulation model generated results that use MAP estimates to the real-world observed values of a specific hydrologic variable (e.g., streamflow). The MAP estimates from the weight-based methods can only be compared using the second approach as these methods do not estimate values at the gauge or non-gauge locations but rather use existing observed values at gauges to obtain MAP.

In the case of surface fitting methods, cross-validation schemes can be used to evaluate MAP estimates. The available rain gauge observations are divided into two datasets. One set of precipitation data, referred to as train dataset, are used for estimation of parameters (coefficients) of the functions that define the surfaces and the other defined as a test dataset, can be used for validation of the estimated precipitation values. In the case of exhaustive cross-validation, the available gauge observations can be divided into several ways of train and test datasets. Different cross-validation schemes that are described next can be used. Leave-p-out cross-validation (LpO CV) uses "p" observations as a test dataset and the remaining is the training dataset. Leave-one-out cross-validation (LOOCV) is a special case of LpOCV with a value of p equal to 1. The holdout cross-validation process uses a split-sample method with dataset partitioned into train and test datasets randomly. In the case of k-fold cross-validation, the available gauge dataset is equally partitioned into k parts or folds and out of these k-folds, k-1 parts are selected as train dataset and one part is selected for validation. Other variants include repeated random subsampling and stratified k-fold cross-validation. Different error and performance measures can be used for the evaluation of estimations based on multiple MAP methods using these cross-validation schemes. The error measures that can be used for evaluation of methods include mean error (ME), root mean squared error (RMSE), mean absolute error (MAE), compound relative error (CRE) and mean relative error (MRE). Performance measures include correlation coefficient, Nash Sutcliffe efficiency criterion (NSEC) and index of agreement. The list of measures provided is by no means exhaustive. The error and performance measures evaluated based on validation test datasets may help to select the best variants of the methods. Since MAP estimates are available as gridded data, they can be used for the hydrologic response evaluation using both lumped and distributed hydrologic models.

The hydrologic response evaluation (HRE) process will involve several approaches using a simulation model results when two types of datasets (gauge-based gridded precipitation products using spatial interpolation and radar or satellite-based precipitation data at the same spatial resolution) are available. The approaches that can be used for HRE are (1) evaluation of parameter set: compare and assess the statistical similarity of parameter sets obtained by calibrating a hydrologic model using the two types of datasets; (2) evaluation of simulation results: compare and assess the statistical similarity of the simulated hydrologic variable (e.g., streamflow) values obtained by model calibration using the two types of datasets; (3) evaluation of simulation results: compare and assess the statistical similarity of simulated hydrologic variable values obtained by a model using the two types of datasets using a specific set of model parameter values; (4) evaluation of simulation results: compare and assess the statistical similarity of simulated hydrologic variable values obtained by a model using the two types of datasets using one parameter set obtained from calibration of model using one of the datasets (gauge-based or interpolated data). The statistical similarity can be evaluated using parametric or nonparametric statistical hypothesis tests. Equifinality issues that arise due to the hydrologic model calibration process may generate similar results with different inputs (i.e., MAP estimates). Multifinality problem is also possible due to sub-optimal calibration process when the hydrologic responses are different even though the same precipitation dataset is used. When hydrologic experts evaluate the results of HRE, there are referred to as Turing tests (1996).

The cross-validation methods can also be used for the evaluation of radar and satellite-based gridded precipitation products. In the case of these gridded precipitation products, split-sample approaches can be used to test the accuracy of these datasets or products by using one dataset (observations from one set of rain gauges) for bias correction and the remaining dataset for validation of the corrected data. Hydrologic-response-based evaluation can also be used for the radar or satellite-based gridded datasets.

8.9 Recommendations for selection of a mean areal precipitation estimation method

Accurate MAP estimates are possible using rain gauges or other sources (i.e., radar or satellite) if advantages and limitations associated with each method are recognized. New developments in methods and sources (e.g., weather radar) for accurate quantitative precipitation estimates (QPEs) are now challenging the attitudes of hydrologic modelers who have confined themselves to the use of rain gauge data for model calibration and validation. The following recommendations can be useful for the selection of the MAP estimation method for hydrologic modeling or design.

1. A comprehensive evaluation of gauge-based measurements needs to be conducted before these datasets can be used in the MAP estimation.
2. Gauge-weighted MAP estimation methods using rain gauge measurements alone are only recommended provided the gauge density is adequate to characterize the spatial variability of precipitation in a region.
3. Gauge-weighted MAP estimation methods that rely on the partitioning of a region by closed polygons, using Euclidean distances to the center of the region/basin, adopting geometric properties of areas or lengths derived by using spatial arrangements of rain gauges are easy to implement for operational hydrology applications. However, some of these methods while being intuitive, simple and replicable, do not consider the spatial and temporal variability of the precipitation while deriving weigths that are used in weighting methods. These methods are rain gauge network-dependent and the application of any one of these methods requires a re-estimate of weights if any changes in network size and orientation are noted.
4. Available region-specific information (topography, the existence of orographic influences on precipitation, large water bodies, etc.) should be used for the selection of the appropriate method. Methods that use region-specific meteorological and climatological information are always recommended for accurate estimation of MAP over other methods.

5. Trend surface or surface fitting models should only be used if there is a reason to believe that precipitation varies over a region in a way that is represented by linear or nonlinear surfaces or fields specified and generated by these models.
6. A computationally intensive method such as kriging should only be used when all the assumptions related to this method are fulfilled and spatial dependence of precipitation in a region can be reasonably modeled using one of the theoretical semi-variogram models.
7. Robustness of the spatial varying precipitation estimates generated using exact or inexact spatial interpolation methods (e.g., inverse distance, trend surface, kriging, regression-based methods) from point measurements (i.e., gauges) with or without auxiliary information (e.g., land slope, aspect, etc.) needs to be evaluated with ground truth using different verification schemes and performance metrics before these methods can be used for MAP estimation.
8. There is no single method for MAP estimation that is appropriate for all regions. Several deterministic and stochastic MAP estimation methods should be evaluated before the best method for a region is selected.
9. Radar and satellite-based gridded precipitation products provide QPEs that can charaterize the spatial variation of precipitation better than rain gauges in a region. MAP estimation methods using these QPEs do not need spatial interpolation methods. However, these products should be scrutinized for errors and biases, and appropriate quality control and bias correction methods should be employed before using these gridded precipitation products for MAP estimation.
10. Precipitation has the strongest causal influence on the hydrologic response from any basin or watershed. Therefore, MAP estimate obtained from any one of the methods discussed in this chapter should not be considered accurate unless the hydrologic response obtained from a well-calibrated and validated lumped hydrologic simulation model using this estimate is close to the hydrologic response observed in real-world situations.
11. Stochastic methods such as kriging and specific regression-based methods can be used for MAP estimates if uncertainty assessments are needed for estimates.
12. Conceptually simple deterministic methods such as inverse distance weighting methods are preferable over computationally intensive stochastic methods such as kriging if the rain gauge network is dense enough to capture the spatial variability of precipitation in the region and when estimates are needed for multiple time steps.
13. Spatial-interpolation or surface fitting methods can be evaluated using local and global variants before selecting the best variant for MAP estimation.
14. The performance and robustness of any method will depend on the temporal resolution at which MAP estimates are derived. Conceptually simple and computationally less intensive deterministic methods such as the inverse distance method may provide equally good estimates as those from stochastic interpolation methods at coarser temporal resolutions (e.g., month, year) compared to finer time scales (day, week).

8.10 Conclusions

Mean areal precipitation (MAP) estimation using deterministic and stochastic methods is elaborated in this chapter. Several conceptually simple weighting methods using point measurements (i.e., rain gauge-based observations) and gridded precipitation estimates from radar and satellite-based sources are discussed. Methods requiring spatial analysis software and computationally intensive stochastic methods are also discussed and limitations and applicability of these methods for accurate estimation of MAP are reported. The widespread availability of quality assessed and quality controlled (QAQCd) radar and satellite-based precipitation estimates are providing the impetus for use of these datasets for MAP estimation. Although these datasets require exhaustive pre-processing and bias corrections, their ability to provide a better spatial characterization of precipitation variability in a region of interest is the main contributing factor for their increasing use in distribued hydrologic modeling applications. Lumped hydrologic modeling efforts primarily aimed at hydrologic design continue to benefit from the emerging gridded precipitation datasets derived from radar and satellite-based sources to accurately estimate MAP. The selection of the best method for MAP requires evaluation of multiple methods considering study region, modeling objectives, and the quality and length of precipitation data available.

References

Abtew, W., Obeysekera, J., Shih, G., 1993. Spatial analysis for monthly rainfall analysis in South Florida. J. Am. Water Resour. Assoc. 29 (2), 179–188.

Ahrens, B., 2006. Distance in spatial interpolation of daily rain gauge data. Hydrol. Earth Syst. Sci. 10, 197–208.

Akin, J.E, 1971. Calculation of areal depth of precipitation. J. Hydrol. 12 (4), 363–376.

Alexander, L.V., et al., 2006. Global observed changes in daily climate extremes of temperature and precipitation. J. Geophys. Res. 111, D05109. doi:10.1029/2005JD006290.

Anderson, E.A, 2002. Calibration of conceptual hydrologic models for use in river forecastingTechnical Report, NOAA Technical Report, NWS 45. Hydrology Laboratory, Silver Spring, Maryland, USA.

ASCE, 1996. Hydrology Handbook, second edition. American Society of Civil Engineers ASCE, New York.

Ball, J.E., Luk, K.C., 1998. Modeling spatial variability of rainfall over a catchment. J. Hydrol. Eng. 3 (2), 122–130.

Bardossy, A., Pegram, G., 2014. Infilling missing precipitation records – a comparison of a new copula-based method with other techniques. J. Hydrol. 519 (Part A), 1162–1170.

Bergman, K.H., 1978. The role of observational errors in optimum interpolation analysis. Bull. Am. Meteorol. Soc. 59 (12), 1603–1611.

Bethlahmy, N., 1976. The two-axis method: a new method to calculate average precipitation over a basin. Hydrol. Sci. Bull. 21 (3), 379–385.

Boots, B.N., 1986Voronoi Thiessen Polygons, Concepts, and Techniques in Modern Geography, 45, Geo Book, Norwich.

Borga, M., Marra, F., Gabella, M., 2022. Rainfall Estimation by Weather Radar. In: Morbidelli, R. (Ed.), Rainfall. Modeling, Measurement and Applications. Elsevier, Amsterdam, pp. 109–134. doi:10.1016/C2019-0-04937-0.
Brassel, K.E., Reif, D., 1979. A procedure to generate Thiessen polygons. Geogr. Anal. 2 (3), 298–303.
Bryan, B.A., Adams, J.A., 2002. Three-dimensional neurointerpolation of annual mean precipitation and temperature surfaces for China. Geogr. Anal. 34 (2), 93–111.
Burrough, P.A., McDonnell, R.A., 1998. Principles of geographical information systems. Oxford University Press, New York, USA.
Buuren, V.S, 2012. Flexible Imputation of Missing Data. CRC Press, Boca Raton, Florida, USA.
Chang, C.L., Lin, Y.T., 2014. A water quality monitoring network design using fuzzy theory and multiple criteria analysis. Environ. Monit. Assess. 186 (10), 6459–6469.
Chang, K.T, 2019. Introduction to Geographic Information Systems. McGraw-Hill, New York.
Chen, F.-W., Liu, C.-W., 2012. Estimation of the spatial rainfall distribution using inverse distance weighting (IDW) in the middle of Taiwan. Paddy Water Environ 10 (3), 209–222.
Chua, S.-H., Bras, R.L., 1982. Optimal estimators of mean areal precipitation in regions of orographic influence. J. Hydrol 57 (1–2), 23–48.
Cho, W., Lee, J., Park, J., Kim, D., 2017. Radar polygon method: an areal rainfall estimation based on radar rainfall imageries. Stochastic Environ. Res. Risk Assess 31, 275–289.
Daly, C., Gibson, W.P., Taylor, G.H., Johnson, G.H., Pasteris, P., 2002. A knowledge-based approach to the statistical mapping of climate. Climate Res. 22, 99–113.
Daly, C., Neilson, R.P., Phillips, D.L., 1994. A statistical topographic model for mapping climatological precipitation over mountainous terrain. J. Appl. Meteorol. 33, 140–158.
Dawdy, D.R., Langbein, W.B., 1960. Mapping mean areal precipitation. Bull. Int. Assoc. Sci. Hydrol. 5 (3), 16–23.
Dingman, S.L., 2015. Physical hydrologyIL. Waveland Press, Inc, Illinois, USA.
France, P.W., 1985. A comparison of various techniques for computation of areal rainfall. J. Inst. Eng. India 66, 74–78.
Franke, R., 1982. Smooth interpolation of scattered data by local thin-plate splines. Comput. Math. Appl. 8 (4), 273–281.
Franke, R., 1985. Thin plate splines with tension. Comput. Aided Geom. Des. 2 (1–3), 87–95.
Gandin, L.S., 1965. Objective analysis of meteorological fields. Translated from the Russian. Israel Program for Scientific Translations, Jerusalem Report, 242 pages.
Goel, S.M., Aldabagh, A.S., 1979. A distance weighted method for computing average precipitation. J. Ins.t of Water Engineers and Science 33, 451–454.
Goly, A., Teegavarapu, R.S.V., 2014. Individual and coupled influences of AMO and ENSO on regional precipitation characteristics and extremes. Water Resour. Res. 50 (6), 4686–4709.
Goovaerts, P., 2000. Geostatistical approaches for incorporating elevation into the spatial interpolation of rainfall. J. Hydrol 228 (1–2), 113–129.
Grimes, D.I.F., Pardo-Igu´zquiza, E., Bonifacio, R., 1999. Optimal areal rainfall estimation using rain gauges and satellite data. J. Hydrol. 222 (1–4), 93–108.
Guillermo, Q., Tabios, G.Q., Salas, J.D., 1985. A comparative analysis of techniques for spatial interpolation of precipitation. J. Am. Water Resour. Asssoc. 21 (3), 365–380.
Habib, E., Lee, G., Kim, D., Ciach, G.J., 2010. Ground-based direct measurement. In: Rainfall: State of the Science, Geophysical Monograph Series, 191. American Geophysical Union, Washington D.C.
Hewitson, B.C., Crane, R.G., 2005. Gridded area-averaged daily precipitation via conditional interpolation. J. Clim. 18 (1), 41–57.

Hutchinson, P., Walley, W.J., 1972. Calculation of areal rainfall using finite element techniques with altitude corrections. Bull. Int. Assoc. Hydrol. Sci. 17 (3), 259–272.
Hutchinson, M.F., 1995. Interpolating mean rainfall using thin plate smoothing splines. Int. J. Geogr. Inf. Sci. 9 (4), 385–403.
Isaaks, H.E., Srivastava, R.M., 1989. An Introduction to Applied Geostatistics. Oxford University Press, New York.
Jain, S.K., Singh, V.P., 2005. Isohyetal method. In: Lehr, J.H., Keeley, J. (Eds.), Water Encyclopedia, 290–292.
Jones, S.B., 1983. The Estimation of Catchment Average Point Rainfall Profiles, Report No. 87. Institute of Hydrology, Wallingford, U.K.
Johnson, D., Smith, M., Koren, V., Finnerty, B., 1999. Comparing mean areal precipitation estimates from NEXRAD and rain gauge networks. J. Hydrol. Eng. 4 (2), 117–124.
Karnieli, A., 1990. Application of kriging technique to areal precipitation mapping in Arizona. Geo. J. 22 (4), 391–398.
Kidd, C., Levizzani, V., 2022. Satellite Rainfall Estimation. In: Morbidelli, R. (Ed.), Rainfall. Modeling, Measurement and Applications. Elsevier, Amsterdam, pp. 135–170. doi:10.1016/C2019-0-04937-0.
Krajewski, W.F., 1987. Co-kriging of radar and rain gauge data. J. Geophys. Res. 92 (8), 9571–9580.
Krige, D.G., 1966. Two-dimensional weighted moving average trend surfaces for ore evaluation. J. S. Afr. Inst. Min. Metall. 66, 13–38.
Kuligowski, R.J., Barros, A.P., 1998. Using artificial neural networks to estimate missing rainfall data. J. Am. Water Resour. Assoc. 34 (6), 1437–1447.
Lal, P.B.B., Al-Mashidani, G., 1978. A technique for determination of areal average rainfall. Hydrol. Sci. Bull 23 (4), 445–453.
Lanza, L.G., Cauteruccio, A., Stagnaro, M., 2022. Rain Gauge Measurements. In: Morbidelli, R. (Ed.), Rainfall. Modeling, Measurement and Applications. Elsevier, Amsterdam, pp. 77–108. doi:10.1016/C2019-0-04937-0.
Lebel, T., Bastin, G., Obled, C., Creutin, J.D., 1987. On the accuracy of areal rainfall estimation: A case study. Water Resour. Res. 23 (11), 2123–2134.
Li, J., Heap, A.D., 2011. A review of comparative studies of spatial interpolation methods in environmental sciences: Performance and impact factors. Ecol. Inform. 6 (3–4), 228–241.
Linacre, E., 1992. Climate data and resources. a reference and guide. Routledge, London.
Linsley Jr., R.K., Kohler, M.A., Paulhus, J.L., 1949. Applied Hydrology. McGraw-Hill, New York, NY.
Ly, S., Charles, C., Degré, A., 2013. Different methods for spatial interpolation of rainfall data for operational hydrology and hydrological modeling at watershed scale: a review. Biotechnol. Agron. Soc. Envir. 17 (2), 392–406.
Mair, A., Fares, A., 2011. Comparison of rainfall interpolation methods in a mountainous region of a tropical island. J. Hydrol. Eng. 16 (4), 371–383.
Mandeville, A.N., Rodda, J.C., 1973. A contribution to the objective assessment of areal rainfall amounts, Results of research on representative and experimental basins, Proceedings of the Wellington Symposium, 2. IAHS, Paris, pp. 120–127.
Masih, I., Maskey, S., Uhlenbrook, S., Smakhtin, V., 2011. Assessing the impact of areal precipitation input on streamflow simulations using the SWAT model. J. Am. Water Resour. Assoc. 47 (1), 179–195.
McCuen, R.H., 2004. Hydrologic Analysis and Design. Prentice-Hall, New Jersey, USA.
Mitasova, H., Mitas, L., 1993. Interpolation by regularized spline with tension: I. Theory and implementation. Math. Geol. 25, 641–655.

Mogheir, Y., de Lima, J.L.M., Singh, V.P., 2013. Entropy and multi-objective based approach for groundwater quality monitoring network assessment and redesign. Water Resour. Manag. 23 (8), 1603–1620.

Moulin, L., Gaume, E., Obled, C., 2009. Uncertainties on mean areal precipitation: assessment and impact on streamflow simulations. Hydrol. Earth Syst. Sci. 13, 99–114.

New, M., et al., 2000. Representing twentieth-century space-time climate variability. Part II: Development of 1901–96 monthly grids of terrestrial surface climate. J. Clim. 13 (13), 2217–2238.

NWS: https://www.weather.gov/abrfc/map#mapx_gridded (Accessed October 2020), 2020.

Noori, J.M., Hussein, H.H., Mustafa, Y.T., 2014. Spatial estimation of rainfall distribution and its classification in duhok governorate using GIS. J. Water Resource Prot. 6, 75–82.

Okabe, A., Boots, B., Sugihara, K., Chiu, S.N., 2000. Spatial Tessellations: Concepts and Applications of Voronoi Diagrams. John Wiley & Sons, Chichester, UK.

Pappas, C., Papalexiou, S.M., Koutsoyiannis, D., 2014. A quick gap filling of missing hydrometeorological data. J. Geophys. Res. Atmos. 119 (15), 1920–1930.

Pathak, C.S., Teegavarapu, R.S.V., 2018. Radar rainfall data estimation and useMOP 139. American Society of Civil Engineers (ASCE), Reston, Virgina.

Paulhus, J.L.H., Kohler, M.A., 1952. Interpolation of missing precipitation records. Mon. Weather Rev. 80 (8), 129–133.

Phillips, D.L., Dolph, J., Marks, D., 1992. A comparison of geostatistical procedures for spatial analysis of precipitation in mountainous terrain. J. Hydrol. 58 (1–2), 119–141.

Raghavan, S., 2003. Radar Meteorology. Springer, Netherlands.

Rainbird, A.F., 1967. Methods of Estimating Average Areal Precipitation. WMO WMO-IHD Report 3, Geneva, Switzerland.

Rogelis, M.C., Werner, M.G.F., 2013. Spatial Interpolation for real-time rainfall field estimation in areas with complex topography. J. Hydrometeorol. 14, 85–104.

Rykiel Jr. E.J., 1996. Testing ecological models: the meaning of validation. Ecol. Modell. 90 (3), 229–244.

Sen, Z., Eljadid, A.G., 2000. Automated average rainfall calculation in Libya. Water Resour. Manage 14 (5), 405–416.

Sen, Z., 1988. Average areal precipitation by percentage weighted polygon method. J. Hydrol. Eng. 3 (1), 69–76.

Sene, K., 2010. Hydrometeorology: Forecasting and Applications. Springer, Netherlands.

Seo, D.-J., Krajewski, W.F., Bowles, D.S., 1990b. Stochastic interpolation of rainfall data from rain gauges and radar using cokriging - 2. Results. Water Resour. Res. 26 (5), 915–924.

Seo, D.-J., Krajewski, W.F., Bowles, D.S., 1990a. Stochastic interpolation of rainfall data from rain gauges and radar using cokriging - 1. Design of experiments. Water Resour. Res. 26 (3), 469–477.

Shaw, E., Beven, K.J., Chappell, N.A., Lamb, R., 2011. Hydrology in Practice. Spon Press, London, U.K.

Shaw, E.M., Lynn, P.P., 1972. Areal rainfall evaluation using two surface fitting techniques. Bull. Int. Assoc. Hydrol. Sci. 17 (4), 419–433.

Shepard, D., 1968. A two-dimensional interpolation function for irregularly spaced data, Proceedings of the Twenty-Third National Conference of the Association for Computing Machinery, 517–524.

Shuttleworth, J.W., 2012. Terrestrial Hydrometeorology. John Wiley and Sons, West Sussex, U.K.

Sibson, R., 1981. A brief description of natural neighbor interpolation. Interpolating multivariate data. *John* Wiley & Sons, New York, pp. 31–36.

Singh V.P., Birsoy Y.K., 1975 *St*udies on Rain-Runoff Modeling: Estimation of Mean Areal Rainfall, WRRI Report No. 061, p. 77.

Singh, V.P., Chowdhury, K., 1986. Comparing some methods of estimating mean aerial rainfall. Water. Resour. Bull. 22 (2), 275–282.

Sokol, Z., 2003. The use of radar and gauge measurements to estimate areal precipitation for several Czech river basins. Stud. Geophys. Geod. 47 (3), 587–604.

Strangeways, I., 2007. Precipitation: Theory, Measurement and Distribution. Cambridge University Press, London, U.K.

Sugawara, M. 1992. On the weights of precipitation stations, Advances in Theoretical Hydrology, J. P. O'Kane, European Geophysical Society Series on Hydrological Sciences, 59–74.

Sumner, G.N., 1988. Precipitation: Process and Analysis. John Wiley, and Sons, Chichester, U.K.

Tabios III, G.O., Salas, J.D., 1985. A comparative analysis of techniques for spatial interpolation of precipitation. Water Resour. Bull. 21, 365–380.

Teegavarapu, R.S.V., Chandramouli, V., 2006. Improved weighting methods, deterministic and stochastic data-driven models for estimation of missing precipitation records. J. Hydrol. 312 (1–4), 191–206.

Teegavarapu, R.S.V., Goly, A., Wu, Q., 2017. Comprehensive framework for assessment of radar-based precipitation data estimates. J. Hydrol. Eng. 22 (5). doi:10.1061/(ASCE) HE.1943-5584.0001277.

Teegavarapu, R.S.V., Meskele, T., Pathak, C., 2012. Geo-spatial grid-based transformation of precipitation estimates using spatial interpolation methods. Comput. Geosci 40, 28–41.

Teegavarapu, R.S.V., Nayak, A., 2017. Evaluation of long-term trends in extreme precipitation: implications of infilled historical data and temporal-window based analysis. J. Hydrol. 550, 614–634.

Teegavarapu, R.S.V., Tufail, M., Ormsbee, L., 2009. Optimal functional forms for estimation of missing precipitation records. J. Hydrol. 374 (1–2), 106–115.

Teegavarapu, R.S.V., 2013a. Climate change-sensitive hydrologic design under uncertain future precipitation extremes. Water Resour. Res. 49 (11), 7804–7814.

Teegavarapu R.S.V., 2018a Design of rainfall monitoring networks, Radar Rainfall Data Estimation and Use, In: C. S. Pathak and R. S. V. Teegavarapu (Eds.), 139, 111-120.

Teegavarapu, R.S.V., 2009. Estimation of missing precipitation records integrating surface interpolation techniques and spatio-temporal association rules. J. Hydroinf. 11 (2), 133–146.

Teegavarapu, R.S.V., 2013b. Floods in a Changing Climate: Extreme Precipitation. Cambridge University Press, London, UK.

Teegavarapu R.S.V., 2018c. Framework for bias analysis of radar data, in Radar rainfall data estimation and use In: C. S. Pathak and R. S. V. Teegavarapu, MOP, 139, 73–93.

Teegavarapu, R.S.V., 2014a. Missing precipitation data estimation using optimal proximity metric-based imputation, nearest neighbor classification and cluster-based interpolation methods. Hydrol. Sci. J. 59 (11), 2009–2026.

Teegavarapu, R.S.V., Aly, A., Pathak, C.S., Ahlquist, J., Fuelberg, H., 2018. Infilling missing precipitation records using variants of spatial interpolation and data-driven methods: use of optimal weighting parameters and nearest neighbour-based corrections. Int. J. Climatol. 38 (2), 776–793.

Teegavarapu, R.S.V., Goly, A., Obeysekera, J., 2013. Influences of Atlantic multi-decadal oscillation on regional precipitation extremes. J. Hydrol. 495, 74–93.

Teegavarapu R.S.V., 2018b. Rain-gauge rainfall data augmentation and radar rainfall data analysis, Radar Rainfall Data Estimation and Use, In: C. S. Pathak, R. S. V. Teegavarapu, (Eds.), 139, 73–93.
Teegavarapu, R.S.V., 2012. Spatial interpolation using nonlinear mathematical programming models for estimation of missing precipitation records. Hydrol. Sci. J. 57 (3), 383–406.
Teegavarapu, R.S.V., 2014b. Statistical corrections of spatially interpolated missing precipitation data estimates. Hydrol. Processes 28 (11), 3789–3808.
Teegavarapu, R.S.V., 2007. Use of universal function approximation in variance-dependent interpolation technique: an application in Hydrology. J. Hydrol. 332 (1–2), 16–29.
Teegavarapu, R.S.V., 2016. Spatial and temporal estimation and analysis of precipitation. Handbook of Applied Hydrology, McGraw Hill, New York, USA, pp. 383–392.
Teegavarapu, R.S.V., 2020b. Precipitation imputation using probability space-based spatial interpolation. J. Hydrol. 584. doi:10.1016/j.jhydrol.2019.124447.
Teegavarapu, R.S.V., 2020a. Evolving adaptive hydrologic design and water resources management in a changing climate: experiences from the U.S.Climate Change-Sensitive Water Resources Management. CRC Press, Florida, USA.
Thiessen, A.J., Alter, J.C., 1911. Precipitation for large areas. Mon. Weather Rev. 39 (7), 1082–1084.
Tobler, W.R., 1970. A computer movie simulating urban growth in the Detroit region. Econ. Geogr. 46, 234–240.
Tsangaratos, P., Dimitrios, R., Andreas, B., 2014. Use of artificial neural network for spatial rainfall analysis. J. Earth Syst. Sci. 123 (3), 457–465.
Wang, F., 2015. Quantitative Methods and Socio-Economic Applications in GIS. CRC Press, Boca Raton. USA.
Whitmore, J.S., Van Eeden, F.J., Harvey, K.J., 1961. Assessment of average annual rainfall over large catchments, Inter-African Conference on Hydrology, C. C. T. A. publication No., 100–107.
Willmott, C.J., Robeson, S.M., 1995. Climatologically aided interpolation (CAI) of terrestrial air temperature. Int. J. Climatol. 15 (2), 221–229.
WMO, 1989. Catalogue of National Standard Precipitation Gauges (B. Sevruk and S. Klemm). Instruments and Observing Methods Report No. 39. Geneva WMO/TD-No. 313a, World Meteorological Organization, Geneva, Switzerland.
WMO, 1983. Guide to Meteorological Instruments and Methods of Observation, Sixth edition. Geneva, Switzerland Part I, Chapter 6, WMO-No. 8, World Meteorological Organization.
WMO, 1984. International Comparison of National Precipitation Gauges with a Reference Pit Gauge (B. Sevruk and W.R. Hamon)Instruments and Observing Methods Report. Geneva No. 17, WMO/TD No. 38, World Meteorological Organization, Geneva, Switzerland.
WMO, 1989b. International workshop on precipitation measurements St Moritz, Switzerland, 3–7 December 1989. In: Sevruk, B. (Ed.), Instruments and Observing Methods Report No. 48, WMO/TD-No. 328. Geneva, Switzerland.
WMO, 1982. Methods of correction for systematic error in point precipitation measurement for operational use (B. Sevruk)Operational Hydrology Report No. 21, WMO-No. 589, World Meteorological Organization. Geneva, Switzerland.
WMO: Papers presented at the workshop on the correction of precipitation measurements (B. Sevruk, ed.) (Zurich, Switzerland, 1–3 April 1985). Instruments and Observing Methods Report No. 25, WMO/TD-No. 104, Switzerland. Geneva, 1986.
WMO: World Meteorological Organization, 1994. Guide to hydrological practices. WMO-No. 168, World Meteorological Organization. Geneva, Switzerland.
Xia, Y., Fabian, P., Winterhalter, M., Zhao, M., 2001. Forest climatology: estimation and use of daily climatological data for Bavaria, Germany. Agric. For. Meteorol. 106 (2), 87–103.

Xing, T., Xuesong, Z., Taylor, J., 2013. Designing heterogeneous sensor networks for estimating and predicting path travel time dynamics: an information-theoretic modeling approach. Transport. Res. B-Meth. 57, 66–90.

Xu, C.-Y., Vandewiele, G.L., 1994. Sensitivity of monthly rainfall runoff models to input errors and data length. Hydrol. Sci. J. 39 (2), 157–176.

Yeggina, S., Teegavarapu, R.S.V., Muddu, S., 2019. Evaluation and bias corrections of gridded precipitation data for hydrologic modelling support in Kabini River basin, India. Theor. Appl. Climatol. 140, 195–1513.

Yeggina, S., Teegavarapu, R.S.V., Muddu, S., 2020. A conceptually superior variant of Shepard's method with modified neighbourhood selection for precipitation interpolation. Int. J. Climatol. 39 (12), 4627–4647.

Zeiger, S., Hubbart, J., 2017. An assessment of mean areal precipitation methods on simulated streamflow: A SWAT model performance assessment. Water 9, 459. doi:10.3390/w9070459.

CHAPTER

9 Ombrian curves advanced to stochastic modeling of rainfall intensity

Demetris Koutsoyiannis, Theano Iliopoulou

Department of Water Resources and Environmental Engineering, National Technical University of Athens, Zographou, Greece

9.1 Introduction

Ombrian relationships, from the Greek word 'όμβρος', i.e. rainfall, are mathematical relationships linking the average rainfall intensity, x, to time scale of averaging, k, and return period, T. These relationships are widely known as intensity-duration-frequency (IDF) curves, even though the term is a misnomer; 'duration' is a misplaced name for time scale of averaging, and 'frequency' reflects an old tradition of confusing frequency with return period. Nonetheless, ombrian curves are an established tool of most hydrological and engineering operations requiring *design storm* estimates, ranging from flood protection and urban drainage design, to construction of highways, bridges, etc. Related methodologies are part of standard hydrological practice (Eagleson, 1970; Chow et al., 1988), and can be traced back in literature as early as in the works of Sherman (1931) and Bernard, 1932. Most countries have performed regionalization analysis of ombrian curves, producing maps for operational uses; such maps are available for the US since 1961 (Hershfield, 1961) and now are available or being updated for most parts of the world (Hailegeorgis et al., 2013; Koutsoyiannis et al., 1998).

The standard modelling procedure consists of fitting a parametric relationship to estimates of (x, k, T), often of the simple power law (Koutsoyiannis, 2021):

$$x = \frac{\lambda T^{\xi}}{k^{\eta}} \tag{9.1}$$

where λ, ξ, η are positive parameters with $\xi \leq \eta \leq 1$. Formulas of this type are often empirical and sometimes supported by fractal representations of rainfall intensity (Veneziano and Furcolo, 2002; Langousis and Veneziano, 2007). Even though the derivation of common formulas is mostly empirical, they have proven useful in practice and been validated by long-term and worldwide hydrological experience. However, there are several reasons that now dictate the need of a solid theoretical basis for their derivation.

First off, the standard empirical approach lacks rigor and entails several theoretical inconsistencies (even dimensional inconsistency, as evident from Eq. (9.1); see

RAINFALL. Modeling, Measurement and Applications. DOI: https://doi.org/10.1016/B978-0-12-822544-8.00003-2

details in Koutsoyiannis, 2021), which become important when one is interested in large return periods. A majority of these shortcomings has been exposed and rectified by Koutsoyiannis et al. (1998) who first connected the derivation of ombrian curves to the theoretical properties of the underlying rainfall process, namely its marginal distribution function and, in particular, its tail index. Linking the properties of the curves to their natural basis which is the parent process is also the only way to comprehend issues of bias and uncertainty in the estimation. Both issues are essential considering the presence of temporal dependence in the rainfall process, which induces bias in the estimation of its quantiles. Estimation uncertainty is further worsened by the usual lack of long-term reliable rainfall data at fine scales. Data scarcity is still a great challenge for many parts of the world (Ayman et al., 2011) and underlines the need to support ombrian curves estimation by a more powerful theoretical approach.

Even without considering the issue of data scarcity, a solid theoretical basis is essential when one is interested in applying simulation. The information conveyed by typical ombrian curves on rainfall frequency is usually not sufficient for studying complex hydrological problems beyond the 'design storm' applications. For instance, hydrosystems in which rainfall is only one of the uncertain components of the system, require thorough modelling of the involved processes in order to determine the overall probability of failure. As a matter of fact, as simulation is increasingly adopted in hydrological problems and the traditional ombrian curves have not kept up with such advances, the framework has even, unjustly, been seen as outdated.

The obvious alternative is to derive the ombrian relationships from simulations by models of the rainfall process, so-called rainfall generators. Yet the challenges involved in the multi-scale rainfall generation often beat the purposes of simplicity and practicality of the traditional curves. Moreover, it is not a given that the synthetic series will preserve the empirical ombrian curves, unless if included in the calibration scheme (Willems, 2000).

To this end, this Chapter presents a two-in-one approach, developed by Koutsoyiannis, 2021, by which curves are themselves advanced to stochastic models of the all-scale rainfall intensity, i.e. *ombrian models*. The approach allows theoretically consistent derivation of ombrian curves, with provision for estimation bias, validity over extended range of scales and capability to be directly used for simulation. It is shown that these advances can be achieved on the basis of simple stochastic characterizations of the parent process, namely of its joint second-order and marginal higher-order properties.

The remaining of the Chapter is structured in 4 sub-sections. In particular, the following sub-section (9.2) sets the requirements for an ombrian model and is devoted to the presentation of the essential stochastic tools for the characterization of the parent process. Section 9.3 introduces two variants of the ombrian model, a full version covering all the range of time scales and a simplified relationship valid for fine timescales. Particular focus is given to the fitting procedure, outlined in Section 9.4, where issues of dependence-induced bias are also addressed. The entire methodology is illustrated in detail by the case study of rainfall in Bologna, in Section 9.5, which stands as a proof of concept of the ombrian model's power. The possibility of taking advantage of possible existence of multi-source data is also highlighted. Further aspects of the ombrian curves are discussed in Section 9.6.

9.2 A stochastic framework for building ombrian models

In this Section, the stochastic concepts used in the ombrian modelling are presented. A stochastic process is an arbitrarily large family of random variables $\underline{x}(t)$ (Papoulis, 1991). To distinguish random variables from regular variables, the former are underlined following the Dutch convention. These variables are indexed by t, which represents time, either discrete (from the set of integers Z, referring to a discrete-time stochastic process), or continuous (from the set of real numbers R, resulting in a continuous-time stochastic process). A continuous-time stochastic process is herein denoted by $\underline{x}(t$, and a discrete-time one by $\underline{x}_\tau$. A realization of stochastic process $\underline{x}(t$, i.e. a timeseries, necessarily referring to discrete time and denoted by x_T.

9.2.1 Basic requirements for an ombrian model

The basic premise of the Chapter is that an ombrian model can be an advance of the classic tool of ombrian curves if it achieves greater modelling power and theoretical consistency but preserves the practical and simple character of the classic curves. Below we outline the basic requirements for such a model as identified by Koutsoyiannis, 2021.

1. A critical prerequisite is that the ombrian model should be applicable in engineering applications without a necessity to perform simulation. Its application should be preferably simple as in traditional IDF curves even if the need for theoretical consistency is compromised to some extent.
2. It is straightforward that, as in every stochastic model, the first and second order properties of the process of interest, i.e. the temporal average of rainfall intensity $\underline{x}^{(k)}$ over any time scale k, should be preserved. Clearly, a constant mean should be preserved for all timescales, although this is often violated in common expressions. An effective methodology to preserve the second-order properties at any scale based on the scaling properties of the variance (climacogram) is outlined in Section 9.2.2.
3. The process's asymptotic variance at $k \to 0$ should be finite; the contrary would imply that the process requires infinite energy to materialize which is absurd for physical processes. In addition, the process's asymptotic variance at $k \to \infty$ should be zero, in order for the process to be ergodic.
4. The model should deal with the intermittence of rainfall occurrences at fine time scales, describing both the probability dry $P_0^{(k)} := P\left\{\underline{x}^{(k)} = 0\right\}$, and the

probability wet, $P_1^{(k)} := \bar{F}^{(k)}(0) = 1 - P_0^{(k)}$ for any time scale k, including for $k \to 0$.

5. The principle modelling focus is on rainfall maxima, and hence it is important to preserve the high-order properties of the process.
6. The tail index of the rainfall intensity distribution should be constant for all time scales. Theoretical justification of this requirement can be found in Koutsoyiannis, 2021.
7. The Pareto distribution constitutes an optimal choice for small time scales due to its simplicity and explicit relationship between the time-averaged intensity and return period. Prevalence of the Pareto distribution for rainfall intensities is also supported by worldwide empirical evidence (Koutsoyiannis, 2004a; Koutsoyiannis and Papalexiou, 2016).

9.2.2 Characterization of second-order properties through climacogram

A comprehensive characterization of a process's second-order properties can be achieved by inspecting the properties of its variance when the process is averaged (or aggregated) over multiple scales. The function of the variance of the averaged process versus the scale of averaging is called the climacogram, while the function of the cumulative process versus the scale is called the cumulative climacogram. The climacogram of a process $\underline{x}(t)$ is defined as:

$$\gamma(k) := \operatorname{var}\left[\frac{\underline{X}(k)}{k}\right] = \frac{\Gamma(k)}{k^2} \tag{9.2}$$

where $\Gamma(k)$ is the cumulative climacogram, and $\underline{X}(k)$ is the process $\underline{x}(t)$ aggregated at timescale k:

$$\underline{X}(k) := \int_0^k \underline{x}(t)\,dt \tag{9.3}$$

or for a discrete-time process, with climacogram γ_κ:

$$\underline{X}_\kappa := \underline{x}_1 + \underline{x}_2 + \ldots + \underline{x}_\kappa \tag{9.4}$$

The discrete time scale κ (integer) is related to the continuous-time one k (real number) by $k = \kappa D$, with D denoting the length of the time step.

The climacogram estimator is the same for discrete- and continuous-time processes and is given as:

$$\underline{\hat{\gamma}}_\kappa \equiv \underline{\hat{\gamma}}(k) := \frac{1}{n}\sum_{\tau=1}^{n}\left(\underline{x}_\tau^{(\kappa)} - \underline{\hat{\mu}}\right)^2 \tag{9.5}$$

where $\underline{\hat{\mu}} := (1/n)\sum_{i=1}^{n} \underline{x}_i$ is the estimator of the true mean.

The climacogram is theoretically equivalent to other second-order properties, namely the autocovariance, autocorrelation and the power-spectrum, but it has superior estimation properties in terms of bias, discretization errors, and sampling uncertainty (Dimitriadis and Koutsoyiannis, 2015). Therefore, it is the basic tool employed here for second-order characterization.

The theoretical climacogram differs among processes with different second-order dependence structure. In case of an independent white-noise process in continuous time, the climacogram is inversely proportional to the time scale:

$$\gamma(k) = \frac{\sigma^2 D}{k} \tag{9.6}$$

where σ^2 is the variance of the process for $k = D$. Notice that the variance of the instantaneous process is infinite. An extension of the white-noise process, again having infinite variance as $k \to 0$ but now exhibiting dependence in time, is the Hurst-Kolmogorov process, which can be defined through its climacogram as:

$$\gamma(k) = \lambda\left(\frac{a}{k}\right)^{2-2H} \tag{9.7}$$

where α and λ are scale parameters, with dimensions $[t]$ and $[x^2]$ and H is the so-called Hurst parameter ranging in the interval (0,1). In the case of $H = 0.5$ the white noise is recovered. For $0.5 < H < 1$ the process is persistent and for $0 < H < 0.5$ antipersistent.

The infinite variance of these processes when the scale tends to zero makes them inappropriate for natural processes, as discussed before. In order to remedy this shortcoming, and improve flexibility of the model to describe the dependence in shorter time scales, the Filtered Hurst-Kolmogorov (FHK) process is developed with several climacogram types (Koutsoyiannis, 2017). The generalized Cauchy-type climacogram (FHK-C) is:

$$\gamma(k) = \lambda_1\left(1+\left(\frac{k}{a}\right)^{2M}\right)^{\frac{H-1}{M}} \tag{9.8}$$

where M is an added dimensionless parameter which controls the local scaling of the process (fractal behaviour), denoted as M in honor of Mandelbrot (Koutsoyiannis et al., 2018). Values of $M < 1/2$ indicate a rough process, while $M > 1/2$ indicate a smooth process.

An alternative flexible type is the composite Cauchy-Dagum-type (FHK-CD) climacogram, which for a rough and persistent process, and for the special case $M = 1 - H$, can be written as:

$$\gamma(k) = \lambda_1\left(1+\frac{k}{a}\right)^{2H-2} + \lambda_2\left(1-\left(1+\frac{a}{k}\right)^{2H-2}\right) \tag{9.9}$$

Both these climacogram models have four parameters and thus great flexibility in capturing the scaling properties of the variance at all scales. More information on the bounds of scaling and on other climacogram-type models are provided in Koutsoyiannis (2017).

Therefore, the empirical climacogram is given by estimating the variance over scales by Eq. (9.5), whereas Eqs. (9.8, 9.9) provide different types of the theoretical climacograms. Because presence of dependence induces downward bias in the estimation of the variance, the two are not directly comparable. To compare them, the bias need to be considered, based on the following equation (Koutsoyiannis, 2003, Koutsoyiannis, 2021):

$$\mathrm{E}\left[\underline{\hat{\gamma}}(k)\right] = \gamma(k) - \gamma(L) \tag{9.10}$$

where L is the length of the observation period.

9.2.3 Assigning empirical return periods using order statistics

Order statistics are a standard tool for dealing with extremes. Below the procedure to apply them for assigning return periods to the data is presented.

Let $\underline{x}$ be a stochastic variable and $\underline{x}_1, \underline{x}_2, \ldots, \underline{x}_n$ be IID copies of it, forming a sample. If we rearrange them by increasing order of magnitude such that $\underline{x}_{(i:n)}$ is the ith smallest of the n, i.e.:

$$\underline{x}_{(1:n)} \le \underline{x}_{(2:n)} \le \cdots \le \underline{x}_{(n:n)} \tag{9.11}$$

then the stochastic variable $\underline{x}_{(i:n)}$ is called the ith order statistic. The minimum and maximum values of a sample are then given respectively by the lowest order statistic, $\underline{x}_{(1:n)} = \min\left(\underline{x}_1, \underline{x}_2, \ldots, \underline{x}_n\right)$ and the highest order statistic $\underline{x}_{(n:n)} = \max\left(\underline{x}_1, \underline{x}_2, \ldots, \underline{x}_n\right)$. It is then well-known that if we define the stochastic variable $\underline{u} := F\left(\underline{y}\right) = F\left(\underline{x}_{(i:n)}\right)$, then its distribution function is the Beta distribution, whose mean is:

$$\mathrm{E}\left[\underline{u}\right] = \mathrm{E}\left[F\left(\underline{x}_{(i:n)}\right)\right] = \frac{i}{n+1} \tag{9.12}$$

Then an estimate of the return period (in time units D) for order statistics $\underline{T}_{(i:n)} := T\left(\underline{x}_{(i:n)}\right)$ is:

$$\frac{T_{(i:n)}}{D} = \frac{1}{1 - \mathrm{E}\left[F\left(\underline{x}_{(i:n)}\right)\right]} = \frac{n+1}{n+1-i} \tag{9.13}$$

which is the well-known Weibull plotting position. This, however, is not recommended for use as it results in high bias in the estimation of the return period of the highest events. There are several other formulae for return period all of which are of the form:

$$\frac{T_{(i:n)}}{D} = \frac{n+B}{n-i+A} \tag{9.14}$$

The parameters A and B depend on the parent distribution of the data. Koutsoyiannis, 2021 developed various parameterizations as the best approximations for a number of distributions and properties of interest. In particular, for distributions belonging to the domain of attraction of EV2, such as the Pareto, the parameters A, B are theoretically proved to be:

$$A = \left(\Gamma(1-\xi)\right)^{-1/\xi}, \quad B = \left(\Gamma(2-\xi)\right)^{-1/\xi} - 1 \tag{9.15}$$

These can be suggested for use (replacing empirical formulae such as the Weibull plotting positions) for assigning return periods to order statistics, assuming independence of the data. Yet since independence is mostly an untenable assumption for natural processes, there is bias involved in the estimation in this case as well.

For an explicit account of dependence in assigning return periods, one could use knowable moments (Koutsoyiannis, 2019). For simplicity however, the following ad-hoc procedure is proposed as an approximation to account for bias for each time scale.

1. A first estimate of the return period of each nonzero value $x_{(i:n)}$ appearing in a sample sorted in ascending order, is obtained based on the independence assumption from Eq. (9.14). This follows the estimation of the coefficients A and B of Eq. (9.15) based on the tail index of the process.
2. The following approximation for a bias correction factor $\Theta(k, L, H)$ is used (Koutsoyiannis, 2021):

$$\Theta(k, L, H) \approx -\frac{\gamma(L)}{2\gamma(k)} \tag{9.16}$$

Accordingly, the empirical return periods are corrected as:

$$T' \approx \min\left(\left(2\Theta + (1-2\Theta)\left(\frac{T\hat{P}_1^{(k)}}{2k}\right)^{(1+\Theta)^2}\right)\frac{2k}{\hat{P}_1^{(k)}}, T\right) \tag{9.17}$$

where $\hat{P}_1^{(k)} = \hat{n}_1 / n$ is the ratio of the non-zero values to the total values at each scale, else the probability wet.

3. This procedure is repeated for all nonzero values $x_{(i:n)}$ for each timescale k, yielding a table of empirical values and associated return periods.

9.3 Building a theoretically consistent ombrian model

9.3.1 All-scale version

The first version of the ombrian model refers to a model valid over the whole range of available scales. To achieve the extension of the typical fine-scale curves to large scales an increase in the complexity of the rainfall's intensity distribution is required. This is described by the following assumptions:

1. At small time scales the rainfall intensity follows a mixed type distribution, with a discrete part at the origin described by the probability dry, and a continuous part following the Pareto distribution with a constant tail index ξ and a state scale parameter $\lambda(k)$ as a function of the timescale:

$$F^{(k)}(x) = 1 - P_1^{(k)}\left(1 + \xi\frac{x}{\lambda(k)}\right)^{-1/\xi} \tag{9.18}$$

2. At larger time-scales the rainfall intensity follows the Pareto-Burr-Feller (PBF) distribution with discontinuity at zero, characterized by an extra parameter $\zeta(k)$ as a function of the timescale:

$$F^{(k)}(x) = 1 - P_1^{(k)}\left(1 + \xi\left(\frac{x}{\lambda(k)}\right)^{\zeta(k)}\right)^{-1/\xi} \tag{9.19}$$

The Pareto distribution is obtained for $\zeta(k) = 1$. The PBF distribution is chosen because, contrary to the Pareto, it becomes bell-shaped for increasing $\zeta(k)$ which is consistent to the behaviour of the rainfall intensity at large time scales (cf. the central limit theorem).

3. The mean of the time-averaged process is constant across all timescales:

$$\mathrm{E}\left[\underline{x}^{(k)}\right] = \mu \tag{9.20}$$

4. The climacogram follows one of the two four-parameter models introduced in Eqs. (9.8, 9.9). Clearly, both Equations satisfy the asymptotic requirements for the variance set in 9.2.1. As $k \to \infty$, $\gamma(k) \to 0$, whereas for $k = 0$, both variances are finite and equal to $\gamma(0) = \gamma_0 = \lambda_1$ in Eq. (9.8) and $\gamma(0) = \gamma_0 = \lambda_1 + \lambda_2$ in Eq. (9.9).
5. The probability wet $P_1^{(k)} = 1 - P_0^{(k)}$ and dry $P_0^{(k)}$ follow the scaling law:

$$\ln P_0^{(k)} = \ln P_0^{(k^*)}\left(k / k^*\right)^{\theta}, \ k \geq k^* \tag{9.21}$$

where k^* is the transition time scale from Pareto to PBF distribution, for which $P_0^{(k^*)} > 0$ and $\zeta(k^*) = 1$ (for continuity of the transition), and θ is a parameter ($0 \leq \theta$

≤ 1). This equation was derived by Koutsoyiannis (2006) from an entropy maximization framework.

The introduction of the two different distributions follows from the need to preserve the shape of the probability of rainfall, which is highly skewed at small timescales but tends to bell-shape at large scales. However, it is noted in Eq. (9.19) that the tail-index of the PBF distribution is not ξ but $\xi/\zeta(k)$ and tends to zero as $k \to \infty$. Thus, at large timescales the constant tail index requirement is violated. An alternative solution would be to replace the PBF with a shape-preserving distribution, yet analytical expressions are too involved and defeat the requirement of practicality. Besides, the violation occurs only at large timescales which are less of interest in applications.

Having assumed the distribution types, it remains to specify the form of the parameters $\lambda(k)$ and $\zeta(k)$ which are derived from the first- and second-order properties, i.e. the mean and the climacogram of the process. For the PBF distribution these are given by:

$$\frac{1}{\zeta(k)} \approx \sqrt{(1-2\xi)\left(P_1^{(k)} \frac{\gamma(k)+\mu^2}{\mu^2} - 1\right)} \tag{9.22}$$

$$\frac{1}{\lambda(k)} \approx \frac{P_1^{(k)}}{\mu}\left(1 + \frac{1}{(1-\xi)(\zeta(k))^2} - \frac{1}{(\zeta(k))^{\sqrt{2}}}\right) \tag{9.23}$$

For the Pareto distribution, $\zeta(k) = 1$, and therefore the probability wet can be explicitly derived from Eq. (9.22) as:

$$P_1^{(k)} = \frac{1-\xi}{1/2-\xi} \frac{\mu^2}{\gamma(k)+\mu^2} \tag{9.24}$$

while in this case Eq. (9.23) can be simplified to:

$$\frac{1}{\lambda(k)} = \frac{P_1^{(k)}}{\mu(1-\xi)} = \frac{\mu}{(1/2-\xi)(\gamma(k)+\mu^2)} \tag{9.25}$$

The special case of $P_1^{(k)} = 1$ denotes the maximum scale till which the Pareto distribution is mathematically feasible, thus $k = k^*_{\max}$, and the following hold:

$$P_1^{(k^*_{\max})} = 1, \quad \frac{\gamma(k^*_{\max})}{\mu^2} = \frac{1}{1-2\xi}, \quad \lambda(k^*_{\max}) = \mu(1-\xi) \tag{9.26}$$

However, in order to preserve the scaling behaviour of the probabilities dry/wet, as specified by Eq. (9.21), the transition scale to the PBF distribution should be chosen much smaller than $k^*_{\max}$, i.e. the Pareto feasibility limit.

On the contrary, the PBF is feasible at any scale, while for large scales in which $P_1^{(k)} = 1$, Eq. (9.22) simplifies to:

$$\frac{1}{\zeta(k)} = \frac{\sqrt{(1-2\xi)\gamma(k)}}{\mu} \tag{9.27}$$

The final version of the ombrian model is obtained by substituting the return period $T = 1/(1 - F^{(k)}(x))$ in the Eq. (9.19) for the PBF:

$$x = \lambda(k)\left(\frac{\left(P_1^{(k)} T / k\right)^{\xi} - 1}{\xi}\right)^{1/\zeta(k)} \tag{9.28}$$

and for the Pareto ($\zeta(k) = 1$)

$$x = \lambda(k)\frac{\left(P_1^{(k)} T / k\right)^{\xi} - 1}{\xi} \tag{9.29}$$

For $\xi = 0$, the PBF distribution switches to the Weibull, and the Pareto to the exponential, i.e.:

$$x = \lambda(k)\left(\ln\left(P_1^{(k)} T / k\right)\right)^{1/\zeta(k)}, \quad x = \lambda(k)\ln\left(P_1^{(k)} T / k\right) \tag{9.30}$$

The final ombrian relationship with its basic properties is summarized in Table 9.1. It is evident that the ombrian relationship is given through the mean, the climacogram, the tail index of the distribution and the probability wet. The full-range model results in a total of seven parameters, depending on the choice of the climacogram model, of four categories: (a) the mean intensity parameter μ with units of $[x]$, i.e. the average of the process, typically mm/h, (b) the intensity scale parameter λ_1, in case of the FHK-C climacogram model, or λ_1, λ_2 in the case of FHK-CD with units of $[x^2]$, (c) the time scale parameter α, in units of time $[t]$, and (d) the dimensionless parameters ξ ($0 < \xi < 0.5$), i.e. the tail index, θ ($0 < \theta < 1$), i.e. the exponent of the expression of probability dry, M ($0 < M < 1$), i.e. the fractal parameter in case of the FHK-C climacogram, and H ($0 < H < 1$), i.e. the Hurst parameter. Note that if the FHK-CD climacogram model (Eq. (9.9)) is used, then the fractal parameter is derived as $M = 1 - H$, and thus it is not an extra parameter.

It is worth noting that typical ombrian curves involve five parameters, yet the gains of including the extra parameters are manifold. In addition to the recovered mathematical and physical consistency, the model yields a better representation of fine scales (through fractal M parameter) and arbitrarily large scales (through H parameter), while precisely preserving the mean, climacogram and probability dry/wet of the process. In principle, this version of the model has the advantage of being valid over all timescales. Yet if only fine timescales are of interest, then a less parameterized and simpler version can be used instead. This is discussed next.

Table 9.1 Ombrian models for the full range of scales and the small scales and their basic properties, i.e. mean, climacogram, probability wet, shape scale parameter, and state scale parameter.

	All-scale ombrian model		Simplified model
	Small scales (Pareto) $k \le k^* \ll k^*_{max}$	**Large scales (PBF) $k \ge k^*$**	**Small scales (Pareto) $k \le \beta$**
x *for* $\xi > 0$	$\lambda(k)\frac{\left(P_1^{(k)}T/k\right)^{\xi}-1}{\xi}$	$\lambda(k)\left(\frac{\left(P_1^{(k)}T/k\right)^{\xi}-1}{\xi}\right)^{1/\zeta(k)}$	$\lambda\frac{\left((T/\beta)^{\xi}-1\right)}{(1+k/\alpha)^{\eta}}$
x *for* $\xi = 0$	$\lambda(k)\ln\left(P_1^{(k)}T/k\right)$	$\lambda(k)\left(\ln\left(P_1^{(k)}T/k\right)\right)^{1/\zeta(k)}$	$\lambda\frac{\ln(T/\beta)}{(1+k/\alpha)^{\eta}}$
		Properties	
$\mathrm{E}\left[\underline{x}^{(k)}\right]$	μ		(inconsistent – not constant)
$\gamma(k)$	$\lambda_1\left(1+(k/\alpha)^{2M}\right)^{\frac{H-1}{M}}$ or $\lambda_1(1+k/\alpha)^{2H-2}+\lambda_2\left(1-(1+a/k)^{2H-2}\right)$		$\lambda(1+k/\alpha)^{2H-2}$
$P_1^{(k)}$	$\frac{1-\xi}{1/2-\xi}\frac{\mu^2}{\gamma(k)+\mu^2}$	$1-\left(1-P_1^{(k^*)}\right)^{\left(\frac{k}{k^*}\right)^{\theta}}$	$\frac{k}{\beta}$
$\frac{1}{\zeta(k)}$	1	$\sqrt{(1-2\xi)\left(P_1^{(k)}\frac{\gamma(k)+\mu^2}{\mu^2}-1\right)}$	(not applicable)
$\frac{1}{\lambda(k)}$	$\frac{P_1^{(k)}}{\mu(1-\xi)}$	$\frac{P_1^{(k)}}{\mu}\left(1+\frac{1}{(1-\xi)(\zeta(k))^2}-\frac{1}{(\zeta(k))^{\sqrt{2}}}\right)$	(not applicable)

9.3.2 Simplified model for small scales

It is possible to simplify the Pareto ombrian relationship (Eq. (9.29)) which is valid for small scales, $k \le k^*_{\max}$, and can be written as:

$$x = \lambda(k)\frac{\left(T / \beta(k)\right)^{\xi} - 1}{\xi} \tag{9.31}$$

where $\beta(k)$ is a function of the time scale with units of time, $\beta(k) := k / P_1^{(k)}$. Then by virtue of Eq. (9.25), the ombrian relationship yields:

$$x = \frac{(1/2 - \xi)\left(\gamma(k) + \mu^2\right)}{\xi\mu}\left(\left(\frac{T}{\beta(k)}\right)^{\xi} - 1\right) \tag{9.32}$$

For small scales we may introduce a series of simplifying assumptions:

1. We assume that $P_1^{(k)} \propto k$, which is an acceptable approximation for small scales, thus that $\beta(k) = \beta =$ constant.
2. Noting that for small scales $\gamma(k) \gg \mu^2$, we neglect the second term in the sum of Eq. (9.23).
3. We assume an FHK-C climacogram (Eq. (9.8)) with a neutral value for $M = 0.5$

   By application of these three assumptions to Eq. (9.32), we get:

$$x = \lambda_1\frac{(1/2 - \xi)}{\xi\mu}\left(1 + \frac{k}{\alpha}\right)^{2H-2}\left(\left(\frac{T}{\beta}\right)^{\xi} - 1\right) \tag{9.33}$$

   which is an ombrian relationship of the form:

$$x = \lambda\frac{b(T)}{a(k)} \tag{9.34}$$

Thus, obtaining the rainfall intensity as a quotient of a function of the time scale and the return period. This facilitates estimation. The function $a(k)$ is given as:

$$a(k) = \left(1 + \frac{k}{\alpha}\right)^{\eta}, \eta := 2 - 2H \tag{9.35}$$

while the parameter λ and the function $b(T)$ are dependent on ξ as follows. For $\xi > 0$:

$$\lambda = \frac{(1/2 - \xi)\lambda_1}{\xi\mu}, \quad b(T) = \left(\frac{T}{\beta}\right)^{\xi} - 1 \tag{9.36}$$

and for $\xi = 0$:

$$\lambda := \frac{\lambda_1}{2\mu}, \quad b(T) = \ln\left(\frac{T}{\beta}\right) \tag{9.37}$$

This simplification results in a total of five parameters, of three categories: (a) λ with units of x, i.e. typically mm/h, (b) the scale parameters α and β with units of time, and (c) the dimensionless parameters ξ ($0 < \xi < 0.5$), i.e. the tail index, and η ($0 < \eta < 1$), which is related to the Hurst parameter of the process. Any value of $\eta < 1$ corresponds to a process with persistence, $H > 0.5$, while $\eta = 1$ (which in reality is never the case) would correspond to $H = 0.5$, a process with purely random behaviour. The model is also summarized in Table 9.1 along with its basic properties.

It can be seen that the simplified version sacrifices some of the requirements set at the beginning, mostly importantly instead of a constant mean, it yields a mean increasing with time-scale (Table 9.1). However, the inconsistencies are negligible if one restricts the range of timescales using as a lower bound the smallest value of the observed data and choosing an upper bound sufficiently below β. Therefore, the model is applicable over this range of observed scales, but if simulation is of interest then the all-scale version should be used instead.

9.4 Model fitting procedure

9.4.1 All-scale version

By assuming an initial parameterization of the all-scale ombrian model we can obtain a theoretical estimate of: (a) the climacogram $\gamma(k)$, (b) the probability wet vs the time scale $P_1^{(k)}$ and (c) the rainfall intensity as a function of the timescale and the return period, i.e. the ombrian model $x(k, T)$, using the equations summarized in Table 9.1. From the empirical series, we also may obtain the empirical estimates of these three relationships as follows. To obtain an estimate of the climacogram $\hat{\gamma}(k)$, we use Eq. (9.5), following the procedure in Section 9.2.1. To estimate the probability wet, we estimate the ratio $\hat{P}_1^{(k)} = \hat{n}_1 / n$ where $\hat{n}_1$ is the number of nonzero observations, and n the total number of observations in the series. Then to assign a return period to each rainfall observation at each scale, thus estimate $x(k, T)$, we use the order statistics method, as presented in Section 9.2.3.

The model is fitted to the empirical estimates, after bias is accounted for, by minimizing the error between the two through a nonlinear solver, available even in any computational (e.g. spreadsheet) environment. Specifically, the error that should be minimized to fit the climacogram adjusted for bias is of the form:

$$E_\gamma := \sum_k w_\gamma(k)\Big(\ln\big(\gamma(k) - \gamma(L)\big) - \ln\hat{\gamma}(k)\Big)^2 \tag{9.38}$$

where $w_\gamma(k)$ is a weighting function of scale. The logarithm is introduced to account for the different orders of magnitude that the climacogram spans. By minimizing Eq. (9.38) we obtain all the climacogram-related parameters.

Likewise, the error for the probability wet is defined as:

$$E_P := \sum_k w_P(k)\left(P_1^{(k)} - \hat{P}_1^{(k)}\right)^2 \tag{9.39}$$

where $w_P(k)$ is the weight, which can be chosen as a function of scale. Since the expression for the probability wet, involves all parameters of the ombrian model, Eq. (9.39) could be used to specify the full version of the model. However, this would give more weight to the representation of the probability dry, than to the extreme values, which are the ones of interest. To this aim, it is better to obtain the parameters directly from the distribution quantiles $x(k, T)$. The total fitting error in this case is given as:

$$E_x := \sum_k \frac{1}{\gamma(k)} \frac{1}{n_k} \sum_T w_x(T)\left(x(k,T) - \hat{x}(k,T)\right)^2 \tag{9.40}$$

where $w_x(T)$ is a weighting function of the return period and n_k is the number of x values at each scale k. The total mean square error over the entire set of return periods is further normalized by the climacogram $\gamma(k)$.

In so doing, we have determined the full parameterization of the ombrian model also accounting for dependence-induced bias. It is also possible to optimize the parameters of the model by formulating an objective function that includes, as a weighted sum, all three errors defined above:

$$E := a_\gamma E_\gamma + a_P E_P + a_x E_x \tag{9.41}$$

where a_γ, a_P, a_x the weights for the three errors.

9.4.2 Simplified version

The simplified version of the ombrian model also allows for a simplified fitting procedure, adjusting the steps previously outlined. In fact, by observing the separability of functions $a(k)$ and $b(T)$ in this version an independent, two-step fitting approach can be used, introduced by Koutsoyiannis (1998). Eq. (9.34) can be expressed as:

$$a(k)x = \lambda b(T) \tag{9.42}$$

We note that the timescale k is not a stochastic variable as it takes values from a fixed set, depending on data availability, whereas $a(k)$ is a deterministic function thereof. The right-hand side of the equation is in fact an expression of the Pareto distribution, independent of timescale k. Substituting Eqs. (9.35, 9.36) in the above equation, yields:

$$\left(1+\frac{k}{\alpha}\right)^\eta x = \lambda\left(\left(\frac{T}{\beta}\right)^\xi - 1\right) \tag{9.43}$$

Now, it is easy to see that for the different timescales k_j the stochastic variables $\underline{y}_j := a(k_j)\underline{x} = (1+k/\alpha)^\eta \underline{x}$ have a common distribution function, with the $\underline{y}_j$ for the different k_j being samples of it. Let then, $\underline{y}_{ji} := a(k_j)\underline{x}_{ji}$ of length $n = \sum_j n_j$ denote the merged sample of all sub-samples $\underline{x}_{ji}$ of size n_j corresponding to timescale k_j. Let also $\underline{r}_{ji}$ denote the rank of each sub-sample $\underline{x}_{ji}$ in the merged sample $\underline{y}_{ji}$ so that the mean rank of each sub-sample is given as $\underline{r}_j = \sum_i \underline{r}_{ji} / n_j$. Replacing all $\underline{r}_{ji}$ with the mean rank value $\underline{r}_j$ we get a sample of n values, with n_1 equal to r_1, n_2 equal to r_2 etc. Then the mean and variance estimators are, respectively:

$$\underline{\bar{r}} := \frac{1}{n} \sum_j n_j \underline{r}_j \tag{9.44}$$

$$\underline{\gamma}_r := \frac{1}{n} \sum_j n_j \left(\underline{r}_j - \underline{\bar{r}} \right)^2 \tag{9.45}$$

If no ties are present among the different ranks, then $\underline{\bar{r}} = (n+1)/2$.

Following the assumption that the samples are from the same distribution, given by the right-hand side of Eq. (9.43), then each $\underline{r}_j$ should be close to the mean $\underline{\bar{r}}$ while the variance should be minimal. Therefore, we can find the parameters α and η as the values that minimize the estimate of the variance γ_r from the observations $\underline{x}_{ji}$. The original values $\underline{y}_{ji}$ could be used as well instead of the ranks, yet the use of the ranks makes the estimation process more robust to outliers. In order to improve the fit to the higher quantile region, we could also use a part of the data of each sample, belonging to the highest 1/2 or 1/3 of the data (Koutsoyiannis, 1998).

Having estimated the α and η parameters, it remains to specify the parameters of the function $b(T)$. Following the same rationale, i.e. of a single distribution function, we merge all k sub-samples into a single sample and we estimate the parameters of the Pareto distribution, which fully determines the form of $b(T)$.

The two-step fitting procedure has an attractive flexibility in using different sources of data. Namely, a reliable determination of parameters α and η requires sub-hourly and sub-daily data, respectively, whereas, on the contrary, the parameters of the function $b(T)$ are better inferred from daily rain-gauge data. Particularly, the most uncertain and critical parameter is the tail-index of the distribution, which requires long timeseries to be reliably estimated. In the absence of long observational rain-gauge records, the tail index of the Pareto should be estimated from regional analysis or be assumed independently of the data, based on local hydrological experience.

9.5 Development of an ombrian model for Bologna in Italy

The all-scale ombrian model is applied to the rainfall of Bologna in Italy, which has one of the longest daily rainfall records worldwide spanning 206 years. The time series of daily observations is available online in the frame of the Global Historical Climatology Network – Daily (GHCN-Daily). Hourly rainfall data from the Dext3r repository are also employed, covering the entire period 1990-2013, with the exception of the missing year 2008. To take advantage of the availability of the two data sources, the ombrian model is fitted to both rainfall series simultaneously.

As a first step, the data of both series are aggregated at larger timescales. Specifically, the hourly rainfall data are aggregated at timescales of 2, 4, 6, 12, 24, 48 and 96 h, thus the modelling scales extend from 1 h to 4 d. The longer daily rainfall data are aggregated at timescales of 2, 4, 8, 16, 32, 64, 128, 182, 365, 730, 1460, and 5840 d, thus in this case, the range of scales extends from 1 d to 16 years. Longer timescales are studied for the daily data due to both their longer length and their higher reliability for the estimation of long-term properties compared to the hourly series. Therefore, the combined series spans from 1 h to 16 years (140 256 h).

The model is fitted according to the procedure outlined in Section 9.4.1. First, the climacogram is graphically inspected in order to choose the most suitable form of the climacogram-models given (Table 9.1). This is shown in Fig. 9.1. It is evident that the behaviour of the variances switches over larger scales, which makes a type FHK-CD climacogram (Eq. (9.9)) more suitable. Next, in order to account for estimation bias, initial theoretical values of the 7 parameters are assumed, namely of μ (mm/h), λ_1 (mm^2/h^2), λ_2 (mm^2/h^2), α (h), H, θ, and ξ. Once the parameters are assumed, the theoretical values of the variance, the probability wet and the empirical quantiles are known for all scales by virtue of Equations shown in Table 9.1. The empirical properties at each timescale, i.e. the variance, the probability wet and the return periods of the non-zero rainfall values, are also estimated. The transition time scale k^* is chosen as 96 h (=4 d) by inspection of the probability wet, in Fig. 9.2.

The model is fitted employing four different optimization procedures with the latter three including all the model parameters, i.e. with optimization targeting in minimizing the (a) error in the climacogram, (b) error in the probability wet, (c) error in the rainfall quantiles and (c) combined error in all previous. Optimization scheme (a), yielding the four parameters of the climacogram model, uses Eq. (9.38) with all scales given equal weight, $w_\gamma(k) = 1$. Optimization scheme (b) is based on Eq. (9.39), assuming equal weights for all scales, $w_P(k) = 1$, whereas optimization scheme (c) is based on Eq. (9.40). In this case, using equal weights would result to a model fit biased in favour of lower return periods, which are more frequent in our dataset. To improve the fit to the higher return periods, which are typically the ones of interest, we use a weighting function increasing with return period, i.e. $w_x(k) \propto \sqrt{T}$. Optimization scheme (d) is done using Eq. (9.41) with weights $a_\gamma = 0.1$, $a_P = 100$, $a_x = 1$. (Note that the chosen high value of a_P counterbalances the fact that E_P is much smaller than the other error components.)

Results from the optimization are shown in Table 9.2. It is interesting to note the high H parameters resulting from all optimization schemes, which yield considerable bias in the climacogram estimation (Fig. 9.1). The ombrian model resulting from the combined optimization is shown in Fig. 9.3, along with the hourly and daily empirical estimates. In this plot, the bias-adjusted results are plotted in order to be comparable to the empirical estimates; therefore, the true theoretical intensity is higher for scales $k > 1000$ h or about 40 d. Overall, the power of the ombrian model is impressive over the whole range of scales spanning 5 orders of magnitude, i.e. from 1 h to 16 years (Fig. 9.3).

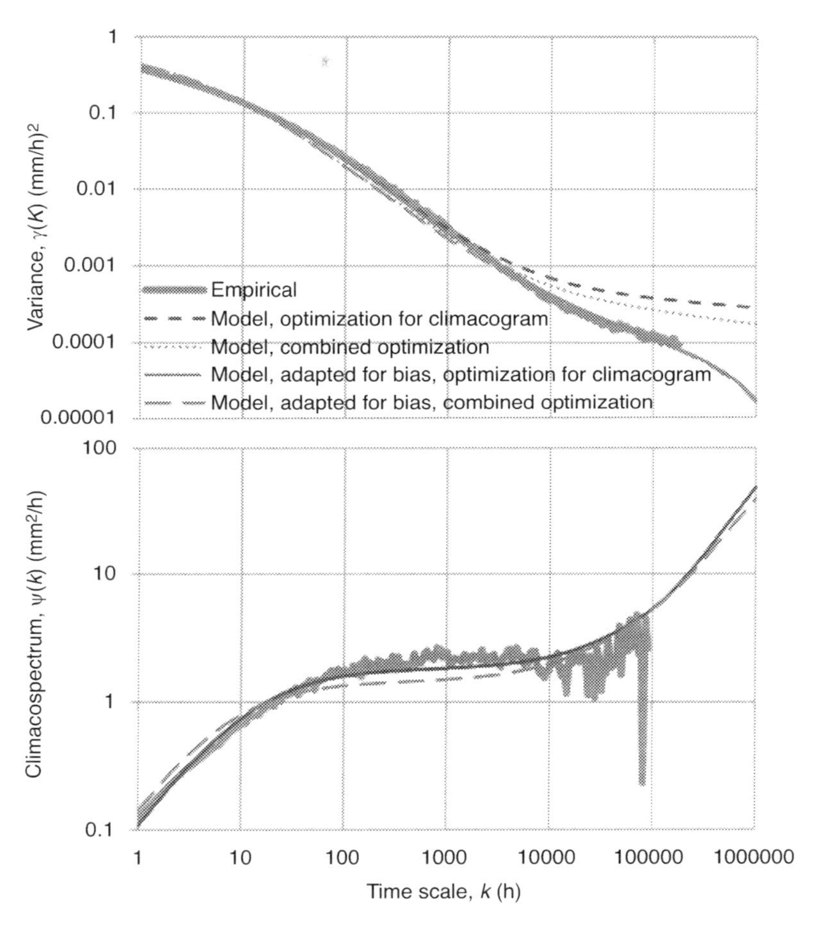

FIG. 9.1

Fitting of the ombrian model to the empirical estimates of the climacogram (Eq. (9.9)) and the climacospectrum for Bologna. The empirical estimates for time scales smaller than or greater than 1000 h (~42 d) are taken from the hourly and daily series, respectively. Note that the climacospectrum is defined through the climacogram as $\boldsymbol{\psi}(\boldsymbol{k}):= \boldsymbol{k}(\gamma(\boldsymbol{k}) - \gamma(2\boldsymbol{k}))/$ ln 2 and is most appropriate for visualization of the process behaviour at small time scales. Source: Koutsoyiannis, 2021.

9.6 Discussion and further aspects

This Section provides an overview of wider research topics within the ombrian modelling framework in light of recent advances in stochastic modelling. Detailed information on these topics can be found in Koutsoyiannis, 2021.

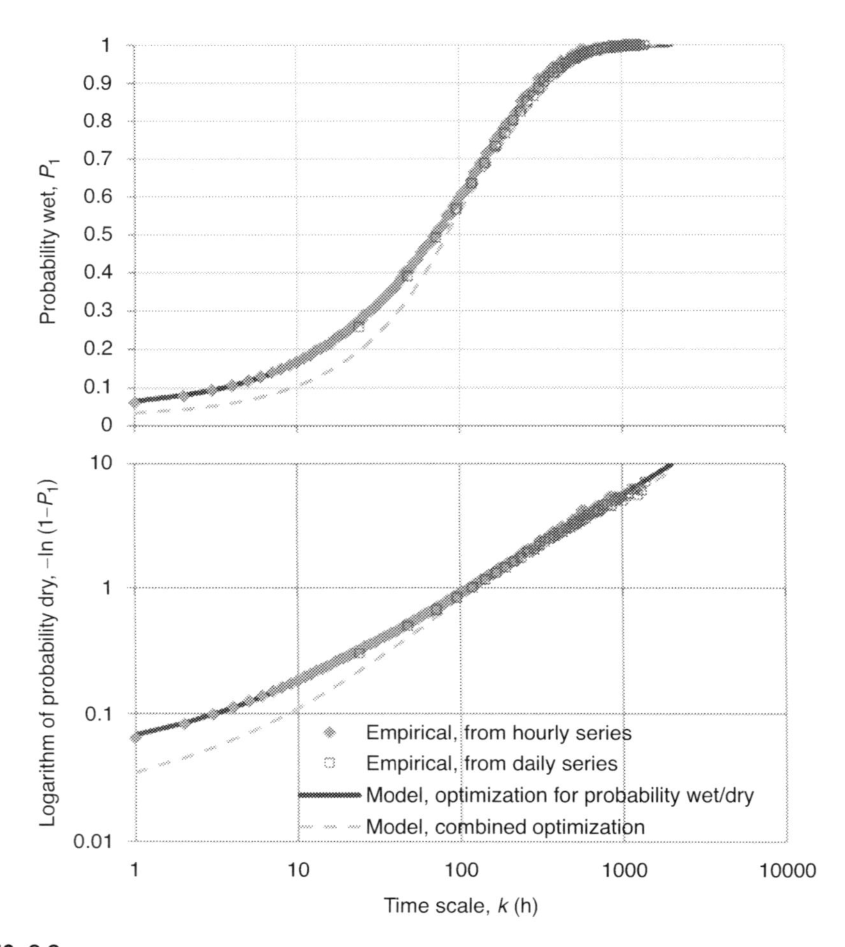

FIG. 9.2

Fitting of the ombrian model (Eqs. (9.21 and 9.24)) to the empirical estimates of probability wet and dry of Bologna. Source: Koutsoyiannis, 2021.

Table 9.2 Parameters of the ombrian model of Bologna from the four optimization schemes (table adapted from Koutsoyiannis, 2021).

Optimization scheme	μ (mm/h)	λ_1 (mm^2/h^2)	λ_2 (mm^2/h^2)	α (h)	H (-)	θ (-)	ξ (-)
(a) Climacogram	-	0.000864	1.51	16.4	0.95	-	-
(b) Probability wet	0.0773	0.00775	0.836	14.15	0.95	0.795	0.121
(c) Quantiles	0.0788	0.00407	1.60	7.70	0.93	0.693	0.125
(d) Combined	0.0823	0.00110	1.43	8.74	0.92	0.787	0.121

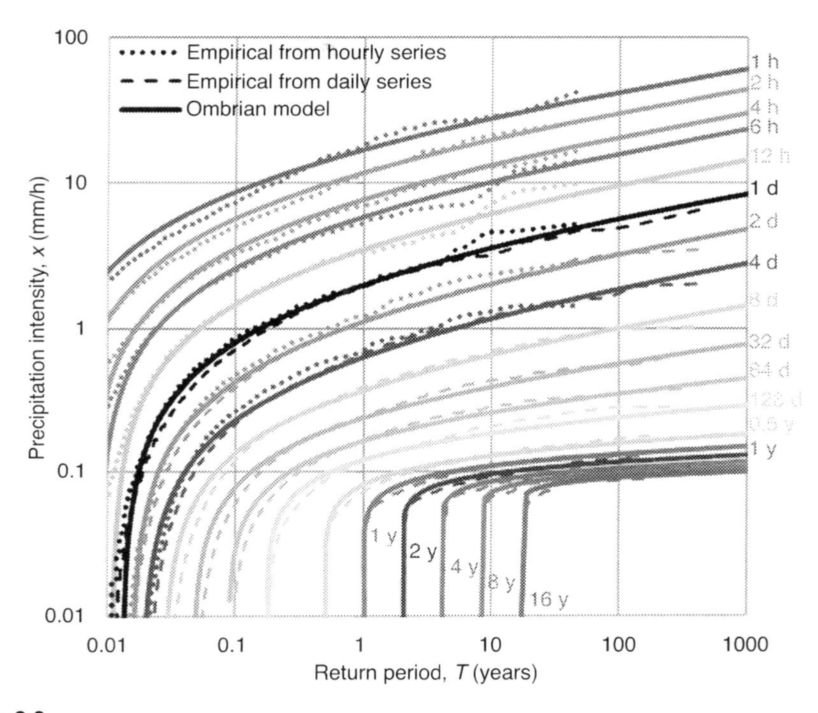

FIG. 9.3

Ombrian curves derived from the ombrian model for Bologna for time scales spanning five orders of magnitude (1 h to 16 years = 140,256 h). The empirical points are estimated from order statistics. The ombrian model results are plotted for bias-adapted variance in order to be comparable with empirical plots (thus, for $k > 1000$ h or about 40 d, the true intensity resulting from the model is higher than what is shown in the graph). The abbreviation "y" stands for year. Source: Koutsoyiannis, 2021.

9.6.1 On the use of all data

It is strongly suggested within the ombrian model approach to exploit and combine all available sources of information, particularly when these are available at different time scales. This approach was first proposed in Koutsoyiannis et al. (1998) on the basis that daily records are usually of higher reliability compared to the shorter sub-hourly series, which are also subjected to larger measurement uncertainty. The exploitation of information from different metering devices is exemplified by the case study of rainfall in Bologna.

Another basic premise of the proposed methodology is to infer the rainfall properties from the whole parent timeseries. This approach may seem at odds with the common practice of using block maxima or a certain amount of values over threshold, yet it is sounder in terms of retained information on the process properties. Series of extremes tend to hide the persistence of the parent process (Iliopoulou and

Koutsoyiannis, 2019), while discarding the body of the distribution, in favour of modelling its tails, has been criticized as wasteful usage of data (Volpi et al., 2019). Current approaches promote the use of the parent process as the natural basis for estimating design quantities in hydrological design. Issues of estimation uncertainty related to classic estimators of higher-order properties (Lombardo et al., 2014) may also be resolved by reliable higher-order estimators of the form of knowable moments (Koutsoyiannis, 2019). It is however often the case that only a part of the series is available for some records, for instance the annual maxima. In such cases, the EV2 distribution which corresponds to a Pareto tail, should be used instead for fitting the data. Yet the final model should again be formulated for the Pareto distribution (Koutsoyiannis et al., 1998).

9.6.2 On the estimation of the tail-index

The tail index of the process is one of the most important properties of extremes and also the hardest to estimate from data. In the Bologna case study, the timeseries was long and could support the estimation from the data, yet this is not the case for typical record lengths < 50 years. In such record lengths, the Gumbel distribution may be falsely supported from the data, even when the true distribution is of the EV2 type (Koutsoyiannis, 2004b, 2004a). This may lead to severe underestimation of risk. It is therefore useful to seek longer timeseries from using data from other stations in the region, or even refer to published results from global or large-scale analyses (Koutsoyiannis, 2004b; Papalexiou and Koutsoyiannis, 2013). The latter analyses provided global-scale evidence of the prevalence of a positive shape parameter, with a mean value of $\xi = 0.15$ and 0.114, respectively. It is also stressed that estimates yielding negative shape parameter, which correspond to a process bounded from above, should be discarded as physically unrealistic, and instead replaced by either the Gumbel distribution, or preferably an EV2 type with a regionally-estimated shape parameter.

9.6.3 On the use of a Hershfield coefficient

The study of a statistical property of a timeseries over multiple scales involves some procedure of aggregation of its values for different scales. Typically, the choice of the starting point for the aggregation is arbitrary, and a change thereof likely results to a different estimate. When studying extremes, it is a common hydrological practice, to either take the maximum estimate resulting from all possible positions of the starting point, or 'inflate' the given estimate by a specific factor, known as the Hershfield coefficient (Hershfield and Wilson, 1957). This practice aims to safer estimates from an engineering point of view. However, it is noted here that when the behaviour of a process is studied in stochastic terms, all realizations are stochastically equivalent and there is no theoretical basis to 'correct' them. In fact, by correcting the series, we distort its stochastic properties by studying, instead of the behaviour of $\underline{x}_\tau^{(k)}$, the behaviour of $\underline{y}_\tau^{(k)} := \max_j \left(\underline{x}_{\tau+j}^{(k)}, j = 0, \ldots k-1 \right)$, which is a different stochastic process.

9.6.4 Area-reduction of point ombrian curves

So far, the estimation of ombrian curves has been presented for the case of point rainfall. However, many hydrological applications, especially the estimation of the streamflow process, require estimates of areal rainfall. To account for the spatial variability of point rainfall within a generally homogeneous climatic region, hydrologists have long used the concept of area-reduction factors (ARF). An ARF is defined as the ratio of areally averaged precipitation depth over a certain area A for a specified return period T and time scale k to the precipitation depth over any point of the area (assumed to be climatically homogeneous) for the same return period and time scale (Flammini et al., 2022). To estimate this ratio, samples of both areal precipitation and point precipitation are needed for several scales and return periods. Moreover, according to this definition, samples of areal precipitation should be derived and stochastically analysed for the entire period, rather than calculating ARF values for isolated events, as sometimes performed in the literature. (A comprehensive review on other empirical approaches and definitions of the ARF concept is provided by Svensson and Jones, 2010 and Flammini et al., 2022).

Extensive investigation on ARF were conducted by the UK by NERC (1975) which resulted to tabulated values of ARF for a wide range of areas (1 to 30 000 km^2) and time scales (1 min to 25 days), ignoring the effect of the return period. Koutsoyiannis and Xanthopoulos (1999, p54) fitted the following relationship to these values:

$$\varphi = \max \left(0.25, 1 - \frac{0.048 A^{0.36 - 0.01 \ln A}}{k^{0.35}} \right) \tag{9.46}$$

where A is the area given in km^2 and k is in h. This relationship has been validated by results from the US by Hershfield and Wilson (1957) for the eastern USA and from the U.S. Weather Bureau (1960) for the western USA. Therefore, it could support ARF estimation in other regions as well.

9.7 Conclusions

Ombrian curves widely known under the misnomer IDF curves are central design tools for a majority of hydrological and engineering tasks. Most of their applications have been based on empirical evidence and hydrological experience. Yet empirically-derived curves entail prominent theoretical inconsistencies and cannot support simulation beyond the range of observed scales. This Chapter presented the traditional tool of ombrian curves and outlined a methodology to traverse their limitations toward building consistent and more powerful stochastic models of rainfall intensity, i.e. *ombrian models*. Going from ombrian curves to models requires understanding the assumptions that are implicit in traditional curves and revisiting thereof through stochastic modelling of the parent process.

Two modelling versions are provided: a simplified relationship valid for small-scales and an all-scale ombrian model, i.e., covering all the range of available scales. Particular emphasis is devoted to the fitting procedure in which issues of bias and data uncertainty are discussed and addressed. It is shown how to account for the effect of dependence-induced bias and further how to combine information from multiple data sources. The entire methodology is illustrated by the case study of rainfall in Bologna, which also stands as a proof of concept of the exceptional performance of the ombrian model from hourly to 16-years scale.

Advancing empirically-derived ombrian curves to theoretically-consistent ombrian models, allows the user to address bias and estimation uncertainty, extrapolate results to longer timescales and perform simulation for complex hydrological systems. These theoretical and practical gains are manifold, while the operational character of traditional ombrian curves is preserved.

References

Ayman, G.A., Mohamed, E., Ashraf, E., Hesham, E., 2011. Developing intensity-duration-frequency curves in scarce data region: an approach using regional analysis and satellite data. Engineering 3 (3), 215–226. doi:10.4236/eng.2011.33025.

Bernard, M.M., 1932. Formulas for rainfall intensities of long duration. Trans. Am. Soc. Civil Eng. 96 (1), 592–606. doi:10.1061/TACEAT.0004323.

Chow, V.T., Maidment, D.R., Mays, L.W., 1988. Applied Hydrology. McGraw-Hill, New York.

Dimitriadis, P., Koutsoyiannis, D., 2015. Climacogram versus autocovariance and power spectrum in stochastic modelling for Markovian and Hurst–Kolmogorov processes. Stoch. Environ. Res. Risk. Assess. 29 (6), 1649–1669. doi:10.1007/s00477-015-1023-7.

Eagleson, P.S.: Dynamic hydrology,1970.

Flammini, A., Dari, J., Corradini, C., Saltalippi, C., Morbidelli, R., 2022. Areal Reduction Factor Estimate for Extreme Rainfall Events. In: Morbidelli, R. (Ed.), Rainfall. Modeling, Measurement and Applications. Elsevier, Amsterdam, pp. 285–306. doi:10.1016/C2019-0-04937-0.

Hailegeorgis, T.T., Thorolfsson, S.T., Alfredsen, K., 2013. Regional frequency analysis of extreme precipitation with consideration of uncertainties to update IDF curves for the city of Trondheim. J. Hydrol. 498, 305–318. doi: 10.1016/j.jhydrol.2013.06.019.

Hershfield, D.M. Rainfall frequency atlas of the United States. Technical paper 40, 1961.

Hershfield, D.M., Wilson, W.T., 1957. Generalizing of rainfall-intensity-frequency data. AIHS. Gen. Ass. Toronto 1, 499–506 1957.

Iliopoulou, T., Koutsoyiannis, D., 2019. Revealing hidden persistence in maximum rainfall records. Hydrol. Sci. J. 64 (14), 1673–1689. doi:10.1080/02626667.2019.1657578.

Koutsoyiannis, D., 2003. Climate change, the Hurst phenomenon, and hydrological statistics. Hydrol. Sci. J. 48 (1), 3–24. doi:10.1623/hysj.48.1.3.43481.

Koutsoyiannis, D., 2004a. Statistics of extremes and estimation of extreme rainfall, 2, empirical investigation of long rainfall records. Hydrol. Sci. J. 49 (4), 591–610, doi:10.1623/hysj.49.4.575.54430.

Koutsoyiannis, D., 2004b. Statistics of extremes and estimation of extreme rainfall, 1, theoretical investigation. Hydrol. Sci. J. 49 (4), 575–590, doi:10.1623/hysj.49.4.591.54424.

Koutsoyiannis, D., 2006. An entropic-stochastic representation of rainfall intermittency: The origin of clustering and persistence. Water Resour. Res. 42 (1), W01401. doi:10.1029/2005WR004175.

Koutsoyiannis, D., 2019. Knowable moments for high-order stochastic characterization and modelling of hydrological processes. Hydrol. Sci. J. 64 (1), 19–33. doi:10.1080/02626667.2018.1556794.

Koutsoyiannis, D., 2021. Stochastics of Hydroclimatic Extremes - A Cool Look at Risk, 1st ed. Kallipos, Athens, p. 333, ISBN: 978-618-85370-0-2. http://www.itia.ntua.gr/en/docinfo/2000/ (Accessed 27 May 2021).

Koutsoyiannis, D., 2017. Entropy production in stochastics. Entropy 19 (11), 581. doi:10.3390/e19110581.

Koutsoyiannis, D., Dimitriadis, P., Lombardo, F., Stevens, S., 2018. From fractals to stochastics: Seeking theoretical consistency in analysis of geophysical data. Advances in Nonlinear Geosciences. Springer, Cham, Switzerland, pp. 237–278.

Koutsoyiannis, D., Kozonis, D., Manetas, A., 1998. A mathematical framework for studying rainfall intensity-duration-frequency relationships. J. Hydrol. 206 (1–2), 118–135. doi:10.1016/S0022-1694(98)00097-3.

Koutsoyiannis, D., Papalexiou, S.M., 2016. Extreme rainfall: Global perspectiveChow's Handbook of Applied Hydrology, 2nd edition. McGraw-Hill, New York.

Koutsoyiannis, D., Xanthopoulos, T., 1999. Engineering Hydrology, 3rd edition. National Technical University of Athens, Athens (in Greek), p. 418. doi:10.13140/RG.2.1.4856.0888.

Langousis, A., Veneziano, D., 2007. Intensity-duration-frequency curves from scaling representations of rainfall. Water Resour. Res. 43 (2). doi:10.1029/2006WR005245.

Lombardo, F., Volpi, E., Koutsoyiannis, D., Papalexiou, S.M., 2014. Just two moments! A cautionary note against use of high-order moments in multifractal models in hydrology. Hydrol. Earth Syst. Sci. 18 (1), 243–255. doi:10.5194/hess-18-243-2014.

NERC (Natural Environment Research Council), 1975. Flood Studies Report. Institute of Hydrology, Wallingford, UK.

Papalexiou, S.M., Koutsoyiannis, D., 2013. Battle of extreme value distributions: a global survey on extreme daily rainfall. Water Resour. Res. 49 (1), 187–201. doi:10.1029/2012WR012557.

Papoulis, A., 1991. Probability, Random Variables, and Stochastic Processes, 3rd edition. McGraw-Hill, New York.

Sherman, C.W., 1931. Frequency and intensity of excessive rainfalls at Boston, Massachusetts. Trans. Am. Soc. Civil Eng. 95 (1), 951–960. doi:10.1061/TACEAT.0004286.

Svensson, C., Jones, D.A., 2010. Review of methods for deriving areal reduction factors. J. Flood Risk Manag. 3 (4), 232–245. doi:10.1111/j.1753-318X.2010.01079.x.

U.S. Weather Bureau, 1960. Generalized Estimates of Probable Maximum Precipitation West of the 105th Meridian, Technical Paper No 38. U.S. Department of Commerce, Washington, DC. https://www.nws.noaa.gov/oh/hdsc/Technical_papers/TP38.pdf.

Veneziano, D., Furcolo, P., 2002. Multifractality of rainfall and scaling of intensity-duration-frequency curves. Water Resour. Res. 38 (6), 42–1 doi:10.1029/2005WR004716.

Volpi, E., Fiori, A., Grimaldi, S., Lombardo, F., Koutsoyiannis, D., 2019. Save hydrological observations! Return period estimation without data decimation. J. Hydrol. 571, 782–792. doi:10.1016/j.jhydrol.2019.02.017.

Willems, P., 2000. Compound intensity/duration/frequency-relationships of extreme precipitation for two seasons and two storm types. J. Hydrol. 233 (1–4), 189–205. doi:10.1016/S0022-1694(00)00233-X.

CHAPTER

Areal reduction factor estimate for extreme rainfall events

10

Alessia Flammini[b], Jacopo Dari[a,b], Corrado Corradini[b], Carla Saltalippi[b], Renato Morbidelli[b]

[a]*National Research Council, Research Institute for Geo-Hydrological Protection, Perugia, Italy*
[b]*Department of Civil and Environmental Engineering, University of Perugia, Perugia, Italy*

10.1 Introduction

Flood frequency estimation is required for many hydrological applications, for example, engineering design of dam spillways as well as of hydraulic structures and flood warning systems aimed at managing natural hazards. Despite streamflow data would provide more reliable estimates of flood frequency, rainfall data are typically more widely available both in space and time. Hence, it is a common hydrological practice to use areal-average rainfall data as input for rainfall–runoff modeling in order to estimate design discharges associated with a specific return period, or recurrence time (Omolayo, 1993).

However, rainfall is generally measured by rain gauges (Lanza et al., 2022) that provide a value at point scale that cannot be directly used in rainfall–runoff procedures where areal depths are required, except for areas smaller than 4 km^2 (Srikanthan, 1995). For larger areas, the upscaling procedure from point to areal-average rainfall depths with specific frequency requires the introduction of a reduction factor, as the extreme local value is not expected to occur simultaneously over the whole area, even in the case of an area with homogeneous characteristics. Starting from 1950s, areal reduction factors (ARFs) have been commonly used to transform point rainfall depth to areal depth characterized by the same probability of exceedance of the local one. The ARF is defined as the ratio between the areal-average rainfall and the point rainfall; several methodologies are available in the literature to express this ratio resulting in different ARFs properties (Wright et al., 2014). Typically, ARFs values are linked to rainfall characteristics (as intensity, duration) and catchment characteristics, that is, size, shape, and location (Svensson and Jones, 2010). The available approaches can be divided into two main categories, namely, empirical and analytical methods, with the former not relying upon specific assumptions on the rainfall process as the latter do. The first effort toward an empirical formulation of an ARF was presented in the Technical Paper N. 29 by the United States Weather Bureau (US WB) (1957-1958) with the development of a set of curves representing the reduction factor as a function of rainfall duration and area extent, to be used for any return period. They were derived using rainfall data time series

RAINFALL. Modeling, Measurement and Applications. DOI: https://doi.org/10.1016/B978-0-12-822544-8.00014-7

with a length ranging from 7 to 15 years in seven stations located in the United States. Storm durations ranging from 30 min to 24 h and areas from 250 to 1000 km^2 were considered. The US WB curves were successively parameterized in a single equation in terms of duration and area by Leclerc and Schaake (1972) and Eagleson (1972). This approach has been one of the most commonly used in the United States (Asquith, 1999) as well as in other areas of the world (often independently of geographical analogies with the US territory) with rainfall data not allowing to generate specific ARFs. Recently, Allen and DeGaetano (2005a) provided a re-evaluation over the US territory of the US WB ARFs on the basis of over 40 years of additional higher quality precipitation data, allowing to investigate the role of the return period, of the geographical location within the US territory, of the spatial density of network, and of the rainfall generating mechanisms. Rainfall data series of 24 h duration and areas up to 35,000 km^2 were considered.

Another empirical approach for estimating ARFs in the United States was proposed in National Oceanic and Atmospheric Administration Technical Report NWS 24 (Myers and Zehr, 1980); it is based on the frequency analysis of annual maximum rainfall at point and areal scales through Chow's (1951, 1964) equation. This approach is currently adopted in the United States and it is also recommended in the Australian design guidelines (Australian Rainfall and Runoff (ARR), 2001) in the versions proposed by the US Weather Service for Chicago (Myers and Zehr, 1980) or Arizona (Zehr and Myers, 1984).

In the United Kingdom (UK), the Natural Environment Research Council (NERC, 1975) developed an empirical methodology for computing reduction factors in a slightly different way from the US WB approach producing ARF values smaller for short durations and small areas (Srikanthan, 1995). This method involves durations ranging between 1 min and 25 days and areas up to 18,000 km^2. Bell (1976) proposed an approach similar to that of NERC (1975), including also the dependence on exceedance probability of rainfall and he obtained decreasing ARF values with increasing return period.

Even if in many countries the official guidelines for estimating ARFs are based on empirical methods, several analytical methods have been developed in the last decades involving assumptions on the dynamics in time and space of rainfall processes, as correlation structure, crossing properties, scaling relationships, and storm characteristics (Rodriguez-Iturbe and Mejía, 1974; Omolayo, 1989; Bacchi and Ranzi, 1996; Sivapalan and Blöschl, 1998; de Michele et al., 2001).

Despite the wide availability of different methodologies, the transposition of ARFs curves to geographical areas different from the ones where they have been developed remains a crucial issue to be addressed. This need arises because accurate local rainfall frequency analysis and characterization of the associated areal rainfall patterns require networks of rain gauges with long time series that often are not available, especially for short durations, that is, subdaily (Omolayo, 1993; Pietersen et al., 2015).

In the last decades, an increasing consciousness of the ARF dependence on specific geographical and climatic features together with the availability of denser rain gauge networks and radar rainfall data with higher spatial and temporal resolutions have determined the development of new approaches to calculate ARFs (Mineo et al., 2018; Ministry of Land, Transport and Maritime Affairs (MLTN), 2011; Lee et al., 2018; Alexander, 2001; Omolayo, 1993).

After an overview on the control factors affecting ARFs estimates in the chapter, a general description of the methodologies available in the literature is presented and advantages and limitations of each one are highlighted. Then a particular focus on the crucial topic of transposition from a region to another of ARFs estimates is deepened. In this context, as a novelty, a new empirical method for estimating specific ARFs in Umbria (Central Italy) is presented. It consists of an ARF empirical relation depending on duration and area obtained by exploiting 6 years of rainfall data with very high temporal resolution (1 min) recorded by the regional rain gauge network located in the upper part of the Tiber basin (about 6600 km^2).

The main objective of the chapter is to provide indications on the choice of the most appropriate methodology to be used for ARFs estimation for all possible combinations of data sets availability and catchment features.

10.2 Main dependencies of areal reduction factors

ARFs are known to be influenced by several factors mainly referring to the geophysical features of the area, to the characteristics of the rainfall events, and to the specific methodology and data set used to calculate the ARFs themselves (Svensson and Jones, 2010). In the following, the abovementioned dependencies are explained.

A first issue influencing the ARF calculation is represented by the physical features of the catchment, namely, its shape and topography. However, several studies showed that such influences are often negligible in the ARFs determination. In principle, basins with an elongated shape are supposed to be characterized by a higher variability of ARFs in case of rainfall isohyets perpendicular to the catchment main direction with respect to circular-shaped basins. Despite this, Veneziano and Langousis (2005) highlighted the scarce impact of the basin shape on the ARFs determination as well as the rareness of catchments whose extension along one direction is much greater than along the other. Another factor potentially affecting the ARFs calculation is the topography; in fact, the orographic features affect the rain gauge networks density, as rain gauges tend to be sparser with increasing altitudes (Prudhomme and Reed, 1999). In addition, procedures aimed at calculating the mean areal rainfall (e.g., Thiessen polygons) do not consider the topography. Under this perspective, Allen and DeGaetano (2005a) proposed a topographic bias adjustment to calculate properly the areal precipitation; nevertheless, the authors found a negligible effect of the proposed adjustment on the ARFs estimation. Finally, the urbanization rate also influences the estimation of reduction factors. Huff (1995) found a higher decrease rate of ARFs associated with urban storms with respect to rural ones for areas greater than 500 km^2. Nevertheless, the study was focused on a limited number of storms in Chicago; hence, it was not possible to properly assess if the results were determined by urbanization only or by other factors.

Rainfall characteristics are crucial factors affecting the ARFs estimation. It is well known that frontal rainfall systems are characterized by a lower spatial variability with respect to convective ones (Corradini et al., 2022). Skaugen (1997) studied frontal and convective events occurred in Norway finding a higher decrease rate of ARFs with increasing area associated with convective rainfalls. The rainfall-generation dynamics are intrinsically linked to seasonality, climate, geographical location, and rainfall duration. Several studies showed that ARFs derived for the warm season are lower than those referring to the cold one (Huff and Shipp, 1969; Allen and DeGaetano, 2005a). Such result corroborates the dependence of reduction factors on the rainfall structure, as convective events are expected to mainly occur during warm seasons. Moreover, Skaugen (1997) pointed out that rainfall extremes associated with frontal systems use to occur over coastal areas, while those associated with convective events are usually observed inland. The influence of the specific area for which ARFs are developed is confirmed by several studies highlighting discrepancies among reduction factors over different countries, for example, United States and Australia in Omolayo (1993), or over different portions of the same country, for example, in the United States in Zehr and Myers (1984) and Asquith and Famiglietti (2000). Rainfall duration is a controversial feature affecting ARFs, as contrasting results can be found in the literature about this topic. Several studies pointed out a higher decreasing rate with increasing area of ARFs associated with short duration events with respect to those for long durations (NERC, 1975; Ramos et al., 2005). On the other side, a low variability of ARFs with respect to different durations was found in studies showing reduction factor developed for the Midwest of the United States (Huff, 1995) and for India (Clark and Rakhecha, 2002). It is noteworthy that these last results can be likely attributed to the limited range of durations investigated in the abovementioned studies (Svensson and Jones, 2010). Finally, the severity of rainfall events, quantified by the associated return period, has an influence on ARFs. In fact, in case of extremely rare events, the asynchrony of point rainfall maxima is exacerbated; hence, ARFs formulations considering rainfall severity show an inverse dependence on the return period. Skaugen (1997) pointed out that differences between ARFs associated with convective and frontal events were more evident for higher return periods. A similar result (i.e., lower ARFs for increasing return periods) was also found by Asquith and Famiglietti (2000) and by Allen and DeGaetano (2005a).

It is noteworthy that ARFs estimates inevitably depend on the type and amount of available data, as well as on the particular methodology adopted. In fact, data sets characterized by a small range of durations do not allow to assess the variation of ARFs with respect to this feature. Similarly, the length of the available time series influences the magnitude of the events that can be characterized, which is expressed through the return period. Another issue influencing the ARFs calculation is the density of rain gauges network, which is strictly linked to the orographic features of the area. Omolayo (1993) found an unexpected behavior of ARFs with increasing areas attributed to the scarce density of stations due to the hilly orography. The study was focused on areas in Australia with size ranging between 100 and 1000 km^2 and a stations density of about 1 rain gauge per 100 km^2. A contrasting result was obtained by Allen and DeGaetano (2005a), who found a negligible influence of investigated stations density, ranging between 1:1800 and 1:545 rain gauge per km^2 over wide flat areas in the United States.

Finally, the use of different methods leads to different ARFs estimates. A description of the main ARFs calculation methods is provided in the following section.

10.3 Different methodologies for estimating areal reduction factors

As previously clarified, ARFs can be determined through analytical methods (if assumptions on the spatio-temporal dynamics of rainfall process are adopted) or empirical methods (if only deterministic data analyses are performed). From another point of view, methods for the estimate of ARFs can be divided into storm-centered and fixed-area approaches, on the basis of the kind of considered area (Omolayo, 1993; Swensson and Jones, 2010). In the storm-centered methods, the involved area is assumed to be the one where rain falls and changes with each storm. For a specific storm, a point characterized by the maximum value of rainfall depth and called storm center is identified. Hence, the ARF is defined as the ratio between the areal rainfall depth referred to an area defined by a selected isohyet and including the storm center (where the rainfall is everywhere greater than or equal to the isohyet value) and the maximum rainfall depth at the storm center. The involved area, as well as the storm center and the duration, change storm by storm and only the average of storm-centered ARFs over many events can be considered representative for a specific region. The storm-centered ARF definition implies a simultaneous evaluation of rainfall at local and areal scale and this makes this approach inappropriate for areal rainfall frequency analysis based on local rainfalls. In fact, extreme local and areal rainfalls are unlikely to be produced by the same storm and generating mechanism. For example, convective rainfalls can produce local critical rainfalls but may be associated with smaller areal depths (Swensson and Jones, 2010). For the aforementioned reason, the storm-centered approaches are expected to produce ARFs not conservative, for example, smaller than those obtained through fixed-area methods (Pietersen et al., 2015; Sivapalan and Blöschl, 1998). In addition, according to Asquith and Famiglietti (2000), storm-centered approaches have not been widely investigated because of the difficulty in including multicentered storms.

Fixed-area (or geographically fixed or centered) approaches are developed to estimate critical areal rainfalls over a specific area (e.g., a catchment) for engineering design purposes. In this case, the ARFs derive from rainfall statistics and are not related to any specific recorded storm. Typically, such approaches are based on series of annual maximum local and areal rainfall depths for specific durations (US Weather Bureau 1957-1958; NERC, 1975) and can also involve frequency analyses of local and areal rainfall extremes (Bell, 1976; Myers and Zehr, 1980).

Empirical methods, based on huge amounts of data and requiring heavy-computational analyses, are often fixed-area (US Weather Bureau 1957-1958; NERC,

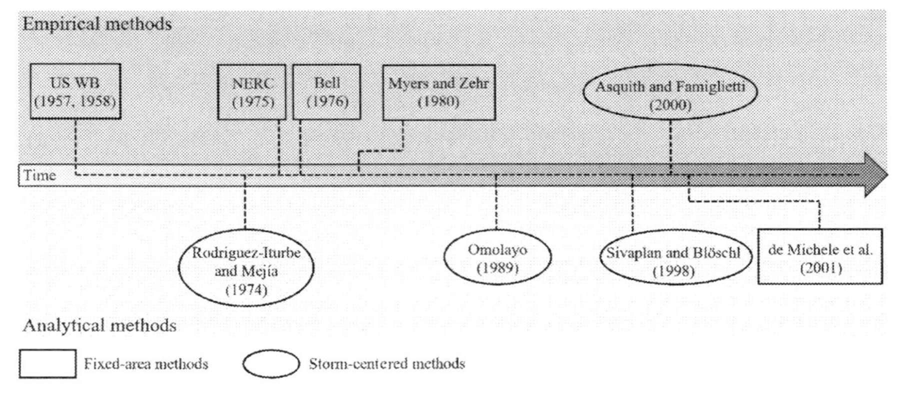

FIG. 10.1

Main empirical and analytical approaches for estimating areal reduction factors available in the literature and widely used. The differentiation between storm-centered and fixed-area methods is also highlighted. Note that the temporal scale is only qualitative.

1975; Bell, 1976; Myers and Zehr, 1980) and usually give probabilistically correct estimates of ARFs. Analytical techniques are more frequently storm-centered, since they are linked to assumptions on the spatial and temporal structure of storms (Rodriguez-Iturbe and Mejìa, 1974; Sivapalan and Blöschl, 1998; Omolayo, 1989). For the same reason, they usually need a limited amount of observed rainfall data. However, it is noteworthy that the assumed theoretical hypotheses may not properly descript the actual physics of the rainfall processes (Swensson and Jones, 2010).

However, storm-centered empirical methods also exist, as for instance that proposed by Asquith and Famiglietti (2000), as well as fixed-area analytical approaches as the one developed by de Michele et al. (2001).

In the following, the most widespread approaches adopted around the world for estimating ARFs, summarized in Fig. 10.1, are shortly described. A critical discussion on the results provided by the different methods, as well as an analysis on the possibility to transfer one method to geographical areas different from those where it has been developed are provided. Lastly, a new empirical approach is proposed.

10.3.1 The US WB approach

The method by US Weather Bureau (1957-1958) published in Technical Paper N. 29 (TP 29) is intuitive to apply although quite laborious. It was developed using rainfall data series of seven stations in eastern United States with length of time series varying from 7 to 15 years and durations from 30 min to 24 h. These data were representative of areas ranging from 250 to 1000 km^2. This approach does not account for return period of rainfall, as expected due to the shortness of the time series. The US WB ARF is defined as the ratio of the mean in time (considering all the n years) of the annual maximum areal-average rainfalls and the spatial mean (considering all the k stations) of the annual maximum point rainfalls averaged for the all the n years:

$$ARF_{US\ WB} = \frac{\frac{1}{n}\sum_{i=1}^{n}\overline{R_i}}{\frac{1}{k}\sum_{j=1}^{k}\left(\frac{1}{n}\sum_{i=1}^{n}R_{ij}\right)} \tag{10.1}$$

where $\overline{R_i}$ is the annual maximum areal-average rainfall in the ith year and R_{ij} the annual maximum point rainfall at jth station and ith year. In order to calculate $\overline{R_i}$, an arithmetic mean was used, as evenly distributed stations were selected. The same was done for determining the spatial mean of the annual maximum point rainfall in the denominator of Eq. (10.1). However, in case of unevenly distributed stations, weighted rainfall estimates can be applied.

The obtained ARFs curves, as shown in Fig. 10.2, were found to be representative of other rainfall data series too and were adopted as a reference in the United States and worldwide.

Leclerc and Schaake (1972) provided a parameterization of the $ARF^{US\ WB}$ values with a single equation:

$$ARF_{US\ WB,p} = 1 - exp\left(-1.1\ D^{0.25}\right) + exp\left(-1.1\ D^{0.25} - 0.01\ A\right) \tag{10.2}$$

where the duration D is expressed in hours and the area A in square miles. The Eq. (10.2) fits properly the experimental curves for D values higher than 3 h, while for smaller durations it overestimates the experimental points. The Eq. (10.2) is also included in Eagleson (1972).

10.3.2 The Flood Studies Report approach

This approach (NERC, 1975), developed in the United Kingdom, does not take return period into account and, from a computational point of view, it can be considered a simplification from a computational point of view of the US WB method. In this case,

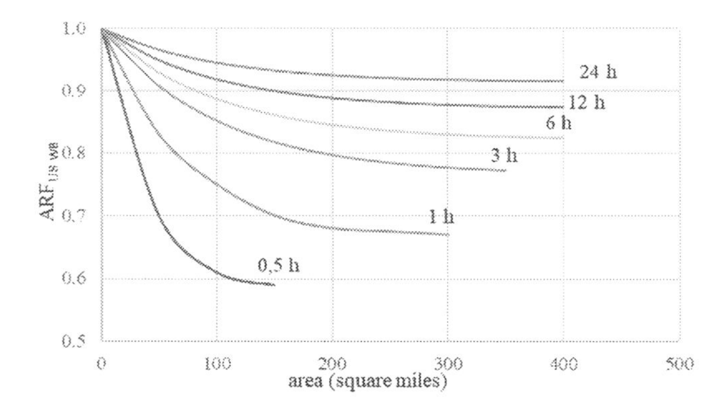

FIG. 10.2

Experimental areal reduction factor according to US Weather Bureau (1957-1958) derived for eastern United States.

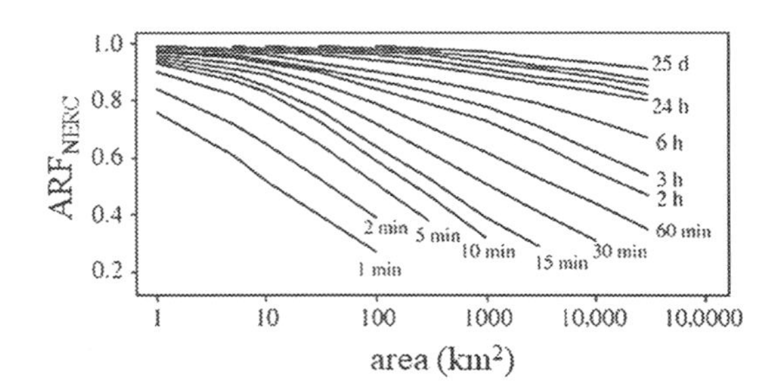

FIG. 10.3

Experimental areal reduction factors according to Flood Studies Report (NERC, 1975) for United Kingdom (diagram derived from tabulated values in NERC, 1975).

labeled with $R_{ij}^{'}$ the point rainfall measurements at station j and year i associated with the annual maximum areal-average rainfall for year i ($\overline{R_i}$), the ARF is defined as:

$$ARF_{NERC} = \frac{1}{n}\sum_{i=1}^{n}\left(\frac{1}{k}\sum_{j=1}^{k}\frac{R_{ij}^{'}}{R_{ij}}\right) \tag{10.3}$$

whereR_{ij} has the same meaning as in Eq. (10.1).

Comparing Eq. (10.3) with Eq. (10.1), it is clear that this approach assumes, with a certain level of approximation, that an average of ratios can appropriately substitute a ratio of averages, which is mathematically not true. The results in terms of ARFs curves of this methodology are given in Fig. 10.3 for a very large range of durations (1 min–25 days) and areas (up to tens of thousands of km^2).

The parameterized formulation proposed by Koutsoyiannis and Xanthopoulos (1999) fits the NERC ARFs:

$$ARF_{NERC,p} = 1 - \frac{0.048\ A^{0.36-0.01\ln(A)}}{D^{0.35}} \tag{10.4}$$

with A and D indicating the area (expressed in km^2) and the duration (expressed in hours), respectively.

10.3.3 The Bell (1976) approach

Re-analyzing the rainfall data used by NERC (1975), Bell (1976) proposed a method that accounts for the return period of the local and areal rainfall. In fact, in the previous methods the averaging operation over all extremes does not allow to consider the specific probability of exceedance of each annual rainfall maximum and to transfer this information to the ARFs curves. To overcome this issue, in the Bell's approach the annual maximum rainfalls for specified duration at local and areal scale were both ranked. The ARF of rank r (i.e., the specific probability of exceedance or return period) was then obtained as the ratio between the annul maximum areal-averaged rainfall with rank r and the spatial mean of annual maximum point rainfalls with the same rank, obtained through the Thiessen polygons interpolation. In other words, frequency distributions were fitted to the areal and point ranked series and ARFs curves for specified return periods were calculated. These curves were found to decrease more rapidly with increasing return periods and to be less conservative with respect to those proposed by NERC (1975) that does not consider the "rarity of rainfall." The inverse dependency of ARFs on return period can be justified by the fact that the more a storm is rare the more it is unlikely to involve a large area simultaneously, thus resulting in lower ARF values.

10.3.4 The National Weather Service approach

Currently, the empirical fixed-area approach proposed by National Oceanic and Atmospheric Administration Technical Report NWS 24 (Myers and Zehr, 1980) is officially adopted in the United States. It is based on frequency analysis of annual maximum point and areal rainfalls; thus, it explicitly considers the rainfall return period. The ARF is defined as:

$$ARF_{NWS} = \frac{\overline{R(T_r, D)}}{R(T_r, D)} \tag{10.5}$$

where $\overline{R(T_r, D)}$ is the areal-average rainfall quantile associated with the return period T_r and duration D over an area A and $R(T_r, D)$ is the equivalent for the point scale. The quantiles (R) for specific T_r of local and areal rainfall maxima are determined according to Chow's (1951, 1964) general equation:

$$R(T_r) = \mu + K_{T_r}\sigma \tag{10.6}$$

with μ and σ representing the mean and the standard deviation of the population of annual rainfall maxima and K_{T_r} is the frequency factor for Gumbel fitting of Fisher–Tippet type I distribution. The estimate of the parameters' distribution is performed through an original procedure that involves averaged series of station-pairs at various distances and for different rainfall durations, based on five-station networks. Thus, the approach minimizes the need of dense networks but requires an isotropic rainfall field.

10.3.5 The annual-maxima-centered approach

This empirical storm-centered approach was developed by Asquith and Famiglietti (2000) to estimate ARFs in Texas, United States. It is based on the ratios between the point annual maximum rainfall with a specific duration and the concurrent local rainfall observed at each surrounding station included in the storm area. Such ratios for each specific return period are plotted against the distance between the considered station-pairs and then a spatial integration over the whole area is performed. This method results in lower theoretical effort with respect to analytical ones that for instance require correlation coefficients estimates; nevertheless, this approach requires a dense network of rain gauges. The resulting ARFs curves remain lower than the US WB ones; this is likely due to the fact that different US areas have been analyzed but also because in general a storm-centered approach, although implying the selection of storms with annual maximum local rainfalls, does not ensure a properly representation of the most critical areal rainfall events.

10.3.6 The Rodriguez-Iturbe and Mejía (1974) approach

The analytical Rodriguez-Iturbe and Mejía (1974) method expresses ARF as function of the correlation coefficient ρd between two rain gauges separated by the characteristic correlation distance d:

$$ARF_{RIM} = \sqrt{\rho_d} \tag{10.7}$$

The quantity d represents the mean distance between two points randomly selected in the area, thus depending on the area itself.

The approach relies upon a specific spatial correlation structure, represented through either an exponential decreasing function either a Bessel-type function, and an isotropic and Gaussian behavior of point rainfall that is, however, unexpected for short-duration and extreme rainfalls. In addition, being a storm-centered method, it should not be able to adequately represent crucial areal rainfalls. However, since this approach considers all precipitation data and not only extreme rainfall events, it is not sure that the ARFs curves are unconservative if compared with fixed-area methods.

10.3.7 The Sivapalan and Blöschl (1998) approach

Sivapalan and Blöschl (1998) proposed an extension of the Rodriguez-Iturbe and Mejía (1974) method with the aim of representing extreme areal-average rainfalls. For the point rainfall intensity, they assumed an exponential distribution, while for the spatial correlation coefficient an isotropic behavior and an exponential trend were considered:

$$\rho_r = exp\left(-\frac{r}{\lambda}\right) \tag{10.8}$$

where r is the distance between two points and λ is the length of the spatial correlation.

Assuming that the areal-average rainfall process is distributed through a gamma function when the local rainfall follows an exponential distribution, the authors fitted the upper tail of the cumulative gamma distribution of areal-average rainfall intensity with a Gumbel function. The parameters of this distribution are linked to the ratio A/λ so to allow the IDF (intension–duration–frequency) curves estimation as a function of the area A. In case of area equal to zero, they are set equal to those of observed point rainfall extremes assumed to be well fitted by a Gumbel distribution, that is not always ensured. It is noteworthy that the method also assumes that the correlation structure of extreme areal-average rainfall is the same as for the parent areal-average process, which could not be true.

The ARFs resulting equation involves the dependence on area, spatial correlation length, return period, and duration. However, the role of the last quantity is very limited, especially for very high return period, this being probably due to its correlation to λ that is the main control factor in this method. In fact, λ is directly linked to storm type and generating rainfall process and is in this method more relevant for estimating ARFs than duration that in any case remains the crucial parameter in hydrological application being the concentration time of a catchment the design criterion.

10.3.8 The Omolayo (1989) approach

The approach proposed by Omolayo (1989) is based on a log-normal distribution in space of rainfall. ARF is expressed as:

$$ARF_{OM} = \exp\left\{K_T\sigma\left(\sqrt{\frac{1+(n-1)\rho}{n}}-1\right)\right\} \tag{10.9}$$

where ρ is the average spatial correlation coefficient, K_T indicates a frequency factor linked to the return period, σ is the standard deviation of log-transformed rainfall, and n represents the number of stations in the considered area.

The dependence of ARF on area is implicit through the spatial correlation coefficient that depends on it. The ARF values increase with ρ and decrease with n, σ, and T. For $\rho = 0$ ARFs reach lower bounds tabulated in Srikanthan (1995) for different values of T, n, and σ. In case of normal distribution of rainfall depth, Eq. (10.9) can be adapted as follows:

$$ARF_{OM.n} = \sqrt{\frac{1+(n-1)\rho}{n}} \tag{10.10}$$

For high values of n, Eq. (10.10) formally provides Eq. (10.7), even if with different meaning of the spatial correlation coefficient.

10.3.9 The de Michele et al. (2001) approach

On the basis of empirical methods results, it is well known that ARF curves show scale-invariant properties both in space and time. This evidence has encouraged researchers to investigate the application of multifractal analysis in estimating ARFs (Veneziano et al., 2006). Under this perspective, de Michele et al. (2001) developed a simple model based on the following scaling relationship:

$$ARF_{dM} = \left[1+\omega\left(\frac{A^{*z}}{D}\right)^{b}\right]^{-\frac{v}{b}} \tag{10.11}$$

where A^{*} is the catchment area reduced by subtracting the rain gauge area, while ω, z, b, and ν are fitted parameters.

The model was calibrated through ARFs curves empirically obtained in Milan (Italy) starting from 8 years long time series of rainfall extremes, without considering return period, in the ranges of 20 min–6 h for duration and 0.25–300 km^2 for area.

The same authors fitted a model also to ARFs by NERC (1975) with durations up to 25 days and areas up to 18,000 km^2. In addition, by analyzing the data used in NERC (1975), they showed a scale invariant behavior of ARFs for areas between 1000 and 10,000 km^2 and durations between 15 min and several hours.

10.3.10 Radar-derived ARFs

The recently increasing availability of data derived from radar (radio detection and ranging) observations has led several scientists to investigate the possibility to formulate radar-derived reduction factors. The radar technology allows the remotely detection of the echoes reflected by the raindrops (Skolnik, 1962; Doviak, 1993; Bringi and Chandrasekar, 2001); starting from the measurement of the backscattered signal, the quantitative estimation of rainfall amounts is possible through processing algorithms often requiring a calibration through gauge-derived measurements. The main advantage in using radar measurements is the good representation of the spatial variability of rainfall (Bacchi and Ranzi, 1996), while the main limitation is represented by the shortness of the available time series (Svensson and Jones, 2010). Durrans et al. (2002) analyzed a 7.5 years data set over a 4 km × 4 km grid covering a portion of the contiguous United States. Durations of 1 h, 2 h, and 4 h were considered. The authors highlighted several issues attributable to the early-life status of the new radar technology, as for instance discrepancies due to updates in the processing algorithm or biases in the estimation of extreme events. Another issue was the shortness of the available time series, limiting the range of return periods to be investigated. Nevertheless, the authors found consistent results with gauge-based ARFs, even if the rate of decrease with increasing area was lower than that obtained with gauge-based measurements. In contrast, Allen and De Gaetano (2005b) found a higher decrease rate for radar-derived ARFs with respect to gauge-derived ones. This study was carried out by exploiting daily radar data from the Global Hydrology Resource Center and adopting a 2 km regular grid over North Carolina and New Jersey. Furthermore, when increasing the spatial density of measurements, a small impact on both gauge- and radar-derived ARFs was pointed out. Finally, differences ranging between 11% and 32% were found in comparisons between radar- and gauge-based reduction factors estimated over an area of 20,000 km^2.

10.4 Comparisons and possibility of transposition of different ARFs approaches

As previously highlighted, the first efforts to develop methodologies for estimating ARFs to be used in design flood determination were made in the United States and the United Kingdom. This element, together with the unavailability of spatially dense and temporally long rainfall series, determined the spread of the US WB (1957-1958) and NERC (1975) approaches around the world without specific adaptation or re-evaluation. For instance, the US WB ARFs were recommended in Australia for many years (Institution of Engineers, Australia, 1987), those of United Kingdom in Norway for a catchment area up to 5000 km^2 (Forland and Kristofferson, 1989). In principle, no theoretical basis has led to this choice except for the assumption that climatically and geographically similar areas (in terms of mean annual rainfall and temperature, general topography, orography effects) are supposed to be characterized by the same ARFs. However, in the last decades several attempts to update traditional methods and to develop new ones have been performed in different areas worldwide finding that the transposition to other geographical zones is not always appropriate. In addition, many comparisons have been made among different ARFs estimate approaches.

For example, Omolayo (1993) re-evaluated the ARFs in a few areas of Australia comparing his results with the US WB curves, which were recommended for many years. He found contrasting results; the US WB reduction factors for durations equal to 1 day were higher than the Australian ones, but successfully transposable to eight Australian urban areas between 200 and 500 km^2. Greater discrepancies were highlighted for 1-h duration reduction factors, showing that the automatic transposition to different areas, even if climatically similar, remains problematic. In addition, he compared the Bell (1976) method, reformulated disregarding the dependence on return period, the US Weather Bureau (1957-1958) method, the NERC (1975) method, and the Rodriguez-Iturbe and Mejia one in eight Australian circular areas around the main cities characterized by significant climatic differences and a rain gauge density of about 1 per 100 km^2. He found that 1-day ARFs derived by the Rodriguez-Iturbe and Mejía method were lower than those obtained by the other methods, with the Bell's approach providing the highest values.

Alexander (1980, 2001) performed a revision of the NERC (1975) approach in South Africa using extreme rainfall series of different durations and provided empirical relations in function of area and concentration time of the catchment. Pietersen et al. (2015) carried out a comparison among different ARFs estimate methods used in South Africa, reaching the conclusion that they have been or transposed from United Kingdom (generally with poor local verification) or locally developed (in this case based on very limited local data).

Allen and DeGaetano (2005a) provided a re-evaluation in the US territory of the US WB ARFs on the basis of over 40 years of additional higher quality precipitation data, including the dependence on return period. Rainfall data series of 24 h duration in two regions, North Carolina of 18,000 km^2 and New Jersey of 3500 km^2, were considered. The revision of the US WB curves was performed by calculating the ARFs according to the following relation:

$$ARF_{A-DeG} = \frac{\overline{R(T_r)}}{\frac{1}{k}\sum_{j=1}^{k} R_j(T_r)} \tag{10.12}$$

that is formally coherent with Eq. (10.1) but involves, instead of annual maximum values of areal-average and local rainfalls, the corresponding quantiles ($\overline{R(T_r)}$ and $R_j(T_r)$ at jth station) for specified return periods determined on the basis of a frequency analysis with beta-P distribution. Such procedure allows to explicitly consider the probability of exceedance in estimating ARFs, being the available data much more consistent than the time series used in TP 29. Due to the aforementioned shortness of rainfall series used, the ARFs curves in TP 29 were interpreted as representative of areal-average rainfall extremes associated with a return period of approximately 2 years. On this basis, for the selected duration of 24 h, the US WB curve was compared with the revised one with return period 2 years in the common area range up to 1000 km^2. The results showed that the US WB curve decreases with area at a faster rate than the re-evaluated one in both study areas, but this deviation is modest considering that more than 40 years of additional data have been exploited. Beyond the 1000 km^2 area limit, the re-evaluated curve continues to decay with an exponential trend reaching the lowest value of 0.81 for an area of 3500 km^2 and of 0.8 in correspondence of an area of 18,000 km^2 over New Jersey and North Carolina, respectively. In addition, the Allen and DeGaetano (2005a) study pointed out, as previously mentioned, the inverse relation existing between ARFs and return period for the considered duration and range of areas. They also explored the possibility to perform spatial interpolations through Thiessen weights or inverse distance weighted averages for determining the numerator of Eq. (10.12) with better perspectives than the unweighted averages used in US Weather Bureau (1957-1958).

Lee et al. (2018) defined ARFs using Eq. (10.12) in relation to the Korean Peninsula, for which the Ministry of Land, Transport and Maritime Affairs (MLMT, 2011) proposed the following parameterization applicable to the major river basins:

$$ARF_{MLTM.p} = 1 - M\exp\left[-\left(aA^{b}\right)^{-1}\right] \tag{10.13}$$

where M, a, and b are parameters depending on the return period and rainfall duration.

Mineo et al. (2018) evaluated ARFs in the Lazio region (Central Italy) through an original empirical approach and compared the obtained values with four widespread empirical methodologies: NERC (1975), as parameterized by Koutsoyiannis and Xanthopoulos (1999), Moisello and Papiri (1986), US Weather Bureau (1957-1958) in the parameterization of Eagleson (1972), and Chow (1964). In the proposed methodology, a single value of ARF is calculated through the relation:

$$ARF_{MRNR} = \frac{\overline{h(A,D)}}{h(D)} \tag{10.14}$$

where $h(D)$ is a heavy rainfall of duration D in a central rain gauge (central with respect to selected areas in circular shape) and $\overline{h(A,D)}$ indicates the arithmetic mean of the rainfall depths with duration D recorded in the rain gauges falling within the considered area. The authors defined as "heavy rainfall" at a specific central rain gauge each observed rainfall event with duration D higher than the 2 years return period extreme. By applying Eq. (10.14) to all the heavy local rainfalls and all the central rain gauges in the study area, a point cloud was obtained and for each duration the relation between the median ARFs and the area was derived. Ten classes of areas ranging from 1 to 1000 km^2 around the selected central rain gauges and durations in the range 0.5–6 h were considered. The return period of 2 years used as threshold to select heavy local rainfalls was considered as precautionary for the design of hydraulic structures. The ARFs showed more sharply decreasing values with increasing areas for shorter durations than for longer ones. Mineo et al. (2018) fitted the derived ARFs curves with an analytical formulation explicitly depending on duration D and area A:

$$ARF_{MRNR,p} = \frac{\gamma(D)^{c(D)}}{\left[\gamma(D)+A\right]^{c(D)}} \tag{10.15}$$

where γ and c are parameters depending on D. As a result of comparisons between the derived ARFs curve and those obtained through the widespread selected methods, the authors highlighted both overestimation and underestimation with major deviations for shorter durations and areas larger than 100 km^2. Specifically, the Eagleson's formulation overestimates the median ARFs, thus providing more conservative values. In any case, the authors concluded that the use of methodologies elsewhere developed can lead to substantial errors in the estimate of design rainfall events.

10.5 A new ARFs empirical formulation in Umbria, central Italy

As original contribution, a new ARFs empirical formulation developed in the Umbria region (Central Italy) is presented.

The Umbria region, with an extension of 8456 km^2, is characterized by a quite complex orography with a mountainous landscape along its eastern side where the Apennines reach up to 2000 m a.s.l., and a hilly morphology in the central and western zones with altitudes ranging from 100 to 800 m a.s.l. The study area mainly falls in the Tiber River basin, which crosses the region from North to South-West. The mean annual rainfall recorded in the last century is about 900 mm, with spatial distribution of values from 650 mm to 1450 mm. The region is equipped with a network of more than 90 rain gauges with a density of about 1 rain gauge every 90 km^2 (see Fig. 10.4).

For the formulation of ARFs, the main sub-basins of the Tiber River Basin, with areas ranging from 667 to 6086 km^2, were considered. In addition, in order to further investigate the ARF-area relation, smaller circular areas from 81 to 577 km^2 were added. Such domains were selected in the region to include the

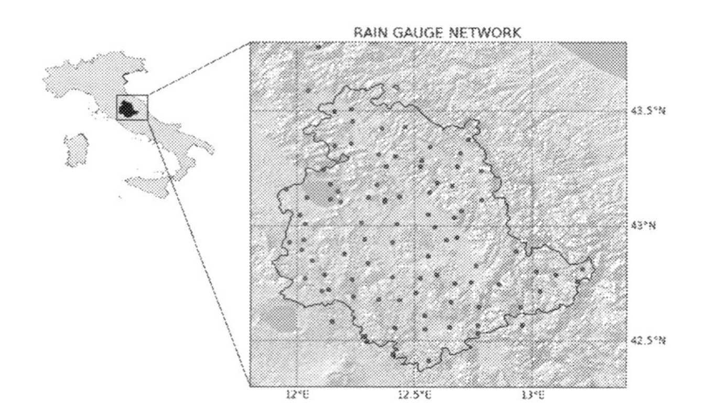

FIG. 10.4

Rain gauge network in Umbria region, Central Italy (coordinate system WGS 84, EPSG 4326).

maximum number of rain gauges. The series of annual maximum rainfalls belonging to the observation period of 2014–2019 with duration in the range 5 min–48 h were extracted; it is noteworthy that, by considering recent years, it was possible to investigate high temporal resolutions (Morbidelli et al., 2020). Considering the limited number of years, the proposed approach does not consider the return period. The ARFs were determined through the following equation, somewhat similar to Eq. (10.3):

$$ARF_{UM} = \frac{1}{n \cdot k} \sum_{h=1}^{n \cdot k} C_h \tag{10.16}$$

where $C_h = \frac{\overline{R_i}}{R_{ij}}$ are all the possible combinations with $i = 1, \ldots, n$ and $j = 1, \ldots, k$, being n the number of available years and k the number of stations included in the considered area.

An example of derived ARFs for duration of 5 min is shown in Fig. 10.5A. In addition, for completeness, the point cloud representing all ratios C_h displaced in function of area and for the same duration is provided in Fig. 10.5B.

A parameterization of ARFs as a function of duration D and area A is also provided with functional form similar to Eq. (10.15):

$$ARF_{UM,p} = \frac{\alpha(D)}{\alpha(D) + A^{0.5}} \tag{10.17}$$

where $\alpha(D)$ assumes the expression:

$$\alpha(D) = aD^b \tag{10.18}$$

with a and b are empirical coefficients.

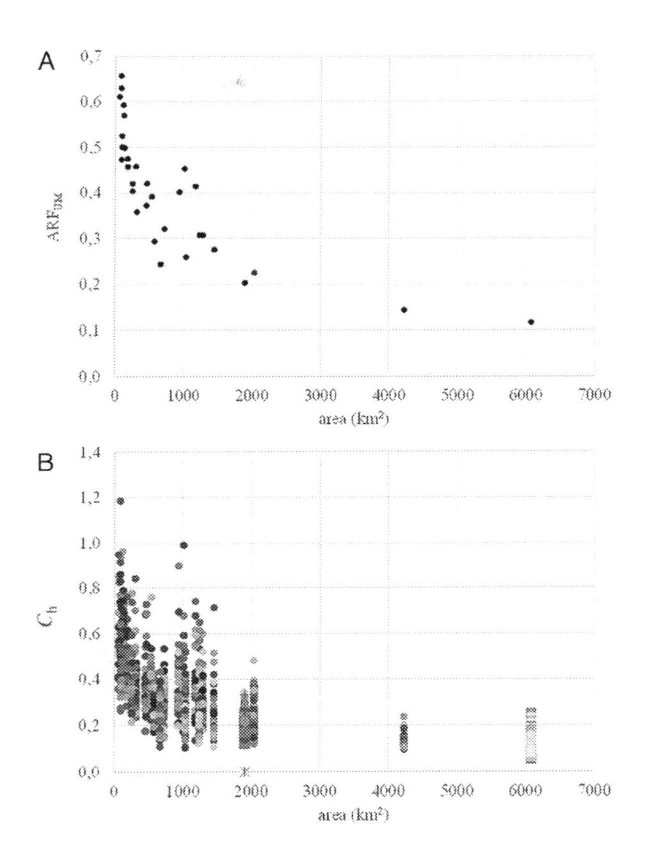

FIG. 10.5

ARFs derived in Umbria region (A) and distribution of ratios C_h of Eq. (10.16) versus area (B) for annual maximum rainfalls of 5-min duration.

The ARFs curves derived by the aforedescribed approach were compared with the following traditional methodologies: (NERC, 1975) as parameterized by Koutsoyiannis and Xanthopoulos (1999), Moisello and Papiri (1986), US Weather Bureau (19857, 1958) in the parameterization of Eagleson (1972), and the one proposed by Mineo et al. (2018). The results for a duration of 1 h are shown in Fig. 10.6. From an overall analysis, the proposed approach is in good accordance with the ARF parameterization by Koutsoyiannis and Xanthopoulos (1999) for very short durations. This is an expected result, since some analogies between the proposed ARF-estimate algorithm and that of NERC (1975) exist. The Moisello and Papiri ARF values are always lower than the corresponding ones calculated through the US WB approach parameterized according to Eagleson (1972), which seem to be always greater than the proposed ones. The Mineo et al. (2018) curves are in the middle. However, the observed differences among the methods tend to reduce with increasing durations, becoming negligible over the value of 24 h.

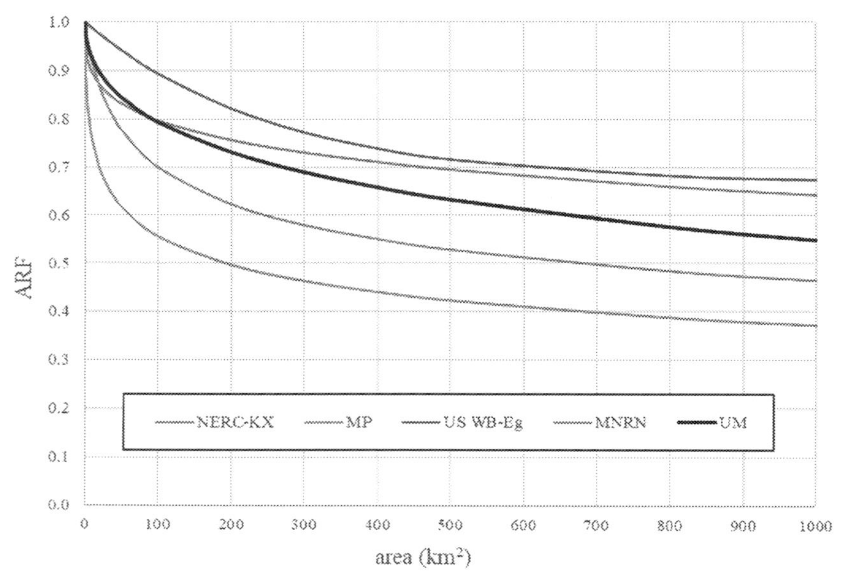

FIG. 10.6

ARFs curves derived according to NERC (1975) as parameterized by Koutsoyiannis and Xanthopoulos (1999), Moisello and Papiri (1986), US Weather Bureau (1957-1958) in the parameterization of Eagleson (1972), Mineo et al. (2018), and the proposed methodology in Umbria region for rainfall duration of 1 h.

10.6 Conclusions

ARFs are essential in the estimate of design rainfall needed in a rainfall–runoff approach for determining design discharge of hydraulic structures. An observed point rainfall event, associated with a specific probability of exceedance, is linked with a uniform spatial distribution of rainfall over an area containing the point measurement to obtain a design rainfall. The upscaling procedure requires a reduction factor that accounts for the nonsimultaneous occurrence of the critical local value of rainfall in all the points of the area, which is supposed to be climatically and orographically homogeneous. In technical applications, a specific value of the ARF is multiplied with a T_r-point rainfall to provide an areal-average rainfall of the same return period. The relationship between point and areal extreme rainfall, and consequently the ARF values, depends on several factors: characteristics of the rainfall, strictly linked with rainfall generating mechanism, as duration, intensity, area extent, seasonality, return period; the characteristics of the area, that is, the catchment features, shape and topography; the characteristics of the data, as length of time series, density of the rain gauge network, time resolution, data type (gauge- or radar-based), strongly related to the adopted methodology. In the literature, two main categories of methodologies to obtain ARFs are available: the classical empirical methods (usually fixed-area) and the analytical methods (usually storm-centered). The latter are based on theoretical assumptions on the rainfall process and generally require a limited amount of data. However, they often are storm-centered, not probabilistically correct, and based on assumptions not always fully realistic. Some of these methods seem to agree with empirical ARFs estimates but within limited temporal and spatial scales. The fixed-area empirical methods, even if usually data-intensive and computationally laborious, remain in many cases the most reliable choice. However, the possibility to transfer ARFs estimated trough such approaches for a specific region/country to other areas on the basis of geographic analogies remains a critical point. Several more or less recent re-evaluations of traditional methods performed in different geographical zones pointed out that a general validation of the transposition procedure is not allowed. On the choice of a specific empirical approach, a particular attention has to be reserved to the consistency of rainfall data. When the rainfall data series are short an empirical method that disregards return period is the only possibility among empirical approaches. In this case, it is ensured that the limited data amount is exploited in the best way to extract an average relationship between local and areal maximum rainfalls. Under this perspective, a new formulation developed in Central Italy is also proposed in this chapter. Alternatively, an analytical approach can be selected, although the limitations are well known. When the length of available rainfall series allows it, empirical methods with an explicit consideration of return period should be preferred, based on the fact that the most common usage of ARFs is as multiplicator factor of a local rainfall extreme with specified return period.

In this context, considering the wide variety of different empirical algorithms proposed by the literature, a further effort from scientific community would be required in order to select the most appropriate approach in relation to the most common ARFs application, namely, the determination of design rainfall and discharge.

References

Alexander, W.J.R., 1980. Depth-Area-Duration-Frequency Properties of Storm Rainfall in South Africa. Department of Water Affairs, Pretoria Technical Report TR103.

Alexander, W.J.R., 2001. Flood Risk Reduction Measures: Incorporating Flood Hydrology for Southern Africa. Department of Civil and Biosystems Engineering, University of Pretoria, Pretoria.

Allen, R.J., DeGaetano, A.T., 2005a. Areal reduction factors for two eastern United States regions with high rain-gauge density. J. Hydrol. Eng. 10 (4), 327–335.

Allen, R.J., DeGaetano, A.T., 2005b. Considerations for the use of radar-derived precipitation estimates in determining return intervals for extreme areal precipitation amounts. J. Hydrol. 315 (1-4), 203–219.

Asquith, W.H., 1999. Areal-Reduction Factors for the Precipitation of the 1-Day Design Storm in Texas. US Geological Survey, Austin, TX, p. 81. Water-Resources Investigations Report 99-4267.

Asquith, W.H., Famiglietti, J.S., 2000. Precipitation areal-reduction factor estimation using an annual-maxima centred approach. J. Hydrol. 230 (1–2), 55–69.

Australian Rainfall and Runoff (ARR), 2001. Design rainfall considerations. In: Pilgrim, D.H. (Ed.), Book 2 of Volume 1 of Australian Rainfall and Runoff. Engineers Australia, Barton, Australia, pp. 1–49.

Bacchi, B., Ranzi, R., 1996. On the derivation of the areal reduction factor of storms. Atmos. Res. 42 (1–4), 123–135.

Bell, F.C., 1976. The Areal Reduction Factor in Rainfall Frequency Estimation. Natural Environment Research Council, UK Institute of Hydrology Report No 35.

Bringi, V.N., Chandrasekar, V., 2001. Polarimetric Doppler Weather Radar: Principles and Applications. Cambridge University Press, New York.

Chow, V.T., 1951. A general formula for hydrologic frequency analysis. Trans. Am. Geophys. Union 32, 231–237.

Chow, V.T., 1964. Statistical and probability analysis of hydrologic data. In: Chow, VT (Ed.), Handbook of Hydrology. McGraw-Hill, New York 8.1–8.42.

Clark, C., Rakhecha, P.R., 2002. Areal PMP distribution of one-day to three-day duration over India. Meteorol. Appl. 9, 399–406.

Corradini, C., Morbidelli, R., Saltalippi, C., Flammini, A., 2022. Meteorological Systems Producing Rainfall. In: Morbidelli, R. (Ed.), Rainfall. Modeling, Measurement and Applications. Elsevier, Amsterdam, pp. 27–48. doi:10.1016/C2019-0-04937-0.

de Michele, C., Kottegoda, N.T., Rosso, R., 2001. The derivation of areal reduction factor of storm rainfall from its scaling properties. Water Resour. Res. 37 (12), 3247–3252.

Doviak, R.J., 1993. Doppler Radar and Weather Observations. Courier Corporation, Chelmsford.

Durrans, S.R., Julian, L.T., Yekta, M., 2002. Estimation of depth-area relationships using radar-rainfall data. J. Hydrol. Eng. 7 (5), 356–367.

Eagleson, P.S., 1972. Dynamics of flood frequency. Water Resour. Res. 8, 878–898.

Forland, E.J., Kristofferson, D., 1989. Estimation of extreme precipitation in Norway. Hydrol. Res. 20 (4-5), 257–276.

Huff, F.A., 1995. Characteristics and contributing causes of an abnormal frequency of flood-producing rainstorms at Chicago. Water Resour. Bull. 31 (4), 703–714.

Huff, F.A., Shipp, W.L., 1969. Spatial correlations of storm, monthly, and seasonal precipitation. J. Appl. Meteorol. 8 (4), 542–550.

Institution of Engineers, Australia, 1987. In: Pilgrim, D.H. (Ed.), Australian Rainfall and Runoff: Guide to Flood estimation. EA Books, Crows Nest, Vol. 1, p. 374.

Koutsoyiannis, D., Xanthopoulos, Th, 1999. Engineering Hydrology, National Technical University of Athens, Athens, Edition 3, 418, doi:10.13140/RG.2.1.4856.0888.

Lanza, L.G., Cauteruccio, A., Stagnaro, M., 2022. Rain Gauge Measurements. In: Morbidelli, R. (Ed.), Rainfall. Modeling, Measurement and Applications. Elsevier, Amsterdam, pp. 77–108. doi:10.1016/C2019-0-04937-0.

Leclerc, G., Schaake, J.C., 1972. Derivation of Hydrologic Frequency Curves. Massachusetts Institute of Technology, Cambridge, MA, p. 151 Report 142.

Lee, J., Park, K., Yoo, C., 2018. Bias from rainfall spatial distribution in the application of areal reduction factor. KSCE J. Civil Eng. 22, 5229–5241.

Mineo, C., Ridolfi, E., Napolitano, F., Russo, F., 2018. The areal reduction factor: a new analytical expression for the Lazio Region in central Italy. J. Hydrol. 560, 471–479.

Ministry of Land, Transport and Maritime Affairs (MLTM), 2011. Improvement and Supplement of Probability Rainfall. Ministry of Land, Transport and Maritime Affairs, Seoul, Korea.

Moisello, U., Papiri, S., 1986. Relazione tra altezza di pioggia puntuale e ragguagliata, XX: Convegno di Idraul. E Costr. Idraul., Atti, Padova. Libreria Progetto, Padova, pp. 615–663 8-10 September 1986.

Morbidelli, R., García-Marín, A.P., Al Mamun, A., Atiqur, R.M., et al., 2020. The history of rainfall data time-resolution in a wide variety of geographical areas. J. Hydrol. 590, 125258.
Myers, V.A., Zehr, R.M., 1980. A Methodology for Point-to-Area Rainfall Frequency Ratios. US Department of Commerce, National Oceanic and Atmospheric Administration, National Weather Service, Washington, DC, USA NOAA Technical Report NWS 24.
Natural Environment Research Council (NERC), 1975. Flood Studies Report. Natl. Environ. Res. Counc., London.
Omolayo, A.S., 1989. PhD Thesis. University of New South Wales, Australia.
Omolayo, A.S., 1993. On the transposition of areal reduction factors for rainfall frequency estimation. J. Hydrol. 145 (1–2), 191–205.
Pietersen, J.P.J., Gericke, O.J., Smithers, J., Woyessa, Y., 2015. Review of current methods for estimating areal reduction factors applied to South African design point rainfall and preliminary identification of new methods. J. S. Afr. Inst. Civil Eng. 57 (1), 16–30.
Prudhomme, C., Reed, D.W., 1999. Mapping extreme rainfall in a mountainous region using geostatistical techniques: a case study in Scotland. Int. J. Climatol. 19, 1337–1356.
Ramos, M.H., Creutin, J.-D., Leblois, E., 2005. Visualization of storm severity. J. Hydrol. 315 (1–4), 295–307.
Rodriguez-Iturbe, I., Mejía, J.M., 1974. On the transformation of point rainfall to areal rainfall. Water Resour. Res. 10 (4), 729–735.
Skaugen, T., 1997. Classification of rainfall into small- and large-scale events by statistical pattern recognition. J. Hydrol. 200 (1–4), 40–57.
Skolnik, M.I., 1962. Introduction to radar, Chap. 2Radar Handbook. McGraw Hill, New York.
Sivapalan, M., Blöschl, G., 1998. Transformation of point rainfall to areal rainfall: intensity–duration–frequency curves. J. Hydrol. 204 (1–4), 150–167.
Srikanthan, R., 1995. A Review of the Methods for Estimating Areal Reduction Factors for Design Rainfall. Cooperative Research Centre for Catchment Hydrology, Australia Report 95/3.
Svensson, C., Jones, D.A., 2010. Review of methods for deriving areal reduction factors. J. Flood Risk Manag. 3 (3), 232–245.
US Weather Bureau, 1957, 1958. Rainfall Intensity-Frequency Regime Parts 1 and 2. U.S. Department of Commerce, Washington, DC, USA Technical Paper No. 29.
Veneziano, D., Langousis, A., 2005. The areal reduction factor: a multifractal analysis. Water Resour. Res. 41 (7).
Veneziano, D., Langousis, A., Furcolo, P., 2006. Multifractality and rainfall extremes: a review. Water Resour. Res. 42 (6).
Wright, D.B., Smith, J.A., Baeck, M.L., 2014. Critical examination of area reduction factors. J. Hydrol. Eng. 19, 769–776.
Zehr, R.M., Myers, V.A., 1984. Depth-Area Ratios in the Semi-Arid Southwest United States. Office of Hydrology, National Weather Service, NOAA, Silver Spring, MD, USA NOAA Technical Memorandum NWS HYDRO-40.

CHAPTER

Analysis of extreme rainfall events under the climatic change

11

Hayley J. Fowler, Haider Ali
School of Engineering, Newcastle University, United Kingdom

11.1 Introduction

Recent studies based on observed data confirm increasing trends in the frequency and intensity of extreme precipitation during the twenty-first century on continental to global scales in a warming climate (Fisher and Knutti, 2016; Westra et al., 2013; Fowler et al., 2021a). This intensification is expected to be more prominent in a warmer future climate, and this is supported by climate model projections (Pfahl et al., 2017; Moustakis et al., 2021). Extreme rainfall events lead to severe damage to societies, posing a challenge to public safety, stormwater infrastructure and the economy (Fadhel et al., 2018). Short-duration, high-intensity rainfall extremes can be more hazardous than longer-duration rainfall extremes as they may lead to intense flash flooding, with shorter catchment response times, and are also responsible for pollution events in sewerage networks (Arnbjerg-Nielsen et al., 2013; Fowler et al., 2021b). They are more relevant in the design of water resource and urban stormwater infrastructure, especially in the mainly concrete-impervious urban areas (Ali et al., 2021). Therefore, knowledge of precipitation changes at hydrologically relevant (sub-hourly to daily) scales and the mechanisms driving these changes are crucial for societal decision-making for climate adaption (Bader et al., 2018; Paschalis et al., 2015). The chapter is structured as follows: we start by defining extreme rainfall events and review methods used to estimate their changes in Section 11.2, followed by a review of studies on understanding observed changes to rainfall extremes and their relationship with temperature variability in Section 11.3. We then review the projected changes to extreme precipitation from climate models in Section 11.4, followed by a summary of the chapter in Section 11.5.

11.2 Rainfall extremes and their analysis

11.2.1 Definition of extreme events

Hydrologic extreme events are natural events that have the potential for a significant adverse impact on lives, infrastructure, or the environment. However, defining

RAINFALL. Modeling, Measurement and Applications. DOI: https://doi.org/10.1016/B978-0-12-822544-8.00017-2

an extreme event is not so simple and the understanding is hindered by a lack of coherence in deciding what makes an extreme event "extreme" (McPhilips et al., 2018). In hydrology, characterization of extreme events is motivated by physical and socioeconomic vulnerability to their impacts (Kharin et al., 2007), which may be because of a single event or compound events which together have strong potential for greater overall impacts (IPCC, 2012; Leonard et al., 2014). Previously, scientists have used different methods to identify, map, characterize, and model extreme rainfall events to understand the factors which govern their magnitude and occurrence (Kunkel, 2003).

11.2.2 Methods

There are many methods available to analyze extreme events using long records of data. Indices can provide useful information on averages, trends, variability and related statistics (Zhange et al., 2011). One of the simplest parameters is to estimate an average (mean) rainfall. However, in climate science, a monthly/daily average smooths out a lot of important information which can be important for impact studies. Therefore, climate studies focusing on change to observed extreme precipitation over time can use the standard set of indices defined by the Expert Team on Climate Change Detection and Indices (ETCCDI) relating to precipitation extremes (Donat et al., 2016a; Sillmann et al., 2013; Peterson and Manton, 2008). This standardization of indices can help in intercomparison between different regional studies in any part of the world and provides seamless merging of index data on the global scale (Sillmann et al., 2013; Zhang et al., 2011). The indices are defined bearing in mind limitations in data availability: data quality and consistent long records.

Sub-daily precipitation extremes associated with atmospheric convection in urban areas may lead to flash flooding; therefore, indices based on daily data may not provide enough information for these locations and can mask some intense sub-daily events. Sub-daily data availability becomes an important issue for conducting such studies; however, with the joint effort of different workshops, meetings, and multiple large community-led projects, some sub-daily datasets are being/have been developed and the focus has shifted to the use of indices derived from sub-daily data as well (Alexander et al., 2019). These important datasets include HadISD (Dunn et al., 2019), the integrated surface database (Smith et al., 2011), and the global sub-daily rainfall dataset (GSDR) (Lewis et al., 2019a). Advances have also been made in relating these indices to remotely sensed products (Alexander et al., 2020) and reanalysis products (Donat et al., 2016a).

11.2.2.1 Indices based on daily precipitation data

Indices based on daily rainfall data generally choose annual maxima or percentiles like 95th, 99th, and even 99.9th percentile as a threshold to define an extreme value. Table 11.1 shows the 10 ETCCDI indices, which are derived from daily precipitation with different measures of frequency (days above fixed thresholds: R10, R20, R95p, R99p), intensity (wettest day: RX1day, RX5day; average daily intensity: SDII,

Table 11.1 The extreme precipitation indices recommended by the ETCCDI (adapted from Zhang et al., 2011).

S.No.	Indicator name	Definition
1	Max 1-day precipitation amount (RX1day)	Monthly maximum 1-day precipitation
2	Max 5-day precipitation amount (RX5day)	Monthly maximum consecutive 5-day precipitation
3	Simple daily intensity index (SDII)	The ratio of annual total precipitation to the number of wet days (≥ 1 mm)
4	Number of heavy precipitation days (R10)	Annual count when precipitation ≥ 10 mm
5	Number of very heavy precipitation days (R20)	Annual count when precipitation ≥ 20 mm
6	Consecutive dry days	Maximum number of consecutive days when precipitation < 1mm
7	Consecutive wet days	Maximum number of consecutive days when precipitation ≥ 1 mm
8	Very wet days (R95p)	Annual total precipitation from days > 95th percentile
9	Extremely wet days (R99p)	Annual total precipitation from days > 99th percentile
10	Annual total wet-day precipitation (PRCPTOT)	Annual total precipitation from days ≥ 1 mm

ETCCDI, Expert Team on Climate Change Detection and Indices.

PRCTOT), and duration (consecutive wet and dry days: CDD, CWD). However, percentile-based thresholds not only hamper intercomparison between studies using different indices and reference periods but may lead to the wrong interpretation of the results, particularly if they are based on a wet-day percentile (described by ETTCDI). For example, for a precipitation intensity distribution with disproportionate changes to different parts, the choice of the wet-day threshold may be very sensitive (Pendergrass, 2018).

11.2.2.2 ***Indices based on sub-daily precipitation data***

Recently, a major international effort to focus on global sub-daily rainfall extremes has been initiated by the INTENSE project (Blenkinsop et al., 2018) by undertaking a task to prepare a quality-controlled sub-daily rainfall dataset (GSDR; Lewis et al., 2019a, 2019b). The INTENSE project aims to provide freely accessible large datasets of sub-daily rainfall indices based on the GSDR dataset through www.climdex.org. These indices are similar to those available at the daily scale which include monthly maximum indices, diurnal cycle indices and frequency threshold indices (Lewis et al., 2019b; Table 11.2).

Table 11.2 Sub-daily precipitation indices proposed by Lewis et al. (2019b) (adapted from Alexander et al., 2019).

S.No.	Index type	Index
1	Monthly maximum indices	Rx1h monthly maximum 1 h precipitation
2		Rx3h monthly maximum 3 h precipitation
3		Rx6h monthly maximum 6 h precipitation
4		Rx1hP percent of the daily total that fell in the monthly maximum 1 h precipitation
5	Diurnal cycle indices	LW1H monthly likely wettest hour within a day
6		LD1H monthly likely driest hour within a day
7		DLW1H dispersion around monthly likely wettest hour within a day
8		S1HII simple hourly precipitation intensity index
9		CW1H maximum length of a wet spell
10	Frequency/threshold Indices	R10mm1h monthly count of hours where precipitation ≥ 10mm
11		R20mm1h monthly count of hours where precipitation ≥ 20mm
12		Rxmm1h annual count of hours when precipitation ≥ n mm, where n is a user-defined threshold
13	General indices	PRCPTOT1h annual total precipitation in wet hours

11.2.3 Extreme value theory

A further way to study the tail of the distribution of non-zero hourly/daily rainfall totals is by focusing only on the part of the distribution exceeding a certain threshold using extreme value theory (EVT) (De Haan et al., 2006). EVT requires large samples from independent and identically distributed data and the hypothesis of convergence to a particular distribution (Coles et al., 2001). In generalized extreme value distribution (GEV; which doesn't make any prior assumption on the distribution of the heavy tail), annual maxima series (AMS) or values exceeding a certain high threshold (peak over threshold) are extracted from the dataset and statistical methods are applied to characterize their distribution (Coles et al., 2001). By fitting this distribution and associated (shape, scale, and location) parameters, the hydrologist can design intensity-duration-frequency curves to estimate the probability of occurrence of events of various magnitudes (AghaKouchak et al., 2018). For example, in the design of large dams, a 1000-year return period is selected which means that the dam should efficiently hold a flood with a recurrence probability of 1 in 1000 years (Salvadori et al., 2011). However, Papalexiou and Koutsoyiannis (2013) argued that the shape parameter is very sensitive to the record length and the geographical location and the best distribution to fit the AMS is the Frechet distribution. The limitations of EVT analysis from short datasets can be overcome by using the metastatistical extreme value (MEV) framework which relaxes some of the assumptions of EVT, and uses the full data record rather than a small subset of its tail by pooling multiple samples through bootstrapping (Marani and Ignaccolo, 2015). MEV has been shown to perform well at both daily and sub-daily timescales (Marra et al., 2018).

This return period-based methodology strongly depends on the assumption of a stationary climate which means that the underlying distribution to estimate return levels (rainfall maxima) does not change with time and can be checked with the Priestley-Subba Rao test (Ali et al., 2017; Cheng et al., 2014). However, due to strong evidence of the increased frequency of extreme events, statistical advances have been made to include nonstationary into the frequency distributions and resilience-based design (Katz, 2013). For example, Ali et al. (2017) used daily dewpoint temperature (DPT) and tropospheric temperature (at 850 hPa height; T850 now onwards) as covariates in the GEV model to estimate the difference between stationary and nonstationary return levels over 23 selected urban areas in India using daily observed records (Fig. 11.1). They found rainfall maxima increased at a majority of locations (13 gauges showed an increase of median 18.3% for 1-day 100 year rainfall maxima) under the assumption of nonstationary atmospheric conditions.

11.3 Observed changes to rainfall extremes

11.3.1 Daily rainfall extremes

With a consistent rise in global mean temperature in the past decades, observed daily rainfall extremes have intensified. This is supported by various studies at regional (Ali et al., 2014), continental (Fischer and Knutti, 2016; Guerreiro et al., 2018), and global (Donat et al., 2016b; Westra et al., 2013; Sun et al., 2021) scales, depending on the availability of the data. The methods used to estimate trends/ changes include a nonparametric linear Man-Kendall test (used in Ali et al., 2014; Sun et al., 2021) and a binning method based on local percentiles (used in Fischer and Knutti, 2016; Guerreiro et al., 2018). However, linear trend techniques may not be appropriate because extreme precipitation nay not follow a linear response to even strong forcing and a nonlinear trend technique may be appropriate instead (e.g., see Fig. 6 in Ali et al., 2014) using a nonlinear method [EEMD] to study trends in extreme rainfall indices). Therefore, the selection of the most suitable methodologies for robust detection of precipitation change is a big challenge.

Globally, Sun et al. (2021) estimated the long term (1950–2018) trends in daily (Rx1day) precipitation extremes over 7593 stations from HadEX2 and found two-thirds (66%) of stations showed increasing trends and close to one-third showed decreasing trends in Rx1day (Fig. 11.2). From these, 9.1% of stations (concentrated in North America, Europe, and South Africa) showed a statistically significant ($p < 0.05$) increasing trend and only 2.1% of stations (over Canadian Prairies, the western United States, and northern China showed a significant ($p < 0.05$) decreasing trend. Dong et al. (2021) used the same data as Sun et al. (2021) and showed an

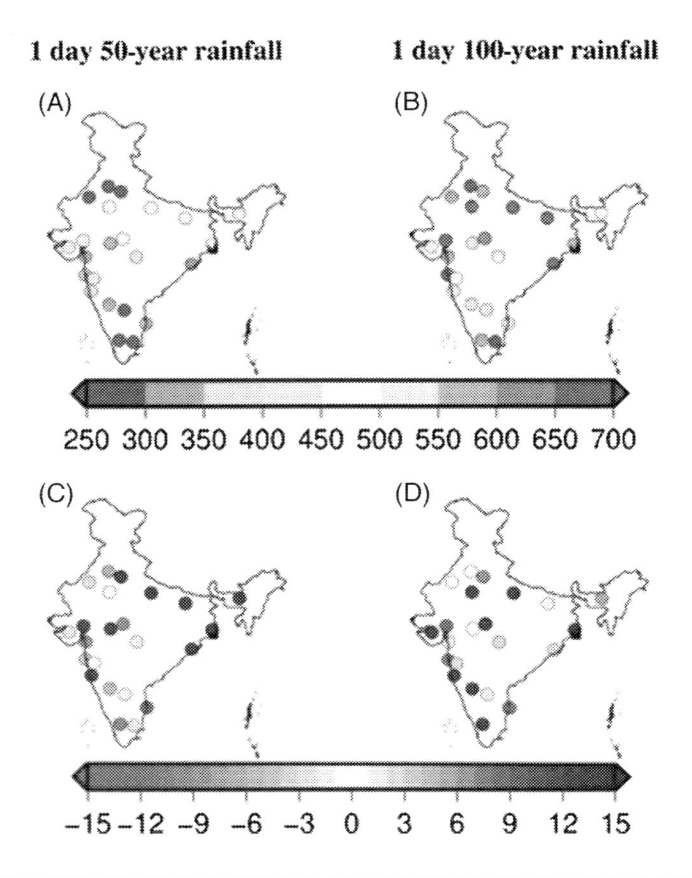

FIG. 11.1

(A) 1 day 50-year rainfall maxima (in mm) for 23 cities across India assuming stationary conditions, (B) same as (A) but for 1 day 100-year rainfall maxima, (C) percentage bias in 1 day 50-year rainfall maxima considering stationary and nonstationary conditions, (D) same as (C) but for 1 day 100 years. DPT and T850 were used as covariate to account nonstationary conditions. Percentage change (= [NS-S]/NS*100) in 1 day 50–100-year rainfall maxima were estimated using the stationary (S) and nonstationary (NS) conditions. Return values were based on GEV model estimated using ismev package in "R". The figure is adapted from Ali and Mishra (2017).

increasing trend in different indices (R99p, R95p, R99pTOT, and R95pTOT) over most locations (see Figure 1 in Dong et al., 2021). The results from these two studies are supported by results from previous global studies using HadEX2 station data comprising over 8000 stations (Westra et al., 2013) and gridded HadEx data using over 6000 stations (Min et al., 2011).

On a continental scale, Fischer and Knutti (2016) reported increases in the frequency of occurrence of heavy precipitation in Europe and the USA in the period 1951–2013 using EOBS gridded observation dataset (See Figure 2 and Figure SF1

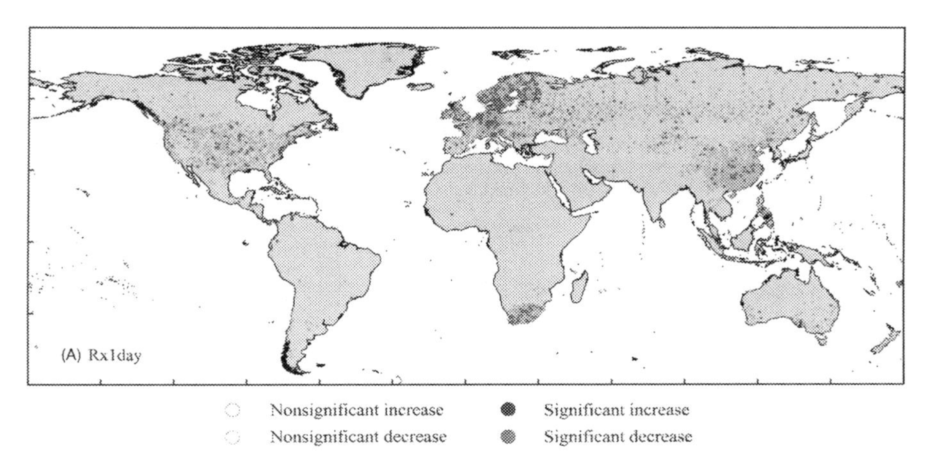

FIG. 11.2 Stations with trends for annual maximum precipitation (Rx1day) over 7293 stations for the period of 1950–2018.

Light blue (*light red*) open dots indicate nonsignificant increasing (decreasing) trends, dark *blue* (*dark red*) dots indicate statistically significant increasing (decreasing) trends. The trends are determined by a non-parametric Mann-Kendall test at the 5% significance level. The figure is adapted from Sun et al. (2021).

respectively, in Fischer and Knutti 2016). Similarly, Sun et al. (2021) showed an intensification in Rx1day over the majority of five continents (Asia, Africa, Europe, North America, and South America) and northwest Australia (see Figure 4 in Sun et al., 2021). The negative trends in Rx1day in southwestern and southeastern Australia have been linked to changes in circulation (Sun et al., 2021). However, Guerreiro et al., (2018) used better quality-controlled GSDR data to study changes in daily precipitation over the Australian continent (period 1966–2013). They showed a continental increase for the first time of around 6.5% per degree warming in daily precipitation during the period 1990–2013 as compared to the period 1966–1989.

On regional scales, Roxy et al. (2017) used daily gridded rainfall data (at 0.25-degree spatial resolution during 1950–2015) and reported a threefold rise in extreme precipitation events over central India. The increasing trends in daily R99 over India are also reported in Ali et al. (2019), Krishnamurthy et al. (2009), and Sen Roy and Balling Jr (2004). Similarly, an increasing trend in extreme precipitation is observed in Bangladesh (Shahid, 2011), South Korea (Park, 2011), southern West Africa (Nkrumah et al., 2019), West China (Wang and Zhou, 2005) and parts of Southeast Asia (Donat et al., 2016a). However, the lack of availability of dense stations (especially in remote areas) still limits such studies at a regional scale.

11.3.2 Sub-daily rainfall extremes

The understanding of changes in sub-daily rainfall extremes is inhibited by the limitations in lack of data, lack of access and quality control checks across a large part

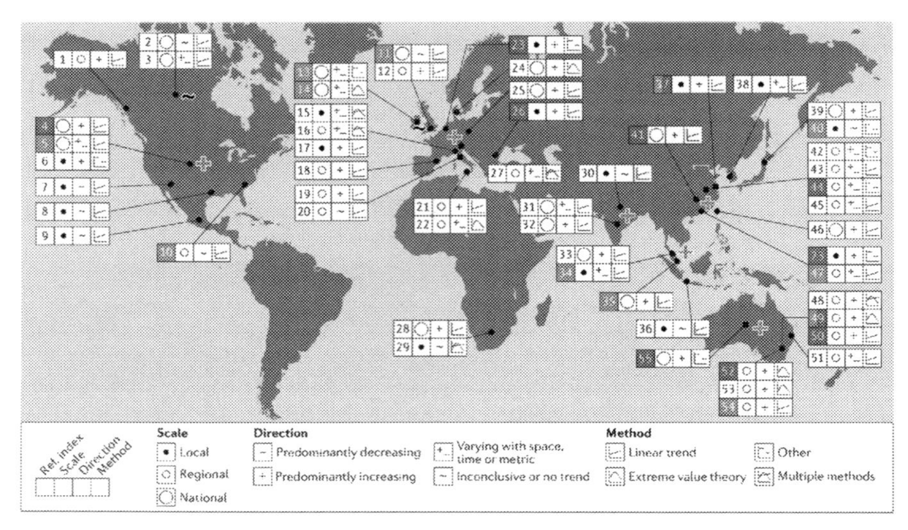

FIG. 11.3 Summary of existing studies of observed changes in the frequency and/or intensity of sub-daily precipitation extremes.

For each study, the details on the spatial scale, change and methods applied are identified. The figure is directly taken from Fowler et al. (2021a) – their Figure 4.

of the globe. Moreover, due to the small network of rain gauges, sub-daily rainfall studies have focused only on local-scale, and not regional-scale, changes (Jun et al., 2020). Recent observational studies evidence increases in the frequency and/or intensity of sub-daily rainfall extremes in, for example, North America (Barbero et al., 2017), Europe (Arnone et al., 2013), Southeast Asia (Syafrina et al., 2015), Australia (Guerreiro et al., 2018), and parts of China (Xiao et al., 2016). Recently, Fowler et al. (2021a) summarized existing studies (mentioned in Table 1 of Fowler et al. 2021a) at a sub-daily scale from rain-gauge observations at different spatial scales (Fig. 11.3). These studies identify the change in sub-daily rainfall extremes in the context of large-scale circulation and modes of variability, increasing urbanization, and the observed relationship between sub-daily rainfall extremes and temperature (also known as temperature scaling).

11.3.3 Temperature scaling

11.3.3.1 Clausius-Clapeyron relationship

The intensification of precipitation extremes with warming can be explained by the increase in moisture availability in the atmosphere due to an increase in saturation vapor pressure with warm air. Though it is not very straightforward, the Clausius-Clapeyron (CC) relationship can be a benchmark to interpret change in extreme precipitation (O'Gorman, 2015). The CC equation determines the equilibrium between two phases of a substance and can be mathematically expressed as the relationship between temperature T and pressure p at the equilibrium as in Eq. 11.1 (Koutsoyiannis, 2012):

$$p = CTe^{\frac{-L}{RT}} \tag{11.1}$$

where CT is an integration constant, L is the enthalpy of vaporisation, and R is the gas constant. The assumption of constant L can be removed by expanding Eq. 11.1 to Eq. 11.2 (derived in Koutsoyiannis, 2012) given by:

$$P = P_0 \exp\left[\frac{\alpha}{RT_0}\left(1-\frac{T_0}{T}\right)\right]\left(\frac{T_0}{T}\right)^{(C_L-Cp)/R} \tag{11.2}$$

where c_p is the specific heat at a constant pressure of the vapour and C_L is the specific heat of the liquid, P_0 is the saturation vapour pressure at a specific temperature T_0. Therefore, Eq. 11.2 can be used to estimate the scaling rate using logarithmic transformation.

In climate change studies, CC scaling, termed "apparent scaling," links surface air temperature (SAT) and atmospheric humidity and states that under the limitation of constant relative humidity, atmospheric moisture will increase at a rate similar to the dependency of vapour pressure on temperature (i.e., 6%–7%/K; CC scaling rate hereafter) (Trenberth et al., 2003). Different methods have been employed to estimate the scaling rate, which include binning methods, quantile regression, and by removing seasonality in temperature discussed in Ali et al. (2018). In addition, scaling estimates are found to become more consistent in space when better scaling variables are used for estimating scaling relationships discussed ahead in the chapter. Fig. 11.4 (based on the binning method) shows the temperature scaling of hourly rainfall intensities with DPT for the Netherlands from the GSDR dataset. The scaling curves show that the dependency of hourly extremes with DPT follow around 2 times CC scaling for higher (99th and 99.9th) percentiles. For more detail on data and methods, please refer to Ali et al. (2021) and Fowler et al. (2021).

11.3.3.2 Factors affecting scaling rates

Previously, the relationship between extreme daily precipitation and daily SAT was found to match the CC scaling rate for Japan (Utsumi et al., 2011), China (Gao et al., 2018), and the Tropics (Wasko et al., 2016). However, many studies have reported deviation in the extreme daily precipitation-daily air temperature relationship from the CC rate, which varies considerably across different regions and seasons (Ali et al., 2018; Zhang et al., 2017). One of the reasons for the deviation in scaling is the peaked structure of the scaling curve, which shows an increase in extreme daily precipitation up to a certain SAT threshold and then a monotonous decrease with higher temperatures (Hardwick Jones et al., 2010; Utsumi et al., 2011).

This behaviour can be because of (1) the cooling effect of extreme precipitation and its accompanying synoptic systems, and (2) a decrease in relative humidity at higher temperatures (Gao et al., 2020). The decrease in relative humidity at higher temperatures may be due to high specific humidity and large saturation deficits at

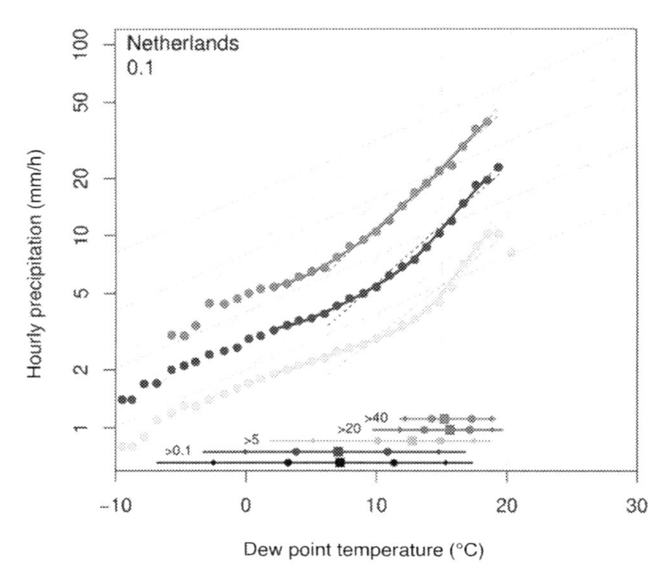

FIG. 11.4 Temperature scaling of hourly rainfall intensities with dewpoint temperature for the Netherlands.

The scaling curves show the dependency of extreme percentiles (95th, *cyan*; 99th, *blue*; 99.9th, *pink*) of the distribution of hourly precipitation (from the GSDR dataset) on daily dewpoint temperature (from HadISD) for the Netherlands (obtained by pooling 29 stations with data from GSDR dataset). Note the logarithmic y axis. The horizontal lines at the bottom show the interquartile ranges of DPT (1%, 5%, 25%, 50%, 75%, 95%, and 99%) for precipitation thresholds of 0.1 mm (*blue*), 5 mm (*yellow*), 20 mm (*red*), 50 mm (*pink*), and overall DPT range, respectively. Dotted lines are the exponential relations given for 1 (*black*) and 2 (dark red) times CC scaling. For methods, please refer to Ali et al. (2021).

higher temperatures, which in turn decreases the specific humidity gradient between the ocean and land, leading to a decrease in moisture transport from the ocean to the land (Gao et al., 2020). This implies that since relative humidity is not stable at higher temperatures, the CC theory may break down at higher temperatures. To overcome these limitations arising from humidity variations, DPT has been proposed for calculating extreme precipitation-temperature sensitivities, since it can show changes to both relative humidity and SAT (Ali and Mishra, 2017; Barbero et al., 2017). Since DPT is the temperature needed by a parcel of air to be cooled at constant saturation vapour pressure (100% relative humidity), a one-degree rise in DPT can be considered as equivalent to a 7% rise in atmospheric moisture content (Lenderink and van Meijgaard, 2008). Fig. 11.5 shows that the peaked structure behaviour in the scaling curve is now transformed to a consistent rising curve when DPT (or T850) is used instead of SAT as a scaling variable. This highlights the importance of considering robust scaling variables while estimating scaling rates.

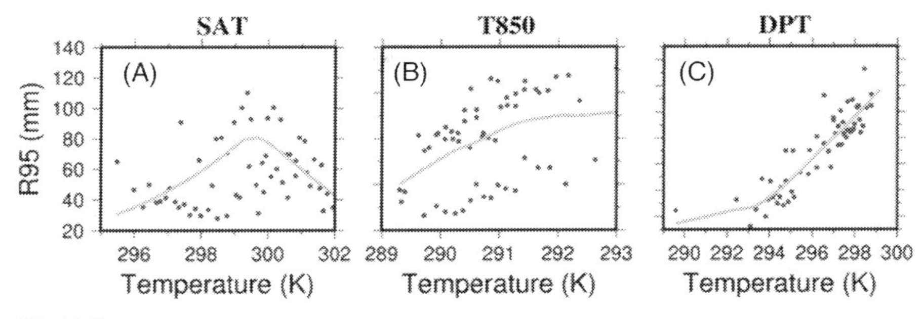

FIG. 11.5

Scaling curves showing the relationship of daily rainfall extremes (R95p) with (A) daily surface air temperature, (B) daily tropospheric temperature (at 850 hPa), and (C) daily dewpoint temperature for stations in the tropical wet zone in India. *Orange* line indicates fitted lines estimated using LOWESS. The figure is adapted from Ali and Mishra (2017). Please refer to Ali and Mishra (2017) for the data and methods.

The other reasons for deviations in scaling rate (negative scaling) may be due to local cooling artefacts, drier surface conditions at higher temperatures, moisture limitations at higher temperatures, temperature seasonality, statistical methods and inappropriate modelling assumptions, and mixing of different rainfall types (as discussed in Ali et al. (2018; 2021)). Moreover, large-scale circulation patterns, local moisture availability through upward motions and moisture convergence, and local-scale dynamics can all influence scaling rates (Ali and Mishra, 2018a; Guerreiro et al., 2018; Pfahl et al., 2017).

11.3.3.3 Observed scaling of sub-daily rainfall extremes

The scaling relation between extreme daily rainfall and the day-to-day variability in temperature approximately follows the CC rate across the globe (Ali et al., 2018; Fowler et al., 2021b). However, sub-daily rainfall extremes have been shown to scale more strongly with temperature than daily rainfall extremes: the scaling rates are observed to be around 2CC scaling in the Netherlands (Lenderink et al., 2011) and even 3CC scaling in Australia (Guerriero et al., 2018). However, global scaling studies based on observed sub-daily rainfall are very limited because of the aforementioned data limitations. A recent global study (Ali et al. 2021) for the first time used a large observational hourly dataset from the GSDR and showed consistent CC scaling at a regional scale and around 2CC scaling at the local scale for sub-daily rainfall extremes. Based on their results, around 60% (of 7088) stations show scaling between CC and 2CC. It is important to note that other dynamical processes, which include changes to large-scale circulation, storm type, cloud size and the spatial extent of storms, and changes to long-term moisture transport patterns have not been considered in their scaling approach, which can change their scaling results (Pfahl et al., 2017). Therefore, further studies are needed to understand the mechanisms (dynamics) of sub-daily rainfall change with temperature.

11.4 Projected changes to rainfall extremes

11.4.1 Changes in rainfall extremes from climate models

Understanding the future response of rainfall extremes to global warming is crucial within the climate change impact debate. Regional climate models (RCMs) and relatively coarser global climate models (GCMs) have shown future increases in the frequency and intensity of daily and multiday rainfall extremes over most land regions under global warming (Blenkinsop et al., 2021; Schwingshackl et al., 2021; Coppola et al., 2021a; b). For instance, Coppola et al. (2021a) used ensembles of GCMs (CMIP6) and RCMs (EURO-CORDEX) and showed a consistent increase in extreme rainfall over the Northern and Central European regions. Similarly, Ali et al. (2019) used an ensemble of GCMs from CMIP5 to show an increase in the frequency of multiday rainfall events over the Indian subcontinent by the end of the 21st century. They showed that the changes in the frequency of 3-day rainfall events exceeding the 20-year return level are projected to increase by around 50% nd 70% respectively from the historic period (1966-2005) under the high emission scenarios (RCP2.6 and RCP8.5 respectively), over all the selected basins (Fig. 11.6). However, GCMs have uncertainties associated with them, leading to the underestimation of extreme convective rainfall and failing to capture processes over complex orography (Ali and Mishra, 2018b) and RCMs are only available for selected regions making it impossible to conduct global studies (Ali et al, 2014; Giorgi et al., 2019). With the advancement in climate modelling and various initiatives, a new framework was launched recently to provide a homogenous ensemble of high resolution (0.25 degree) projections over the globe: CORDEX-CORE (Gutowski et al., 2016). Coppola et al. (2021b) used ensembles of these RCMs to show a future increase in wet extremes over the La Plata basin in South America, the Congo basin in Africa, northeast Europe, eastern North America, and India.

RCMs, and even GCMs, due to their coarser resolution, do not accurately capture extreme rainfall intensities as they fail to reproduce small-scale clouds and other processes that drive extreme rainfall, especially convective storms (Zhang et al., 2017). The issue in GCMs (especially CMIP5 and CMIP6 used in most of the recent studies) in representing the storm dynamics responsible for extreme precipitation cannot be fixed at their current spatial resolution (Wehner et al., 2019). Moreover, for sub-daily rainfall extremes, GCMs fail to produce extreme rainfall statistics, particularly in warm convective seasons, and model performance decreases with coarser temporal resolution (Fowler et al., 2021b). Nevertheless, changes (increases) at sub-daily timescales are projected to be stronger than at the daily scale, especially for the Tropics (Ali and Mishra, 2018b; Morrison et al., 2019). Therefore, the focus has shifted to very high-resolution climate models with improved representations of extreme rainfall intensity termed convection-permitting models (CPMs), where the deep convection parameterisation scheme is switched off and the models directly simulate deep convection (Fowler et al., 2021b; Kendon et al., 2014, 2017, 2020).

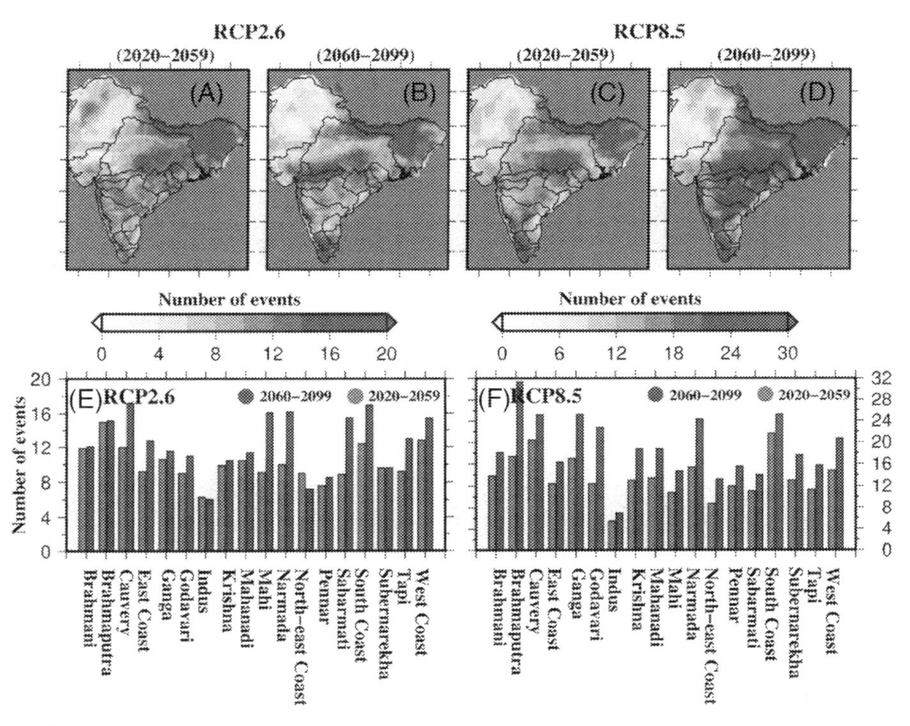

FIG. 11.6

Change in the frequency of 3-day extreme rainfall events in (A) 2020–2059 and (B) 2060–2099, exceeding the 20-year return level threshold based on the historic period (1966-2005) using an ensemble of five GCMs for the RCP2.6 scenario, (C) and (D) similar to (A) and (B), respectively, but for the RCP8.5 scenario, (E) and (F) are median change in the frequency of 3-day extreme rainfall for 2020–2059 (*red*) and 2060–2099 (*blue*) periods for each basin, for RCP2.6 and RCP8.5 respectively. This figure is taken directly from Ali et al. (2019).

11.4.2 Convection-permitting models

CPMs simulate storm processes, including cloud convection and local feedbacks, and produce better representations of sub-daily extreme rainfall, including the diurnal cycle and duration characteristics at a resolution of less than 4 km (Fowler et al., 2021b; Prein et al., 2017). They efficiently reproduce small-scale (1–10 km^2) clouds and other processes controlling rainfall, even over mountains (Blenkinsop et al., 2021; Kendon et al., 2017). Studies based on CPM multimodel projections show more intense storms with warming over the USA (Prein et al., 2017), Europe (Chan et al., 2020), the UK (Kendon et al., 2020), and Africa (Kendon et al., 2019). CPMs also show increased intensity of future daily rainfall extremes in winter over orography and due to increases in embedded convection in winter storms, which does not occur in coarser-resolution models (Kendon et al. 2021). CPMs produce rainfall that is too intense but with more realistic representation of the rate of change of intensity with return period using EVT. However, there is some difficulty in quantifying future uncertainties since there are limited CPM runs to date and these simulations involve significant amounts of computational time and costs due to their high level of detail (Blenkinsop et al., 2021; Fowler et al., 2021b). Despite these limitations, CPMs are a significant advance in climate modelling and their simulations should be expanded to the global scale to merge dynamical changes together with convective process changes – providing more robust future projections of changes to sub-daily (and daily) rainfall extremes (Kendon et al., 2017, 2019).

11.5 Conclusions

This chapter reviews previous work on changes in observed rainfall extremes at different durations and their projected future changes from climate model outputs. Trend studies based on the ETCCDI indices have shown increases in the frequency and intensity of observed rainfall extremes and the changes are greater for short-duration extremes than at daily or longer durations (based on scaling studies). Moreover, studies based on EVT theory suggest an increase in precipitation maxima for both stationary and nonstationary atmospheric conditions. A strong relationship has been observed between rainfall extremes and temperature (apparent scaling) suggesting more sensitivity (scaling rates) of sub-daily rainfall extremes to warming than daily rainfall extremes. These scaling rates become more consistent (at around the CC rate) when robust methods and scaling variables are considered in estimating scaling.

Despite their importance, global sub-daily (including sub-hourly) studies are still very limited because of the lack of observational records due to the poor density of rain gauge network and data sharing across the majority of the globe (Morbidelli et al., 2022). With advances in climate modelling, different regional and global climate models (RCMs and GCMs, respectively) have been used to study the changes to daily rainfall extremes in the future. Notwithstanding their limitations in producing extreme rainfall at a finer resolution (especially convective rainfall), they suggest an increase in frequency and intensity of extreme daily rainfall in the future for the majority of the globe. To overcome these limitations in GCMs and RCMs, convection-permitting climate models have been recently developed which accurately capture extreme rainfall intensities at finer spatial and temporal resolutions (≤4 km, 1h). The limited regional studies so far based on CPMs suggest more intense rainfall extremes with future warming, highlighting the importance of running CPMs on a global scale.

This chapter highlights the importance of more global studies on rainfall extremes (especially at sub-daily timescales) to understand both observed and future changes in extreme rainfall and the mechanisms, which govern these changes. This has strong implications for the urban flooding and the design of urban stormwater infrastructure in the context of climate change impacts and adaptation.

References

AghaKouchak, A., Ragno, E., Love, C., & Moftakhari, H. (2018). Projected changes in California's precipitation intensity-duration-frequency curves. Energy Commision of Satet of California: Sacramento, CA, USA.

Alexander, L.V., Fowler, H.J., Bador, M., Behrangi, A., Donat, M.G., Dunn, R., Venugopal, V., 2019. On the use of indices to study extreme precipitation on sub-daily and daily timescales. Environ. Res. Lett. 14 (12), 125008.

Alexander, L.V., Bador, M., Roca, R., Contractor, S., Donat, M.G., Nguyen, P.L., 2020. Intercomparison of annual precipitation indices and extremes over global land areas from in situ, space-based and reanalysis products. Environ. Res. Lett. 15 (5), 055002.

Ali, H., Mishra, V., 2017. Contrasting response of rainfall extremes to increase in surface air and dewpoint temperatures at urban locations in India. Sci. Rep. 7 (1), 1–15.

Ali, H., Mishra, V., Pai, D.S., 2014. Observed and projected urban extreme rainfall events in India. J. Geophys. Res. Atmos. 119 (22), 12–621.

Ali, Haider, Mishra, Vimal, 2018a. Contributions of dynamic and thermodynamic scaling in subdaily precipitation extremes in India. Geophys. Res. Lett. 45 (5), 2352–2361.

Ali, H., Mishra, V., 2018b. Increase in subdaily precipitation extremes in india under 1.5 and 2.0 C warming worlds. Geophys. Res. Lett. 45 (14), 6972–6982.

Ali, H., Fowler, H. J., Lenderink, G., Lewis, E., Pritchard, D. (2021). Consistent large-scale response of hourly extreme precipitation to temperature variation over land. Geophys. Res. Lett., 48(4), e2020GL090317.

Ali, H., Modi, P., Mishra, V., 2019. Increased flood risk in Indian sub-continent under the warming climate. Weather Clim. Extremes 25, 100212.

Ali, Haider, Fowler, Hayley J., Mishra, Vimal, 2018. Global observational evidence of strong linkage between dew point temperature and precipitation extremes. Geophys. Res. Lett. 45 (22), 12–320.

Arnbjerg-Nielsen, K., Willems, P., Olsson, J., Beecham, S., Pathirana, A., Bülow Gregersen, I., Nguyen, V.T.V., 2013. Impacts of climate change on rainfall extremes and urban drainage systems: a review. Water Sci. Technol. 68 (1), 16–28.

Arnone, E., Pumo, D., Viola, F., Noto, L.V., Loggia, G.L., 2013. Rainfall statistics changes in Sicily. Hydrol. Earth Syst. Sci. 17 (7), 2449–2458.

Bader, D.A., Blake, R., Grimm, A., Hamdi, R., Kim, Y., Horton, R., Rosenzweig, C., 2018. Urban climate science. In Rosenzweig, C., Solecki, W., Romero-Lankao, P., Mehrotra, S., Dhakal, S., and Ali Ibrahim, S. (eds.),Climate Change and Cities: Second Assessment Report of the Urban Climate Change Research Network. Cambridge University Press, New York, 27–60.

Barbero, R., Fowler, H.J., Lenderink, G., Blenkinsop, S., 2017. Is the intensification of precipitation extremes with global warming better detected at hourly than daily resolutions? Geophys. Res. Lett. 44 (2), 974–983.

Blenkinsop, S., Muniz Alves, L., Smith, A.J.P., 2021. Climate change increases extreme rainfall and the chance of floods. In: Le Quéré, C., Liss, P., Forster, P. (Eds.), Critical Issues in Climate Change Science. https://doi.org/10.5281/zenodo.4779119.

Blenkinsop, S., Fowler, H.J., Barbero, R., Chan, S.C., Guerreiro, S.B., Kendon, E., & Tye, M.R. (2018). The INTENSE project: using observations and models to understand the past, present and future of sub-daily rainfall extremes. Advances in Science and Research, 15, 117–126.

Chan, S.C., Kendon, E.J., Berthou, S., Fosser, G., Lewis, E., Fowler, H.J., 2020. Europe-wide precipitation projections at convection permitting scale with the Unified Model. Clim. Dyn. 55 (3), 409–428.

Cheng, L., AghaKouchak, A., Gilleland, E., Katz, R.W., 2014. Non-stationary extreme value analysis in a changing climate. Clim. Change 127 (2), 353–369. 10.1007/s10584-014-1254-5.

Coles, S., Bawa, J., Trenner, L., Dorazio, P., 2001An Introduction to Statistical Modeling of Extreme Values208. Springer, London, p. 208.

Coppola, E., Nogherotto, R., Ciarlo', J.M., Giorgi, F., van Meijgaard, E., Kadygrov, N., Wulfmeyer, V, 2021. Assessment of the European climate projections as simulated by the large EURO-CORDEX regional and global climate model ensemble. J. Geophys. Res. Atmos. 126 (4), e2019JD032356.

Coppola, E., Raffaele, F., Giorgi, F., Giuliani, G., Xuejie, G., Ciarlo, J.M., Rechid, D., 2021. Climate hazard indices projections based on CORDEX-CORE, CMIP5 and CMIP6 ensemble. Clim. Dyn., 1–91.

De Haan, L., Ferreira, A., Ferreira, A., 2006Extreme Value Theory: An Introduction21. Springer, New York.

Donat, M.G., Alexander, L.V., Herold, N., Dittus, A.J., 2016a. Temperature and precipitation extremes in century-long gridded observations, reanalyses, and atmospheric model simulations. J. Geophys. Res. Atmos. 121 (19), 11–174.

Donat, M.G., Lowry, A.L., Alexander, L.V., O'Gorman, P.A., Maher, N., 2016b. More extreme precipitation in the world's dry and wet regions. Nat. Clim. Change 6 (5), 508–513.

Dong, Q., Wang, W., Kunkel, K.E., Shao, Q., Xing, W., & Wei, J. (2021). Heterogeneous response of global precipitation concentration to global warming. International Journal of Climatology, 41, E2347-E2359.

Dunn, R.J.H, 2019. HadISD version 3: monthly updates Technical Note HCTN_103_2019P.

Fadhel, S., Rico-Ramirez, M. A., & Han, D. (2018). Sensitivity of peak flow to the change of rainfall temporal pattern due to warmer climate. Journal of Hydrology, 560, 546–559.

Fischer, E.M., Knutti, R., 2016. Observed heavy precipitation increase confirms theory and early models. Nat. Clim. Change 6 (11), 986–991.

Fowler, H.J., Lenderink, G., Prein, P., Westra, S., Allan, R.P., Ban, N., Barbero, R., Berg, P., Blenkinsop, S., Do, H.X., Guerreiro, S., Haerter, J.O., Kendon, E., Lewis, E., Schaer, C., Sharma, A., Villarini, G., Wasko, C., Zhang, X., 2021a. Anthropogenic intensification of short-duration rainfall extremes. Nat. Rev. Earth Environ. doi:10.1038/s43017-020-00128-6.

Fowler, HJ., Ali, H., Allan, R.P, Ban, N., Barbero, R., Berg, P., Blenkinsop, S., Cabi, N.S., Chan, S., Dale, M., Dunn, R.J.H., Ekström, M., Evans, J.P., Fosser, G., Golding, B., Guerreiro, S.B., Hegerl, G.C., Kahraman, A., Kendon, E.J., Lenderink, G., Lewis, E., Li, X.-F., O'Gorman, P.A., Orr, H.G., Peat, K.L., Prein, A.F., Pritchard, D., Schär, C., Sharma, A., Stott, P.A., Villalobos-Herrera, R., Villarini, G., Wasko, C., Wehner, M.F., Westra, S., Whitford, A., 2021b. Towards advancing scientific knowledge of climate change impacts on short-duration rainfall extremes. Phil. Trans. Roy. Soc. A., 20190542. doi:10.1098/rsta.2019.0542.

Gao, Xichao, et al., 2018. Temperature dependence of hourly, daily, and event-based precipitation extremes over China. Sci. Rep. 8 (1), 1–10.

Gao, X., Guo, M., Yang, Z., Zhu, Q., Xu, Z., Gao, K., 2020. Temperature dependence of extreme precipitation over Mainland China. J. Hydrol., 124595.

Giorgi, F., Raffaele, F., & Coppola, E. (2019). The response of precipitation characteristics to global warming from climate projections. Earth System Dynamics, 10(1), 73–89.

Guerreiro, S.B., Fowler, H.J., Barbero, R., Westra, S., Lenderink, G., Blenkinsop, S., & Li, X.F. (2018). Detection of continental-scale intensification of hourly rainfall extremes. Nature Climate Change, 8(9), 803–807.

Gutowski, J.W., Giorgi, F., Timbal, B., Frigon, A., Jacob, D., Kang, H.S., Tangang, F., 2016. WCRP coordinated regional downscaling experiment (CORDEX): a diagnostic MIP for CMIP6. Geosci. Model Dev. 9 (11), 4087–4095.

Intergovernmental Panel on Climate Change, 2012. Managing the risks of extreme events and disasters to advance climate change adaptation. In: Field, C.B., Barros, V., Stocker, T.F., Qin, D., Dokken, D.J., Ebi, K.L. et al., (Eds.), A Special Report of Working Groups I and II of the Intergovernmental Panel on Climate Change. Cambridge University Press, Cambridge, England/New York, NY.

Jones, Hardwick, Westra, Rhys Seth, Sharma, Ashish, 2010. Observed relationships between extreme sub-daily precipitation, surface temperature, and relative humidity. Geophys. Res. Lett. 37 (22).

Jun, C., Qin, X., Tung, Y.K., De Michele, C., 2020. On the statistical analysis of rainstorm events between historical (1777–1907) and modern (1961–2010) periods in Seoul, Korea. Int. J. Climatol. 40 (4), 2078–2090.

Katz, R.W., 2013. Statistical methods for nonstationary extremes. Extremes in a Changing Climate. Springer, Dordrecht, Netherlands, pp. 15–37. https://doi.org/10.1007/978-94-007-4479-0_2.

Kendon, E.J., Roberts, N.M., Fowler, H.J., Roberts, M.J., Chan, S.C., Senior, C.A., 2014. Heavier summer downpours with climate change revealed by weather forecast resolution model. Nat. Clim. Change 4 (7), 570–576.

Kendon, E.J., Ban, N., Roberts, N.M., Fowler, H.J., Roberts, M.J., Chan, S.C., Wilkinson, J.M., 2017. Do convection-permitting regional climate models improve projections of future precipitation change? Bull. Am. Meteorol. Soc. 98 (1), 79–93.

Kendon, E.J., Stratton, R.A., Tucker, S., Marsham, J.H., Berthou, S., Rowell, D.P., Senior, C.A., 2019. Enhanced future changes in wet and dry extremes over Africa at convection-permitting scale. Nat. Commun. 10 (1), 1–14.

Kendon, E.J., Roberts, N.M., Fosser, G., Martin, G.M., Lock, A.P., Murphy, J.M., Tucker, S.O., 2020. Greater future UK winter precipitation increase in new convection-permitting scenarios. J. Clim. 33 (17), 7303–7318.

Kendon, E.J., Prein, A.F., Senior, C.A., Stirling, A., 2021. Challenges and outlook for convection-permitting climate modelling. Philos. Trans. R. Soc. A 379 (2195), 20190547.

Kharin, V.V., Zwiers, F.W., Zhang, X., & Hegerl, G.C. (2007). Changes in temperature and precipitation extremes in the IPCC ensemble of global coupled model simulations. Journal of Climate, 20(8), 1419–1444.

Krishnamurthy, C.K.B., Lall, U., Kwon, H.H., 2009. Changing frequency and intensity of rainfall extremes over India from 1951 to 2003. J. Clim. 22 (18), 4737–4746.

Koutsoyiannis, D., 2012. Clausius–Clapeyron equation and saturation vapour pressure: simple theory reconciled with practice. Eur. J. Phys. 33 (2), 295.

Kunkel, K.E., Easterling, D.R., Redmond, K., & Hubbard, K. (2003). Temporal variations of extreme precipitation events in the United States. Geophysical research letters, 30(17): 1895–2000.

Lenderink, G., van Meijgaard, E., 2008. Increase in hourly precipitation extremes beyond expectations from temperature changes. Nat. Geosci. 1, 511–514.

Lenderink, G., Mok, H.Y., Lee, T.C., Van Oldenborgh, G.J., 2011. Scaling and trends of hourly precipitation extremes in two different climate zones-Hong Kong and the Netherlands. Hydrol. Earth Syst. Sci., 15(9), 3033–3041.

Leonard, M., Westra, S., Phatak, A., Lambert, M., van den Hurk, B., McInnes, K., et al., 2014. A compound event framework for understanding extreme impacts. Wiley Interdiscip. Rev. Clim. Change 5 (1), 113–128. doi10.1002/wcc.252.

Lewis, E., Fowler, H., Alexander, L., Dunn, R., McClean, F., Barbero, R., Blenkinsop, S., 2019a. GSDR: a global sub-daily rainfall dataset. J. Clim. 32 (15), 4715–4729.

Lewis, E., Guerreiro, S., Blenkinsop, S., Fowler, H.J., 2019b. Quality control of a global sub-daily precipitation dataset and derived extreme precipitation indices. Geophys. Res. Abst. 21.

Marani, M., Ignaccolo, M., 2015. A metastatistical approach to rainfall extremes. Adv. Water Resour. 79, 121–126.

Marra, F., Nikolopoulos, E.I., Anagnostou, E.N., Morin, E., 2018. Metastatistical extreme value analysis of hourly rainfall from short records: estimation of high quantiles and impact of measurement errors. Adv. Water Resour. 117, 27–39.

McPhillips, L.E., Chang, H., Chester, M.V., Depietri, Y., Friedman, E., Grimm, N.B., & Shafiei Shiva, J. (2018). Defining extreme events: A cross-disciplinary review. Earth's Future, 6(3), 441–455.

Min, S.K., Zhang, X., Zwiers, F.W., Hegerl, G.C., 2011. Human contribution to more-intense precipitation extremes. Nature 470 (7334), 378–381.

Morbidelli, R., Corradini, C., Saltalippi, C., Flammini, A., 2022. Time Resolution of Rain Gauge Data and its Hydrological Role. In: Morbidelli, R. (Ed.), Rainfall. Modeling, Measurement and Applications. Elsevier, Amsterdam, pp. 171–216. doi:10.1016/C2019-0-04937-0.

Morrison, A., Villarini, G., Zhang, W., Scoccimarro, E., 2019. Projected changes in extreme precipitation at sub-daily and daily time scales. Glob. Planet. Change 182, 103004.

Moustakis, Y., Papalexiou, S.M., Onof, C.J., Paschalis, A., 2021. Seasonality, intensity, and duration of rainfall extremes change in a warmer climate. Earths Future 9 (3).

Nkrumah, F., Vischel, T., Panthou, G., Klutse, N.A.B., Adukpo, D.C., Diedhiou, A., 2019. Recent trends in the daily rainfall regime in southern West Africa. Atmosphere 10 (12), 741.

O'Gorman, Paul A, 2015. Precipitation extremes under climate change. Curr. Clim. Change Rep. 1 (2), 49–59.

Papalexiou, S.M., & Koutsoyiannis, D. (2013). Battle of extreme value distributions: A global survey on extreme daily rainfall. Water Resources Research, 49(1), 187–201.

Park, J.S., Kang, H.S., Lee, Y.S., Kim, M.K., 2011. Changes in the extreme daily rainfall in South Korea. Int. J. Climatol. 31 (15), 2290–2299.

Paschalis, A., Fatichi, S., Katul, G.G., Ivanov, V.Y., 2015. Cross-scale impact of climate temporal variability on ecosystem water and carbon fluxes. J. Geophys. Res. Biogeosci. 120 (9), 1716–1740.

Pendergrass, A.G. (2018). What precipitation is extreme? Science, 360(6393), 1072–1073.

Peterson, T.C., Manton, M.J., 2008. Monitoring changes in climate extremes: a tale of international collaboration. Bull. Am. Meteorol. Soc. 89 (9), 1266–1271.

Pfahl, S., O'Gorman, P.A., Fischer, E.M., 2017. Understanding the regional pattern of projected future changes in extreme precipitation. Nat. Clim. Change 7 (6), 423–427.

Prein, A.F., Rasmussen, R.M., Ikeda, K., Liu, C., Clark, M.P., Holland, G.J., 2017. The future intensification of hourly precipitation extremes. Nat. Clim. Change 7 (1), 48–52.

Roxy, M.K., Ghosh, S., Pathak, A., Athulya, R., Mujumdar, M., Murtugudde, R., Rajeevan, M., 2017. A threefold rise in widespread extreme rain events over central India. Nat. Commun. 8 (1), 1–11.

Roy, S.Sen, Jr, Balling, R., C, 2004. Trends in extreme daily precipitation indices in India. Int. J. Climatol. 24 (4), 457–466.

Salvadori, G., De Michele, C., Durante, F., 2011. On the return period and design in a multivariate framework. Hydrol. Earth Syst. Sci. 15 (11), 3293–3305 10.5194/hess-15-3293-2011.

Schwingshackl, C., Sillmann, J., Vicedo-Cabrera, A.M., Sandstad, M., Aunan, K., 2021. Heat stress indicators in cmip6: estimating future trends and exceedances of impact-relevant thresholds. Earths Future 9 (3), e2020EF001885.

Shahid, S., 2011. Trends in extreme rainfall events of Bangladesh. Theor. Appl. Climatol. 104 (3), 489–499.

Sillmann, J., Kharin, V.V., Zhang, X., Zwiers, F.W., Bronaugh, D., 2013. Climate extremes indices in the CMIP5 multimodel ensemble: part 1. Model evaluation in the present climate. J. Geophys. Res. Atmos. 118 (4), 1716–1733.

Smith, A., Lott, N., Vose, R., 2011. The integrated surface database: Recent developments and partnerships. Bull. Am. Meteorol. Soc. 92 (6), 704–708.

Sun, Q., Zhang, X., Zwiers, F., Westra, S., Alexander, L.V., 2021. A global, continental, and regional analysis of changes in extreme precipitation. J. Clim. 34 (1), 243–258.

Syafrina, A.H., Zalina, M.D., Juneng, L., 2015. Historical trend of hourly extreme rainfall in Peninsular Malaysia. Theor. Appl. Climatol. 120 (1), 259–285.

Trenberth, Kevin E., et al., 2003. The changing character of precipitation. Bull. Am. Meteorol. Soc. 84 (9), 1205–1218.

Utsumi, Nobuyuki, et al., 2011. Does higher surface temperature intensify extreme precipitation? Geophys. Res. Lett. 38 (16).

Wang, Y., Zhou, L., 2005. Observed trends in extreme precipitation events in China during 1961–2001 and the associated changes in large-scale circulation. Geophys Res Lett 32:L09707.

Wasko, C., Parinussa, R.M., Sharma, A., 2016. A quasi-global assessment of changes in remotely sensed rainfall extremes with temperature. Geophys. Res. Lett. 43 (24), 12–659.

Wehner, M., Lee, J., Risser, M., Ullrich, P., Gleckler, P., Collins, W.D., 2019. Evaluation of extreme sub-daily precipitation in high-resolution global climate model simulations. Philos. Trans. R. Soc. A 379 (2195), 20190545.

Westra, S., Alexander, L.V., Zwiers, F.W., 2013. Global increasing trends in annual maximum daily precipitation. J. Clim. 26 (11), 3904–3918.

Xiao, C., Wu, P., Zhang, L., Song, L., 2016. Robust increase in extreme summer rainfall intensity during the past four decades observed in China. Sci. Rep. 6 (1), 1–9.

Zhang, X., Alexander, L., Hegerl, G.C., Jones, P., Tank, A.K., Peterson, T.C., Zwiers, F.W., 2011. Indices for monitoring changes in extremes based on daily temperature and precipitation data. Wiley Interdiscip. Rev. Clim. Change 2 (6), 851–870.

Zhang, X., Zwiers, F.W., Li, G., Wan, H., Cannon, A.J., 2017. Complexity in estimating past and future extreme short-duration rainfall. Nat. Geosci. 10 (4), 255–259.

CHAPTER

12 Rainfall regionalization techniques

Pierluigi Claps, Daniele Ganora, Paola Mazzoglio

Department of Environment, Land and Infrastructure Engineering, Politecnico di Torino, Torino, Italy

12.1 Introduction

The need for reliable design rainfall in whatever point in an area is becoming more and more relevant to support design and management of stormwater drainage, combined sewer overflows and flood control systems, particularly under the threat of climate change. When a long and complete extreme rainfall time series is available in a site of interest, the intensity – duration – frequency (IDF) relationships can be reliably built, applying usual frequency analysis techniques, to produce estimations of a design rainfall for a given return period (Koutsoyiannis and Iliopoulou, 2022). The most commonly used data are the annual maximum rainfall depths recorded in time intervals between 1 and 24 h. In ungauged locations, the estimation of rainfall quantiles requires the use of techniques able to transfer the information from other gauged locations. This goal is generally intended as the main objective of regionalization methods.

In today's observation networks, then, rain gauges are unevenly distributed and often characterized by short and fragmented records (Kidd et al., 2017; Libertino et al., 2018). This is also due to the response of various monitoring agencies to the above-mentioned technical needs, which often results in the installation of new rain gauges. New stations produce increased information in the spatial dimension, but the new records will inevitably be too short as compared to the other historical ones. These circumstances require careful reassessment of consolidated techniques of rainfall regionalization to profitably exploit short and fragmented records.

The "traditional" regionalization methods are based on the concept of hydrologic similarity (Hosking and Wallis, 1997) which allows to obtain estimates of rainfall statistic in ungauged sites by pooling information collected at gauged sites that are "similar" to each other. Regionalization is based on a "space-for-time substitution" approach: time (i.e. the "missing" time series of the ungauged sites) is filled with "spatially close" data (i.e., the time series of nearby gauged sites) and this can happen by grouping stations in appropriate homogeneous regions (Blöschl, 2011; Grimaldi et al., 2011). The same paradigm, however, is also successfully applied to increase the information content in gauged locations, e.g. when the number of local observations is particularly low with respect to the return period of interest (D. Faulkner, 1999; Naghettini and Pinto, 2017).

RAINFALL. Modeling, Measurement and Applications. DOI: https://doi.org/10.1016/B978-0-12-822544-8.00013-5

In "traditional," as well as in recent, regionalization methods, the most widely adopted operational procedure is the index-flood approach, originally introduced by Dalrymple (1960), in which the quantile h_T of the rainfall depth related to the return period T comes as the product of an index variable h_i with a dimensionless frequency quantile K_T, also known as the "growth factor":

$$h_T = h_i \times K_T \tag{12.1}$$

h_i is intended as a position parameter and is generally taken as the mean or the median of the distribution: h_i has, therefore, a very local relevance. This means that with a barely sufficient length of the time series at the site of interest (say 5–10 years), the index value can be directly computed on observations (Svensson and Jones, 2010 and references therein).

The rainfall depth h_T must be associated to a duration d, i.e., the time interval over which annual maxima are computed and recorded. Eq. 12.1 can therefore be written as:

$$h_{d,T} = h_{i,d} \cdot K_d(T) \tag{12.2}$$

where d ranges between a few minutes and a few days. Usually, the duration is considered varying between 1 and 24 hours, which is the interval where most systematic data is available. Considering the 1–24 hours interval, the index rainfall $h_{i,d}$ is then represented by a function (called the average depth-duration curve), almost everywhere represented with the power law:

$$h_{i,d} = ad^n \tag{12.3}$$

where a is the scale factor and n is a scaling exponent. The coefficient a represents the best unbiased linear estimation of the 1-h average rainfall extreme, given that $h_{i,1} \cong a$.

A practical approach in dealing with the estimation of quantiles for different durations is to consider the growth curves $K_d(T)$ as extracted from the same probability distribution for all the durations. Burlando and Rosso (1996) refer to this circumstance as the "simple scaling" paradigm. Exceptions and limitations of this hypotheses (Marani, 2003) will not be considered in this chapter mainly because, as highlighted by Svensson and Jones (2010), this hypothesis is largely predominant in regional analyses adopted in various countries.

The main consequence of the adoption of the "simple scaling" is that a single growth curve can be defined for the whole homogeneous region, whatever the duration of interest. This leads to a very compact form of the family of Depth – Duration - Frequency (DDF) curves which represents the main tool used to derive a design rainfall (Grimaldi et al., 2011).

Application of the index-flood methodology within a traditional regional frequency analysis (RFA) is then based on three fundamental steps: (1) the data reliability check, (2) the definition of the homogeneous region and of its representative regional growth curve $K(T)$, and (3) the evaluation of the index-rainfall curve $h_{i,d} = f(d)$ for the ungauged site of interest. The latter step, if no direct measure is locally available to estimate h_i, can be pursued through various interpolation approaches, as discussed later in this chapter.

The analysis of the three steps (1) to (3) will form the structure of the chapter. Reference to the data availability and preparation represents a relative novelty in the presentation of rainfall frequency analyses. This is justified by the nature of the most recent approaches, that tend to answer to the increasing importance of correctly attributing return periods to new record-breaking rainstorms, that are sometimes recorded in stations with short time series. Inclusion of isolated record-breaking values in regional analyses can also help to answer questions about the significance of the trends in amount and frequency of extreme rainfall, that is one of the possible outcomes of global warming.

12.2 Variables to be regionalized, data preparation, and data scarcity

12.2.1 Regionalized variables

In a RFA, the selection of the variable to be regionalized requires some discussion, as it depends on the availability of raw data versus pre-processed data, and can have an impact on the dataset preparation. If a complete sequence of observations at the gauging maximum resolution (e.g., 1–2 minutes for electronic devices, or a continuous line for analogic devices) is available, it is possible to compute the "complete duration series" (CDS) by applying a moving average of a given width (d) to the whole record; the CDS will then include all the measured rainfall depth in a year aggregated at the duration d, and will allow the widest possible range of analysis. From the complete sequence one can indeed extract the peak values but also evaluate other occurrence measures, like the inter-event waiting time or the seasonality of events. If the time series contains only those events that exceed a fixed threshold, for whatever duration, it is named partial duration series or peaks-over-threshold (POT) series. The most popular selection of data for the analysis of rainfall extremes is, however, the block-maxima selection, that rely on series that contains only the highest values that occur in a fixed period; if the period is one year, the sample represents the annual maximum series (AMS).

As initially suggested by Gumbel (1954), AMS is the data set most commonly used in probabilistic analyses. POT is sometimes preferred when the objective is to increase the time series length, especially while dealing with heavy-tailed distributions (Madsen et al., 1997). However, it must be considered that the POT method introduces discretional terms in the choice of the threshold value, as clarified by Claps and Laio (2003). CDS advantages are described by Marani and Zanetti (2015) and can be taken into account when critical high-return period estimation is the objective. However, when using CDS, more evidence is still required on the selection of the minimum set of necessary data, given a target return period (Requena et al., 2019).

To apply analyses on long records, however, one must consider that for older records only AMS may be available, and this issue may limit the operational options.

Whatever the available observations are, the definition of the variables to use for the derivation of the regional dimensionless frequency curve $K(T)$ is not a trivial task, similarly to what happens in the case of flood peaks. The main alternatives in literature are to proceed towards regionalization of rainfall quantiles (Svensson and Jones, 2010) for one duration within the simple scaling assumption (Soltani et al., 2017) or for different durations under the multiscaling approach (i.e., when $K_d(T)$ changes with the duration d, Burlando and Rosso, 1996). However, many authors assume that it is preferable to regionalize the probability distribution parameters (Svensson and Jones, 2010 and reference therein) or the sample ordinary moments, or the L-moments (Modarres and Sarhadi, 2012; Ngongondo et al., 2012; Smithers and Schulze, 2001). In all cases, a thorough understanding of the dependence of the variance of the regionalized parameters on the record length is necessary, especially for the estimation of high-return period quantiles.

Our evaluation is that the average practitioner would not easily distinguish the differences in the outcomes when different variables have been chosen for the regionalization. Differences can emerge in "critical applications," where the aim is the estimation of high-return period design rainfall, or in the evaluation of rainfall spatial variability in regions with few data and high hydro-climatic variability (e.g., mountainous regions). In both cases, however, marked differences can arise from the intervention of individual (significant) records, even if included in short time series (Libertino et al., 2018)

12.2.2 Data preparation and data scarcity

As mentioned by CSAGroup (2019), practitioners are now more and more concerned with data and methods that can emphasize the evolution of rainfall hazard due to climate change. For this reason, the management of large databases of rainfall measurements requires to consider and manage the presence of gaps in the time series, the use of recent but short records, and the possible role of rain gauges relocation.

In the literature of rainfall extremes two main approaches are applied to deal with the data fragmentation issue. The first one is draconian and is based on the definition of a minimum acceptable threshold of record length: in this approach, to achieve the data homogeneity required to proceed with the RFA, only the time series longer than a given threshold will be considered in the analysis. Although this approach can be considered as precautionary, one can end up discarding important information included in some short records, and this can affect the reliability of the RFA (Ouali et al., 2016; Libertino et al., 2018).

The second approach tries to preserve the available information in relation to its influence on each parameter to be estimated (Laio et al., 2011). In essence, even if short time series cannot provide great influence on higher-order moments, their data can be confidently used to improve the estimation of the first order parameters.

Considering that this second approach may lead to errors if based on non-robust assumptions (Teegavarapu and Nayak, 2017) significant research efforts are being recently addressed on data augmentation techniques, using also ancillary information. A brief review is reported below.

Most of the methodologies proposed to retain information from short records are based on interpolation or data reconstruction. For instance, Pappas et al. (2014) use time series autocorrelation to deal with sporadic time gaps. However, low autocorrelation or frequent and systematic missing data can make the method not applicable. Ouali et al. (2016) propose a conditional quantile regression model that, even if not address a data reconstruction, can be ascribed to the "data augmentation" category. The "patched kriging" (PK) approach, proposed by Libertino et al. (2018), reconstructs missing values using contemporary observations at nearby sites through a sequential application of the ordinary kriging. Homogeneous annual maxima series at each location can also be obtained with a bootstrap procedure, that accounts for all the measurements in nearby stations (Uboldi et al. 2014). In this case, data relevance decreases as the distance between the rain gauge and the estimation point increases. However, the results of this application end up being highly sensitive to the presence of outliers in the observed series, and can produce quantiles that can deviate substantially from those computed from local samples even of consistent length.

Data augmentation in rainfall frequency analysis from ancillary data are the last frontier, both for applications to areas affected by data scarcity, and for the increase in the spatial detail of the statistical analyses. In the first case, some studies investigate the possibility of evaluating statistics using hydrometeorology attributes rather than inadequately observed precipitation measurements (Satyanarayana and Srinivas, 2008). This approach, preliminarily tested in India over the summer monsoon affected regions, suggests using both large-scale atmospheric variables that affect the precipitation in the study region, and location attributes as additional features in the regionalization. Over large areas, satellite data can certainly provide additional insights in RFA (Qamar et al., 2017) since global-scale rainfall datasets like those acquired by TRMM and GPM constellations have homogeneous and reasonably long time series. However, additional work is required to take enough advantage of this information, considering its low resolution in space and some inherent inaccuracies (Zorzetto and Marani, 2020).

Data scarcity is, however, a relative concept, as it can depend on the required details of the spatial rainfall analysis. Urban hydrological applications, for instance, tend to be data-hungry for the need to reproducing rainfall estimates at high resolution. In this context, Weather Radar data can provide interesting support to the regional frequency analyses, as witnessed by recent literature (Goudenhoofdt et al., 2017; Ochoa-Rodriguez et al., 2019; Kašpar et al., 2021). As seen from the above applications, new and interesting developments in rainfall frequency analysis can therefore come from the inclusion of additional information from other sources, providing advances already seen with the use of data assimilation in rainfall nowcasting applications (Li et al., 2018).

12.3 Regional methods

12.3.1 Fixed region and region-of-influence methods

In regional statistical methods a homogeneous region is a domain where some statistical properties of the variable under study can be considered constant. For instance, in the index-flood framework, regions are used to define a unique growth curve $K(T)$ valid for every site of interest falling into the region. Construction of the actual regional growth curve derives from the information common to a pooled sample. The sample can be built using all the nondimensional (i.e., normalized by the index-value) series observed in stations falling within the homogeneous region. When the curve is directly built using all nondimensional data together, the method is called "station-year." This way to define a unique $K(T)$ curve is quite practical but doesn't take explicitly into account the different uncertainties deriving from the different lengths of the individual series. More accredited approaches are based on the computation of the regional parameters of the $K(T)$ curve, using weighted averages to account for the different record lengths. When a predetermined probability distribution is selected (e.g., the TCEV, Rossi and Villani, 1994) different pooled samples (or regions) for different order moments can be devised, according to a hierarchical scheme. More in general, regional L-moments (Hosking and Wallis, 1997) of different order, all related to the same pooled sample, or related to different pooled samples according to the order, can be regionalized (Alila, 1999). In this way, different probability distributions representing the growth function can be built, if necessary, in different areas.

The procedure for delineating homogeneous regions has been a field where creativity has been applied to the highest extent, both for rainfall and for flood variables (Reed et al., 1999) and has evolved in time. A very first approach is based on the delineation of fixed and non-overlapping geographical regions (see Fig. 12.1 as a reference), obtained through the split of the entire study area in separate areas that can be considered similar from morphological or climatological points of view (Cole, 1966). To minimize the adoption of subjective delineation criteria, as highlighted by Acreman and Wiltshire (1989), some authors have suggested the development of delineation procedures that still create fixed regions but in non-geographic spaces. Examples are regions in the space of some descriptors (generally geomorphological ones, like latitude/longitude, elevation, distance from the sea, aspect).

The "delineation" of the regions can be assisted by various cluster analysis techniques, able to suggest a preliminary configuration of the regions. Cluster analysis is a class of methods designed to classify groups with similar behaviors within large collections of data (Modarres and Sarhadi, 2012; Naghettini and Pinto, 2017; Ngongondo et al., 2012; Satyanarayana and Srinivas, 2008). These methods organize measurements (or stations parameters) into groups (clusters) that present a high level of internal similarity, while showing evident dissimilarity with all the other groups. The entire dataset is split into several fixed and non-overlapping clusters, based on combination of values of climatic, morphological or geographic characteristics (Fig. 12.1). Once these regions are delineated, further adjustments may be required to ensure the homogeneity of all the regions, to make the whole process non-ambiguous (Ganora and Laio,

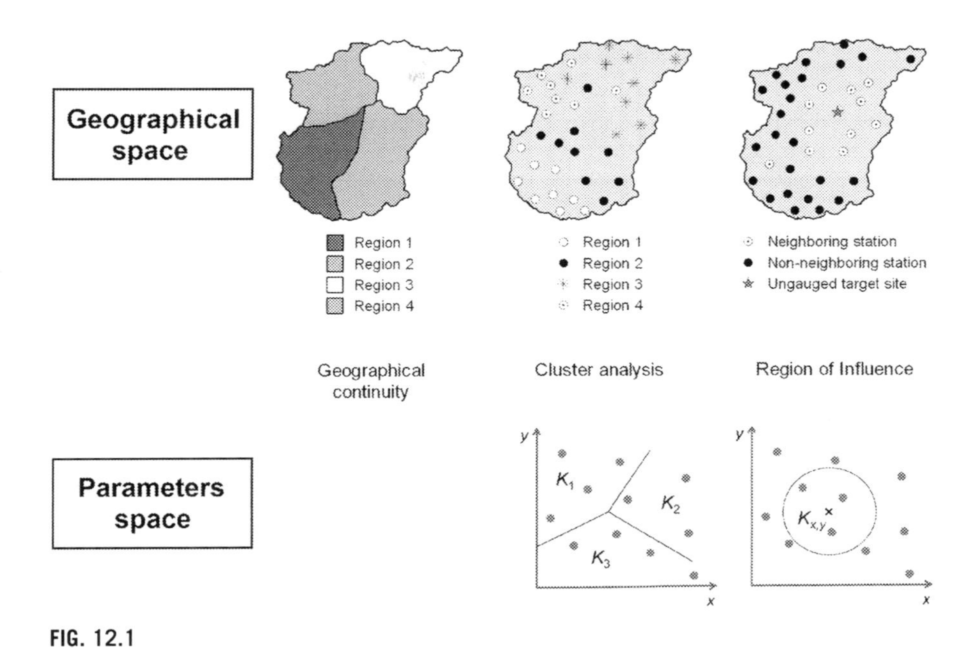

FIG. 12.1

Differences between geographical contiguity (fixed and non-overlapping regions), cluster analysis, and region of influence.

2016). Conceptual and operational difficulties can also emerge with geographically delineated regions, where, as in cluster analyses, inconsistencies can emerge when the point (area) of interest falls in proximity of the boundary of two adjacent regions. In these cases, small variations in the configuration of the regions (due to, e.g., subjective placement of the edge or effects of different delineation algorithms) would cause the point (area) to potentially fall in different regions (with potentially very different regional parameters), hampering the robustness of the estimates.

A major response to this significant drawback has been given by the region of influence (ROI) approach, in which the regions are no longer fixed (in whatever space). In ROI, a new and individual "region" is defined for each ungauged site (Burn, 1990) so that the desired amount of information is acquired by the "surrounding" stations, granting a sort of continuity among estimates while moving the point of interest in the space (Fig. 12.1). This technique has gained a high level of refinement over the years, and represents the base of different national-scale applications, as reported in Svensson and Jones (2010). Not particularly simple is the way in which the ROI techniques deal with the areal rainfall estimations. This task, despite being very frequent in the practitioner's work, ends up to be very difficult to manage as a mere extension of the quantile extraction in a given location.

In all the above-mentioned techniques the regionalized curve can be "donated" to ungauged sites only if the statistical homogeneity within the data-pooling group is respected. Statistical homogeneity (or heterogeneity) is commonly evaluated with specific tests (Hosking and Wallis, 1997), in which discordant data, outliers and heterogeneities can be identified by comparing the sample L-moment ratios among different stations. These ratios are the dimensionless L-variance, L-skewness and L-kurtosis (Hosking and Wallis, 1997). Discordancy measures identify samples with statistical characteristics too different from those of the whole group (Hosking and Wallis, 1997; Naghettini and Pinto, 2017). Application of these tests is done continuously during the region delineation procedure: large regions produce larger samples, but generally less homogeneous; small regions have fewer stations, and may not reach a sufficient number of data (see the UK FORGEX method, reviewed in Svensson and Jones, 2010). The main problem in this interactive delineation/test process is that it can converge with different and equally likely solutions.

To limit subjectivity, the use of ancillary climatological information has been again proposed. For instance, Schaefer (1990), Alila (1999) and Gaál et al. (2008) identify climatologically homogeneous subregions on the basis of the mean annual precipitation. Pooling criteria based on similar climatological conditions have demonstrated to be successful in supporting delineation, and possibly underlie that the extremes are produced by similar heavy rainfall generating mechanisms (Persiano et al., 2020).

12.3.2 Regionless methods

Starting from the name of the methodology, the "RFA" requires a definition of "regions," in which pooled data achieve a "critical mass" required to support statistical estimation ungauged sites. Looking to the objective closer, the amount of necessary data reduces consistently if the return period T is small, or if only regions for lower order moments are considered (Rossi and Villani, 1994). If the index rainfall estimation is of interest, its spatial variability is so high that the definition of homogeneous regions loses significance. For this reason, many regional studies propose that the index rainfall in ungauged locations is estimated through interpolation methods (Trefry et al., 2006; Faulkner and Prudhomme, 1998). The development of automated GIS techniques has subsequently led to an increased, up to becoming dominant, interest in interpolation techniques, involving also higher order parameters. This has motivated the definition of "regionless methods," that include regression methods, interpolation methods and geostatistical methods.

Regression methods mainly involve the assessment of the spatial variability of low-order moments (Prudhomme and Reed, 1998, 1999) as dependent on geomorphological features. Parameters in ungauged sites are related to geographical/climatological covariates (Prudhomme and Reed, 1998, 1999; Van De Vyver, 2012). Reasonably good results can be obtained by linear models: an application performed over South of France (Carreau et al., 2013) showed that non-parametric non-linear interpolation of GEV parameters does not outperform simple linear interpolations in the investigated areas. Complex topography can play a significant role in the rainfall spatial variability: to account for this influence, regression methods need to consider morphological covariates these external variables to minimize prediction errors (Beguería and Vicente-Serrano, 2006). In the cited works, however, a geographic-based explanation of possible model inadequacies is lacking.

The quite popular category of interpolation methods includes several techniques available from long time in the literature, ranging from Thiessen polygons to inverse distance weighting (IDW), to geostatistical models. The simplest interpolation approach is based on Thiessen polygons: the record of the closest rain gauge is substantially assigned to the ungauged location. The method is based on the drawing of a polygon of influence around every rain gauge, with boundaries at a distance halfway between each pair of rain gauges (Goovaerts, 2000). In practice, in each polygon the parameters values are assumed equal to these of its central station. Even if this methodology is mainly used to estimate areal rainfall, it has been applied to several cases, as in the South of France (Creutin and Obled, 1982), in the Central Continental United States (Tabios and Salas, 1985) and in the South of Portugal (Goovaerts, 2000).

With the IDW technique, an estimate in an ungauged site is obtained using a weighted average of the measurements collected in the nearby locations (Tabios and Salas, 1985; Goovaerts, 2000). The method assumes that the influence of each actual measurement on the prediction value in a site of interest decreases quadratically with the respective distance. Like in the Thiessen interpolation, the IDW considers distance as the only relevant predictor criterion. Fig. 12.2 shows an example of spatial interpolation, performed with IDW, of 24-h average annual maximum rainfall depths. The example is related to the Piemonte Region, described later. The limits of this technique emerge with the "bull's eye effect," an artifact that overweighs observed values around the measurement points and that is more evident as the network density decrease. Bull's eye effect decreases significantly using a geostatistical approach like kriging.

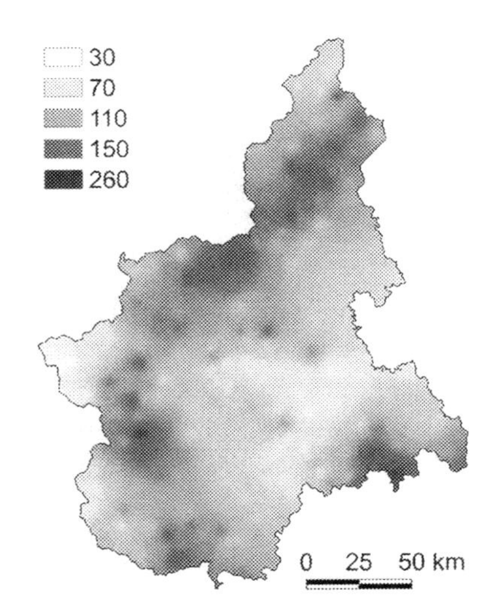

FIG. 12.2

IDW interpolation of 24 h annual maximum rainfall depths measured over Piemonte in 2010.

12.3.3 Geostatistical techniques

Kriging is an interpolation technique widely used in hydrology, originally developed by Krige and by Matheron (Matheron, 1963) and more generally referred to as geostatistics. This technique takes into account the spatial correlation between neighboring observation to predict values in unsampled locations (Goovaerts, 1999). In the geostatistical techniques, the spatial dependence of rainfall data, that is seen as a problem in RFA (because indicates regional inhomogeneity), is indeed an advantage. In fact, if the target is a homogeneous region, intersite dependence acts like reducing the number of independent sites available in the region. And is understandable that the use of very similar observations coming from two correlated stations merely duplicate the information, reducing the value of adding stations to a group (Svensson and Jones, 2010). On the other hand, the use of methods that properly consider spatial correlation, like kriging, helps to obtain estimations with a correct degree of confidence as a function of the distance from the actual sources of information.

In kriging, the spatial variability of the observations is represented by a variogram, defined as the half of the spatial variance of the variable Z in two points with a spacing h.

$$\gamma(h) = \frac{1}{2} E\left\{\left[Z(x+h) - Z(x)\right]^2\right\} \tag{12.4}$$

where x is the location and E is the expected value (Blöschl, 2011). The kriging technique is often defined as a best linear unbiased estimator (BLUE): the linearity condition implies that the unknown value $\hat{z}_0$ in position x_0 can be evaluated as a weighted average of the nearby measurements:

$$\hat{z}(x_0) = \sum_{i=1}^{n} \lambda_i z(x_i) \tag{12.5}$$

where λ_i is the interpolation weight in position x_i and n is the number of measurements included in the interpolation.

In the ordinary kriging (OK), measurements over a specific region are used for the evaluation of a sample variogram, to which a theoretical variogram is fitted (Fig. 12.3). This approach relies on the assumption that the mean value is constant, but unknown. However, the complexity of the rainfall field, as a result of orographic effects, can result in heterogeneity and anisotropy, which may not be adequately represented in ordinary kriging (Prudhomme and Reed, 1999). If a strong trend can be recognized in the data, the global mean may not be representative of the local value: therefore, the ordinary kriging should not be applied over the entire domain.

Several approaches can allow to take into account the presence of a significant trend: for instance, the trend can be removed to allow a safer application of the ordinary (or simple) kriging on the detrended set of observations (Prudhomme and Reed, 1999). This methodology is generally called detrended kriging or regression kriging (RK). RK can be applied if the correlation coefficient of the trend is high (Ahmed and De Marsily, 1987). Moreover, variables must have a significant number of common measurement locations to allow the building of a regression model (Ahmed and

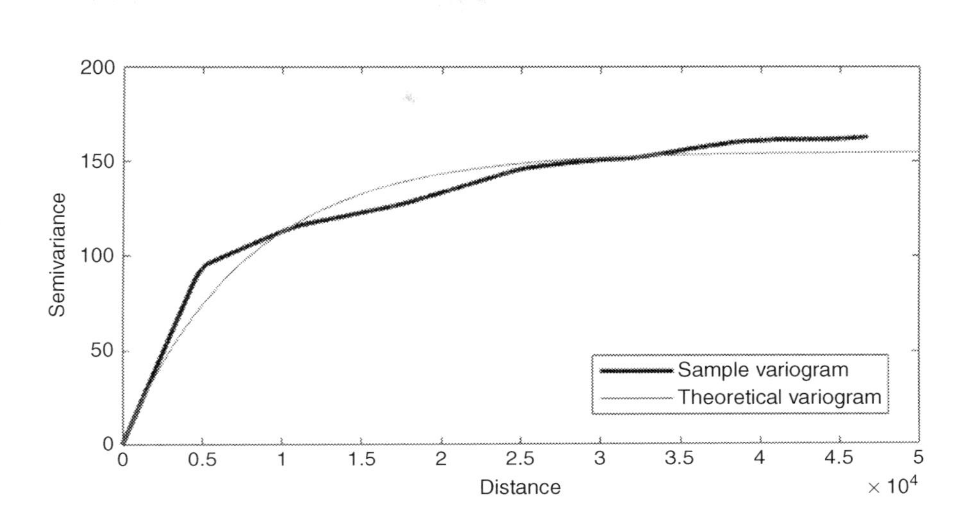

FIG. 12.3

Sample and theoretical variograms. An exponential expression is used for the theoretical curve.

De Marsily, 1987). Consequently, RK might perform poorly if the data sample is small and non-representative, if the relation between the target variable and predictors is non-linear, and if the points do not represent the whole domain or represent only the central part of it (Hengl, 2007).

A rather different answer to the presence of trend is to model the spatial dependence by restricting the analyses to smaller moving windows around the location where prediction is required, so that an ordinary kriging can be applied. The moving window should be small enough to consider the sub-areas as stationary but, in the same way, large enough to include a sufficient number of sampled points (Grimes and Pardo-Igúzquiza, 2010; Fouedjio, 2017).

The kriging with external drift (KED) is another model, based on a geostatistical regression, that takes into account the correlation between an observed (dependent) variable and an additional information (independent variable), often called "drift" or "external drift" (Varentsov et al., 2020). In the rainfall interpolation field, the elevation is often considered as the drift. Within the KED, the drift is not likely to be constant in any direction (Hudson and Wackernagel, 1994) and is modelled together with the variable (i.e., it is the opposite situation compared to RK or residual kriging). In other words, while with RK the deterministic (regression) and stochastic (kriging) predictions are done separately, in KED both components are predicted simultaneously (Hengl, 2007). The second variable should be sampled at a large number of points in space and it should give a good image of the underlying structure of the first (low sampled) variable, with which it is assumed to be strongly correlated (Ahmed and De Marsily, 1987). If the correlation between the target and predictors is significant, the KED or RK prediction errors will be lower than the ordinary kriging error (Hengl et al., 2003).

Cokriging is another geostatistical technique, developed to improve the estimation of a variable using the information on another spatially correlated variable with a better (spatial and temporal) availability (Ahmed and De Marsily, 1987). It is an extension of kriging when multivariate data and a multivariate variogram (or a covariance model) are available (Wackernagel, 1998). The dataset may not cover all variables at all sample locations: the measurements available for different variables in a specific domain may be located either in the same or in different sample points for each variable (Ahmed and De Marsily, 1987). Depending on how the measurements of the different variables are scattered in space it is possible to distinguish between isotopic and heterotopic datasets (Wackernagel, 1998). In entirely heterotopic data, the variables are measured at different points (i.e., they have no sample locations in common). In a partial heterotopic data situation, variables share some sample location. In this case, both the cross variogram and the covariance function model can be inferred using the isotopic subset of the data (Wackernagel, 1998). An isotopic condition is the one characterized by data available for each variable at all sampling points. For a stable cross-variogram, a strong correlation between the primary and secondary variables is required (Prudhomme and Reed, 1999; Grimes and Pardo-Igúzquiza, 2010). For data with spatially correlated residuals, cokriging is a better estimation technique when the correlation coefficient is high (>0.7) (Ahmed and De Marsily, 1987). In this case, cokriging should be preferred to other techniques. If the correlation coefficient is low, cokriging provides results similar to KED or RK. When residuals of the estimation are spatially uncorrelated, cokriging generally introduces less bias but does not give the best estimate (Ahmed and De Marsily, 1987). According to Ahmed and De Marsily (1987), kriging combined with linear regression (RK) is a better estimator for all values of the correlation coefficient, but with a slightly larger bias than KED (it comes second), and cokriging (third). If the number of auxiliary variables is low and they are not available at all grid nodes, cokriging should be used to improve the prediction accuracy (Hengl et al., 2003). If auxiliary information is available at all grid-nodes and they are correlated with the target variable, KED should be used (Hudson and Wackernagel, 1994; Hengl et al., 2003).

Usually, kriging techniques are applied to interpolate distribution parameters or precipitation quantiles (Das, 2019). In this regard, it is worth mentioning that the presence of statistical noise in the observed field (due to measurement or sampling errors) can affect the optimal nature of the ordinary kriging. When spatial interpolation of the field is the only desired result, ordinary kriging often represents a quick and reasonably accurate predictor. However, when the 'observations' are statistical values, e.g. parameters of a distribution, the noise in the data can affect the results and the variance of predictor residuals is of interest. To provide an answer to this need, the kriging for uncertain data (KUD) approach has been proposed (De Marsily, 1986; Deidda et al., 2021; Furcolo et al., 1995; Todini, 2001). In KUD, as local data is assumed to be known with uncertainty, interpolation around all observation points are less sensitive to the local sample values than it happens in the ordinary kriging. As a consequence, the less the variance in local samples (due to large record lengths), the more the sample values become confirmed by the kriging application.

Among the few alternatives to the above paradigms in the framework of the RFA, we can refer to the PK methodology (Libertino et al., 2018). This approach is based on the sequential application, year by year, of the ordinary kriging equation to all the values recorded in the region of interest in each year. This technique has been already cited with reference to the data augmentation issue.

However, a bias correction undertaken in the cited paper make this method a complete RFA approach, that has been adopted at the scale of Piemonte region (Claps et al., 2015), becoming quite operational. In this method, to make the newly assembled dataset useful for design rainfall estimation in ungauged sites, the kriging variance in each year, as a measure of the estimation uncertainty, is used to weight the contribution of each rainfall observation in the estimation of the sample L-moments of the reconstructed series. This can somewhat preserve the spatial correlation of higher-order L-moments.

This method is among the very few RFA approaches that aim to include all the available measurements in the analysis, regardless of possible station relocation. In Fig. 12.4, that summarizes the PK workflow, one can notice how the influence of the different accuracies of data propagates in the reconstructed rainfall data cube.

Geostatistical techniques generally outperform other interpolations methods (Tabios and Salas, 1985; Goovaerts, 2000). For instance, comparison between IDW interpolation and geostatistical techniques highlighted that kriging is superior in the estimation (Das, 2019). The improvements in the prediction become extremely relevant in case of gauging networks with low or variable space density. This circumstance can be quite important, because the monitoring agencies end up adapting the spatial density of stations in relation to the known spatial variability of the phenomena.

Considering a multivariate approach, sparsely observations of the primary attribute can be completed by secondary attributes that are more densely sampled. For instance, a secondary variable can be a digital elevation model or weather-radar data. In mountainous regions, often characterized by uneven station density and poor rainfall data representativity, interpolation methods, even sophisticated as kriging, can still suffer from the lack of data (Prudhomme and Reed, 1999). However, a comparison between three multivariate geostatistical algorithms (simple kriging with varying local means, KED, and cokriging) that incorporate a digital elevation model into the rainfall prediction, showed smaller prediction errors than univariate techniques based on ordinary kriging or inverse square distance or Thiessen polygons (Goovaerts, 2000).

All the above considered, better prediction performances can reasonably derive from considering the relations between rainfall and the elevation (i.e., the orographic effects on precipitation). Indeed, some rainfall frequency analyses explicitly include adjustment factors for specific topographic conditions, even in the homogeneous regions paradigm (Svensson and Jones, 2010). For example, in Canada (Svensson and Jones, 2010), adjustment factors are applied in mountainous areas, depending on the type of relief (onshore coastal or noncoastal mountains); in Germany, the gradient and the orientation of the slope of the terrain are taken into account during the spatial interpolation proposed in the KOSTRA model (Bartels et al., 1997; Svensson

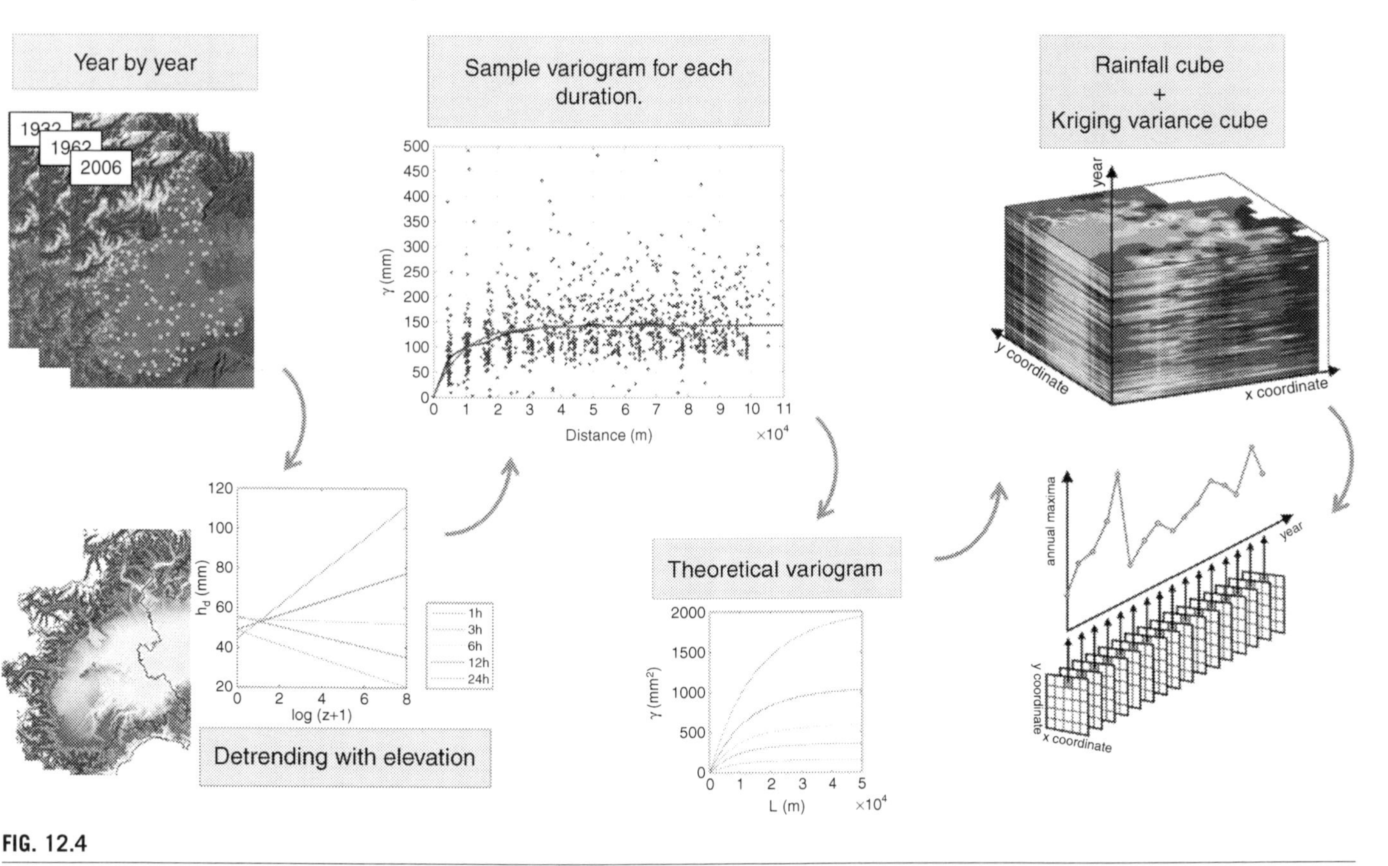

FIG. 12.4

Patched kriging workflow (Libertino et al., 2018). Last step demonstrates the sampling, in a given location, of a reconstructed time series.

and Jones, 2010); similar parameters are accounted for in the United States, where the PRISM method (Daly et al., 2002) allows continuous time reconstruction of missing data. In conclusion, where orographic effects play a major role, geostatistical approaches of the "multivariate" type appear to be well organized to apply selectively in the space the relations found over a large-scale analysis (Prudhomme and Reed, 1999).

12.4 Methods adopted in practice

While this chapter intends to resume and compare the performances of available regionalization techniques, it is equally important to look at their impact, which is well assessed when they become national methods. Svensson and Jones (2010) have thoroughly described the rainfall frequency estimation methods adopted in Canada, Sweden, France, Germany, United States, South Africa, New Zealand, Australia, and United Kingdom. In many nations, more established methods are still preferred to new approaches.

In France, for instance, maps of rainfall estimates are provided for duration from 6 min to 10 days for return period up to 100 years, but no regionalization technique is applied. In Sweden, different techniques are adopted, ranging from at-site estimates to the definition of homogeneous regions. In United States and in South Africa, growth curves for homogeneous fixed regions are evaluated using a cluster analysis based on topographic and climatological parameters. A different situation applies to Canada: while Environment Canada "Rainfall Frequency Atlas for Canada" is the only official source of information regarding IDF curves on a national scale, several institutions have performed similar research on smaller areas (CSAGroup, 2019).

The ROI approach is used in New Zealand, UK and Australia. In Germany, a spatial interpolation method is suggested, with an explicit dependence on seasonality (i.e., the methods applied during summer, winter, or during the entire year are different).

The national method of Italy, not mentioned in Svensson and Jones (2010), has been so far based on the methodology suggested in the Valutazione delle Piene (VAPI) project, applied in a series of publications referred to different macroregions. The VAPI methodology (Rossi and Villani, 1994) is based on a hierarchical approach in the selection of homogeneous regions, where higher-order parameters are considered uniform in areas larger than the homogeneous regions of mid and low-order parameters. The VAPI is one of the few RFA based on a four-parameters distribution, the two component extreme value (TCEV) distribution. Parameter combinations allow to have three levels of regions that are fixed and non-overlapping (Rossi and Villani, 1994). Being comparted in regional reports and lacking computer tools for supporting the application, actual application of VAPI is still a challenging task for practitioners.

The provision to the end-users of rainfall quantiles for different durations and return periods is sometimes undertaken: some countries have limited the technical objective to return periods in the order of 100 years (e.g., Canada; CSAGroup, 2019), other countries have decided to extend the analyses to a 10,000 years return period (e.g., Sweden, France, and United Kingdom; Svensson and Jones, 2010).

12.4.1 National atlases

In the past decades, several national agencies dealt with the production of rainfall or hydrological atlases. The contents of these collections of information varies from place to place, depending on the purposes of each atlas. In the United States, the NOAA Atlas 14 contains precipitation estimates for different durations and selected return periods. In Canada, DDF parameters are available in a map format for durations between 5 and 24 h, while estimates for longer durations are available upon request (Svensson and Jones, 2010). The Hydrological Atlas of Switzerland (https://hydrologicalatlas.ch/), accessible through a WebGIS application, contains hydrological information like the mean precipitation depths, maps of extreme precipitation for a 1-day duration and different return periods, and annual, seasonal and monthly rainfall totals for individual years. The Hydrological Atlas of Austria (Fürst et al., 2009) contains an even more comprehensive collection of information linked to the hydrological cycle (precipitation, evapotranspiration, snow and glacier, river and lakes, groundwater, mass balance, water management, and water and environment). In Germany, the KOSTRA-DWD (coordinated heavy precipitation regionalization and evaluation of the DWD) initiative provides maps of extreme rainfall depths for different durations and return periods.

12.5 Considerations on applicability and evolution of the regional frequency analyses

As seen for Canada, but frequent in other areas, different regionalization method can be found in the same areas, posing problems to non-expert users that need to choose among different regionalization techniques and different probability distributions (Svensson and Jones, 2010). If needed, model performances can be evaluated using model selection criteria, like the Akaike information criterion, the Bayesian information criterion, and the Anderson-Darling criterion. However, several studies have pointed out that the identification of the "best model" is a difficult task (Di Baldassarre et al., 2009; Velázquez et al., 2011). Conversely, by averaging the results obtained with different models, instead of using a model selection criterion, one can address the consistency in rainfall estimates and can reduce the uncertainties in the estimation of quantiles for large return period.

If different methods relate to different amount of data, before accepting the changes a thorough assessment of differences is often required by the practitioners. In this context, a case study can classify the issue. The case study is related to the Piedmont region, in Italy, a region of about 25,000 km^2 whose geographical position is displayed in Fig. 12.5. Historical rainfall extremes are available since 1913 and

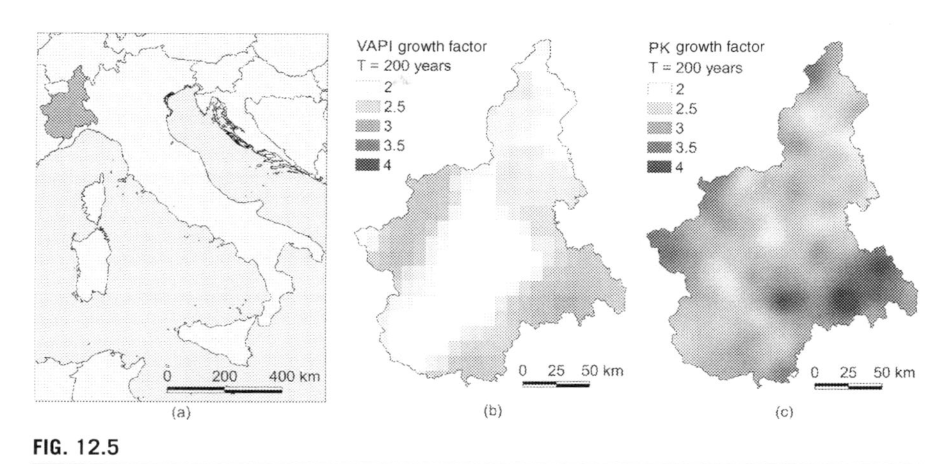

FIG. 12.5

Area of interest (a) VAPI growth factor related to the TCEV distribution, (b) patched kriging growth factor related to a GEV distribution (c).

various regional frequency analyses have been conducted on this area over the years. We investigate here the results of the application of two regionless methods. The first one is an evolution of the VAPI procedure based on the adoption of the TCEV probability distribution but following a regionless approach using the kriging for uncertain data (Furcolo et al., 1995). The second one is based on the PK (Libertino et al., 2018). In the second method, a GEV distribution is adopted. A comparison between the results of the two methods would allow to recognize the effects of two changes: one is the increase in data availability, as the second method has been applied on series with up to 30 more years of observation; the second change is in the method, from which we expect less significant variations. The comparison has been made between the respective growth factors $K(T)$ computed in the region related to a 200-years return period. Fig. 12.5b and c represent the spatial variability of the mentioned growth factor and confirm the higher data content of the PK analysis, also in terms of the resolution of the maps (PK pixels size are 250×250 meters while VAPI maps are characterized by a 9×11.5 km-size cells).

The absolute differences ΔK found between the two factors are represented in Fig. 12.6a.

Fig. 12.6b shows the relative difference $\Delta K/K$, evaluated as the ratio between ΔK and the PK growth factor. The higher (positive) differences are visible in the southern part of the region, close to Liguria region and to the sea. The increase in variability and of the $K(T)$ non-dimensional quantiles displayed by the PK procedure is however consistent with the occurrence of recent severe events in those areas.

The example presented here suggests the need of undertaking a transition between the results of an outdated analysis and those of an updated one. In many cases, old and recent results of RFA coexist in the application environment, and a practitioner should be prepared to have both in his toolbox.

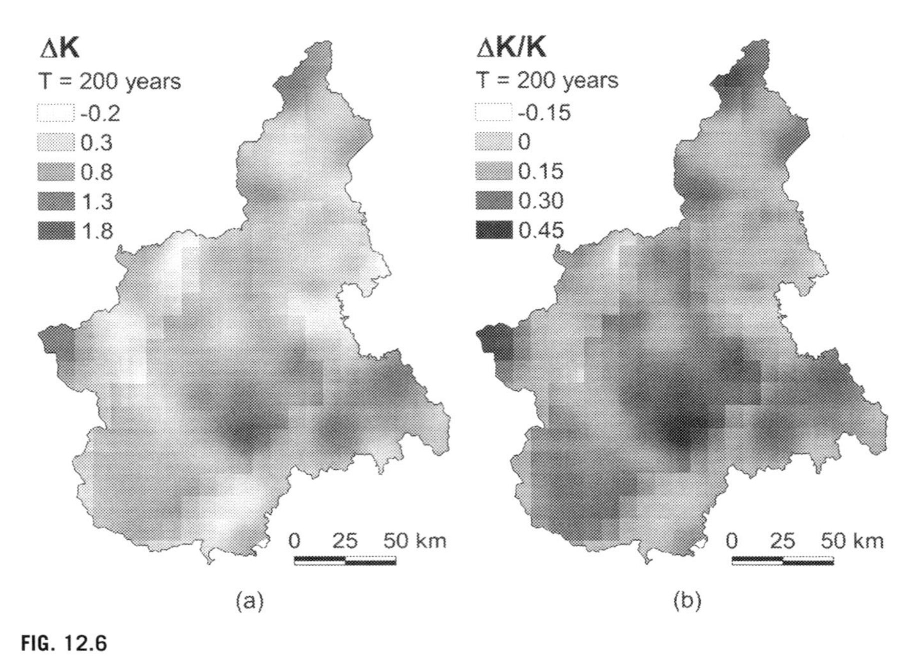

FIG. 12.6

Absolute (a) and relative (b) differences between the growth factors evaluated with the VAPI and patched kriging methods.

The possibility of having simple GIS-based tools for the frequency analysis could indeed be important to facilitate their applicability by practitioners, who also sometimes still manage data in different formats, both digital and paper, and must address areal estimates for a whole river basin. A recent example is the MultiRain tool that has been developed for the estimation of the design rainfall over Piemonte region (NW Italy) on the basis of a multimodel approach (Grasso et al., 2020).

In this tool, a GIS-based procedure allows to obtain depth-duration-frequency estimates, the growth factor and the design rainfall at a point and over an area, using different regional frequency analyses. A similar tool (Schardong et al., 2020), named IDF_CC tool, has been developed to facilitate the use of historical data for the development of IDF curves in Canada. This tool considers historical observed sub-daily data, historical daily data obtained with climate model experiments, and future model representative concentration pathway scenarios at daily resolution, to extract yearly maxima that can be fitted with a generalized extreme value distribution.

In both cases, that are not the only one available, the goal is to transfer to the end user a standardized and computer-assisted procedure, with the aim of reducing the errors in the actual application of the rainfall frequency analyses.

12.6 Conclusions

This chapter has analyzed the state of the art of rainfall RFA, developed to estimate design rainfall in ungauged sites and over basin areas. To pursue this important technical goal, rain gauges data have been considered as grouped data and are analyzed in various ways. While for many years the debate has been addressed to the definition of homogeneous regions, this approach has evolved toward the consideration of regionless methods.

As suggested also by Svensson and Jones (2010) and by Blöschl (2011), there is not a "best" way to perform an RFA: every approach has its pros and cons, and different methods have been adopted nationwide. As of today, the majority of the national RFA approaches are still based on the delineation of homogeneous regions, while the most advanced geostatistical techniques, even those widely documented in the literature, are still not used.

However, while regionless methods become closer to the application field, the directions of most promising developments in this field are those which take more advantage of the strong increase of remote-sensing data. As the requirements of design rainfall estimates compatible with climate change issues and urban-scale flooding increase, the rain gauge – radar assimilation is likely to become the next frontier of rainfall frequency analysis.

References

Acreman, M., Wiltshire, S., 1989. The regions are dead; long live the regions. Methods of identifying and dispersing with regions for flood frequency analysis. In: Roald, K., Nordseth, K., Hassel, K.A. (Eds.). FRIENDS in Hydrology, 187. IAHS Publ, Wallingford, Oxfordshire, UK, pp. 175–188.

Ahmed, S., De Marsily, G., 1987. Comparison of geostatistical methods for estimating transmissivity using data on transmissivity and specific capacity. Water Resour. Res. 23 (9), 1717–1737. https://doi.org/10.1029/WR023i009p01717.

Alila, Y., 1999. A hierarchical approach for the regionalization of precipitation annual maxima in Canada. J. Geophys. Res. Atmos. 104 (D24), 31645–31655. https://doi.org/10.1029/1999JD900764.

Bartels, H., G. Malitz, S. Asmus, F.M. Albrecht, B. Dietzer, T. Gunther, H. Ertel. 1997. Starkniederschlagshohen Fur Deutschland (KOSTRA-Atlas).

Beguería, S., Vicente-Serrano, S.M., 2006. Mapping the hazard of extreme rainfall by peaks over threshold extreme value analysis and spatial regression techniques. J. Appl. Meteorol. Climatol. 45 (1), 108–124. https://doi.org/10.1175/JAM2324.1.

Blöschl, G., 2011. Scaling and regionalization in hydrology. In: Wilderer, P. (Ed.), Treatise on Water Science. Elsevier, Oxford, UK, pp. 519–535. https://doi.org/10.1016/B978-0-444-53199-5.00113-5.

Burlando, P., Rosso, R., 1996. Scaling and multiscaling models of depth-duration-frequency curves for storm precipitation. J. Hydrol. 187 (1–2), 45–64. https://doi.org/10.1016/S0022-1694(96)03086-7.

Burn, D.H., 1990. Evaluation of regional flood frequency analysis with a region of influence approach. Water Resour. Res. 26 (10), 2257–2265. https://doi.org/10.1029/WR026i010p02257.

Carreau, J., Neppel, L., Arnaud, P., Cantet, P., Philippe, C., 2013. Extreme rainfall analysis at ungauged sites in the south of France: comparison of three approaches. Journal de La SFdS 154 (2), 119–138.

Claps, P., Laio, F., 2003. Can continuous streamflow data support flood frequency analysis? An alternative to the partial duration series approach. Water Resour. Res. 39 (8), 1–12. https://doi.org/10.1029/2002WR001868.

Claps, P., F. Laio, P. Allamano, A. Libertino, M. Iavarone. 2015. Attività Di Ricerca Nell'ambito Del Progetto Strada 2.0 Modulo CAPPIO (CAPitalizzazione Azione Di Caratterizzazione Delle PIOgge Estreme). http://www.idrologia.polito.it/~claps/Papers/RELAZIONE_prog_STRADA_Piemonte.pdf.

Cole, G., 1966. An application of the regional analysis of flood flowsRiver Flood Hydrology, 39–57. https://doi.org/10.1680/rfh.44944.0003.

Creutin, J., Obled, C., 1982. Objective analyses and mapping techniques for rainfall fields: an objective comparison. Water Resour. Res. 18 (2), 413–431. https://doi.org/10.1029/WR018i002p00413.

CSAGroup. 2019. Technical guide. Development, interpretation, and use of rainfall intensity-duration-frequency (IDF) information: guideline for Canadian water resources practitioners.

Dalrymple, T., 1960. Flood-frequency analyses. Manual of hydrology: part 3. Flood-flow techniques. Usgpo 1543-A 80 http://pubs.usgs.gov/wsp/1543a/report.pdf.

Daly, C., Gibson, W.P., Taylor, G.H., Johnson, G.L., Pasteris, P., 2002. A knowledge-based approach to the statistical mapping of climate. Clim. Res. 22 (2), 99–113. https://doi.org/10.3354/cr022099.

Das, S., 2019. Extreme rainfall estimation at ungauged sites: comparison between region-of-influence approach of regional analysis and spatial interpolation technique. Int. J. Climatol. 39 (1), 407–423. https://doi.org/10.1002/joc.5819.

Di Baldassarre, G., Laio, F., Montanari, A., 2009. Design flood estimation using model selection criteria. Phys. Chem. Earth 34 (10–12), 606–612. https://doi.org/10.1016/j.pce.2008.10.066.

De Marsily, G., 1986. Quantitative Hydrogeology: Groundwater Hydrology for Engineers. Academic Press, Orlando.

Deidda, R., Hellies, M., Langousis, A., 2021. A critical analysis of the shortcomings in spatial frequency analysis of rainfall extremes based on homogeneous regions and a comparison with a hierarchical boundaryless approach. Stoch. Environ. Res. Risk Assess. https://doi.org/10.1007/s00477-021-02008-x.

Faulkner, D.S., C. Prudhomme. 1998. Mapping an index of extreme rainfall across the UK. Hydrol. Earth Syst. Sci. 2 (2–3): 183–94. https://doi.org/10.5194/hess-2-183-1998.

Faulkner, D., 1999. Flood Estimation Handbook Volume 2: Rainfall Frequency Estimation. Institute of Hydrology, Wallingford, UK.

Fouedjio, F., 2017. Second-order non-stationary modeling approaches for univariate geostatistical data. Stoch. Environ. Res. Risk Assess. 31 (8), 1887–1906. https://doi.org/10.1007/s00477-016-1274-y.

Furcolo, P., Villani, P., Rossi, F., 1995. Statistical analysis of the spatial variability of very extreme rainfall in the mediterranean area, U.S. - Italy Research Workshop on the Hydrometeorology, Impacts, and Management of Extreme Floods. Perugia (Italy).

Fürst, J., Godina, R., Peter Nachtnebel, H., Nobilis, F., 2009. The hydrological atlas of Austria - comprehensive transfer of hydrological knowledge and data to engineers, water resources managers and the public. IAHS-AISH Publication 327, 36–44.

Gaál, L., Kyselý, J., Szolgay, J., 2008. Region-of-influence approach to a frequency analysis of heavy precipitation in Slovakia. Hydrol. Earth Syst. Sci. 12, 825–839. https://doi.org/10.5194/hess-12-825-2008.

Ganora, D., Laio, F., 2016. A comparison of regional flood frequency analysis approaches in a simulation framework. Water Resour. Res. 52, 5644–5661. https://doi.org/10.1002/2016WR018604.

Goovaerts, P., 1999. Using elevation to aid the geostatistical mapping of rainfall erosivity. Catena 34 (3–4), 227–242. https://doi.org/10.1016/S0341-8162(98)00116-7.

Goovaerts, P., 2000. Geostatistical approaches for incorporating elevation into the spatial interpolation of rainfall. J. Hydrol. 228 (1–2), 113–129. https://doi.org/10.1016/S0022-1694(00)00144-X.

Goudenhoofdt, E., Delobbe, L., Willems, P., 2017. Regional frequency analysis of extreme rainfall in Belgium based on radar estimates. Hydrol. Earth Syst. Sci. 21, 5385–5399. https://doi.org/10.5194/hess-21-5385-2017.

Grasso, S., Libertino, A., Claps, P., 2020. Multirain: a GIS-based tool for multi-model estimation of regional design rainfall for scientists and practitioners. J. Hydroinform. 22 (1), 148–159. https://doi.org/10.2166/hydro.2019.016.

Grimaldi, S., Kao, S.C., Castellarin, A., Papalexiou, S.M., Viglione, A., Laio, F., Aksoy, H., Gedikli, A., 2011. Statistical hydrology. In: Wilderer, P. (Ed.), Treatise on Water Science. Elsevier, Oxford, UK, pp. 479–517. https://doi.org/10.1016/B978-0-444-53199-5.00046-4.

Grimes, D.I.F., Pardo-Igúzquiza, E., 2010. Geostatistical analysis of rainfall. Geogr. Anal. 42 (2), 136–160. https://doi.org/10.1111/j.1538-4632.2010.00787.x.

Gumbel, E.J., 1954. Statistical Theory of Extreme Values and Some Practical Applications: A Series of Lectures. US Government Printing Office, District of Columbia.

Hengl, T., 2007. A Practical Guide to Geostatistical Mapping of Environmental Variables. Office for Official Publications of the European Communities, Luxembourg.

Hengl, T., Heuvelink, G., Stein, A., 2003. Comparison of kriging with external drift and regression-kriging. Technical Note 17. https://doi.org/10.1016/S0016-7061(00)00042-2.

Hosking, J.R.M., Wallis, J.R., 1997. Regional Frequency Analysis: An Approach Based on L-Moments. Cambridge University Press, Cambridge, UK. https://doi.org/10.1017/CBO9780511529443.

Hudson, G., Wackernagel, H., 1994. Mapping temperature using kriging with external drift: theory and an example from Scotland. Int. J. Climatol. 14 (1), 77–91. https://doi.org/10.1002/joc.3370140107.

Kašpar, M., Bližňák, V., Hulec, F., Müller, M., 2021. High-resolution spatial analysis of the variability in the subdaily rainfall time structure. Atmos. Res. 248, 105202. https://doi.org/10.1016/j.atmosres.2020.105202.

Kidd, C., Becker, A., Huffman, G.J., Muller, C.L., Joe, P., Skofronick-Jackson, G., Kirschbaum, D.B., 2017. So, how much of the earth's surface is covered by rain gauges? Bull. Am. Meteorol. Soc. 98 (1), 69–78. https://doi.org/10.1175/BAMS-D-14-00283.1.

Koutsoyiannis, D., Iliopoulou, T., 2022. Ombrian Curves Advanced to Stochastic Modelling of Rainfall Intensity. In: Morbidelli, R. (Ed.), Rainfall. Modeling, Measurement and Applications. Elsevier, Amsterdam, pp. 261–284. doi:10.1016/C2019-0-04937-0.

Laio, F., Ganora, D., Claps, P., Galeati, G., 2011. Spatially smooth regional estimation of the flood frequency curve (with uncertainty). J. Hydrol. 408 (1–2), 67–77. https://doi.org/10.1016/j.jhydrol.2011.07.022.

Li, Z., Ballard, S.P., Simonin, D., 2018. Comparison of 3D-var and 4D-var data assimilation in an NWP-based system for precipitation nowcasting at the met office. Q. J. R. Meteorol. Soc. 144 (711), 404–413. https://doi.org/10.1002/qj.3212.

Libertino, A., Allamano, P., Laio, F., Claps, P., 2018. Regional-scale analysis of extreme precipitation from short and fragmented records. Adv. Water Res. 112, 147–159. https://doi.org/10.1016/j.advwatres.2017.12.015.

Madsen, H., Rasmussen, P.F., Rosbjerg, D., 1997. Comparison of annual maximum series and partial duration series methods for modeling extreme hydrologic events: 1. at-site modeling. Water Res. Res. 33 (4), 747–757. https://doi.org/10.1029/96WR03848.

Marani, M., 2003. On the correlation structure of continuous and discrete point rainfall. Water Resour. Res. 39 (5), 1–8. https://doi.org/10.1029/2002WR001456.

Marani, M., Zanetti, S., 2015. Long-term oscillations in rainfall extremes in a 268 year daily time series. Water Resour. Res. 51, 639–647. https://doi.org/10.1002/2014WR015885.

Matheron, G., 1963. Principles of geostatistics. Econ. Geol. 58 (8), 1246–1266. https://doi.org/10.2113/gsecongeo.58.8.1246.

Modarres, R., Sarhadi, A., 2012. Statistically-based regionalization of rainfall climates of Iran. Glob. Planet. Change 75 (1–2), 67–75. https://doi.org/10.1016/j.gloplacha.2010.10.009.

Naghettini, M., Pinto, E.J.A., 2017. Regional Frequency Analysis of Hydrologic Variables. Fundamentals of Statistical Hydrology. Springer, Cham. https://doi.org/10.1007/978-3-319-43561-9_10.

Ngongondo, C.S., Xu, C.Y., Tallaksen, L.M., Alemaw, B., Chirwa, T., 2012. Regional frequency analysis of rainfall extremes in Southern Malawi using the index rainfall and L-moments approaches. Stoch. Environ. Res. Risk Assess. 25 (7), 939–955. https://doi.org/10.1007/s00477-011-0480-x.

Ochoa-Rodriguez, S., Wang, L.P., Willems, P., Onof, C., 2019. A review of radar-rain gauge data merging methods and their potential for urban hydrological applications. Water Resour. Res. 55 (8), 6356–6391. https://doi.org/10.1029/2018WR023332.

Ouali, D., Chebana, F., Ouarda, T.B.M.J., 2016. Quantile regression in regional frequency analysis: a better exploitation of the available information. J. Hydrometeorol. 17 (6), 1869–1883. https://doi.org/10.1175/JHM-D-15-0187.1.

Pappas, C., Papalexiou, S.M., Koutsoyiannis, D., 2014. A quick gap filling of missing hydrometeorological data. J. Geophys. Res. Atmos. 119 (15), 9290–9300. https://doi.org/10.1002/2014JD021633.

Persiano, S., Ferri, E., Antolini, G., Domeneghetti, A., Pavan, V., Castellarin, A., 2020. Changes in seasonality and magnitude of sub-daily rainfall extremes in Emilia-Romagna (Italy) and potential influence on regional rainfall frequency estimation. J. Hydrol: Reg. Stud. 32. https://doi.org/10.1016/j.ejrh.2020.100751.

Prudhomme, C., Reed, D.W., 1998. Relationships between extreme daily precipitation and topography in a mountainous region: a case study in Scotland. Int. J. Climatol. 18 (13), 1439–1453. https://doi.org/10.1002/(SICI)1097-0088(19981115)18:13<1439::AID-JOC320>3.0.CO;2-7.

Prudhomme, C., Reed, D.W., 1999. Mapping extreme rainfall in a mountainous region using geostatistical techniques: a case study in Scotland. Int. J. Climatol. 19 (12), 1337–1356. https://doi.org/10.1002/(SICI)1097-0088(199910)19:12<1337::AID-JOC421>3.0.CO;2-G.

Qamar, M.U., Azmat, M., Shahid, M.A., Ganora, D., Ahmad, S., Cheema, M.J.M., Faiz, M.A., Sarwar, A., Shafeeque, M., Khan, M.I., 2017. Rainfall extremes: a novel modeling approach for regionalization. Water Resour. Manag. 31 (6), 1975–1994. https://doi.org/10.1007/s11269-017-1626-5.

Reed, D.W., Jakob, D., Robinson, A., Faulkner, D., Stewart, E.J., 1999. Regional frequency analysis: a new vocabulary, Hydrological Extremes: Understanding, Predicting, Mitigating. Proceedings of the IUGG 99 Symposium. Birmingham. *IAHS Publ 255*. pp. 237–243.

Requena, A.I., D.H. Burn, P. Coulibaly. 2019. Pooled frequency analysis for intensity–duration–frequency curve estimation. Hydrol. Process. 33 (15): 2080–94. https://doi.org/10.1002/hyp.13456.

Rossi, F., Villani, P., 1994. Regional flood estimation methods. In: Rossi, G., Harmancioglu, N., Yevjevich, V. (Eds.), Coping with Flood. Kluwer Academic Publishers, The Netherlands, pp. 135–169.

Satyanarayana, P., Srinivas, V.V., 2008. Regional frequency analysis of precipitation using large-scale atmospheric variables. J. Geophys. Res. Atmos. 113 (24), 1–16. https://doi.org/10.1029/2008JD010412.

Schaefer, M.G., 1990. Regional analyses of precipitation annual maxima in washington state. Water Resour. Res. 26 (1), 119–131. https://doi.org/10.1029/WR026i001p00119.

Schardong, A., Simonovic, S.P., Gaur, A., Sandink, D., 2020. Web-based tool for the development of intensity duration frequency curves under changing climate at gauged and ungauged locations. Water 12 (5). https://doi.org/10.3390/W12051243.

Smithers, J.C., Schulze, R.E., 2001. A methodology for the estimation of short duration design storms in South Africa using a regional approach based on L-moments. J. Hydrol. 241 (1–2), 42–52. https://doi.org/10.1016/S0022-1694(00)00374-7.

Soltani, S., Helfi, R., Almasi, P., Modarres, R., 2017. Regionalization of rainfall intensity-duration-frequency using a simple scaling model. Water Resour. Manag. 31 (13), 4253–4273. https://doi.org/10.1007/s11269-017-1744-0.

Svensson, C., Jones, D.A., 2010. Review of rainfall frequency estimation methods. J. Flood Risk Manag. 3, 296–313. https://doi.org/10.1111/j.1753-318X.2010.01079.x.

Tabios, G.Q., Salas, J.D., 1985. A comparative analysis of techniques for spatial interpolation of precipitation. J. Am. Water Resour. Assoc. 21 (3), 365–380. https://doi.org/10.1111/j.1752-1688.1985.tb00147.x.

Teegavarapu, R.S.V., Nayak, A., 2017. Evaluation of long-term trends in extreme precipitation: implications of in-filled historical data use for analysis. J. Hydrol. 550, 616–634. https://doi.org/10.1016/j.jhydrol.2017.05.030.

Todini, E., 2001. A bayesian technique for conditioning radar precipitation estimates to rain-gauge measurements. Hydrol. Earth Syst. Sci. 5 (2), 187–199. https://doi.org/10.5194/hess-5-187-2001.

Trefry, C.M., Watkins Jr, D.W., Johnson, D., 2006. Regional rainfall frequency analysis for the state of Michigan. J. Hydrol. Eng. 10 (6), 437–449. https://doi.org/10.1061/(ASCE)1084-0699(2005)10:6(437).

Uboldi, F., Sulis, A.N., Lussana, C., Cislaghi, M., Russo, M., 2014. A spatial bootstrap technique for parameter estimation of rainfall annual maxima distribution. Hydrol. Earth Syst. Sci. 18 (3), 981–995. https://doi.org/10.5194/hess-18-981-2014.

Varentsov, M., Esau, I., Wolf, T., 2020. High-resolution temperature mapping by geostatistical kriging with external drift from large-eddy simulations. Mon. Weather Rev. 148 (3), 1029–1048. https://doi.org/10.1175/MWR-D-19-0196.1.

Velázquez, J.A., Anctil, F., Ramos, M.H., Perrin, C., 2011. Can a multi-model approach improve hydrological ensemble forecasting? A study on 29 French catchments using 16 hydrological model structures. Adv. Geosci. 29, 33–42. https://doi.org/10.5194/adgeo-29-33-2011.

Van de Vyver, H., 2012. Spatial regression models for extreme precipitation in Belgium. Water Resour. Res. 48 (9), 1–17. https://doi.org/10.1029/2011WR011707.

Wackernagel, H., 1998. Multivariare Geostatistics. Springer-Verlag Berlin Heidelberg. https://doi.org/10.1007/978-3-662-03550-4.

Zorzetto, E., Marani, M., 2020. Extreme value metastatistical analysis of remotely sensed rainfall in ungauged areas: spatial downscaling and error modelling. Adv. Water Resour. 135. https://doi.org/10.1016/j.advwatres.2019.103483.

CHAPTER

13 Rainfall and development of floods

Carla Saltalippi[a], Corrado Corradini[a], Jacopo Dari[a,b], Renato Morbidelli[a], Alessia Flammini[a]

[a]Department of Civil and Environmental Engineering, University of Perugia, Perugia, Italy
[b]National Research Council, Research Institute for Geo-Hydrological Protection, Perugia, Italy

13.1 Introduction

Prerequisite for flood development in a given river section is the occurrence of significant rainfall on its watershed. Floods in rivers with very large watersheds are commonly generated by widespread rainfalls associated with frontal systems (Corradini et al., 2022), while in small catchments they can also be produced by convective systems.

The flood hydrograph involves two components (Chow, 1988): the direct flow produced by effective rainfall and the baseflow that has generally a minor role.

Primary importance in flood studies is therefore ascribed to the knowledge of the time evolution of effective rainfall (effective hyetograph) starting from the observed rainfall. Its determination requires the assessment of the rainfall losses due to interception, evaporation, surface retention, and infiltration. Rainfall interception by vegetation, mainly due to woodland, can in any case be neglected in flood studies. The evaporation/evapotranspiration process, very important for a few hydrological investigations for example, agricultural water management and droughts, can generally be disregarded, particularly for floods produced by continuous rainfalls. The role of the surface retention loss depends on the depressions on the ground through their storage capacity. The downward water flow through the soil surface represents the infiltration loss that can produce a substantial reduction of the observed rainfall associated to moderate floods and can significantly affect also heavy flood events. Flood formation is primarily due to the transformation of observed rainfall to runoff (Ogden and Warner, 2017). The main contribution derives from the effective rainfall-direct runoff transformation, with a spatiotemporal variability of the effective rainfall as a result of that commonly characterizing both observed rainfall (Corradini et al., 2022; Borga et al., 2022) and infiltration (Govindaraju and Goyal, 2022).

Flood monitoring in terms of discharge at selected river sections and rainfall observed through their watersheds is crucial for many hydrological applications that require an extended database of these quantities.

RAINFALL. Modeling, Measurement and Applications. DOI: https://doi.org/10.1016/B978-0-12-822544-8.00015-9

The availability of numerous observed rainfall-runoff events allows to realize different types of mathematical models for simulation and prediction of flood events. These are commonly classified as black-box models, conceptual models and models based on integrated forms of the primitive equations for movement and mass conservation of surface and subsurface water. The conceptual models, which are widely used in the hydrological practice, can be characterized by different levels of complexity linked with the simplifications involved in the schematization of both the physical processes and the watershed configuration. The last choice identifies the model distribution level and influences the representation of the rainfall spatial distribution that is the main input of each flood model. Commonly the watershed is divided into different areas, each considered at a given time, subjected to a uniform rainfall rate. These flood models generally require a calibration, through the choice of a few incorporated quantities and parameters, to be performed using the complete knowledge of hyetograph and hydrograph of each calibration event. The models can be then adapted for real-time flood forecasting, with rainfall known until the time origin of the flow forecast and later forecasted with a given lead time.

The availability of continuous rainfall data for many tenths of years would represent the optimal condition to investigate by a statistical approach extreme rainfalls of different durations, that are crucial in the determination of the flood hydrograph for designing hydraulic structures. In this context, for a given return period, design hyetographs referred to observed and effective rainfall can be deduced. The latter is then used as input to a rainfall-runoff model providing direct flow component of the design hydrograph for the hydraulic structure of interest.

This chapter first deals with the formation and separation of the flood hydrograph through the effective hyetograph associated to a specific rainfall-runoff event. On this basis the main structure of typical rainfall-runoff models for simulating single flood events is highlighted in general terms. Then, the specific structure of an adaptive rainfall-runoff model for real-time flood forecasting is also examined.

Finally, through a synthetic statistical analysis of extreme rainfalls, a classical procedure for determining the design hydrograph of hydraulic structures is presented.

13.2 Formation and separation of the flood hydrograph

The watershed, associated to a given river section, is constituted by drainage areas that contribute to the hydrograph formation at the outlet.

A significant rainfall event, usually represented as a function of time by a histogram (hyetograph), produces a hydrological response at the watershed outlet. This response, denoted as hydrograph, is the continuous evolution in time of the stream discharge, $Q_E(t)$ (water volume per time unit, $[L^3T^{-1}]$) (Dingman, 2015). It is noteworthy that, given a rainfall event occurring on a watershed, the volume of water associated to the hydrograph is only a fraction of the total input. A remaining water amount through the infiltration process increases the soil water content and could

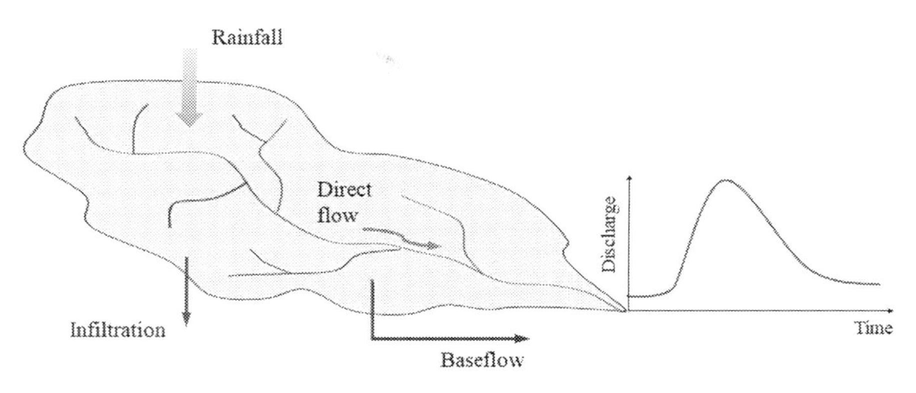

FIG. 13.1

A schematic representation of the main mechanisms producing through the rainfall-runoff process the flood hydrograph at the basin outlet.

produce groundwater recharge that supplies at long times the river flow as baseflow (Fig. 13.1). A typical hydrograph shape is shown in Fig. 13.2.

At some time after the beginning of the rainfall the discharge begins to increase rapidly from the pre-event rate to a well-defined peak discharge, $Q_E(t_p)$, with t_p time to peak. This part of the hydrograph is called the rising limb. This stage is determined by the effective rainfall starting from the time to ponding (Govindaraju and Goyal, 2022), that is the time when soil surface reaches saturation. After the peak the flow decreases more slowly to a value close to the pre-event one. This is the recession limb. The hydrograph is the trace of a flood wave crossing a point of the stream

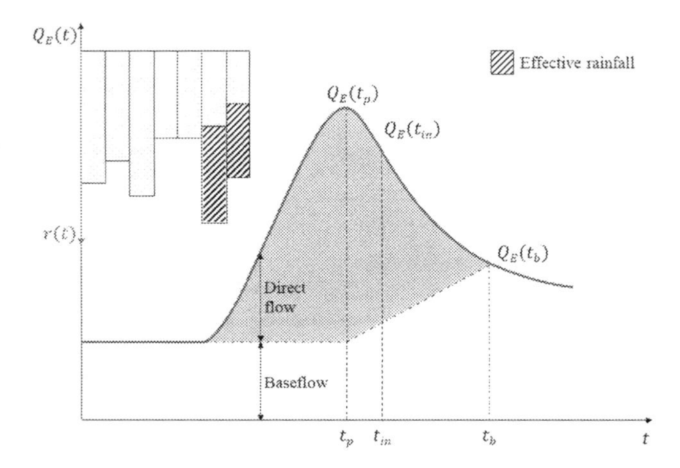

FIG. 13.2

Example of a graphical procedure of hydrograph separation into baseflow and direct flow, given a rainfall-runoff event: $r(t)$ is the areal-average rainfall rate over the watershed and $Q_E(t)$ is the discharge observed at the basin outlet. For other symbols see text.

network. Typically, in humid regions there is a nonzero flow, slowly decreasing, before the rising limb. This discharge is supplied by groundwater as a result of saturated soil with a water table higher than the free surface of the stream.

The discharge evolution until the base time, t_b (see Fig. 13.2), is determined by both baseflow and direct flow. The latter, generated by the effective rainfall, identifies the water moving on the soil surface and in a narrow sub-surface layer that reaches as lateral inflow the channel network and then the watershed outlet. The direct flow coming from different areas of the watershed crosses its outlet section at different times depending on the distance to be covered. Specifically, the travel time to the watershed outlet is determined by the movement over the soil surface and in the stream network. This element together with the typical form of a natural basin and its channel network causes the hydrograph shape of Fig. 13.2. After t_b, the river discharge is supplied only by groundwater while direct flow has run out. Groundwater contribution consists of water entered the watershed in previous rainfall events and transferred to the streams with significant delay. It is also expected to be increased after t_p by the water infiltrated during the current rainfall event starting from its beginning. This contribution to discharge is also called baseflow. On this basis, it is clear that it is not easy to identify all the sources of the streamflow during rainfall events whose effects occur at very different times (Dingman, 2015).

On the other hand, in order to develop methods for flood forecasting, it is of practical interest to define the stream response that is associated to a given event. This objective can be obtained through a procedure known as separation of hydrograph into the baseflow, $B(t)$, and direct flow, $Y_E(t)$, components. In this approach, the observed total streamflow $Q_E(t)$ is given by:

$$Q_E(t) = B(t) + Y_E(t) \tag{13.1}$$

In the absence of more detailed information, a graphical heuristic technique can be used along the following steps (Fig. 13.2):

1. The pre-event flow trend (i.e., baseflow) is projected until the time to peak, t_p;

2. After t_p the baseflow is connected by a straight line that intersects the observed flow at the point (t_b, $Q_E(t_b)$), with t_b base time identified as described below: a. Assumed the trend of the two straight lines as baseflow, the direct flow is obtained as difference of $Q_E(t)$ and $B(t)$.

The identification of t_b requires:

1. To find the flex point on the recession limb corresponding to the inflection time, t_{in};

2. To fit the recession limb for $t \geq t_{in}$ by an exponential function with a single parameter α, as:

$$Q_E(t) = Q_E(t_{in}) \exp\left[-\alpha\left(t - t_{in}\right)\right] \tag{13.2}$$

where α remains constant until t_b, after which it assumes a different value due to a change of curvature,

1. To plot the hydrograph on semi-logarithmic paper ($\log[Q_E(t)]$ versus t) for $t \geq t_{in}$;

2. To identify a couple of straight lines well-fitting the two recession curve parts on semi-logarithmic paper and intersecting at time t_b, that can be so assessed. The base time is assumed to be the time when direct flow stops to contribute to discharge.

13.3 A modeling framework for flood simulation and real-time flood forecasting

Simulation by a rainfall-runoff model of an observed flood hydrograph relies upon the knowledge of the associated hyetograph usually derived by a rain gauge network (Lanza et al., 2022), if possible integrated by remote sensing techniques (Kidd and Levizzani, 2022; Borga et al., 2022). The model could be identified as lumped, semi-distributed and distributed according to the representation used for both the basin configuration and the involved physical processes. Fig. 13.3 shows a semi-distributed approach incorporating a distributed representation of a watershed with different sub-basins and minor regions draining directly into the main river (herein denoted as regions of lateral inflow), both with the physical process that could be approximated by a lumped conceptual formulation. More detailed representations of the watershed channel network and associated drainage areas would allow to increase the model

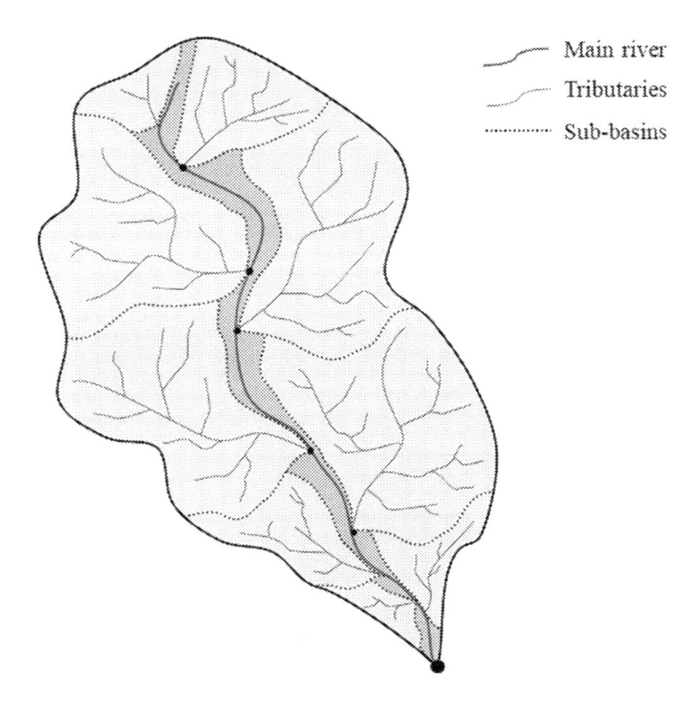

FIG. 13.3

A semi-distributed representation of the watershed with different sub-basins and minor regions draining directly into the main river.

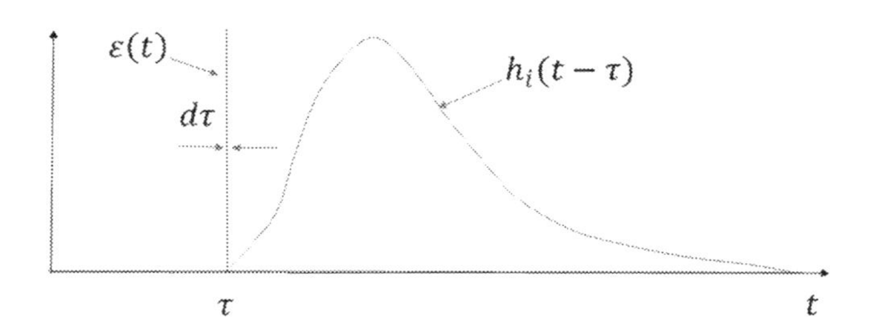

FIG. 13.4

A typical sub-basin response, $h_i(t-\tau)$, as a function of time, to an effective unit rainfall, $\varepsilon(t)$, of δ– Dirac type.

distribution level (Beven and Kirkby, 1979; Woolisher et al., 1990; Melone et al., 1998). The direct runoff at a time t at the basin outlet can be computed by the contributions, herein denoted as $Y_i(t)$, $i = 1\ldots, n$, of the effective rainfall-direct runoff transformation in the n sub-areas then routed to the outlet.

For a quantitative formulation of $Y_i(t)$, for $i = 1\ldots, n$ the technique of the unit hydrograph is widely used in the hydrological practice (Singh and Jain, 2017). It relies upon an instantaneous unit hydrograph (IUH) expressed by the implicit function $h_i(t-\tau)$, defined as the sub-area response to a spatially uniform effective rainfall rate, $\varepsilon(\tau)$, of unit depth, R_u, with time distribution of δ-Dirac type, $\delta(t-\tau)$, expressed by:

$$\varepsilon(t) = R_u \delta(t-\tau) \quad \text{with} \quad \int_{-\infty}^{+\infty} \varepsilon(t)dt = R_u \text{ at the time } \quad t = \tau \tag{13.3}$$

Fig. 13.4 shows that $\varepsilon(t) = 0$ at each time except in an infinitesimal time interval dt when $\varepsilon \rightarrow \infty$. The shape of $h_i(t-\tau)$, in $[L]^3[T]^{-1}$, is that expected for the sub-basins considering their typical geometrical configurations and the structure of their channel network.

The lateral inflow typically concerns a narrow and almost continuous zone along the main river, that can be split up into regions on the basis of the distance from the basin outlet. The shape of $h_i(t-\tau)$ for each region can be approximated by a decreasing curve associated to a fictitious single reservoir. For a real temporal distribution of a spatially uniform effective rainfall rate, $E_i(t)$, the direct runoff at the outlet of the i-th area, Y_{i0}, can be expressed as:

$$Y_{i0}(t) = \frac{1}{R_u} \int_0^t E_i(\tau) h_i(t-\tau) d\tau \tag{13.4}$$

Eq. (13.4) derives from the assumption that each area is a linear and stationary operator and $E_i(\tau)$ is obtained from the observed areal-average rainfall rate, $r_i(\tau)$, as:

$$E_i(\tau) = r_i(\tau) - I_i(\tau) - \frac{dV_i}{d\tau} \tag{13.5}$$

where $I_i(\tau)$ is the areal-average infiltration rate (Govindaraju and Goyal, 2022) and $\frac{dV_i}{d\tau}$ is the areal-average surface retention loss. The direct runoff hydrograph $Yi_0(t)$ experiences an appreciable variation due to the transfer process towards the basin outlet. Its contribution at the basin outlet becomes:

$$Y_i(t) = \frac{1}{V_0}\int_0^t \left(\frac{1}{R_u}\int_0^\tau \left[r_i(\tau') - I_i(\tau') - \frac{dV_i}{d\tau'} \right] h_i(\tau - \tau')\, d\tau' \right) g_i(t - \tau; x_i)\, d\tau \quad (13.6)$$

where g_i denotes a transfer function for a unit water volume with time distribution of δ–Dirac type depending on the length, x_i, of the main river reach between the basin and i-th sub-area outlets. The solution of Eq. (13.6) requires an explicit representation of h_i and g_i. Among the variety of available formulations, the Nash model and the geomorphologic model (Rodríguez-Iturbe and Valdes, 1979; Gupta et al., 1980; Corradini and Singh, 1985) are those mainly adopted for h_i of a sub-basin. For g_i, a kinematic or diffusive model could be chosen on the basis of the slope of the main river. The discharge at the basin outlet, $Q(t)$, can be expressed as:

$$Q(t) = \sum_{i=1}^{n} Y_i(t) + B \quad (13.7)$$

where B is the baseflow that, as shown in the previous section, can be approximated by the discharge observed at the basin outlet before the increasing limb of the flood hydrograph. The model framework for simulating the formation of single flood events involves some quantities/parameters that should be assessed in advance and then adjusted through calibration events. Complete measurements of the spatiotemporal rainfall distribution and discharge at the basin outlet are required. A separate set of rainfall-runoff events should be used for model testing.

Then, the model can be extended to perform real-time flood forecasting under the condition that real-time observations of the spatiotemporal rainfall field and discharge at the basin outlet are provided by a suitable experimental network. The discharge forecasted at a time t_c for a lead time T is given by:

$$Q(t_c + T) = B + \frac{1}{F_c(t_c)}\sum_{i=1}^{n}\frac{1}{V_0}\int_0^{t_c+T}\left(\frac{1}{R_u}\int_0^\tau E_i(\tau')h_i(\tau - \tau')\,d\tau'\right) g_i(t_c + T - \tau; x_i)\,d\tau \quad (13.8)$$

Until $\tau = t_c$, $E_i(\tau')$ is obtained from the observed rainfall, while for $t_c < \tau \le t_c + T$ rainfall prediction models (Grabowski, 2022) should be used. In the absence of a reliable model for rainfall forecast, $E_i(\tau') = 0$ for $t_c < \tau \le t_c + T$. The correction factor $F_c(t_c)$ can be deduced by an adaptive component of the model (Corradini et al., 1987) that corrects the discrepancy typically existing between $Q(t_c)$ computed by the model and the observed discharge at the same time.

13.4 A short description of the implicit functions incorporated in the real-time flood forecasting model framework

A simple approach for expressing $h_i(t)$ for each sub-basin is based on the use of the dimensionless IUH, $h^*(t^*)$, which represents the IUH scaled by the peak flow, $h_{pi,}$ for its ordinate and by the average time the effective rainfall spends in the sub-basin, L_i, for the time basis. Considering that the change of $h_i(t)$ from one sub-basin to another may be primarily produced by the variation of L_i, a very reduced variability of $h^*(t^*)$ is generally expected. Corradini et al. (1995) investigated this topic in a variety of watersheds located in Central Italy and proposed the dimensionless geomorphologic IUH shown in Fig. 13.5. Given $h^*(t^*)$, $h_i(t)$ is computed as:

$$h_i(t) = h_{pi}\ h^*\left(\frac{t}{L_i}\right) \tag{13.9}$$

where h_{pi} can be expressed through the conservation equation for the effective rainfall as:

$$\int_0^{+\infty} h_i(t)dt = L_i h_{pi} \int_0^{+\infty} h^*\left(t^*\right)dt^* = A_i R_u \tag{13.10}$$

$$h_{pi} = \frac{A_i R_u}{L_i \int_0^{+\infty} h^*\left(t^*\right)dt^*} \tag{13.11}$$

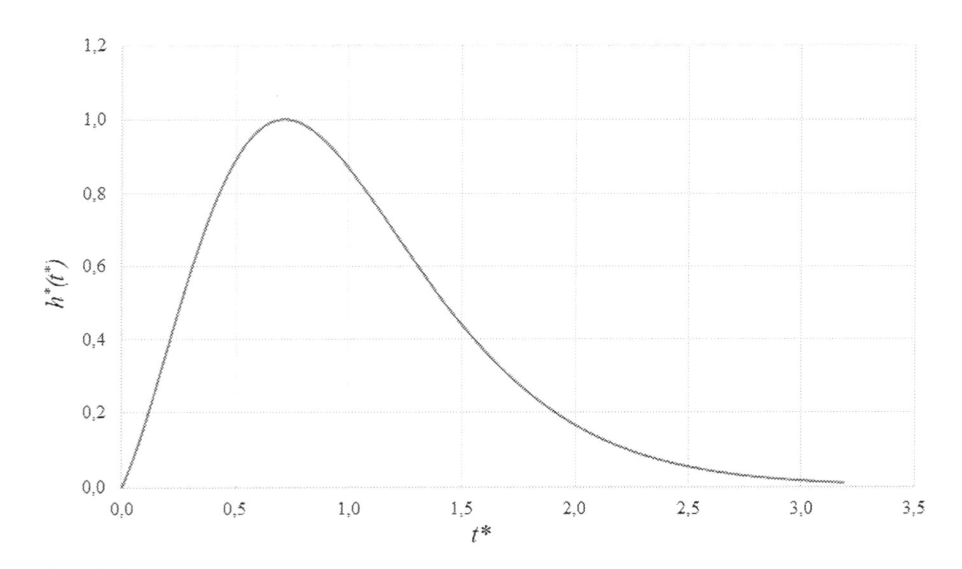

Fig. 13.5

Dimensionless instantaneous unit hydrograph (from Corradini et al., 1995).

and Eq. (13.9) becomes:

$$h_i(t) = \frac{A_i R_u}{L_i \int_0^{+\infty} h^*(t^*)\,dt^*} h^*\left(\frac{t}{L_i}\right) s \tag{13.12}$$

where A_i is the sub-basin area. The sub-basin lag for significant flood events is commonly deduced by the empirical relationship:

$$L_i = \beta \, A_i^{\alpha} \tag{13.13}$$

with the parameters α and β usually considered to be constant for watershed located in the same geographic area. With L_i in hours and A_i in Km^2 α is frequently assumed as 0.33 or 0.38 (Singh, 1988), while $\beta = 1.19$ was proposed by Melone et al. (2002) for hilly watersheds in Central Italy. Finally, $h_i(t) = h_i(t - \tau)$ because each sub-basin is considered to be a stationary operator.

A very simple approach can be used for expressing h_i associated to the regions of lateral inflow. Each element is roughly considered as a linear reservoir, with storage coefficient K to be determined by calibration, that allows to estimate $h_i(t - \tau)$ as:

$$h_i(t-\tau) = \frac{A_i R_u}{K} \exp\left[-\frac{(t-\tau)}{K}\right] \backslash n \tag{13.14}$$

For small slopes of the main river and considerable floods the transfer function $g_i(t - \tau, x_i)$ can be approximated by the linearized diffusive model as:

$$g_i(t-\tau, x_i) = \frac{x_i V_0}{\left[4\pi D(t-\tau)^3\right]^{\frac{1}{2}}} \exp\left\{-\frac{\left[C(t-\tau)-x_i\right]^2}{4D(t-\tau)}\right\} \tag{13.15}$$

where C is the absolute celerity and D is the diffusivity, both assessed in advance through the knowledge of the involved average discharge and the geometry of the main river (Reed, 1984).

The surface retention loss $\frac{dV_i}{d\tau'}$ in the i-th region is frequently represented in the hydrologic modeling by a fictitious reservoir of capacity V_D and V_i expressed as:

$$V_i(\tau) = V_D \left\{1 -- \exp\left[-\frac{1}{V_D}\int_0^{\tau}(r_i - I_i)\,d\tau'\right]\right\} \tag{13.16}$$

where V_D is a parameter representing the depression storage capacity and $V_i(\tau)$ is the cumulative contribution of rainfall to the depression storage up to the time τ.

The areal-average infiltration rate, $I_i(\tau)$, can be expressed by a variety of models widely described by Govindaraju and Goyal (2022). This is a process strictly linked to the spatiotemporal rainfall distribution, and with hydraulic soil properties (Corradini et., 1997; Morbidelli et al., 2006) as the areal-average value of the initial soil water content, θ_{in}, and the saturated hydraulic conductivity and its coefficient of variation.

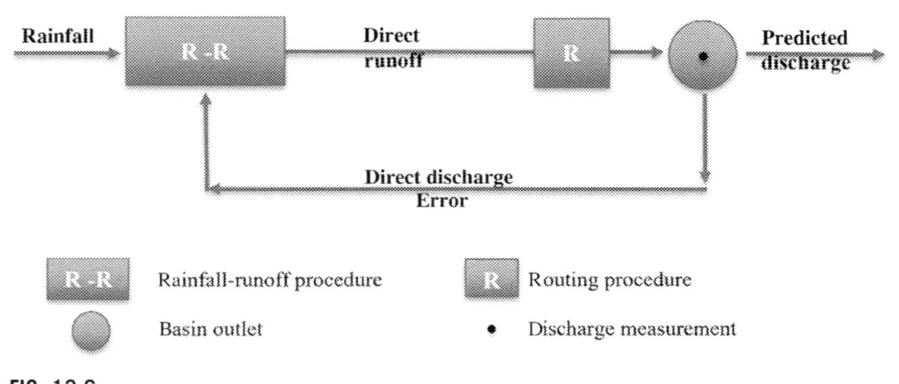

FIG. 13.6

Schematization of the adaptive flood forecasting model.

The last two quantities can be approximately assessed in advance on the basis of the soil structure and soil use, while θ_{in} also depends on the previously observed rainfall.

For the sake of simplicity, the quantity θ_{in}, assumed constant through the basin, is used herein, as an example, to implement the adaptive model component that allows to compute $F_c(t_c)$ of Eq. (13.8). At the time origin of flow forecasting, it is adapted through a minimization of the objective function $G(\theta_{in})$ defined by the differences between observed discharge, Q_E, and computed value, Q, in the last two times of forecasting separated by a time interval of Δt as:

$$G(\theta_{in}) = \sum_{m=0}^{1} \left[Q_E(t_c - m\Delta t) - Q(t_c - m\Delta t; \theta_{in}) \right]^2 \tag{13.17}$$

subjected to the constraint to have within Δt the same trend for computed and observed discharge, that is:

$$\frac{Q_E(t_c) - Q_E(t_c - \Delta t)}{Q(t_c) - Q(t_c - \Delta t)} > 0 \tag{13.18}$$

The last two relationships typically produce the discrepancy $Q(t_c) \neq Q_E(t_c)$.

This is removed by the correction factor $F_c(t_c)$ that changes the computed direct runoff term in Eq. (13.8). Fig. 13.6 shows a schematization of the adaptive flood forecasting model.

13.5 Rainfall-runoff for designing extreme floods

The design of a hydraulic structure to be realized at a specific river cross section requires the estimate of the maximum streamflow expected in the life time of the structure. The design discharge represents the critical event for the structure (Dingman, 2015); for larger discharge values its failure is expected to occur. The higher the cost of a structure, the higher will be the assumed expected life time. The concern of life time is usually reformulated in terms of return period, T_r, that determines the average number of periods (typically years) after which an exceedance of the design streamflow is expected. Mathematically, T_r can be expressed as:

$$T_r = \frac{1}{1 - P(X')} \tag{13.19}$$

where X' is the critical value of the random variable X, namely the annual maximum streamflow, and $P(X')$ is the nonexceedence probability of X':

$$P(X') = \{X \ values \leq X'\} \in [0,1] \tag{13.20}$$

while $1 - P(X')$ is the exceedance probability.

On the basis of the relation (13.19), the statistical analysis of extreme values of X becomes the procedure to identify the design flood event.

The nonexceedance probability distribution $P(X')$ can be estimated through theoretical models. Among the available probability distributions, the most commonly used are the Gumbel and the log-normal ones (Singh and Zhang, 2017). Once a value of T_r has been fixed and the analytical function of the probability distribution has been selected, from Eq. (13.19) the critical value X' can be derived. This approach requires a large amount of data of annual maximum streamflow, at least a few tenths of years, at the specific river cross section of interest. Since usually this procedure cannot be applied for data scarcity, an indirect approach involving extreme rainfall data, typically available over the watershed, and a rainfall-runoff model has to be used. This approach relies upon the simplifying supposition that a flood event with specific value of T_r is determined by a critical rainfall event characterized by the same probability of exceedance. This assumption completely disregards the initial soil conditions, but can be considered acceptable for design events. In this context, the aim of this indirect approach is to estimate the design rainfall, evenly distributed in space, to be used as an input of a rainfall-runoff model. The computed hydrograph can be assumed as the design hydrograph.

The determination of the design rainfall requires that an inference analysis of the annual maximum local rainfalls of appropriate durations is performed for each rain gauge located in the basin and characterized by sufficiently long time series of data.

For different durations, D_r, estimation of the rainfall depths, R_{D_r}, with a given nonexceedance probability is called depth-duration-frequency (DDF) analysis. An equivalent procedure that substitutes rainfall depths with rainfall intensities is called intensity-duration-frequency (IDF) analysis.

Specifically, the DDF curves express annual maxima of rainfall as function of duration and probability of non-exceedance (or T_r). At the local scale (or rain gauge scale), the procedure to determine the DDF curves consists of (Kottegoda and Rosso, 2008):

1. Extracting annual maximum time series that is, the largest value in each available year for each fixed duration, preferably from rainfall data with a

very small-time resolution (Morbidelli et al., 2017; Morbidelli et al., 2022), obtaining a sample of the random variable 'annual maximum rainfall of specified duration' to be checked for the homogeneity and stationarity of the series.

2. Adopting a theoretical model for the probability distribution function (typically the Gumbel one), estimating its parameters on the basis of sample values through one of the available techniques (moments, L-moment, maximum likelihood methods), and checking the assumed distribution through tests as Pearson, Anderson-Darling, and Kolmogorov-Smirnov.
3. Deriving the value of annual maximum rainfall over the fixed duration for a specific value of T_r and corresponding probability of nonexceedance.
4. Repeating all previous steps for different durations.
5. Fitting the point cloud in the plot annual rainfall maxima versus duration for the given T_r by a curve that can be represented through different equations. One of the most widely used is known as Montana Curve (Di Baldassarre et al., 2006a):

$$R_{D_r}(T_r) = aD_r^{\ b} \tag{13.21}$$

where a and b are parameters depending on T_r. A generalization of Eq. (13.21) in terms of intensity, $r_{D_r}(T_r) = \frac{R_{D_r}(T_r)}{D_r}$, is the following (Chow, 1988):

$$r_{D_r}(T_r) = \frac{cD_r^{\ m}}{D_r^{\ e} + f} \tag{13.22}$$

where c, e, f, and m are coefficients linked with T_r.

A typical trend of local DDF curves for different values of T_r is shown in Fig. 13.7.

From the estimated DDF curves at each rain gauge located in the basin and/or close to its boundaries (if any), a single set of DDF curves representative for the whole watershed can be obtained. In fact, an areal-reduction factor has to be applied to the last curves in order to upscale the information from point to areal scale (Flammini et al., 2022).

The duration of the design rainfall for the watershed can be determined in many different ways. For example, the rainfall duration could be assumed equal to the base of the watershed IUH, identified as the time period when the IUH discharge is greater than 10% of its peak value. This choice assures that the response to an instantaneous uniform effective rainfall is practically completely incorporated.

The total depth of the design rainfall derived from the areal DDF with selected T_r can be differently distributed in time. The alternating block method is a simple and widely used approach that considers n successive time intervals Δt within the total duration $D_r = n\Delta t$ (Chow, 1988). The depth for the durations Δt, $2\Delta t$, $3\Delta t \ldots n\Delta t$ is taken out from the DDF curve. The differences between successive rainfall depth values represent the rainfall amounts to be included in the design hyetograph. These

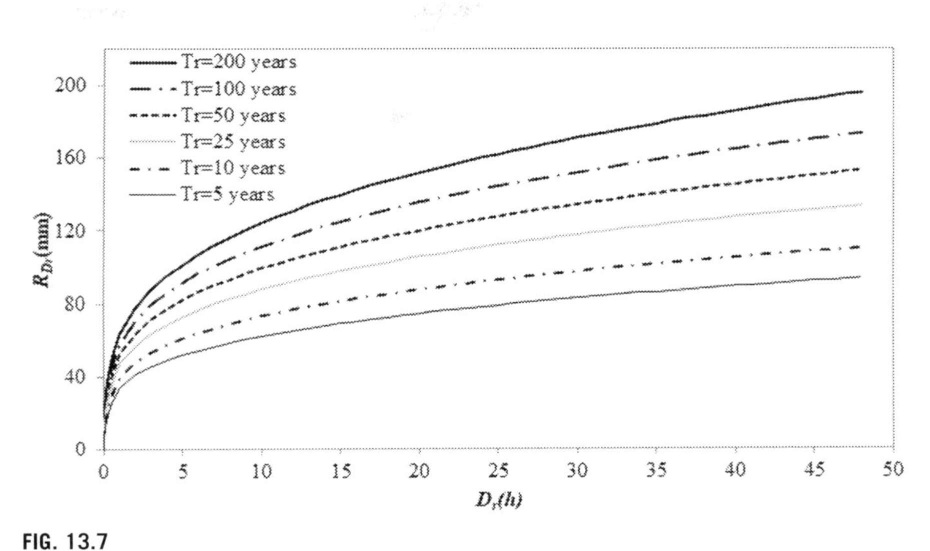

FIG. 13.7

Example of depth-duration-frequency (*DDF*) curves derived for San Biagio station, Umbria, Central Italy. T_r is the return period, D_r stands for rainfall duration, and R_{D_r} for rainfall depth with given nonexceedance probability.

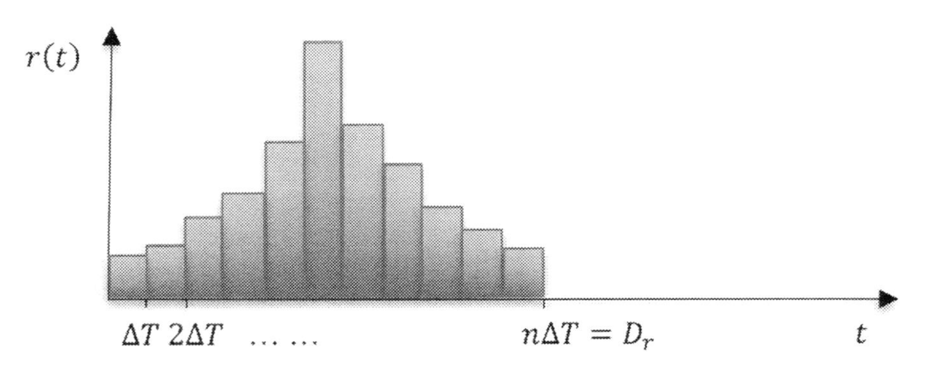

FIG. 13.8

Example of alternating block method to derive a design rainfall hyetograph (Chow, 1988).

quantities, or blocks, are reordered into a time sequence with the maximum at the center of the duration D_r and the remaining ones arranged in descending order alternately to the right and left of the central block to form the design hyetograph (see Fig. 13.8).

The obtained design rainfall can be then used as a spatially uniform input in a hydrological model of rainfall-runoff type to derive the design hydrograph.

13.6 Conclusions

Understanding the quantitative link between heavy rainfall and flood generation is a basic element in the hydrological practice. The described rainfall-runoff modeling satisfies this objective using a flexible framework that allows to include representations with different approximation level of each incorporated physical process. For example, changing the basin map scale it is possible to increase the detail in the representation of the drainage channel network and the number of homogeneous areas that constitute the watershed. This choice can be made on the basis of the basin dimension, the expected variability of rainfall and that of the main quantities influencing the infiltration loss. In principle a greater model distribution level is closer to the physical reality but typically requires a difficult calibration of a larger number of quantities/parameters.

In the application of the model to extreme events the same design hyetograph over all the watershed sub-areas is used because of the approach adopted to derive the empirical relationships available for areal reduction (Flammini et al., 2022).

The model in the version for real-time flood forecasting can be used to provide forecasts which, in the absence of reliable forecast of rainfall, can be considered to be acceptable for lead times T much lower than the basin lag. These forecasts of discharge can have in any case a significant applicative role for large basins.

Finally, selecting a previously observed flood event and performing flow forecasts at a time origin t_c for given values of T, but including also rainfall observed between t_c and $t_c + T$ in the computations, it is possible to assess the accuracy of the model framework independently of the lack of rainfall forecasts.

References

Beven, K., Kirkby, M.L., 1979. A physically based variable contributing area model of basin hydrology. Hydrol. Sci. Bull. 24 (1), 43–69.

Borga, M., Marra, F., Gabella, M, 2022. Rainfall estimation by weather radar. In: Morbidelli, R (Ed.), Rainfall. Modeling, Measurement and Applications. Elsevier, Amsterdam, pp. 109–134. doi:10.1016/C2019-0-04937-0.

Chow, V.T., Maidment, D.R., Mays, L.W, 1988. Applied Hydrology. McGraw-Hill, New York.

Corradini, C., Singh, V.P, 1985. Effect of spatial variability of effective rainfall on direct runoff by a geomorphologic approach. J. Hydrol. 81 (1-2), 27–43.

Corradini, C., Melone, F., Singh, V.P, 1987. On the structure of a semi-distributed adaptive model for flood forecasting. Hydrol. Sci. J. 32 (2), 227–242.

Corradini, C., Melone, F., Singh, V.P, 1995. Some remarks on the use of the GIUH in the hydrological practice. Nord. Hydrol. 26 (4-5), 297–312.

Corradini, C., Melone, F., Smith, R.E, 1997. A unified model for infiltration and redistribution during complex rainfall patterns. J. Hydrol. 192 (1-4), 104–124.

Corradini, C., Morbidelli, R., Saltalippi, C., Flammini, A, 2022. Meteorological systems producing rainfall. In: Morbidelli, R (Ed.), Rainfall. Modeling, Measurement and Applications. Elsevier, Amsterdam, pp. 27–48. doi:10.1016/C2019-0-04937-0.

Di Baldassarre, G., Brath, A., Montanari, A, 2006. Reliability of different depth-duration-frequency equations for estimating short-duration design storms. Water Resour. Res. 42 (12), W12501.

Dingman, S.L, 2015. Physical Hydrology. Waveland Press, Long Grove.

Flammini, A., Dari, J., Corradini, C., Saltalippi, C., Morbidelli, R, 2022. Areal reduction factor estimate for extreme rainfall events. In: Morbidelli, R (Ed.), Rainfall. Modeling, Measurement and Applications. Elsevier, Amsterdam, pp. 285–306. doi:10.1016/C2019-0-04937-0.

Govindaraju, R.S., Goyal, A, 2022. Rainfall and Infiltration. In: Morbidelli, R (Ed.), Rainfall. Modeling, Measurement and Applications. Elsevier, Amsterdam, pp. 367–396. doi:10.1016/C2019-0-04937-0.

Grabowski, W., 2022. Rainfall Modeling. In: Morbidelli, R (Ed.), Rainfall. Modeling, Measurement and Applications. Elsevier, Amsterdam, pp. 49–76. doi:10.1016/C2019-0-04937-0.

Gupta, V.K., Waymire, E., Wang, C.T, 1980. A representation of an instantaneous unit hydrograph from geomorphology. Water Resour. Res. 16 (5), 855–862.

Kidd, C, Levizzani, V, 2022. Satellite Rainfall Estimation. In: Morbidelli, R (Ed.), Rainfall. Modeling, Measurement and Applications. Elsevier, Amsterdam, pp. 135–170. doi:10.1016/C2019-0-04937-0.

Kottegoda, N.T., Rosso, R, 2008. Applied Statistics for Civil and Environmental Engineers. Wiley-Blackwell Ed., Hoboken, New Jersey, US.

Lanza, L., Cauteruccio, A., Stagnaro, M, 2022. Rain gauge measurements. In: Morbidelli, R (Ed.), Rainfall. Modeling, Measurement and Applications. Elsevier, Amsterdam, pp. 77–108. doi:10.1016/C2019-0-04937-0.

Melone, F., Corradini, C., Singh, V.P, 1998. Simulation of the direct runoff hydrograph at basin outlet. Hydrol. Process 12 (5), 769–779.

Melone, F., Corradini, C., Singh, V.P, 2002. Lag prediction in ungauged basins: an investigation through actual data of the upper Tiber river valley. Hydrol. Process 16 (5), 1085–1094.

Morbidelli, R., Corradini, C., Govindaraju, R.S, 2006. A field-scale infiltration model accounting for spatial heterogeneity of rainfall and soil saturated hydraulic conductivity. Hydrol. Process 20 (7), 1465–1481.

Morbidelli, R., Saltalippi, C., Flammini, A., Cifrodelli, M., Picciafuoco, T., Corradini, C., Casas-Castillo, M.C., Fowler, H.J., Wilkinson, S.M, 2017. Effect of temporal aggregation on the estimate of annual maximum rainfall depths for the design of hydraulic infrastructure systems. J. Hydrol. 554, 710–720.

Morbidelli, R., Corradini, C., Saltalippi, C., Flammini, A, 2022. Time resolution of rain gauge data and its hydrological role. In: Morbidelli, R (Ed.), Rainfall. Modeling, Measurement and Applications. Elsevier, Amsterdam, pp. 171–216. doi:10.1016/C2019-0-04937-0.

Ogden, F.L., Warner, G.S, 2017. Watershed runoff, streamflow generation, and hydrologic flow regimens. In: Singh, VP (Ed.), Handbook of Applied HydrologySecond Edition. McGraw Hill, New York, US, p. 12 chapter 49.

Reed D.W. A Review of British Flood Forecasting Practice. Tech. Report No. 90, Institute of Hydrology, Wallingford, England, p. 113, 1984.

Rodríguez-Iturbe, I., Valdes, J.B, 1979. The geomorphologic structure of hydrologic response. Water Resour. Res. 15 (6), 1435–1444.

Singh, V.P, 1988. Hydrologic Systems: Rainfall-Runoff Modeling 480. Prentice Hall, Englewood Cliffs, New Jersey.

Singh, V.P., Zhang, L, 2017. Frequency distributions. In: Singh, VP (Ed.), Handbook of Applied HydrologySecond Edition. McGraw Hill, p. 11 chapter 21.

Singh, V.P., Jain, S.K, 2017. Rainfall-runoff modeling. In: Singh, VP (Ed.), Handbook of Applied Hydrology Second Edition, McGraw Hill, New York, US, p. 8 chapter 59.

Woolhiser, D.A., Smith, R.E., Goodrich, D.C., 1990. KINEROS, A Kinematic Runoff and Erosion Model: Documentation and User Manual, RS-77. US Dept. of Agriculture, Agricultural Research Service, Washington, DC, p. 130.

CHAPTER

Rainfall and infiltration 14

Rao S. Govindaraju, Abhishek Goyal
Lyles School of Civil Engineering, Purdue University, West Lafayette, IN, United States

14.1 Introduction

Infiltration plays a major role in the partitioning of rainfall into surface runoff and subsurface water which governs the transport of pollutants through the vadose zone and groundwater recharge. Hillel (1980) and Brutsaert (2005) defined infiltration as the entry of water into the soil surface and its subsequent movement away from the soil surface within the unsaturated zone. Understanding the infiltration process is critical to modeling unsaturated and groundwater flow, and the associated transport of pollutants in surface and subsurface waters.

There are myriad challenges that one faces in modeling infiltration—from challenges in assessing the variation in parameter values due to spatial heterogeneity to the uncertainty about the physics at different scales. For example, soil saturated hydraulic conductivity (K_s) can vary by as much as twelve orders of magnitude when compared between compacted clays and gravel (Kutílek and Nielsen, 1994).

The spatio-temporal evolution of infiltration rate under natural conditions cannot be readily deduced by direct measurements at any scale of interest in applied hydrology, therefore the use of infiltration models that allow transfer of knowledge of measurable quantities over multiple scales is of fundamental importance.

The first set of models focus on the point- or local-scale infiltration. Numerous local infiltration models for vertically homogeneous soils with constant initial soil water content and horizontal surfaces have been proposed (Green and Ampt, 1911; Kostiakov, 1932; Horton, 1933, 1939; Philip, 1957a, 1957b, 1957c; Holtan, 1961; Smith and Parlange, 1978; Parlange et al. 1982; Dagan and Bresler, 1983; Swartzendruber, 1987; Broadbridge and White, 1988; Smith et al., 1993; Corradini et al., 1994; Corradini et al., 1997; Wang et al. 2017). Furthermore, extended forms of the Green-Ampt model (Mein and Larson, 1973; Chu, 1978), the Philip model (Chow et al., 1988), and the Smith-Parlange model (Parlange et al., 1982) are used for isolated storms and for pre- and post-ponding conditions. However, these become inapplicable for complex rainfall patterns involving rainfall hiatus periods. Modifications were offered by Mls (1980), Péschke and Kutílek (1982), Verma (1982), using the time compression approximation (Reeves and Miller, 1975) for post-hiatus rainfall producing immediate ponding. However, the last approach neglects the soil water

RAINFALL. Modeling, Measurement and Applications. DOI: https://doi.org/10.1016/B978-0-12-822544-8.00007-X

redistribution process which is particularly important when long periods with light or no rainfall occur (Smith et al., 1993). Models combining infiltration and redistribution are better suited for complex rainfall patterns which are frequent in nature.

In applied hydrology, upscaling of point infiltration models to the field scale is required to estimate the areal-average infiltration. This is a complex task because of spatial heterogeneity of soil hydraulic properties (Nielsen et al., 1973; Warrick and Nielsen, 1980; Greminger et al., 1985; Sharma et al., 1987; Loague and Gander, 1990), and requires computationally expensive simulations. For example, Smith and Hebbert (1979), Maller and Sharma (1984), Saghafian et al. (1995), Corradini et al. (1998) used distributed/semi-distributed rainfall-runoff models with local-scale infiltration models and Monte Carlo (MC) simulations of K_s, but these methods may be computationally expensive for practical considerations. Sivapalan and Wood (1986) also used MC simulations to approximate simplified field-scale solution of infiltration rate considering K_s as a lognormal random variable and assuming constant rainfall rates, however the averaging procedure was applied in space over a single realization and the behavior of the resulting errors was not specified. Woolhiser et al. (1996) incorporated the effects of runon phenomena in areas of low K_s that were then followed by areas of higher K_s downstream by solving the kinematic wave equation with the Smith-Parlange infiltration model, using a numerical finite-difference scheme.

For practical applications, it is often desirable to use simplified upscaled models to parameterize semi-empirical approaches, and to serve as a benchmark to validate semi-analytical models. Along these lines Govindaraju et al. (2001) developed a semi-analytical/conceptual model for estimating the expected areal-average infiltration into vertically homogeneous soils with a spatial lognormal distribution of K_s, under the conditions of negligible runon. The model was extended by Govindaraju et al. (2006) to include the coupled effects of spatial variation of K_s as a lognormal distribution and spatial variation of rainfall rate, r, modeled by a uniform distribution. Furthermore, Morbidelli et al. (2006) formulated a more complete mathematical model for the expected areal-average infiltration that combines the aforementioned semi-analytical approach with a semi-empirical/conceptual component to represent of the runon process.

Since the actual infiltration occurs mostly over significantly sloping surfaces (Beven, 2002; Fiori et al., 2007), there has been a wide interest to study the impacts of slope on infiltration. However, there has been little consensus in the conclusions from various theoretical (Philip, 1991; Chen and Young, 2006; Wang et al., 2018) and experimental (Nassif and Wilson, 1975; Sharma et al., 1983; Poesen, 1984; Cerdà and García-Fayos, 1997; Fox et al., 1997; Chaplot and Le Bissonais, 2000; Janeau et al., 2003; Assouline and Ben-Hur, 2006; Essig et al., 2009; Ribolzi et al., 2011; Patin et al., 2012; Lv et al., 2013; Morbidelli et al., 2015; Morbidelli et al., 2016) investigations. The process is confounded by factors such as rainfall intensity, microtopography, vegetation, soil texture, and vertical and horizontal heterogeneity in soil properties, suggesting an improved understanding (Morbidelli et al., 2019) and modeling of infiltration over sloping surfaces is required.

To isolate the effects of rainfall over infiltration, researchers (Sidiras and Roth, 1987; Boers, et al., 1992; Gupta et al., 1993; Gupta et al., 1994; Morbidelli et al.,

2017) used rainfall simulators to exert control on the applied water. Although the controlled experiments can potentially generate more detailed data compared to experiments using natural rainfall, the process could be expensive. Careful considerations have to be given to the rain drop size, the height of fall and the allowable range of intensities. In addition, Goyal et al. (2019) argued that infiltration measurements obtained from rainfall-runoff experiments are fundamentally limited by the nature of rainfall. For an input rainfall rate, r, calibration methods would only lead to a conditional estimation of model parameters as a fraction of the field with K_s greater than r would essentially remain unresolved (or unidentifiable).

A comprehensive overview of the progress made with physically based models in hydrology, in last 50 years, was published by Paniconi and Putti (2015). It focuses on how the interplay between theory, experiments, and modeling has contributed to advancing the state-of-the-art.

14.2 Rainfall-infiltration process

Considering a horizontally homogeneous one-dimensional vertical system with its upper boundary at the ground surface and a gravity drainage lower boundary, the water movement in the vertical direction is governed by soil water flow and continuity equations. The flow rate per unit cross-sectional area, q, is described by Darcy's equation, proposed by Buckingham (1907), as:

$$q = -K\left(\frac{\partial \psi}{\partial z} - 1\right), \tag{14.1}$$

where K is the hydraulic conductivity, ψ the soil water matric capillary head and z the vertical soil depth assumed positive vertically downward.

The continuity equation, with a constant water density and porosity as well as absence of sinks and sources, is expressed by:

$$\frac{\partial \theta}{\partial t} = \frac{\partial q}{\partial z}, \tag{14.2}$$

where θ is the volumetric water content and t is the time. Substituting Eq. (14.1) in Eq. (14.2) leads to the well-known and widely used one-dimensional version of Richards' equation:

$$\frac{\partial \theta}{\partial t} = \frac{\partial}{\partial z}\left(D(\theta)\frac{\partial \theta}{\partial z}\right) - \frac{dK}{d\theta}\frac{\partial \theta}{\partial z}, \tag{14.3}$$

where $D(\theta)$ is the soil water diffusivity. The initial condition at time $t = 0$ for $z \geq 0$ is $\psi = \psi_i$, and the upper boundary conditions at the soil surface are:

$$z = 0, \qquad 0 < t \leq t_p, \qquad q_0 = r; \tag{14.4a}$$

$$z = 0, \qquad t_p < t \le t_r, \qquad \theta_0 = \theta_s; \tag{14.4b}$$

$$z = 0, \qquad t_r < t, \qquad q_0 = 0; \tag{14.4c}$$

where r is the rainfall rate, t_p the time to ponding, t_r the duration of rainfall; hereafter the subscripts i and s denote initial and saturation quantities, respectively, and the subscript 0 stands for quantities at the soil surface. The lower boundary conditions at a depth z_b which is not reached by the wetting front is $\psi(z_b) = \psi_i$ for $t > 0$. The soil water hydraulic properties can be represented by the following parameterized forms (Smith et al., 1993):

$$\psi = \psi_b \left[\left(\frac{\theta - \theta_r}{\theta_s^* - \theta_r} \right)^{-c/\lambda} - 1 \right]^{1/c} + d, \tag{14.5a}$$

$$K = K_s^* \left[1 + \left(\frac{\psi - d}{\psi_b} \right)^c \right]^{-(b\lambda + a)/c}, \tag{14.5b}$$

where θ_s^* and K_s^* are used as scaling quantities, ψ_b is the air entry head, θ_r is the residual volumetric water content, c, λ and d are empirical coefficients, $b = 3$ and $a = 2$ according to Burdine's method (Brooks and Corey, 1964). For particular values of the parameters, Eqs. (14.5a–b) reduce to the equations proposed by Brooks and Corey (1964) and van Genuchten (1980). For two-layered soils, two additional conditions are required at the interface between the two layers:

$$\psi_1(Z_C) = \psi_2(Z_C) = \psi_C, \tag{14.6a}$$

$$K_1 \left[\left(\frac{\partial \psi_1}{\partial z} \right)_{Z_C} - 1 \right] = K_2 \left[\left(\frac{\partial \psi_2}{\partial z} \right)_{Z_C} - 1 \right], \tag{14.6b}$$

where the subscripts 1, 2 and C denote variables in the upper layer, lower layer and at the interface, respectively, and Z_C is the interface depth.

14.3 Point infiltration methods

Due to the highly non-linear nature of the flow process in the unsaturated zone of soils, there is a near equality between the infiltration relations for the inflow rate under rainfall and that under unlimited water supply at the surface (Smith, 2002; Smith and Goodrich, 2005). This allows one to use infiltration models to predict both the onset of ponding under rainfall and the infiltration behavior for post-ponding periods (Smith and Goodrich, 2005).

Infiltration models can be broadly categorized into two types: empirical models that have no rigorous physical basis and models based on relationships derived from solutions of equations of water transport in a porous medium (Rumynin, 2015).

Among the numerous models proposed to describe infiltration (see Section 14.1), empirical models of Kostiakov (1932) and Horton (1933; 1939) and physical/analytical models proposed by Green and Ampt (1911), Philip (1957a, 1957b, 1957c), Parlange et al. (1982) are widely recognized in the scientific literature. These models have been extensively used in their original versions with vertically homogeneous soils and constant initial soil water content, as well as in their extended formulations in applied hydrology, and also as local models to support the development of infiltration approaches at the field scale.

A two-layer approximation, with each layer treated as being homogeneous, is frequently used to set up models of infiltration for natural soils. The process of formation of a sealing layer was examined by Mualem and Assouline (1989) and Mualem et al. (1993), while that of disruption was considered by Bullock et al. (1988), Emmerich, 2003, Morbidelli et al. (2011). On the other hand, vertical profiles with a more permeable upper layer are frequently found in practice and can be also used, for example, as a first approximation in the representation of infiltration into homogeneous soils with grassy vegetation (Morbidelli et al., 2014).

14.3.1 Empirical equations

Many empirical expressions have been developed to describe the infiltration rate as a function of time, *t*, using monotonically decreasing functions based on exponential or power law decays. The coefficients of these equations are evaluated by fitting to experimental data and may offer limited physical interpretation (Haverkamp et al., 1988; Kutílek et al., 1988).

Kostiakov (1932) used a power law of the form

$$f_c = \eta t^{-\alpha}, \tag{14.7}$$

where f_c is the soil infiltration capacity, and η and α are empirical coefficients. The coefficient $\alpha \in (0, 1)$, hence, $f_c \to 0$ as $t \to \infty$. Eq. (14.7) has been modified by researchers such as Mezencev (1948), Smith (1972), Furman et al. (2006), Parhi et al. (2007), Valiantzas et al. (2009) to overcome the shortcoming by including various initial and boundary conditions. A modified form of Eq. (14.7) is (Smith, 1972):

$$f_c = \begin{cases} f_f + \eta t^{-\alpha}, & t \le t_p, \\ f_f + \eta\left(t - t_p\right)^{-\alpha}, & t > t_p, \end{cases} \tag{14.8a, b}$$

where f_f is the value of f_c at saturation.

Horton (1933, 1939) represented the infiltration capacity using an exponential function as

$$f_c = f_f + \left(f_i - f_f\right)\exp(-\gamma t), \tag{14.9}$$

where f_i represent the initial value of f_c, and γ is the decay constant. As $t \to \infty$, f_f becomes approximately equal to K_s.

14.3.2 Green-Ampt model

In the understanding of groundwater flow, it is often required to describe the temporal dynamics of a one-dimensional vertical wetting front advancing downward into a soil following a rainfall event (Brutsaert, 2005). The Green-Ampt model relies on fundamental physics but makes strong assumptions about the soil hydraulic properties and the shape of the wetting front to achieve an exact analytical solution. It represents infiltration into homogeneous soils under the conditions of continuously saturated soil surface and uniform initial soil moisture as:

$$f_c = K_s\left[1 - \frac{\psi(\theta_s - \theta_i)}{F}\right], \tag{14.10}$$

where F is the cumulative depth of infiltrated water. To express infiltration as a function of time, this equation can be solved after the substitution $f_c = dF/dt$, and integrating to obtain (Chow et al., 1988):

$$F = K_s t - \psi(\theta_s - \theta_i)\ \ln\left[1 - \frac{F}{\psi(\theta_s - \theta_i)}\right]. \tag{14.11}$$

Eq. (14.11) assumes ponded conditions at the soil surface and no limitation in supply of water. Under conditions of a constant rainfall rate $r > K_s$, which begins at the time $t = 0$, surface saturation is reached at a time $t_p > 0$. For $t \le t_p$ the infiltration rate q_0 is equal to r and subsequently to the infiltration capacity. Mein and Larson (1973) formulated this process through Eq. (14.10) as:

$$r - K_s = -\frac{K_s\psi(\theta_s - \theta_i)}{\int_0^{t_p} r\,dt}, \quad t = t_p, \tag{14.12}$$

which determines t_p as:

$$t_p = -\frac{K_s\psi(\theta_s - \theta_i)}{r(r - K_s)}. \tag{14.13}$$

Furthermore, Eq. (14.11) becomes:

$$F = F_p - \psi(\theta_s - \theta_i)\ \ln\left[\frac{F - \psi(\theta_s - \theta_i)}{F_p - \psi(\theta_s - \theta_i)}\right] + K_s(t - t_p),\ t > t_p, \tag{14.14}$$

where F_p is the cumulative depth of infiltrated water at $t = t_p$.

Eq. (14.14) can be solved at each time, for example, by successive substitutions of F which is then substituted in Eq. (14.10) to obtain the corresponding value of f_c. Alternatively, utilizing series expansions and simplification by neglecting higher terms, the following explicit approximation to F is obtained (Govindaraju et al., 2001):

$$F = F_p + \left[2K_s\psi(\theta_s - \theta_i)\right]^{1/2}\left(t^{1/2} - t_p^{1/2}\right) + \frac{2}{3}K_s(t - t_p)$$
$$+ \frac{1}{18}\left[\frac{2K_s^3}{\psi(\theta_s - \theta_i)}\right]^{1/2}\left(t^{3/2} - t_p^{3/2}\right), \quad t > t_p. \tag{14.15}$$

Selkar and Assouline (2017) provided an explicit and accurate (within 1% of the exact implicit solution of vertical Green and Ampt infiltration) solution for the position of the wetting front in time based on approximating the term describing early time behavior by means of the sum of gravitational flow and the exact solution for capillary imbibition:

$$q = K_s + \frac{\omega K_s + \sqrt{\dfrac{\psi(\theta_s - \theta_i)K_s}{2t}}}{1 + \omega\dfrac{K_s t}{\psi(\theta_s - \theta_i)} + \sqrt{\dfrac{2K_s t}{\psi(\theta_s - \theta_i)}}}, \tag{14.16}$$

where ω is a fitting parameter modulating the gravitational component and approximated by 2/3.

14.3.3 Philip model

Philip (1957a, 1957b, 1957c, 1969) proposed an approximate analytical series solution of Richards' equation under the condition of vertically homogeneous soil, constant initial moisture content and saturated soil surface with immediate ponding. For early to intermediate times, the series solution may be restricted to the first two terms. The infiltration capacity can be expressed as:

$$f_c = \frac{1}{2}St^{-1/2} + A, \tag{14.17}$$

where S is the sorptivity, depending on soil properties and initial moisture content, and A is a quantity varying between 1/3 K_s and 2/3 K_s (Talsma and Parlange, 1972).

The sorptivity, S, is defined as:

$$S = \int_{\theta_i}^{\theta_s} \zeta \, d\theta, \tag{14.18}$$

where ζ is the independent variable resulting from the Boltzmann transformation:

$$\zeta = zt^{-1/2}. \tag{14.19}$$

The quantity S may also be approximated through the parameters of the Green-Ampt equation (see Youngs, 1964) as:

$$S = \left[2(\varphi - \theta_i) K_s |\psi|\right]^{1/2}, \tag{14.20}$$

where φ is the soil porosity.

It may be noted that the form of Eq. (14.17) is very similar to that of Eq. (14.8a). In fact, the equations become equivalent for $\alpha = ½$, $\eta = S/2$, and $f_f = A$.

For $t \to \infty$, Eq. (14.17) is replaced by $f_c = K_s$, and integration yields the cumulative infiltration:

$$F = St^{1/2} + At. \tag{14.21}$$

Philip's model was extended for applications to less restrictive conditions. For constant rainfall rate $r > K_s$, surface saturation occurs at a time $t_p > 0$, and following Chow et al. (1988), infiltration can be described through an equivalent time origin for potential infiltration after ponding, t_o, as:

$$t_p = \frac{S^2\left(r - \frac{A}{2}\right)}{2r(r - A)^2}, \tag{14.22}$$

$$t_o = t_p - \frac{1}{4A^2}\left[\left(S^2 + 4AF_p\right)^{1/2} - S\right]^2, \tag{14.23}$$

$$f_c = \frac{1}{2}S(t - t_o)^{-1/2} + A, \quad t > t_p, \tag{14.24}$$

For unsteady rainfall under the condition of a continuously saturated surface for $t > t_p$, the infiltration process can be represented adopting a similar procedure as in Chow et al (1988).

14.3.4 Parlange-Lisle-Braddock-Smith model

Parlange et al. (1982) introduced a quasi-exact solution obtained through an analytical integration of Richards' equation expressed for the entire time range but limited to non-ponded conditions, as:

$$f_c = K_s\left\{1 + \frac{\beta}{\exp\left[\frac{\beta F'}{G(\theta_s - \theta_i)}\right] - 1}\right\}, \tag{14.25}$$

where $F' = F - Kit$ is the cumulative dynamic infiltration rate with K_i as the value of K at θ_i, β a parameter linked with the behavior of hydraulic conductivity and soil water diffusivity, D, as functions of θ, and G is the integral capillary drive defined by

$$G = \frac{1}{K_s} \int_{\theta_i}^{\theta_s} D(\theta) d\theta. \tag{14.26}$$

The value of β usually varies between 0.8 to 0.85 (Smith, 2002) for a range of soils. Eq. (14.25) reduces to the Green-Ampt equation for $\beta = 0$. It can be applied to determine t_p and f_c for any rainfall pattern, and for $t > t_p$, can be rewritten under the condition of surface saturation as:

$$\left[(1-\beta)K_s - K_i\right](t - t_p) = F' - F'_p - \frac{K_s G(\theta_s - \theta_i)}{K_d} \ln \left\{ \frac{\exp\left[\frac{\beta F'}{G(\theta_s - \theta_i)}\right] - 1 + \frac{\beta K_s}{K_d}}{\exp\left[\frac{\beta F'_p}{G(\theta_s - \theta_i)}\right] - 1 + \frac{\beta K_s}{K_d}} \right\}, \tag{14.27}$$

where $K_d = K_s - K_i$ and $F'_p = F'(t_p)$. The quantities F'_p and t_p are the values of F' and t, respectively, at which Eq. (14.25) with $f_c = r(t_p)$ is first satisfied.

14.4 Rainfall-infiltration process at field scale

Even though several models describe infiltration into vertically-homogeneous soils at the point scale fairly accurately, upscaling of these models to the field scale has been complicated even in the absence of runon. Natural watersheds exhibit spatial heterogeneity in topography, surface roughness, vegetation, and soil infiltration characteristics. Field measurements of soil properties (Nielsen et al., 1973; Sharma et al., 1987; Loague and Gander, 1990, Morbidelli et al., 2017) and infiltration characteristics (Grah et al., 1983; Flammini et al., 2018) have shown that these properties exhibit a high degree of spatial variability. This behavior is particularly evident in K_s, which is shown to manifest large variations for even seemingly uniform areas (Nielsen et al., 1973; Russo and Bresler, 1981; Morbidelli et al., 2017). Assouline and Mualem (2002, 2006) demonstrated the combined impact of impervious areas and spatial variability in soil properties on infiltration. Accounting for field spatial variability reduces the ponding times (Smith and Hebbert, 1979; Woolhiser and Goodrich, 1988), and early surface runoff in heterogeneous fields when compared to homogeneous ones. Also, parameters relevant for infiltration, as well as experiments for direct infiltration measurements, are typically observed and performed at the point scale. As a result, the point-scale infiltration models are limited as they do not explicitly account for the heterogeneity commonly encountered at larger scales that significantly impacts the responses of a field to a rainfall event (Sharma et al., 1979; DeRoo et al., 1992).

In addition to the heterogeneity of soil hydraulic properties, experimental studies show that rainfall also exhibits spatial variability (Goodrich et al., 1995; Krajewski et al., 2003; Corradini et al., 2022), and the variability effects of K_s and r on spatially-averaged infiltration and surface runoff are not entirely decoupled. Govindaraju et al. (2006) and Morbidelli et al. (2006) demonstrated that a uniform probability distribution function (PDF) of rainfall rate allows for a reasonable representation of spatial variability of rainfall over a field.

Various approaches have been proposed to upscale the infiltration-related parameters, such as similarity scaling (Miller and Miller, 1956; Nielsen et al., 1998; Zhu and Mohanty, 2006; Montzka et al., 2017), aggregation, and Bayesian upscaling (Jana and Mohanty 2012a, 2012b, 2012c). This chapter will be restricted to physical upscaling in which PDFs are used to upscale local infiltration to the field scale. These models represent a useful representation of field-scale hydrologic behavior without having to resort to expensive computational schemes. Three models of varying complexity and application domain are presented here. A detailed review of the various scaling approaches can be found in Vereecken et al. (2019).

14.4.1 A semi-empirical approach

Smith and Goodrich (2000) developed a semi-empirical field-scale infiltration model reflecting the random spatial variability of K_s. They assumed a lognormal PDF of K_s with a mean value $E[K_s]$ and a coefficient of variation $\mathrm{CV}(K_s)$. Employing the Parlange et al. (1982) model coupled with Latin Hypercube sampling through a large number of simulations performed for many values of $\mathrm{CV}(K_s)$ and rainfall rate, the authors developed the following relation for the scaled areal-average infiltration rate, I_e^*, linked with the corresponding scaled cumulative depth, F_e^*:

$$I_e^* = 1 + (r_e^* - 1)\left\{1 + \left[\frac{r_e^* - 1}{\beta}(\mathrm{e}^{\beta F_e^*} - 1)\right]^{C_c}\right\}^{-1/C_c}, \quad r_e^* > 1, \tag{14.28}$$

where $I_e^* = I_e / K_e$, $F_e^* = F_e / \left[G(\theta_s - \theta_i)\right]$, and $r_e^* = r / K_e$, with K_e being the areal effective value of K_s determined by the PDF of K_s and r. Here, I_e and F_e denote the areal expected infiltration rate and cumulative infiltration, respectively.

The parameter c_c describing the curvature of the ensemble I_e^* relation in the region of ponding development is given by

$$c_c \cong 1 + \frac{0.8}{\left[\mathrm{CV}(K_s)\right]^{1.3}}\left[1 - \exp\left(-0.85\left(r_b^* - 1\right)\right)\right], \tag{14.29}$$

where $r_b^* = r / E[K_s]$.

As $\mathrm{CV}(K_s) \to 0$, c_c becomes large and $K_e \to E[K_s]$, and with r_e^* becoming large, Eq. (14.28) approaches the behavior of the scaled soil infiltration capacity rate given by Smith et al. (1993).

In contrast to the Green-Ampt equation, Eq. (14.28) describes the variation of actual (ensemble) infiltration rate I_e, implying Eq. (14.28) can be used throughout a temporally variable rainfall with changes in r_e^* as the storm progresses ($r_e^* > 1$), using ensemble F_e^* as the continuous independent variable.

14.4.2 A semi-analytical/conceptual model

Govindaraju et al. (2001) formulated a semi-analytical model, based on the extended form of the Green-Ampt model, to estimate the expected field-scale infiltration rate. The heterogeneity of K_s was incorporated as a spatially auto-correlated random variable with a lognormal PDF. Using the cumulative infiltration amount, F, as the independent variable, the field-scale infiltration rate, $E[I(F)]$, under a rainfall rate, r, invariant with time, t, is given by

$$E[I(F)] = r[1 - M(K_c, 0)] + \frac{\psi(\theta_s - \theta_i) + F}{F} M(K_c, 1), \tag{14.30}$$

where K_c denotes the maximum value of K_s leading to surface saturation and is determined by:

$$K_c = \frac{Fr}{\psi(\theta_s - \theta_i) + F} = F_c r. \tag{14.31}$$

The first and second terms on the right-hand side of Eq. (14.30) represent the contributions of the non-ponded and ponded areas, respectively. The term $M(K_c, \xi)$ is the ξ-th partial moment of K_s over $[0, K_c]$, and is given by (Chen et al., 1994)

$$\begin{aligned} M(K_c, \xi) &= \int_0^{K_c} k^{\xi} f_{K_s}(k)dk \\ &= \exp\left(\xi \mu_Y + \frac{\sigma_Y^2 \xi^2}{2}\right) \cdot \left[1 - \frac{1}{2} erfc\left(\frac{\ln K_c - \mu_Y}{\sqrt{2}\sigma_Y} - \frac{\sigma_Y \xi}{\sqrt{2}}\right)\right], \end{aligned} \tag{14.32}$$

where $f_{K_s}(k)$ is the PDF of K_s, and μ_Y and σ_Y are the mean and standard deviation, respectively, of the Gaussian random variable $Y = \ln(K_s)$, and $erfc(.)$ is the complementary error function.

The expected time to infiltrate a depth of water, F, over the field may be written in the following generalized form applicable to unsteady rainfall as well (Govindaraju et al., 2001):

$$\begin{aligned} E[t] = t_0 &+ \frac{F}{r}(1 - M(K_c, 0)) - \frac{F_0}{r} + \left\{F + \psi(\theta_s - \theta_i)\ln\left[\frac{\psi(\theta_s - \theta_i)}{\psi(\theta_s - \theta_i) + F}\right]\right\} M(K_c, -1) \\ &+ \psi(\theta_s - \theta_i)\sum_{i=1}^{\infty} \frac{M(K_c, i)}{(i+1)r^{i+1}}, \end{aligned} \tag{14.33}$$

where, considering a stepwise rainfall rate represented in successive pulses, t_0 is the time at the beginning of a rainfall pulse of intensity r and F_0 is the cumulative infiltration amount at t_0.

The infinite series in the last term of Eq. (14.33) converges rapidly and can be approximated by the first five terms for most applications. Eq. (14.33) is an implicit relation between t and F; therefore, for a given expected time, $E[t]$, F must be found out using numerical techniques.

This model was extended by Govindaraju et al. (2006) to also incorporate heterogeneity of r, assuming a uniform PDF. The quantity $E[I(F)]$, is estimated through the averaging procedure over the ensemble of two-dimensional realizations of K_s and r, and can be written as:

$$E\left[I(F)\right]=\int_{0}^{\infty}\int_{K_c}^{\infty} r f_r(r) f_{K_s}(k)\, dr\, dk + \int_{0}^{\infty}\int_{0}^{K_c}\left[1+\frac{\psi(\theta_s-\theta_i)}{F}\right] k f_r(r) f_{K_s}(k)\, dr\, dk, \tag{14.34}$$

where $f_r(r)$ is the PDF of r.

Eq. (14.34) can be expressed as:

$$\begin{aligned}E\left[I(F)\right]=&\frac{1}{2RF_c^2}\left\{M\left[(r_{\min}+R)F_c,2\right]-M\left[r_{\min}F_c,2\right]\right\}\\&-\frac{r_{\min}^2}{2R}\left\{M\left[(r_{\min}+R)F_c,0\right]-M\left[r_{\min}F_c,0\right]\right\}\\&+\left(r_{\min}+\frac{R}{2}\right)\left\{1-M\left[(r_{\min}+R)F_c,0\right]\right\}\\&+\frac{1}{F_c}\left(\frac{r_{\min}+R}{R}\right)\left\{M\left[(r_{\min}+R)F_c,1\right]-M\left[(r_{\min}F_c),1\right]\right\}\\&-\frac{1}{F_c}\left(\frac{1}{RF_c}\right)\left\{M\left[(r_{\min}+R)F_c,2\right]-M\left[(r_{\min}F_c),2\right]\right\}+\frac{1}{F_c}M\left[r_{\min}F_c,1\right],\end{aligned} \tag{14.35}$$

with $r_{\min}$ and $r_{\min}+R$ as the lower and upper bounds of the uniform PDF of r, respectively.

To relate time to F, an implicit relation between the expected value of t, $E[t]$, and F is provided as:

$$\begin{aligned}E[t]=&\frac{F}{E[r]}\left\{1-M\left[E[r]F_c,0\right]\right\}+\left\{F+\psi(\theta_s-\theta_i)\ln\left[\frac{\psi(\theta_s-\theta_i)}{\psi(\theta_s-\theta_i)+F}\right]\right\}\\&\times M\left[E[r]F_c,-1\right]+\psi(\theta_s-\theta_i)\sum_{j=1}^{\infty}\frac{1}{(j+1)\left(E[r]\right)^{j+1}}M\left[E[r]F_c,j\right].\end{aligned} \tag{14.36}$$

To provide a more complete analysis, Morbidelli et al. (2006) extended the model by incorporating the runon effect, as additional empirical term:

$$E\left[\tilde{I}(t)\right]\cong E\left[I(t)\right]+E[r]a_M\left(\frac{t}{t_p}\right)^{b_M}\exp\left(-c_M\frac{t}{t_p}\right), \tag{14.37}$$

where t_p is the time to ponding associated with $E[r]$ and $E[K_s]$. The parameters a_M, b_M and c_M are expressed by:

$$a_M = 2.8\left[\text{CV}(r)+\text{CV}(K_s)\right]^{0.36}, \tag{14.38a}$$

$$b_M = 5.35-6.32\left[\text{CV}(r)\,\text{CV}(K_s)\right], \tag{14.38b}$$

$$c_M = 2.7+0.3\left[\frac{E[r]/E[K_s]}{\text{CV}(r)\,\text{CV}(K_s)}\right]^{0.3}, \tag{14.38c}$$

where Eq. (14.38a) holds for $\theta_i << \theta_s$, and for $\theta_i \to \theta_s$ we have $a_M \to 0$.

The solution of the model even in the conditions of coupled spatial variability of r and K_s is fairly simple and requires little computational effort. The additional empirical term in Eq. (14.37) could be adapted for unsteady rainfall events following the guidelines in Morbidelli et al. (2006).

The model for coupled heterogeneity of r and K_s was validated by comparison with the results derived from MC sampling and using a combination of the extended Green-Ampt formulation at the local scale with the kinematic wave approximation (Singh, 1996) that is required to represent runon. It was shown that the model produced accurate estimates of $E[\tilde{I}]$ over a clay loam soil and a sandy loam soil, and the spatial heterogeneity of both r and K_s can be neglected only when $E[r] >> K_s$ or for storm durations much greater than t_p.

14.4.3 An explicit approximation of the upscaled Green and Ampt infiltration equation

Craig et al. (2010) provided a direct method to upscale the Green-Ampt infiltration solution, while addressing the variability in initial saturation, porosity, and ψ, besides the spatial variability in K_s. For homogeneous initial conditions, they proposed the mean infiltration rate, $E[I]$, with the infiltration rate a function of K_s, as:

$$E[I(t)] = \int_0^{\infty} I(X(t), K_s).f_{K_s}(k)dk, \tag{14.39}$$

where X represents a dimensionless time parameter given by:

$$X = \frac{1}{1+1/(r/\psi)t}. \tag{14.40}$$

An approximation to the Green-Ampt equation:

$$I(t) = \min\left(r, \frac{K_s}{X}\right)+\varepsilon(X, K_s), \tag{14.41}$$

with ε as a dimensionless error term, was substituted in Eq. (14.39) to approximate the $E[I(t)]$ as:

$$E\left[I(t)\right]=\frac{r}{2}erfc\left(\frac{\ln(rX)-\mu_Y}{\sqrt{2}\sigma_Y}\right)+\frac{1}{2X}\exp\left(\mu_Y+\frac{\sigma_Y^2}{2}\right)erfc\left(\frac{\ln(rX)-\mu_Y}{\sqrt{2}\sigma_Y}\right)$$
$$+r\int_0^{X(t)}\varepsilon\left(X(t),K_s\right).f_{K_s}(k)dk. \tag{14.42}$$

The last term in Eq. (14.42) has to evaluated numerically using the approximation based on Barry et al. (2005):

$$\varepsilon\approx 0.3632\left(1-X\right)^{0.484}\left(1-\frac{K_s}{rX}\right)^{1.74}\left(\frac{K_s}{rX}\right)^{0.38}. \tag{14.43}$$

The semi-analytical upscaling approach, which neglects the effects of runon or spatial correlation, was evaluated against MC simulations and produced results accurate to 3% of the computationally intensive 'exact' cases, which appears to be acceptable for many practical purposes.

14.5 Case study

To better highlight the complexities involved in the rainfall infiltration process, a case study using multiple rainfall-runoff experiments is presented. It investigates the effects on the variability of K_s revealed by different rainfall-runoff experiments. The issue is addressed by developing a consolidated posterior estimate of the range of PDFs of K_s, given the observations of areal-average infiltration rates from multiple experiments. The approach is useful for assessing the properties of the K_s random field which are required in field-scale infiltration models described in section 14.4.

14.5.1 Methodology

An MC model was formulated based on the simplified semi-analytical infiltration model developed by Govindaraju et al. (2001) as given by Eqs. (14.30)–(14.33). It could be hypothesized that the deterministic model, with fixed μ_Y and σ_Y, is one of the numerous models that could fit the experimental observations because, for each experiment at any time t, the expected areal-average infiltration observation, $(E[I])_{obs}$, could be corrupted with a Gaussian measurement error with zero mean and variance v as:

$$E\left[I\right]=\left(E\left[I\right]\right)_{\text{obs}}+\mathcal{N}\left(0,v\right). \tag{14.44}$$

The quantities μ_Y and σ_Y were chosen as independent uniformly distributed random variables with ranges [−5, 5] and (0, 5] (with K_s value expressed in mm/h), respectively. A total of 10,000 MC simulations consisting of 100 realizations each for both parameters were performed.

For an experiment h of a total of H experiments, the model simulated expected infiltration rate at time $t = \tau$ for the n-th of N MC realizations is given by

$$m_{h\tau n} = E_{\tau}\left[I_h \mid \mu_{Y,n}, \sigma_{Y,n}\right], \tag{14.45}$$

and $M_{hn} = \{m_{h\tau n}: \tau = \{1, ..., T_h\}\}$ denotes the vector of simulated infiltration rates for experiment h and parameter set $(\mu_{Y,n}, \sigma_{Y,n})$ for $h = \{1,..., H\}$ and $n = \{1,..., N\}$.The likelihood of a model given the observations is

$$L_{hn}(\mu_{Y,n}, \sigma_{Y,n} \mid M_{h,n}) = \prod_{\tau=1}^{T_h} N(m_{h\tau n} \mid o_{h\tau}, v), \tag{14.46}$$

where $\mathcal{N}(m_{h\tau n} \mid o_{h\tau}, v)$ is the Gaussian PDF with parameters $o_{h\tau}$ and v, evaluated at $m_{h\tau n}$.

Normalizing the likelihood in Eq. (14.46) gives the joint PDF of μ_Y and σ_Y for each experiment.

Due to the limitations of the data from experiments mentioned in section 14.1, multiple values of these two parameters could yield similar results in terms of goodness of fit with experimental data because of non-uniqueness, and the acceptable parameter space could be different for different experiments. Hence, for a consolidated estimate of K_s over the study area, the probability distributions obtained from multiple events were summarized using an information-based measure (Hill and Miller, 2011):

$$\Omega(x) = \frac{f_1(x)...f_J(x)}{\int_{-\infty}^{\infty} f_1(\ell)...f_J(\ell)d\ell}, \tag{14.47}$$

where $f_1,\ldots,f_J$ are J input PDFs and Ω is the consolidated distribution.

The consolidation of multiple distributions gleaned from each rainfall-runoff experiment into a single distribution leads to information loss; however, with Eq. (14.47), the loss is minimized (in terms of Shannon information) when compared to other forms of consolidation, such as averaging probabilities, averaging input data, etc.

14.5.2 Study area, soil and rainfall characteristics

As noted in Goyal et al. (2019), an 8.1 m × 8.7 m closed plot (Fig. 14.1), located close to University of Perugia, Italy, was constructed with an artificially laid silty loam soil (USDA classification) evenly packed to a thickness of 70 cm. It was carefully meshed to create a spatially uniform grain size distribution. A 15-cm layer of gravel laid beneath the soil to aid drainage, and the base and the sides were made impermeable to prevent water fluxes. Table 14.1 lists the main characteristics of the soil, incorporated in the functional forms expressing the hydraulic properties (Smith et al., 1993). The top surface of the study plot, having a gradient of 4%, was exposed to natural weather variations.

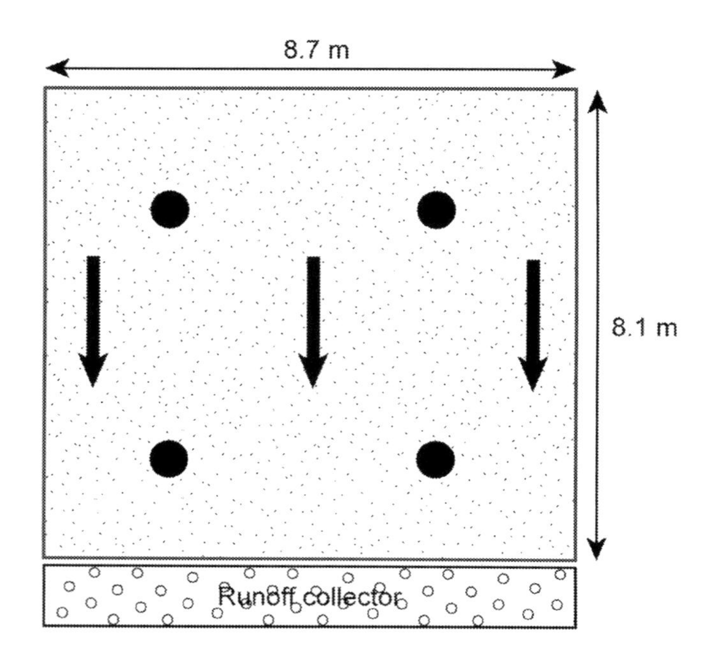

FIG. 14.1

Schematic representation of the experimental plot. The four circles indicate the locations of soil moisture measurements collected from a time domain reflectometer. The initial soil moisture content for an experiment was evaluated as the spatial average of soil moisture values obtained from the four locations. (Source: Goyal et al., 2019).

Table 14.1 Main characteristics of the study soil. (Source: Goyal et al., 2019).

Soil components	Clay		Silt		Sand	
Weight (%)	28		57		15	
Hydraulic properties	θ_r	θ_s	ψ_b (mm)	λ	b	d (mm)
	0.070	0.36	–500	0.2	5	50

The experiments were performed under natural rainfall conditions during the autumn-winter period of 2013/2014 (Flammini et al., 2018) when the soil surface was likely not affecited by the formation of sealing layers and cracks. A typical rainfall event is shown in Fig. 14.2. No lag was assumed between the rainfall occurrence and runoff generation during any time interval; hence, for an interval with no observed runoff, entire rainfall depth in the time period was considered to have infiltrated into the soil. Sixteen natural rainfall-runoff events were equally divided for model calibration and validation, to have two sets with similar characteristics in terms of

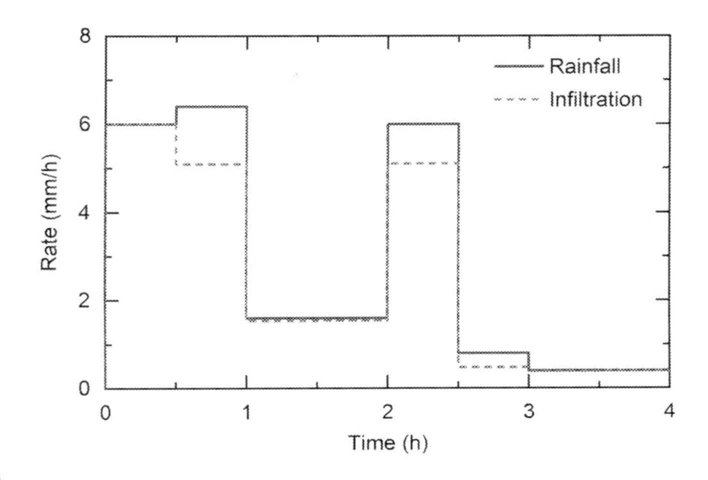

FIG. 14.2

Rainfall and infiltration rates observed during the event of November 2, 2013 (cal_4 of Table 13.2) in the study plot. (Source: Goyal et al., 2019).

observed total rainfall, surface runoff, and hydraulic parameters. Table 14.2 lists the characteristics of these rainfall-runoff events and summarizes the hydraulic characteristics of the study soil. θ_s for the soil was found to be 0.36.

14.5.3 Inference from rainfall-runoff experiments

In general, the addition of observations to a model should lead to a narrower distribution of the parameters. Fig. 14.3 shows the joint PDF of μ_Y and σ_Y at different time-steps for the calibration experiments cal_1, cal_2, and cal_4. The distributions remained mostly flat at the first time-step, $t = 0.5$ h. For experiments cal_4 and cal_5, large rainfall rates occurred in the first half-hour, thereby, constraining the distribution.

It is useful to note that a stable posterior distribution, which is typically governed by the highest rainfall rate for an experiment, was achieved and further data points on infiltration did not constrain the K_s distribution any further. The stable distribution indicates the most constrained parameter distribution obtained from the experiment. The stabilization times for the different runs are given in Table 14.3. For the experiments cal_1, cal_2, cal_3, and cal_8, the joint distribution stabilized as soon as the maximum rainfall intensity occurred for that experiment. For cal_4, cal_5, cal_6, and cal_7, the maximum rainfall rates were followed by rates of similar magnitudes, thereby, extending the effect of the highest rainfall rate, and the distribution stabilized only after all the contiguous high rainfall events ceased to occur.

The distribution shapes for K_s are also dictated by factors such as the time distribution of the rainfall during the event, antecedent moisture conditions, and the nature of the measurement error. Thus, the data from each rainfall event might lead to a different conclusion about the nature of variability in K_s. Not only was there non-uniqueness (equifinality) when dealing with data from a single rainfall-runoff

Table 14.2 Characteristics of the selected rainfall-runoff events. ψ was determined as a numerical solution of the net capillary drive (Melone et al., 2006), which is a function of $\theta_i - \theta_S$, and expressed through the functional forms of the soil hydraulic properties (Smith et al., 1993). (Source: Goyal et al., 2019).

Event date	ID	Total rainfall (mm)	Rainfall duration (h)	Peak rainfall intensity (mm/h)	θ_i	ψ (mm)	Observed surface runoff (mm)
Calibration events							
09/29/13	cal_1	10.60	4.0	15.20	0.189	702.57	1.02
10/09/13	cal_2	7.40	2.0	13.20	0.292	664.08	2.36
10/12/13	cal_3	8.40	4.5	12.00	0.309	627.84	1.77
11/02/13	cal_4	11.60	4.0	6.40	0.289	651.72	1.34
11/10/13	cal_5	21.00	3.5	11.60	0.307	647.03	6.32
11/11/13	cal_6	18.40	7.0	7.60	0.319	618.57	6.58
02/26/14	cal_7	21.00	9.0	6.00	0.304	649.09	2.21
03/04/14	cal_8	5.80	2.5	6.00	0.320	606.09	0.56
Validation events							
10/05/13	val_1	11.00	2.0	13.60	0.238	699.64	1.87
11/04/13	val_2	5.00	2.0	8.80	0.306	633.76	0.66
11/11/13	val_3	9.60	4.0	6.40	0.334	558.99	3.07
11/15/13	val_4	10.20	8.0	6.40	0.314	628.86	1.00
11/21/13	val_5	20.20	4.5	9.20	0.319	607.65	5.77
03/04/14	val_6	7.40	4.0	5.20	0.328	593.41	0.66
04/04/14	val_7	17.00	8.0	7.60	0.299	650.01	1.29
05/02/14	val_8	9.80	4.5	5.60	0.301	648.09	0.60

experiment, but also the posterior distributions of acceptable parameter combinations varied between experiments adding another layer of uncertainty. With natural field variability in soil hydrologic properties, there will likely be some portions of the field where the water supply is limiting and others where soil properties are limiting infiltration, and this control is expected to change as the storm progresses in time. Moreover, even for a rainfall event of practically infinite duration, some portions of the field may never reach ponding conditions, and consequently, some portion of the field never gets truly resolved.

Fig. 14.4A shows the consolidated PDF obtained using Eq. (14.47), considering the stable distributions from individual experiments as the inputs, and the 95% credible region (95-CR) is shown in Fig. 14.4B. Only a fraction of the initial set of parameters is found to be consistent with all experiments with 462 parameter combinations (out of an initial 10,000 combinations) containing 95% of the total mass of

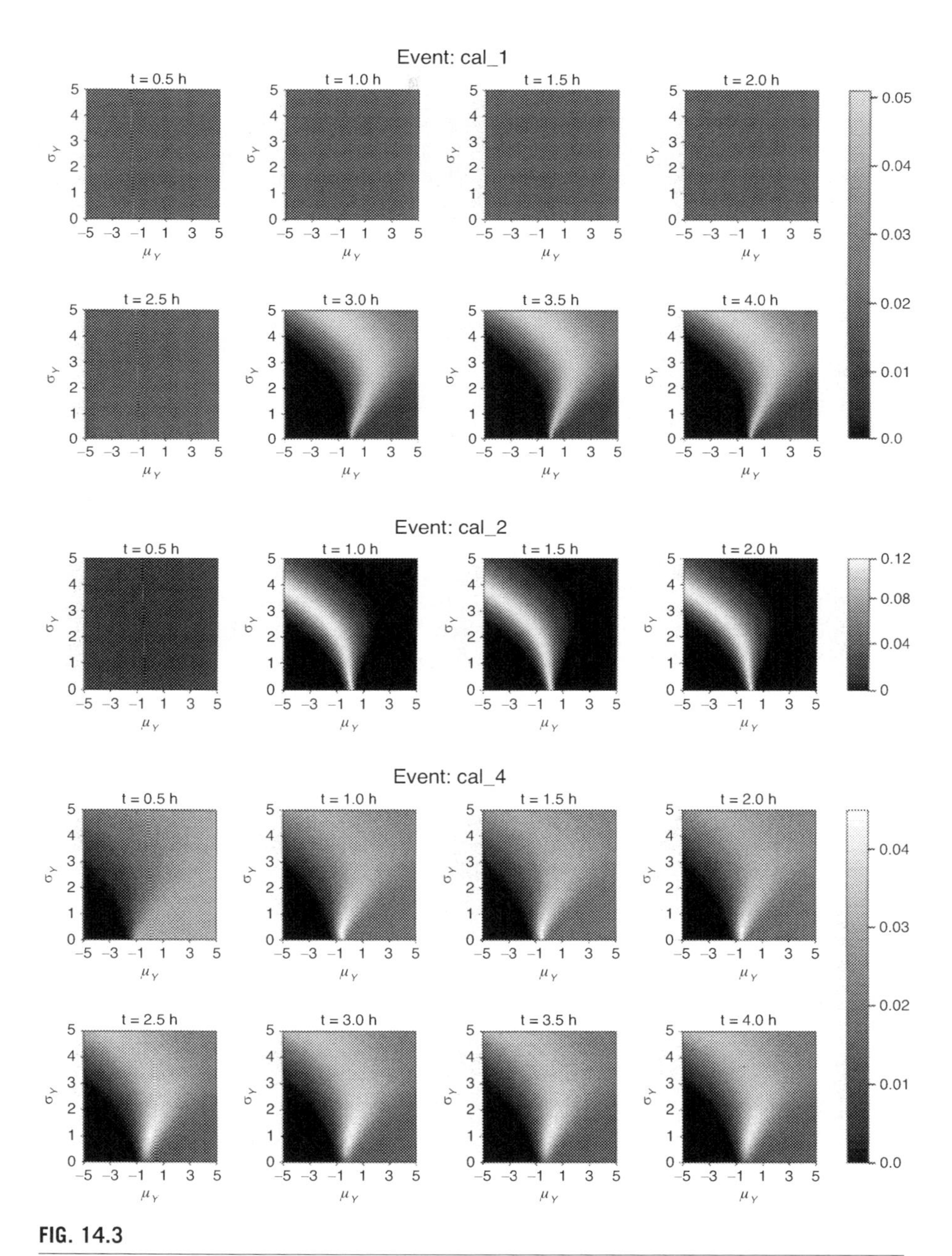

FIG. 14.3

The posterior distributions obtained after each time-step for calibration experiments cal_1, cal_2, and cal_4. Color bars indicate the PDF values; lighter colors show higher values. (Source: Goyal et al., 2019).

Table 14.3 Stabilization times of parameter distributions for the calibration experiments. (Source: Goyal et al., 2019).

Experiment ID	Experiment run time	Distribution stabilization time	Time when maximum rainfall intensity occurred
	(h)	(h)	(h)
cal_1	4.0	3.0	3.0
cal_2	2.0	1.0	1.0
cal_3	4.5	1.5	1.5
cal_4	4.0	2.5	1.0
cal_5	3.5	2.5	1.5
cal_6	7.0	5.0	2.5
cal_7	9.0	7.5	1.5
cal_8	2.5	1.5	1.0

the consolidated distribution, indicating the improvement in estimation of distributional properties achieved over multiple rainfall-runoff experiments.

Even with the refinement of the parameter space, as shown in Fig. 14.4A–B, it needs to be noted that the uncertainty around the mean of K_s (μ_{K_s}) is still considerable, which is evident from the descriptive statistics shown in Table 14.4. Although the most likely estimates of μ_{K_s}, σ_{K_s}, and CV(K_s) agreed with the ranges suggested in the literature (Mohanty et al., 1994; Smettem and Clothier, 1989) for similar soil types, the median estimate of CV(K_s) was over three times higher than the suggested values. The range of estimates of μ_{K_s} and σ_{K_s} emphasized the existence of a wide array of possible lognormal PDFs that could describe the infiltration observations over the study area from multiple rainfall events.

The parameter sets in the 95-CR were used to gauge the performance of the model using the eight rainfall-runoff validation events (Table 14.2). The performance of each validation event, as shown in Fig. 14.5, indicates that after the maximum rainfall rate had occurred over the field the posterior distribution stabilized, and the observations typically fell within the credible region, implying that consolidation of individual experimental outcomes is a step in the right direction.

14.6 Conclusions

Infiltration plays a fundamental role in the understanding of groundwater and subsurface flows. Due to the natural variability of soil hydraulic properties, the estimate of infiltration at different spatial scales is still a complex problem. The understanding of infiltration is further complicated due to the spatio-temporal patterns of rainfall, and the requirements of computationally expensive MC simulations. Even with

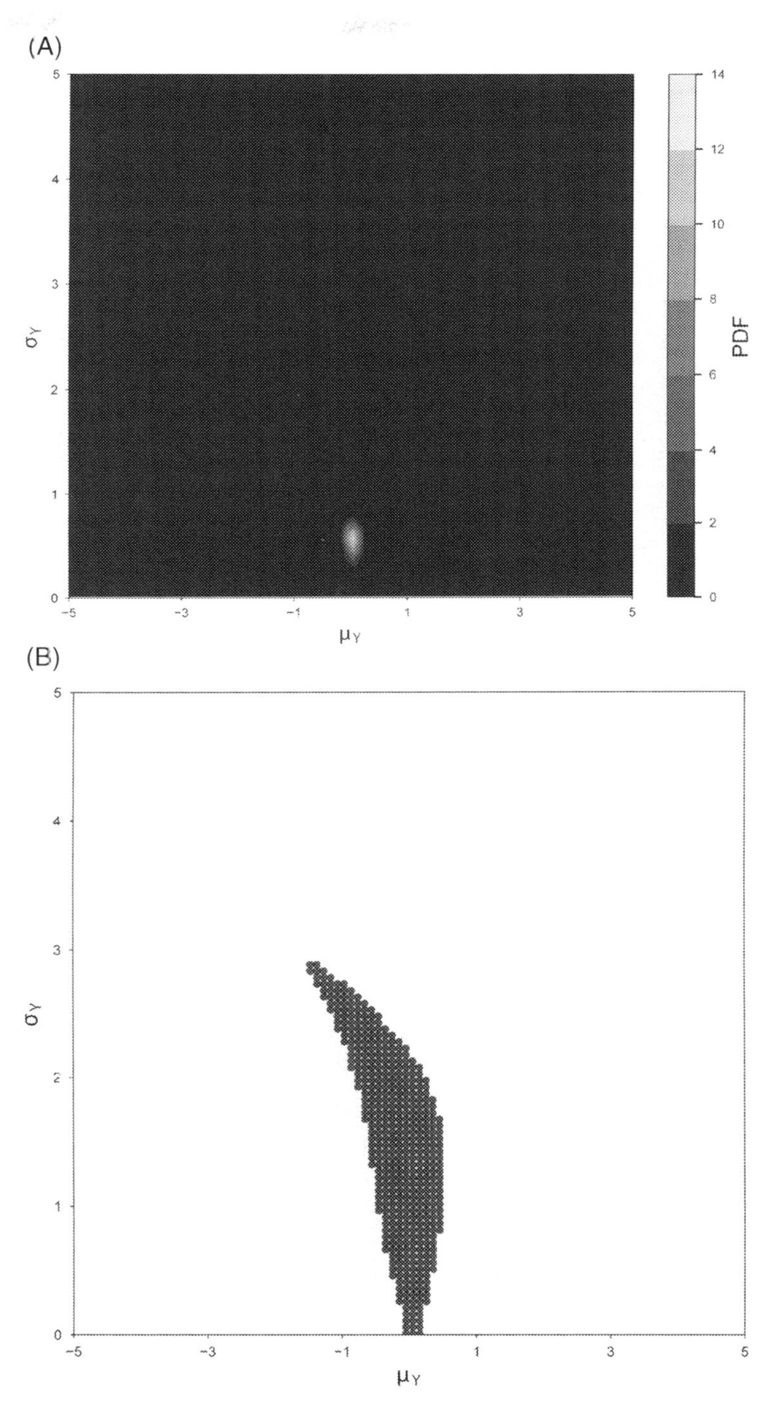

FIG. 14.4

(A) The consolidated distribution and (B) the 95% credible region of the consolidated distribution, obtained from the calibration stable distributions. (Source: Goyal et al., 2019).

Table 14.4 The range of estimates obtained for the mean (μ_{K_S}) and standard deviation (σ_{K_S}) of K_s, and the coefficient of variation (CV(K_s)), calculated using the parameters contained in the 95% credible region. The most likely estimate corresponds to the estimates obtained from the most likely parameter combination, (μ_Y, σ_Y) = (0.0505, 0.6148), in the consolidated distribution. (Source: Goyal et al., 2019).

Parameter	Minimum estimate	Maximum estimate	Median estimate	Most likely estimate
$\propto_{K_s}$ (mm/h)	0.87	16.32	2.98	1.27
σ_{K_s} (mm/h)	0.01	1041.20	8.93	0.86
CV(K_S)	0.01	63.80	3.02	0.68

infiltration theories for inclined surfaces (Philip, 1991; Chen and Young, 2006), the effects of the slope angle are not representative of results obtained in controlled laboratory experiments. In this chapter, several local-scale infiltration models were first presented, which were then used to upscale infiltration and soil properties, using simplified approaches. An important issue to be addressed when areal estimates are involved is concerning the determination of $E[K_s]$, CV(K_s), $E[r]$ and CV(r) together with the corresponding quantities for soil moisture content which influence the infiltration process even though in a limited way (Morbidelli et al., 2012). Furthermore, the infiltration rates obtained from different local-scale instruments are plagued with measurement and systematic errors, thereby, introducing new uncertainties in the results of the upscaled methods.

One major challenge for estimating rainfall-infiltration due to the spatial heterogeneity lies in the non-uniqueness of infiltration parameters–for a single rainfall event and in between different rainfall events. For a given rainfall event, at any time t, only a fraction of the probability distribution function of K_s may be resolved. Future studies that investigate different ways of consolidating infiltration information from independent experiments and exploring simpler ways of assessing the nature of uncertainty in field-scale infiltration are required.

14.7 Recent developments and challenges

Some topics of continued interest are listed below.

14.7.1 Uncertainty quantification

Flow and transport processes are characterized by incomplete knowledge about physical processes and model parameterization. Quantification of the uncertainties arising from these limitations—measurement, parametric or structural—is a necessary step

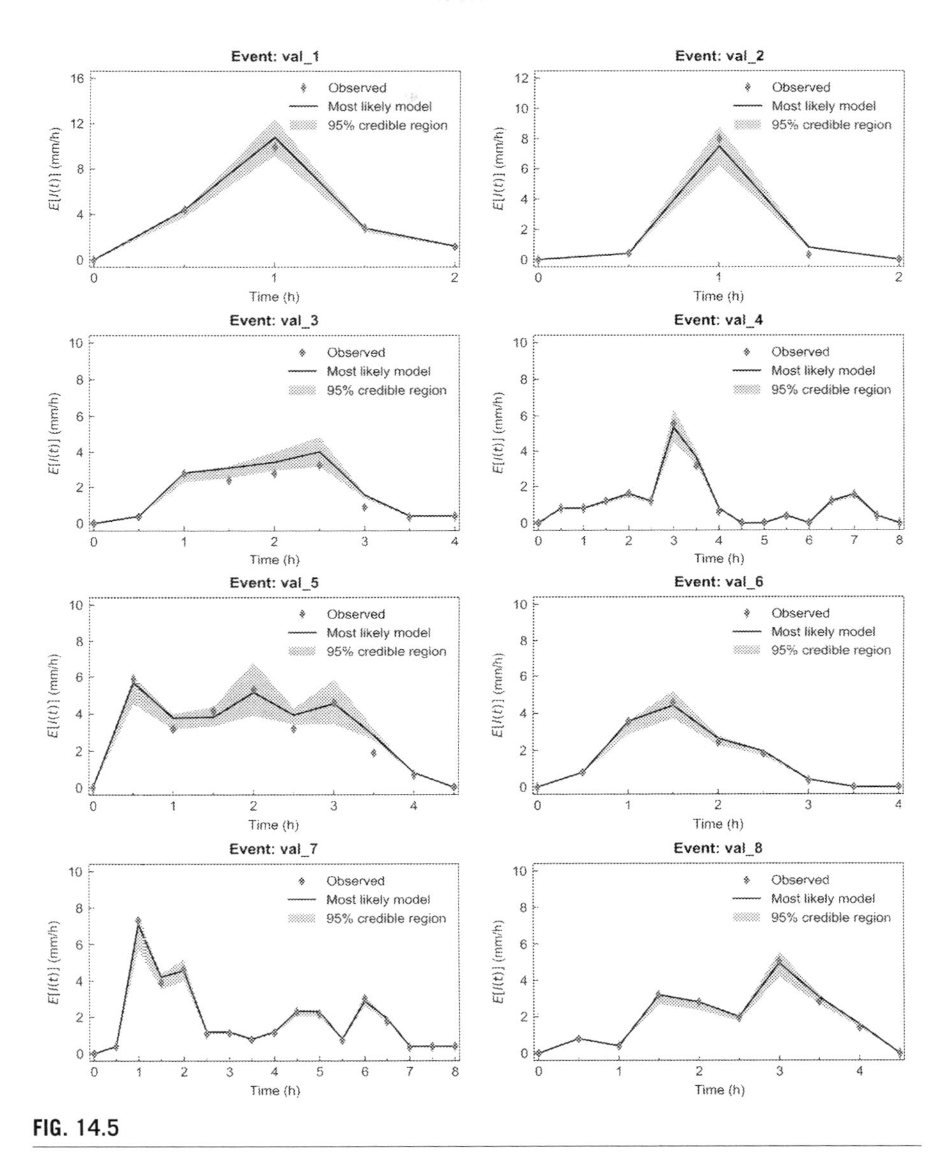

FIG. 14.5

Evaluation of the model performance using eight validation rainfall-runoff events. (Source: Goyal et al., 2019).

to develop a better understanding of these processes. As described in section 14.5, measurements obtained from a rainfall-runoff experiment, will not be representative of the entire field, as a result, the infiltration model parameters may not be identified uniquely. Consolidating information from multiple experiments provides a more realistic estimate of the identifiable parameter space.

14.7.2 Benchmarking model results

Due to the expensive nature of field experiments and the reasons given in section 14.7.1, identification of true values of the infiltration parameters (either point estimates or their probability distributions) remains a challenge. Morbidelli et al. (2017) used rainfall simulators in homogeneous laboratory settings and over a field site to assess the variability in results from different instruments, and their utility in estimating plot-scale variation of K_s and field-scale runoff. They found that the different instruments for estimating point-scale values (Guelph permeameter, double ring infiltrometer, and CSIRO permeameter) showed substantial inter- and intra-instrument variability, thereby highlighting the need for methods to reconcile parameter estimates from different measurement instruments. Data assimilation from multiple instruments is as yet an unresolved problem.

References

Assouline, S., Ben-Hur, M., 2006. Effects of rainfall intensity and slope gradient on the dynamics of interrill erosion during soil surface sealing. Catena 66 (3), 211–220.

Assouline, S., Mualem, Y., 2002. Infiltration during soil sealing: the effect of areal heterogeneity of soil hydraulic properties. Water Resour. Res. 38 (12), 22.1–22.9.

Assouline, S., Mualem, Y., 2006. Runoff from heterogeneous small bare catchments during soil surface sealing. Water Resour. Res. 42 (12), W12405.

Barry, D.A., Parlange, J.Y., Li, L., Jeng, D.S., Crapper, M., 2005. Green Ampt approximations. Adv. Water Resour. 28 (10), 1003–1009.

Beven, K.J., 2002. Rainfall-Runoff Modelling. John Wiley & Sons, New Jersey.

Boers, T.M., van Deurzen, F., Eppink, R.E., Ruytenberg, R.E., 1992. Comparison of infiltration rates measured with an infiltrometer, a rainulator and a permeameter for erosion research in SE Nigeria. Soil Tech. 5 (1), 13–26.

Broadbridge, P., White, I., 1988. Constant rate rainfall infiltration: a versatile nonlinear model. 1. Analytic solution. Water Resour. Res. 24 (1), 145–154.

Brooks, R.H., Corey, A.T., 1964. Hydraulic Properties of Porous MediaHydrology Paper3. Colorado State University, Fort Collins, CO, USA.

Brutsaert, W., 2005. Hydrology: An introduction. Cambridge University Press, Cambridge.

Buckingham, E., 1907. Studies on the movement of soil moisture. U.S. Department of Agriculture, Bureau of Soils Bull. 38, 1–61.

Bullock, M.S., Temper, W.D., Nelson, S.D., 1988. Soil cohesion as affected by freezing, water content, time and tillage. Soil Sci. Soc. Am. J. 52 (3), 770–776.

Cerdà, A., García-Fayos, P., 1997. The influence of slope angle on sediment, water and seed losses on badland landscapes. Geomorphology 18 (2), 77–90.

Chaplot, V., Le Bissonais, Y., 2000. Field measurements of interrill erosion under different slopes and plot sizes. Earth Surf. Process Landf. 25 (2), 145–153.

Chen, Z., Govindaraju, R.S., Kavvas, M.L., 1994. Spatial averaging of unsaturated flow equations under infiltration conditions over areally heterogeneous fields: 1. Development of models. Water Resour. Res. 30 (2), 523–533.

Chen, L., Young, M.H., 2006. Green-Ampt infiltration model for sloping surfaces. Water Resour. Res. 42 (7), (W07420.

Chow, V.T., Maidment, D.R., Mays, L.W., 1988. Applied Hydrology. McGraw-Hill, New York.

Chu, B.T., Parlange, J.Y., Aylor, D.E., 1978. Edge effects in linear diffusion. Acta. Mech. 21, 13–27.

Corradini, C., Melone, F., Smith, R.E., 1994. Modeling infiltration during complex rainfall sequences. Water Resour. Res. 30 (10), 2777–2784.

Corradini, C., Melone, F., Smith, R.E., 1997. A unified model for infiltration and redistribution during complex rainfall patterns. J. Hydrol. 192 (1–4), 104–124.

Corradini, C., Morbidelli, R., Melone, F., 1998. On the interaction between infiltration and Hortonian runoff. J. Hydrol. 204 (1–4), 52–67.

Corradini, C., Morbidelli, R., Saltalippi, C., Flammini, A., 2022. Meteorological Systems Producing Rainfall. In: Morbidelli, R. (Ed.), Rainfall. Modeling, Measurement and Applications. Elsevier, Amsterdam, pp. 27–48. doi:10.1016/C2019-0-04937-0.

Craig, J.R., Liu, G., Soulis, E.D., 2010. Runoff–infiltration partitioning using an upscaled Green–Ampt solution. Hydrol. Proc. 24 (16), 2328–2334.

Dagan, G., Bresler, E., 1983. Unsaturated flow in spatially variable fields. 1. Derivation of models of infiltration and redistribution. Water Resour. Res. 19 (2), 413–420.

DeRoo, A.P.J., Hazelhoff, L., Heuvelink, G.M.B., 1992. Estimating the effects of spatial variability of infiltration on the output of a distributed runoff and soil erosion model using Monte Carlo methods. Hydrol. Processes 6 (2), 127–143.

Emmerich, W.E., 2003. Season and erosion pavement influence on saturated soil hydraulic conductivity. Soil Sci. 168 (9), 637–645.

Essig, E.T., Corradini, C., Morbidelli, R., Govindaraju, R.S., 2009. Infiltration and deep flow over sloping surfaces: comparison of numerical and experimental results. J. Hydrol. 374 (1–2), 30–42.

Fiori, A., Romanelli, M., Cavalli, D.J., Russo, D., 2007. Numerical experiments of streamflow generation in steep catchments. J. Hydrol. 339 (3–4), 183–192.

Flammini, A., Morbidelli, R., Saltalippi, C., Picciafuoco, T., Corradini, C., Govindaraju, R.S., 2018. Reassessment of a semi-analytical field-scale infiltration model through experiments under natural rainfall events. J. Hydrol. 565 (October), 835–845.

Fox, D.M., Bryan, R.B., Price, A.G., 1997. The influence of slope angle on final infiltration rate for interrill conditions. Geoderma 80 (1–2), 181–194.

Furman, A., Warrick, A., Zerihun, D., Sanchez, C., 2006. Modified Kostiakov infiltration function: Accounting for initial and boundary conditions. J. Irrig. Drain. Eng. 132 (6), 587–596.

Goodrich, D.C., Faurès, J.M., Woolhiser, D.A., Lane, L.J., Sorooshian, S., 1995. Measurement and analysis of small-scale convective storm rainfall variability. J. Hydrol. 173 (1–4), 283–308.

Govindaraju, R.S., Morbidelli, R., Corradini, C., 2001. Areal infiltration modeling over soils with spatially-correlated hydraulic conductivities. J. Hydrol. Eng. 6 (2), 150–158.

Govindaraju, R.S., Corradini, C., Morbidelli, R., 2006. A semi-analytical model of expected areal-average infiltration under spatial heterogeneity of rainfall and soil saturated hydraulic conductivity. J. Hydrol. 316 (1–4), 184–194.

Goyal, A., Morbidelli, R., Flammini, A., Corradini, C., Govindaraju, R.S., 2019. Estimation of field-scale variability in soil saturated hydraulic conductivity from rainfall-runoff experiments. Water Resour. Res. 55 (9), 7902–7915.

Grah, O.J., Hawkins, R.H., Cundy, T.W., 1983. Distribution of infiltration on a small watershed, Proceedings of ASCE Irrigation and Drainage Speciality Conference on Advances in Irrigation and Drainage: Surviving External Pressures, Jackson, Wyoming, 44–54.

Green, W.A., Ampt, G.A., 1911. Studies on soil physics: 1. The flow of air and water through soils. J. Agric. Sci 4 (1), 1–24.
Greminger, P.J., Sud, Y.K., Nielsen, D.R., 1985. Spatial variability of field measured soil-water characteristics. Soil Sci. Soc. Am. J. 1 (49), 1075–1082.
Gupta, R.K., Rudra, R.P., Dickinson, W.T., Patni, N.K., Wall, G.J., 1993. Comparison of saturated hydraulic conductivity measured by various field methods. Trans of the ASAE 36 (1), 51–55.
Gupta, R.K., Rudra, R.P., Dickinson, W.T., Patni, N.K., Wall, G.J., 1994. Spatial and seasonal variations in hydraulic conductivity in relation to four determination techniques. Canadian Water Resour. J. 19 (2), 103–113.
Haverkamp, R., Kutilek, M., Parlange, J.Y., Rendon, L., Krejca, M., 1988. Infiltration under ponded conditions: 2. Infiltration equations tested for parameter time-dependence and predictive use. Soil Sci. 145 (5), 317–329.
Hill, T.P., Miller, J., 2011. How to combine independent data sets for the same quantity. Chaos 21 (3), 1–8.
Hillel, D., 1980. Fundamentals of Soil Physics. Academic Press, San Diego.
Holtan, H.N., 1961. A Concept for Infiltration Estimates in Watershed Engineering, 41–51.
Horton, R.E., 1933. The role of infiltration in the hydrological cycle. Trans. Am. Geophys. Union 14 (1), 446–460.
Horton, R.E., 1939. Analysis of runoff plat experiments with varying infiltration capacity. Trans. Am. Geophys. Union 20 (4), 693–711.
Jana, R., Mohanty, B., 2012a. On topographic controls of soil hydraulic parameter scaling at hillslope scales. Water Resour. Res. 48 (2), W02518.
Jana, R., Mohanty, B., 2012b. A topography-based scaling algorithm for soil hydraulic parameters at hillslope scales: field testing. Water Resour. Res. 48 (2), W02519.
Jana, R., Mohanty, B., 2012c. A comparative study of multiple approaches to soil hydraulic parameter scaling applied at the hillslope scale. Water Resour. Res. 48 (2), W02520.
Janeau, J.L., Bricquet, J.P., Planchon, O., Valentin, C., 2003. Soil crusting and infiltration on steep slopes in northern Thailand. Eur. J. Soil Sci. 54 (3), 543–553.
Kostiakov, A.N., 1932. On the dynamics of the coefficients of water percolation in soils and on the necessity of studying it from a dynamic point of view for purpose of amelioration. In: Fauser, O. (Ed.), Transactions of the 6th Commission of the International Society of Soil Science, Russia. Martinus Nijhoff, Groningen, the Netherlands, pp. 17–21.
Krajewski, W.F., Ciach, G.J., Habib, E., 2003. An analysis of small-scale rainfall variability in different climatic regimes. Hydrol. Sci. J. 48 (2), 151–162.
Kutílek, M., Krejča, M., Haverkamp, R., Rendon, L.P., Parlange, J.Y., 1988. On extrapolation of algebraic infiltration equations. Soil Tech. 1 (1), 47–61.
Kutílek, M., Nielsen, D.R., 1994. Soil Hydrology: Textbook for Students of Soil Science, Agriculture, Forestry, Geoecology, Hydrology, Geomorphology and Other Related Disciplines. Catena Verlag, Cremlingen-Destedt, Germany.
Loague, K., Gander, G.A., 1990. R-5 revisited: 1. Spatial variability of infiltration on a small rangeland catchment. Water Resour. Res. 26 (5), 957–971.
Lv, M., Hao, Z., Liu, Z., Yu, Z., 2013. Conditions for lateral downslope unsaturated flow and effects of slope angle on soil moisture movement. J. Hydrol. 486 (12 April), 321–333.
Maller, R.A., Sharma, M.L., 1984. Aspects of rainfall excess from spatially varying hydrological parameters. J. Hydrol. 67 (1–4), 115–127.
Mein, R.G., Larson, C.L., 1973. Modeling infiltration during a steady rain. Water Resour. Res. 9 (2), 384–394.

Melone, F., Corradini, C., Morbidelli, R., Saltalippi, C., 2006. Laboratory experimental check of a conceptual model for infiltration under complex rainfall patterns. Hydrol. Proc. 20 (3), 439–452.

Mezencev, V, 1948. Theory of formation of the surface runoff. Meteorol. Hidrol. 3, 33–40.

Miller, E.E., Miller, R.D, 1956. Physical theory for capillary flow phenomena. J. Appl. Phys. 27, 324–332.

Mls, J, 1980. Effective rainfall estimation. J. Hydrol. 45 (3–4), 305–311.

Mohanty, B.P., Ankeny, M.D., Horton, R., Kanwar, R.S, 1994. Spatial analysis of hydraulic conductivity measured using disc infiltrometers. Water Resour. Res. 30 (9), 2489–2498.

Montzka, C., Herbst, M., Weihermüller, L., Verhoef, A., Vereecken, H, 2017. A global data set of soil hydraulic properties and sub-grid variability of soil water retention and hydraulic conductivity curves. Earth Syst. Sci. Data 9 (2), 529–543.

Morbidelli, R., Corradini, C., Govindaraju, R.S, 2006. A field-scale infiltration model accounting for spatial heterogeneity of rainfall and soil saturated hydraulic conductivity. Hydrol. Proc. 20 (7), 1465–1481.

Morbidelli, R., Corradini, C., Saltalippi, C., Flammini, A., Rossi, E, 2011. Infiltration-soil moisture redistribution under natural conditions: experimental evidence as a guideline for realizing simulation models. Hydrol. Earth Syst. Sci. 15 (9), 2937–2945.

Morbidelli, R., Corradini, C., Saltalippi, C., Brocca, L, 2012. Initial soil water content as input to field-scale infiltration and surface runoff models. Water Resour. Manag. 26 (7), 1793–1807.

Morbidelli, R., Corradini, C., Saltalippi, C., Flammini, A., Dari, J., Govindaraju, R.S, 2019. A new conceptual model for slope-infiltration. Water 11 (4), 678.

Morbidelli, R., Saltalippi, C., Flammini, A., Rossi, E., Corradini, C, 2014. Soil water content vertical profiles under natural conditions: matching of experiments and simulations by a conceptual model. Hydrol. Proc. 28 (17), 4732–4742.

Morbidelli, R., Saltalippi, C., Flammini, A., Cifrodelli, M., Picciafuoco, T., Corradini, C., Govindaraju, R.S, 2016. Laboratory investigation on the role of slope on infiltration over grassy soils. J. Hydro. 543 (Part B), 542–547.

Morbidelli, R., Saltalippi, C., Flammini, A., Cifrodelli, M., Picciafuoco, T., Corradini, C., Govindaraju, R.S, 2017. In situ measurements of soil saturated hydraulic conductivity: Assessment of reliability through rainfall-runoff experiments. Hydrol. Proc. 31 (17), 3084–3094.

Morbidelli, R., Saltalippi, C., Flammini, A., Cifrodelli, M., Corradini, C., Govindaraju, R.S, 2015. Infiltration on sloping surfaces: Laboratory experimental evidence and implications for infiltration modelling. J. Hydrol. 523 (April), 79–85.

Mualem, Y., Assouline, S, 1989. Modeling soil seal as a nonuniform layer. Water Resour. Res. 25 (10), 2101–2108.

Mualem, Y., Assouline, S., Eltahan, D, 1993. Effect of rainfall-induced soil seals on soil water regime: wetting processes. Water Resour. Res. 29 (6), 1651–1659.

Nassif, S.H., Wilson, E.M, 1975. The influence of slope and rain intensity on runoff and infiltration. Hydrol. Sci. Bull. 20 (4), 539–553.

Nielsen, D.R., Biggar, W., Erh, K.T, 1973. Spatial variability of field-measured soil-water properties. Hilgardia 42 (7), 215–260.

Nielsen, D.R., Hopmans, J.W., Reichardt, K, 1998. An emerging technology for scaling field soil-water behavior. In: Sposito, G. (Ed.), Scale Dependence and Scale Invariance in Hydrology. Cambridge Univ Press, Cambridge, pp. 136–166.

Paniconi, C., Putti, M, 2015. Physically based modeling in catchment hydrology at 50: survey and outlook. Water Resour. Res. 51 (9), 7090–7129.

Parhi, P.K., Mishra, S.K., Singh, R, 2007. A modification to Kostiakov and modified Kostiakov infiltration models. Water Resour. Manage. 21 (11), 1973–1989.

Parlange, J.Y., Lisle, I., Braddock, R.D., Smith, R.E., 1982. The three-parameter infiltration equation. Soil Sci. 133 (6), 337–341.

Patin, J., Mouche, E., Ribolzi, O., Chaplot, V., Sengtaheuanghoung, O., Latsachak, K.O., Soulileuth, B., Valentin, C., 2012. Analysis of runoff production at the plot scale during a long-term survey of a small agricultural catchment in Lao PDR. J. Hydrol. 426–427 (21 March), 79–92.

Péschke, G., Kutílek, M., 1982. Infiltration model in simulated hydrographs. J. Hydrol. 56 (3–4), 369–379.

Philip, J.R., 1957a. The theory of infiltration: 1. The infiltration equation and its solution. Soil Sci. 83 (5), 345–358.

Philip, J.R., 1957b. The theory of infiltration: 2. The profile at infinity. Soil Sci. 83 (6), 435–448.

Philip, J.R., 1957c. The theory of infiltration: 4. Sorptivity algebraic infiltration equation. Soil Sci. 84 (3), 257–264.

Philip, J.R., 1969. Theory of infiltration. In: Chow, W.T. (Ed.). Advances in Hydroscience, 5. Academic Press, New York, pp. 215–296.

Philip, J.R., 1991. Hillslope infiltration: planar slopes. Water Resour. Res. 27 (1), 109–117.

Poesen, J., 1984. The influence of slope angle on infiltration rate and Hortonian overland flow volume. Z. Geomorphol. 49, 117–131.

Reeves, M., Miller, E.E., 1975. Estimating infiltration for erratic rainfall. Water Resour. Res. 11 (1), 102–110.

Ribolzi, O., Patin, J., Bresson, L., Latsachack, K., Mouche, E., Sengtaheuanghoung, O., Silvera, N., Thiébaux, J.P., Valentin, C., 2011. Impact of slope gradient on soil surface features and infiltration on steep slopes in northern Laos. Geomorphology 127 (1–2), 53–63.

Russo, D., Bresler, E., 1981. Soil hydraulic properties as stochastic processes: 1. An analysis of field spatial variability. Soil Sci. Soc. Am. J. 45 (5), 682–687.

Rumynin, V.G., 2015. Overland Flow Dynamics and Solute Transport. Springer International Publishing, Switzerland.

Saghafian, B., Julien, P.Y., Ogden, F.L., 1995. Similarity in catchment response: 1. Water Resour. Res. 31 (6), 1533–1541.

Selker, J.S., Assouline, S., 2017. An explicit, parsimonious, and accurate estimate for ponded infiltration into soils using the Green and Ampt approach. Water Resour. Res. 53 (8), 7481–7487.

Sharma, K., Singh, H., Pareek, O., 1983. Rainwater infiltration into a bar loamy sand. Hydrol. Sci. J. 28 (3), 417–424.

Sharma, M.L., Seely, E., 1979. Spatial variability and its effect on infiltration. Proc. Int. Hydrol. Water Resour. Symp., 69–73.

Sharma, M.L., Barron, R.J.W., Fernie, M.S., 1987. Areal distribution of infiltration parameters and some soil physical properties in lateritic catchments. J. Hydrol. 94 (1–2), 109–127.

Sidiras, N., Roth, C.H., 1987. Infiltration measurements with double-ring infiltrometers and a rainfall simulator under different surface conditions on an oxisol. Soil Tillage Res. 9 (2), 161–168.

Singh, V.P., 1996. Kinematic Wave Modeling in Water Resources: Surface Water Hydrology. John Wiley & Sons, New York.

Sivapalan, M., Wood, E.F., 1986. Spatial heterogeneity and scale in the infiltration response of catchments. In: Gupta, V.K., RodríguezIturbe, I., Wood, E.F. (Eds.), Scale Problems in Hydrology. D. Reidel Publishing, Dordrecht, pp. 81–106.

Smettem, K.R.J., Clothier, B.E., 1989. Measuring unsaturated sorptivity and hydraulic conductivity using multiple disc permeameters. J. Soil Sci. 40 (3), 563–568.
Smith, R.E., 1972. The infiltration envelope: results from a theoretical infiltrometer. J. Hydrol. 17 (1–2), 1–22.
Smith, R.E., Parlange, J.Y., 1978. A parameter-efficient hydrologic infiltration model. Water Resour. Res. 14 (3), 533–538.
Smith, R.E., Hebbert, R.H.B., 1979. A Monte Carlo analysis of the hydrologic effects of spatial variability of infiltration. Water Resour. Res. 15 (2), 419–429.
Smith, R.E., Corradini, C., Melone, F., 1993. Modeling infiltration for multistorm runoff events. Water Resour. Res. 29 (1), 133–144.
Smith, R.E., Goodrich, D.C., 2000. Model for rainfall excess patterns on randomly heterogeneous area. J. Hydrol. Eng. 5 (4), 355–362.
Smith, R.E., 2002. Infiltration Theory for Hydrologic Applications. American Geophysical Union, Washington, DC, USA.
Smith, R.E., Goodrich, D.C., 2005. Rainfall excess overland flow. In: Anderson, M.G. (Ed.), Encyclopedia of Hydrological Science. Wiley, pp. 1708–1718.
Swartzendruber, D., 1987. A quasi solution of Richard equation for downward infiltration of water into soil. Water Resour. Res. 23 (5), 809–817.
Talsma, T., Parlange, J.Y., 1972. One dimensional vertical infiltration. Aust. J. Soil Res. 10 (2), 143–150.
Van Genuchten, M.T., 1980. A closed-form equation for predicting the hydraulic conductivity of unsaturated soils. Soil Sci. Soc. Am. J. 44 (5), 892–898.
Valiantzas, J.D., Pollalis, E.D., Soulis, K.X., Londra, P.A., 2009. Modified form of the extended Kostiakov equation including various initial and boundary conditions. J. Irrig. Drain. Eng. 135 (4), 450–458.
Vereecken, H., Weihermüller, L., Assouline, S., Simunek, J., Verhoef, A., Herbst, M., Archer, N., Mohanty, B., Montzka, C., Vanderborght, J., Balsamo, G., Bechtold, M., Boone, A., Chadburn, S., Cuntz, M., Decharme, B., Ducharne, A., Ek, M., Garrigues, S., Goergen, K., Ingwersen, J., Kollet, S., Lawrence, D.M., Li, Q., Or, D., Swenson, S., deVrese, P., Walko, R., Wu, Y., Xue, Y., 2019. Infiltration from the pedon to global grid scales: an overview and outlook for land surface modeling. Vadose Zone J. 18 (1), 1–53.
Verma, S.C., 1982. Modified Horton's infiltration equation. J. Hydrol. 58 (3–4), 383–388.
Wang, J., Chen, L., Yu, Z., 2018. Modeling rainfall infiltration on hillslopes using flux-concentration relation and time compression approximation. J. Hydrol. 557 (February), 243–253.
Wang, K., Yang, X., Liu, X., Liu, C., 2017. A simple analytical infiltration model for short-duration rainfall. J. Hydrol. 555 (December), 141–154.
Warrick, A.W., Nielsen, D.R., 1980. Spatial variability of soil physical properties in the field. In: Hillel, D. (Ed.), Applications of Soil Physics. Academic Press, New York, pp. 319–344.
Woolhiser, D.A., Goodrich, D.C., 1988. Effect of storm rainfall intensity patterns on surface runoff. J. Hydrol. 102 (1–4), 335–354.
Woolhiser, D.A., Smith, R.E., Giraldez, J.V., 1996. Effects of spatial variability of saturated hydraulic conductivity on Hortonian overland flow. Water Resour. Res. 32 (3), 671–678.
Youngs, E.G., 1964. An infiltration method measuring the hydraulic conductivity of unsaturated porous materials. Soil Sci. 97 (5), 307–311.
Zhu, J., Mohanty, B.P., 2006. Effective scaling factor for transient infiltration in heterogeneous soils. J. Hydrol. 319 (1–4), 96–108.

CHAPTER

Rainfall and erosion/ sediment transport

15

J.V. Giráldez[a,b], O. Castro-Orgaz[a], J.A. Gómez[b], A.M. Laguna[c]

[a]*University of Córdoba, Department of Agronomy, Córdoba, Spain*

[b]*Institute for Sustainable Agriculture, CSIC, Department of Agronomy, Córdoba, Spain*

[c]*University of Córdoba, Department of Applied Physics, Córdoba, Spain*

The rain one of the most important sources of the water cycle but it creates some problems to the soil due to the impact of the raindrops that dislodge the aggregates, compact part of them and detach particles which are later dispersed with the splash. Furthermore, the raindrop impact on the surface runoff flow induces a shear strength of the water near the bed and the dispersal of fine sediments suspended in the rill and sheet flow.

The purpose of this chapter is the exploration of the aspects of soil erosion caused by the rain to detect some of the relevant aspects that might deserve more attention by the research in the near future.

15.1 The erosive power of the rainfall

The erosivity of the rain has been usually attributed to the kinetic energy of drops reaching the surface, although the momentum can be considered as an alternative. In this section the two components of both magnitudes, mass and velocity of the raindrops, are analyzed.

15.1.1 Estimation of the distribution of the number and mass of the raindrops

Wischmeier and Smith (1958) made one of the first attempts to relate rain properties with the aggressivity, or erosivity as they called it, based on the empirical equation of the distribution of the number of raindrops per unit volume of air of Laws and Pearson (1943), and of the terminal velocity of Gunn and Kinzer (1949) for the formulation of the kinetic energy, k_e

$$k_e = \frac{\pi \rho_w}{12} \int_0^{D_{max}} v_t^2(D) D^3 N(D) dD \tag{15.1}$$

RAINFALL. Modeling, Measurement and Applications. DOI: https://doi.org/10.1016/B978-0-12-822544-8.00006-8

where ρ_w is the water density, v_t the terminal velocity of the raindrops, D their diameter, which varies between 0 and a maximum value, D_{max}, and $N(D)$ dD is their number per unit of air volume in the atmosphere. The integration of the elemental kinetic energy of raindrops yielded an equation simplified to a linear expression relating its value per unit of rain depth to the logarithm of the rain intensity. This equation was incorporated into the Universal Soil Loss Equation in its successive versions (Wischmeier and Smith 1965; Wischmeier and Smith 1978, and the revised, RUSLE, Renard et al., 1997), in the form of the erosivity factor, R, as the event-integrated product of the computed kinetic energy and the maximum intensity in a 30-minutes period, what is called the I_{30} term. In the more recent version of the USLE, the RUSLE, the linear logarithmic equation has been replaced by an exponential equation (Brown and Foster, 1987). Nearing et al. (2017) gave many interesting details of the evolution of the erosivity factor.

The EI_{30} product of the erosivity factor was not based on sound physical principles. Wischmeier and Smith (1958) adopted the I_{30} index combined with the kinetic energy through a multiple linear regression of the measured soil loss in experimental plots against the kinetic energy alone, with the I_{30}, the precedent 30-day cumulative rainfall, and with the accumulated rainfall energy since the last tillage operation. Wischmeier (1959) checked the convenience of combined index with several soil types, confirming the good fits to the experimental data.

Nevertheless, a proper revision of the EI_{30} has not been made in posterior works. The convenience of the I_{30} has been discussed by several scientists as Hudson (1965 § 9.1) and D'Odorico et al. (2001). D'Odorico et al. (2001) have shown that the change from a 30-minutes interval to a 1 hour-interval does not change much the efficiency of the erosivity factor in the soil loss estimation. Likewise, Dunkerley (2019), after a detailed examination of high-resolution rainfall hyetographs from two Australian locations, concluded that the 30-minutes interval does not properly represents the most intense periods of rain. The results found by Dunkerley (2019) motivated him to propose another index such as the duration of the rain events whose intensity were above a certain threshold as 20 mm h^{-1}, given the inadequacy of a 30-minutes interval to capture a very intense rain pulse, and to the increasing accuracy of modern rain gauges to detect the intense rain bursts. Kinnell et al. (1994) proposed the $I_X E_A$ index developed for soils where sheet erosion was the predominant water erosion form. The $I_X E_A$ index consisted of an integration of the product of the kinetic energy of the rain and the excess of infiltration rate assumed to be a constant fraction of the net infiltration rate. With such an expression the $I_X E_A$ index is an alternative version of the EI_{30}. Nevertheless, Kinnell (1983) confirmed the observations of Ekern and Muckenhirn (1948) and Ekern (1951) first, and later by Hudson (1965), that the soil loss was related to the product of the rainfall energy and the intensity of the effective, or net rainfall, detracting the soil infiltration rate.

The computation of the kinetic energy of the rainfall was later explored by other research teams using more sophisticated techniques like the optical spectropluviometer (Salles et al. 1998).

In an attempt to summarize the available experimental information needed to estimate the kinetic energy, Sempere-Torres et al. (1994) presented a general formulation for the characterization of raindrop size distribution based on the number of raindrops per unit of air volume whose diameters were in the range *(D, D+dD)*, $N(D, M_i)$, where M_i was the *i*-th moment of the raindrop diameter distribution function

$$N(D, M_i) = M_i^{\alpha} g\left(D M_i^{-\beta}\right) = M_i^{\alpha} g(x_1) \quad (15.2)$$

The $g(x_1)$ function included any of the several expressions chosen to characterize the raindrops population (Sempere-Torres et al., 1994, Table 1). The normalized relationship of Sempere-Torres et al. (1998) can estimate rain-related properties such as liquid water content, and with the terminal raindrop velocity as a function of the drop diameter, the kinetic energy. A further refinement was advanced by Testud et al. (2001), whom, trying to improve the fit of the normalized equations to the observed raindrop size distributions, applied a second normalization to the Sempere-Torres et al. (1998) expressions. Lee et al. (2004) generalized the double-moment normalization of Testud et al. (2001), re-escribing Eq (15.2) as

$$N(D) = \left(M_i^{j+1} M_j^{-i-1}\right)^{1/(i-j)} h\left[D\left(M_i M_j^{-1}\right)^{1/(j-i)}\right] \quad (15.3)$$

where the arguments are the respective *i*-th and *j*-th moments of the drop size distribution, M_i, and M_j, and the normalizing function is now $h(x_2)$. The double-moment relationship fits the measured data better than the single-moment renormalization. In this framework, many probability functions have been used for the description of the observed raindrop distribution, like the exponential function by Marshall and Palmer (1948), the lognormal function by Feingold and Levin (1986), the gamma function by Ulbrich (1983) and the Weibull function by Mualem and Assouline (1986) and Assouline (2020), among other ones. More recently, Cugerone and de Michele (2015) recovered the Johnson system bounded or SB distribution of Johnson (1949) for the reduced, bounded, range of the raindrops.

Although the adoption of these probability distribution functions for the characterization of the raindrop size has not been exempt of criticism (Ignaccolo and de Michele 2014) due to failure of some goodness of fit tests, their use is widely accepted (e.g. Thurai and Bringi 2018). Alternatively, the statistical moments can be good estimators of the variation of the raindrop size as, for instance, Ignaccolo and de Michele (2014) suggested.

The representation of the raindrop size distribution with the statistical functions must be accepted without forgetting that there are some restrictive hypotheses implicitly assumed, as Jameson and Kostinski (2001) have clearly remarked. The raindrop distribution can change in time within a rain event (Lavergnat and Golé, 1998; D'Adderio et al., 2018), and it depends on the measuring instrument as indicated by Peters et al. (2005).

The use of the probability-based distribution functions of raindrops with relevance for the estimation of rainfall erosivity implies that some effects might be overlooked, as the break-up and collisions of them during their travel through the atmosphere, and the occurrence of great, or super-large raindrops. Testik and Barros (2007) explored in a thorough review the mechanisms involved in the drop interaction within a rain, essentially bouncing effects after collisions between drops, coalescence of drops forming a larger unit, and break up in smaller drops. There was some discussion about whether fragmentation, as suggested by Villermaux and Bossa (2009), or collisions supported by Barros et al., (2010), were more relevant for the evolution of raindrop size. The observations of Testik and Rahman (2017) with a high-speed optical disdrometer remarked the increasing importance of the collision theory.

Another aspect that has not been properly considered in the adoption of probability distribution functions to the description of raindrop size is the occurrence of super-large raindrops. The drops of great size have been observed by Hobbs and Rangno (2004) in tropical environments, in some cases induced by condensation onto giant smoke particles. The most relevant property of the super-large raindrops is their terminal velocity which will presented below. The rain interception by vegetation induces the formation of secondary water drops, called ‘gravity drops’ by Moss and Green (1987) with great erosive effects.

15.1.2 Estimation and measurement of the rain drops velocity

The terminal velocity of raindrops, their velocity as they reach the ground, is other factor of the rain kinetic energy. After the theoretical analysis of Spilhaus (1948), the precise measurements of the terminal velocity of the raindrops by Gunn and Kinzer (1949) opened the way to a new line of research. The movement of raindrops are explained with the interaction of the gravity and buoyancy force, F_g, and the drag force, F_d. The gravity-buoyancy force, assuming spherical raindrops with the drop diameter, D, the water and air density, ρ_w and ρ_a and the acceleration of the gravity, g, is

$$F_g = \frac{1}{6}(\rho_w - \rho_a)\pi D^3 g \tag{15.4}$$

The drag force, with, the velocity, v, the drag coefficient, c_D, and the cross-sectional area normal to the flow, A, is

$$F_d = \frac{1}{2} c_D A \rho_a v^2 \tag{15.5}$$

At the equilibrium, the velocity reaches a terminal value, v_t. The value of the drag coefficient is then

$$c_D = \frac{\rho_w - \rho_a}{\rho_a} \frac{\pi D^3 g}{3 A v_t^2} \tag{15.6}$$

The data of Gunn and Kinzer (1949) fitted by the Best equation (1950) were a reference for later works. The drag coefficient is usually expressed as a function of the Reynolds number, R_e

$$R_e = \frac{v_t \rho_a D}{\mu_a} \tag{15.7}$$

where D is a scale reference for the raindrop, as, for instance, its diameter in the case of a spherical drop, and μ_a the dynamic viscosity of the air. Mitchell (1996) used the Best number, X, introduced by Pruppacher and Klett (1978)

$$X = c_D R_e^2 \tag{15.8}$$

to express the raindrop terminal velocities and its relationship with the drag coefficient, as shown in Fig. 15.1 with the Gunn and Kinzer (1949) data.

Beard (1976) separated the raindrops in three classes attending not only the diameter but the value of the Reynolds number, the ratio between inertial and viscous forces which influences the air resistance to the advance of the falling drop into the atmosphere. The three classes of drops, slightly modified by Testik and Barros (2007) are: I. cloud drops (diameters below 0.25 mm); II, small raindrops (diameters between 0.25 and 1 mm); and III, large raindrops (diameters greater than 1 mm).

Several simplified relationships between the raindrop terminal velocity and diameter such as Best (1950), Uplinger (in Ferrier 1994), and Gossard et al. (1992) have been adopted for the estimation of the kinetic energy of the rain as a function of the intensity (Salles et al. 2002, van Dijk et al. 2002b). The Uplinger equation is not well suited for the larger drop diameters because the terminal velocity decreases in that region. Inspired in the work of van Boxel (1998), Furbish et al. (2007, Appendix A)

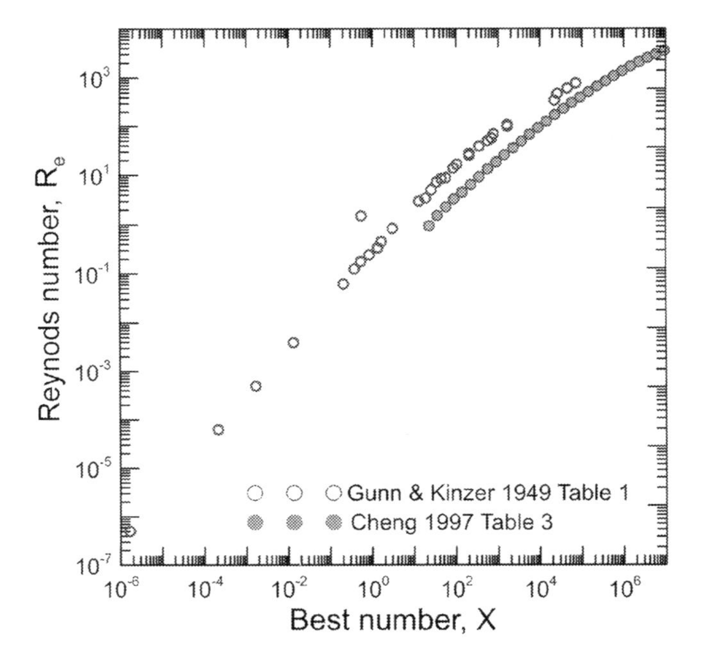

FIG. 15.1

Comparison of the Best and Reynolds number computed with the terminal velocity of raindrops of Gunn and Kinzer (1949, Table 1), and the settling velocity of solid particles of Cheng (1997, Table 15.3).

presented a simplified equation for the ratio of the fall and terminal velocities of raindrops descending from short distances like plant canopies.

Nevertheless, some anomalous values of the raindrop speed have been recorded. Montero-Martínez et al. (2009) observed that drops of medium size reached a terminal velocity higher than the expected value for their diameter. They called super-terminal drops to those exceeding their expected terminal velocity by at least a 30%. Among the possible explanations of their origen they suggested that these drops were fragments of a larger one with differently size moving with the velocity of the parental one. Villermaux and Eloi (2011) justified the high terminal velocity of a fraction of drops produced in the breakup of the parental ones with their fragmentation model. In any case the number of observations of super-terminal raindrops has increased recently (Bringi et al. 2018; Das et al., 2020; Larsen et al., 2014). Possibly, the occurrence of super-terminal drops should be considered in the estimation of the kinetic energy of the rain.

The use of cloud resolving models, like the Weather Research and Forecasting model, WRF of Skamarock et al. (2005) adopted by Nissan and Toumi (2013a), can improve the estimation of the kinetic energy of the rain, including the presence of large drops. With the same model Nissan and Toumi (2013b) explored the influence of aerosols on the kinetic energy of the rain using the cloud microphysics of the WRF model. Nissan and Toumi (2013b) modeled the behavior of a stationary supercell and a moving warm and moist cloud over a hill. Their results showed that as the concentration of aerosols increases in the supercell due to updrafts inside, it stimulates the cloud droplets generating large drops, until the water vapor is consumed. After this concentration threshold is reached, the kinetic energy decreases. In the cloud moving over the hill the rainfall and the kinetic energy decreases with the aerosol concentration. The conclusion was that the presence of aerosols in the atmosphere can accentuate the erosivity of the rain. Therefore, given the measured concentration of aerosols in the atmosphere (Hidy, 2019), the rain erosivity might increase in the near future.

There are several research questions to solve: How can the kinetic energy of the rain be properly expressed considering its temporal evolution during the rain event, and its spatial variability? Is the maximum rain intensity in any time interval like the EI_{30} necessary to express the erosivity of the rain? How can be separated the effect of the raindrop impact from that of the runoff shear stress?

15.2 Raindrop impact on the soil surface

Salles et al. (2000) measured the impact of single raindrops on the surface of two sandy soils with small clay content with an optical spectro pluviometer (Salles and Poesen, 1999), relating the detached mass with the products of several powers of the mass and the velocity of the drops. The Salles et al. (2000) results indicated a slight advantage of the momentum over the kinetic energy as a mechanical erosivity index. A similar conclusion was obtained by Furbish et al. (2007) in their trials of high-speed imaging of drop impacts on dry sand. The best index found by Salles et al. (2000) to describe the mass of soil detachment, was the product of the fourth power of raindrop diameter by the velocity what be considered as the product of the momentum by the diameter of the drops. In their experiments Salles et al. (2000) detected a certain threshold that depended on the content of fine particles in the soil, in a similar range of particle diameter of the Shields diagram used for the estimation of the threshold values of the shear stress at the incipient movement of bed particles in water flumes, (Garcia, 2000), and in wind erosion processes (Shao and Lu, 2000). The effective diameter size range, roughly from 0.2 to 1 mm, contains silts and fine sand, with small weight, and reduced cohesion forces between particles. Therefore, the soils with a great content of particles of that interval are very susceptible to the water erosion whether by rain or by runoff flow.

In this section several aspects of the impact are examined like the formation of surface crusts, the influence of slope and wind on the splashing of solid particles, the different size of particles dispersed and the modification of raindrops by plant canopies.

15.2.1 Formation of soil surface crusts

The raindrop impact has been carefully described in several articles (Ghadiri and Payne, 1981), with an overview by Yarin (2006) for the general process, and van der Meer (2017) for the splash on a porous media. As Furbish et al. (2007) explained, the water drop that falls on a soil with medium to large pores partly infiltrates generating fluid pressure wave, a shock wave. The depth of penetration of the raindrop into the sand layer of the experiments of Furbish et al. (2007) was between 1 and 2 times the representative diameter of the layer particles. The central part of the impact area becomes a crater, leading to a crust by the downwards displacement of fine particles. This crust is of the erosion, or structural type as independently defined by Shainberg (1985), and Valentin and Bresson (1992), and described in the field by Bielders et al. (1996) in a case of West Africa. Moss (1991a) recognized the presence of tree distinct components in the structural crusts: (i) the vesicles, cavities filled with air, in the compacted layer; (ii) silt layers not necessarily continuous above the compacted layer; and (iii) small patches of coarse particles above the previous layers. The coarse particles can prevent the formation of crusts as Moss (1991b) suggested. There are several environmental factors such as the surface roughness, possibly associated with the presence of the coarse particles mentioned by Moss (1991b), and the occurrence of hydrophobic substances in the soil. Terry and Shakesby (1993) and Terry (1998) discussed the crust formation in water repellents soils.

In the external part of the impact area several processes develop (Yarin 2006), as the dispersal of particles by the splash. In the horizontal surfaces the particles are displaced individually or in aggregates (Mouzai and Bhouhadeff, 2003), following a negative exponential function of the distance, proposed by van Dijk et al. (2002a). The distance of the rainsplash depends on the particle size with the maximum values for the intermediate diameters as Leguédois et al. (2005) in aggregated soils, and Furbish et al. (2007) in dry sand have observed. The movement of sand grains or aggregate fragments induced by the raindrop impact was characterized as a stochastic advection-dispersion process by Furbish et al (2009) who proposed two Fokker-Planck equations for the description of the particle concentration and for the evolution of the land surface, respectively, with a slope-dependent drift. An important result of this work is the study of the small mounds generated under the isolated plant canopies, with an initial 'harvesting' in the authors' words of the soil particles, what is a very common feature of arid zone landscapes (e.g. Parsons et al. 1992 and the ring-growth patterns of Ravi et al. 2010). A more recent exploration of the stochastic processes involved in the rain splash phenomena by Sochan et al. (2019) separated two stages in the displacement of the particles. The particles splashed in the first stage traveled longer that the particles later moved. The ejection of soil particles by rain splash is more effective in hydrophobic media than in the otherwise similar hydrophilic media as reported Ahn et al. (2013). An alternative model based of a Markov chain was proposed by Wright (1987).

15.2.2 Slope influence on rainsplash

The slope exerts a great influence of rain splash displacements, generating an asymmetric distribution as Wright (1986) explained due to difference between path angles and initial velocities. As a consequence, downslope mass of particle is greater than the analog upslope mass. Gabet and Dunne (2003) developed a simple splash model based on what they denominated rain power, a function of the time derivative of the kinetic energy of the raindrops, including a correction for the fraction of vegetation covered area, and applying it to an integrated value of the elemental threads of flow across a slope. In a further step, Dunne et al. (2010) elaborated a model for describing the particle splash solving the ballistic equation to calculate the number of particles whose path ends on the slope plane, similar to the model of Reeve (1982) and of Furbish et al. (2007). A weighted integration for these numbers over the whole plane gave the mass of particles splashed upslope and downslope. The dependency of the mass transported by the splash on the hillslope angle followed the same pattern predicted by the Poesen (1985) model, tested in the field by Poesen (1986).

15.2.3 Influence of the wind on the rainsplash transport

The wind increases the soil detachment as Mazurak and Mosher (1968) indicated. The horizontal component of the wind velocity enhances the shear stress generated during the impact of raindrops on the soil aggregates and other particles, inducing what de Lima et al. (1992) called splash saltation. Erpul et al. (2002) and Cornelis et al. (2004) measured the increase of the impact pressure, the mass of detached particles, and the travel distance of these particles. In both works the wind effect was characterized with the shear velocity, that is the conversion of the shear stress into velocity terms, estimated by fits of the logarithmic relationship between the wind velocity and the elevation. In a posterior work Erpul et al. (2005) expressed the kinetic energy of the wind-driven raindrops as a potential exponent of the horizontal component of the wind velocity. Similar results were obtained by Marzen et al. (2015) emphasizing the importance of the splash creep, whose relevance was remarked by Terry (1998). This form of splash transport when the particles slide or roll, observed in the high-speed films of Mazurak and Mosher (1968), is similar to the contact load defined by Moss et al. (1979). In a later work, Marzen et al. (2016) explored the influence of different factors in the mass of soil detached in their controlled trials in laboratory, namely, the erosive agent, rain and wind separately, or combined; the type of soil of a sandy or a loamy textural class; the slope, upslope, plane or downslope conditions; the surface roughness, smooths, with transversal rills with respect to the wind direction. The most relevant factor, by large, was the erosive agent, with the wind-driven splash as the treatment that displaced the greatest soil mass of the whole set of them. The laboratory results were tested under field conditions with the portable wind and rainfall simulator of Fister et al. (2012) in different environments by Marzen et al. (2017). The wind-driven rainsplash was the more effective erosive agent in the trials in the three environments, semiarid, Mediterranean conditions, with silty loam soils; humid conditions with clayey soils, and humid conditions in the coast, with sandy soils, compared to windless rainsplash.

15.2.4 Rainsplash transport of the different soil particle sizes

The influence of the raindrops on the transport of sediments in water have been explored in detail by the work of the CSIRO researchers. Moss et al. (1979) conducted several experiments to assess this effect in shallow flows, under 1 cm, in laboratory conditions, applying artificial rain with different rates, over a runoff flow, under several slopes from .001 to .298, in an experimental setup described by Walker et al. (1977, 1978). The rainfall intensities were adjusted to the surface flow rates to produce the total water discharge equivalent to that of a rainfall rate of 100 mmh^{-1}. The soil was a poorly graded sand. The raindrop-stimulated transport consisted of a combination of the saltation, the contact load, defined by the authors as an initial motion to induce the rolling of the solid particles, and suspension. Their results suggested that the concurrent effect of the rain impact and the water flow shear is more erosive than the isolated rain splash. At the same time, the rainless treatment generated sediments at the 0.01 slope in the first stage, a small load rate, about 0.1 $gm^{-2}s^{-1}$, but the sediment loads of almost all the treatments under rainfall were similar. As Moss et al. (1979) explained, the overland flow was able to mask the contribution of the rain either by direct impact or by splash transport. Nevertheless, the type of the displaced particles was different. Moss et al. (1979) used the enrichment factor as a transport index. The enrichment factor is the ratio of the masses of the different size fractions in the sediment load and in the bed or parent material. In general, the enrichment factor values were greater in the case of fine particles like clay, under the intense rains and steep slopes than in the other sizes of particles. In all cases the values decreased with time from the initial stages. A similar trend was detected with load rate. For the particles of coarse sand size, the changes of the enrichment factor with time were small under the different conditions of the trial. The sediment load was essentially bed load. In all cases the load increased with the slope, due to the increment of energy of flowing water.

However, the water depth protects the soil over a threshold that depends on the size of falling raindrops, about 2 diameters according to the results of Moss and Green (1983), with the greater effectiveness for the increasing drop diameters. The airsplash can be very active from the bare soil surfaces within the water covered areas contributing to the sediment load of the runoff. Moss and Green (1983) observed the penetration of the small water drops, with a diameter of 1.2 mm that contributed to the displacement of soil particles in suspension.

15.2.5 The impacts of the gravity drops

Raindrops of small diameters can coalesce in the plant canopies producing the gravity drops mentioned earlier. These drops are greater than the original raindrops, in the measurements of Moss and Green (1987, Table 15.1), the average diameter was 5.3 mm, more than the 0.5 mm of the fine mist and the 2.7 mm of rain applied. Park et al. (1983) proposed a simple model to determine the velocity of the raindrops as function of the elevation of the falling point in the canopy, using the second law of Newton

$$m\frac{dv}{dt} = F_g - F_d \tag{15.9}$$

Assuming spherical raindrops with the drag coefficient mentioned above (15.6), neglecting the air density

$$\frac{dv}{dt} = g\left[1-\left(\frac{v}{v_t}\right)^2\right] \tag{15.10}$$

Table 15.1 Conditions in experiments* by Iwagaki (1955). Rainfall durations tested are t_r = 10, 20, and 30 s.

	Coordinates of the reach, m		Net rainfall rate, *r*
S_0	Usptream	Downstream	cm s^{-1}
0.020	0	8	0.1080
0.015	8	16	0.0638
0.010	16	24	0.0800

*(Series B, cascade of planes)

A simple integration of this equation from $t = 0$, $v = 0$, leads to an implicit expression of the velocity as a function of time

$$t = \frac{v_t}{2g} \ln\left(\frac{1 + v/v_t}{1 - v/v_t}\right) \tag{15.11}$$

Changing the variable from time to elevation z, with the equation of velocity $dz = v\,dt$, the final expression for the velocity of the raindrop falling from the elevation h is an alternative to equation (A2) of Furbish et al. (2007)

$$\frac{v}{v_t} = \left[1 - exp\left(-\frac{2gh}{v_t^2}\right)\right]^2 \tag{15.12}$$

Therefore, a simple computation reveals that the kinetic energy of a gravity drop with a diameter of 5.3 mm falling from an elevation of 0.4 m is 2,770 times that of a mist drop of a diameter of 0.5 mm, and 0.973 times that of a drop of a diameter of 2.7 mm falling at terminal velocity on the ground.

The airsplash produced by the gravity drops were in some cases about half of the effect caused by raindrops reaching the ground at the terminal velocity. The rain-flow transport of sediment increased as well with the gravity drops. Moss and Green (1987) presented the potential effects of these drops falling from different layers of the plant canopy. Nevertheless, their experiments did not include real plants, except for a single, small branch of a tree. In field observations, the craters formed by gravity drops falling the canopies sometimes detected by micro-fairy chimneys when a small stone or leaf protect the soil surface. The erosive effect of the gravity drops can be even more intense when the wind moves the branches of the trees as Moss and Green (1987) detected in their experiment. Summarizing part of his trials Moss (1988) discussed the three main mechanisms of water erosion: (1) airsplash that acts in the bare soil and is enhanced by the slope and wind, causing random displacements of particles, except for wind action or terrain slopes, or due to movements from spots of high activity to other ones with low activity, (2) overland water flow that can degenerate into rills and gullies, and (3) rainflow transport, that complements the splash effect. The three modes can be very erosive, reducing the aggregates to primary particles. Confirming the suggestion of Loch (1984), Moss (1988) explained the importance of the rill flow transporting more effectively the sediments, with greater particles than those transported by sheet flow. Moore and Burch (1986) applied the Yang (1973) total load equation to describe the experimental results of Loch and Donnollan (1983), considering the soil aggregates instead of the primary particles. In the experiments of Loch (1984) the representative size of the particles carried by rill flow were twice the size of the corresponding particles moved by sheet flow. The coexistence of flooded areas with rill flow and non-flooded interrill areas with rain splash might enhance the overall soil loss rate as Moss observed. Moss (1989) acknowledged the beneficial effects of plant covers in the soil protection.

15.3 Overview of soil erosion models

The raindrops falling over the surface might affect the runoff flow, as Glass and Smerdon (1967) saw in the hydraulic flume of a laboratory, reducing the velocity. A related effect was observed by Izzard (1946) in hydrographs generated by simulated rain, in which a sudden increase, pip, occurred when the simulated rain ended. The reduction of the velocity of flow by the raindrop impact is limited to small flow rates. Yu and McNown (1964) explained this phenomenon by a sudden decrease of the Reynolds number, as can be seen in the Glass and Smerdon (1967) Table 1. Li et al. (1975), reproduced the pip of the hydrograph of Izzard (1946) considering the rainfall intensity in a modified expression of the Darcy-Weisbach friction factor used in a nonlinear kinematic wave model for runoff flow. Beuselinck et al. (2002), observed an increase of the sediment transport of the flow under raindrops impact what can be attributed to the detachment effect mentioned by Moss et al. (1979).

In this section a brief survey of one of the most general soil erosion models based on the kinematic approximation to the St-Venant equations followed by a simple comparison of this approach with the more complete dynamic wave solution of those equations.

15.3.1 Soil erosion models that incorporate the rainfall effects

The development of the soil erosion model of Rose et al. (1983a) might be used as framework to present new developments on the influence of the rainfall on soil erosion. The model consists of the mass conservation of water in the control volume, a land strip in a hillslope, the momentum conservation of water, usually reduced to the kinematic wave approximation, and mass conservation for the different groups of sediments separated in size intervals, with the possibility of deposit over the soil surface and re-entrainment by the water flow or by the raindrop impact. An important component of the Rose model is the recovery of the bottom withdrawal method to measure the soil aggregate settling velocities (Lowell and Rose 1988). With this method is more easy to consider the movement of soil aggregates as Moore and Burch (1986) indicated. The method was already described by Stanley and Parker (1943) but Lowell and Rose (1988) improved its interpretation. An example of the size-intervals and their associated settling velocities can be found in Rose et al. (1983b, Table 2).

The erosive processes generating sediments in the Rose et al. (1983a) model were rainfall and runoff entrainment and a part of them were lost from the water flow by deposition. The rain entrainment was formulated by the product of a coefficient depending on the water depth over the soil surface and a power of the rainfall intensity. The runoff entrainment was formulated as stream power, defined as the product of the shear stress per unit water depth, by the volumetric flux rate per unit width, over a threshold value, and corrected by an efficiency factor. The sediment deposition was equal to the product of its concentration and the corresponding settling velocity affected by another correction factor for every size class. Later Proffitt et al. (1989) and Hairsine and Rose (1991) introduced the influence of the freshly deposited sediments which implied the characterization of the deposited layer, and the addition of the re-detachment processes by the rain or the runoff (Hairsine and Rose 1992a, 1992b).

An interesting result of the comparison of the splash-cup and flume techniques of Proffitt et al. (1989) was the observation that the soil loss rate is decreasing with time from the initial values until a 'true' equilibrium rate is attained, with similar results for each soil type using both techniques. The reduction of sediment concentration with time, measured by Proffitt and Rose (1991a, 1991b) at, approximately, the time taken to reach the 'true' equilibrium values of the detachment rate, was attributed to the accumulation of the sediment over the soil surface. The settling velocity distribution of the sediment's characteristic of a finer soil than the soil used in the experiments evolved with time towards a distribution curve closer to that of the original soil. The results of Proffitt et al. (1991), the time-decreasing sediment concentrations and the evolution of settling velocity distributions to that of the original soil, were approximated by the improvement of the initial Rose (1983a) erosion model trough the inclusion of the time variable sediment layer suggested by Hairsine et al. (1999). In this work (Hairsine et al. 1999, Appendix), the settling velocity of soil particles was characterized with the Newton second law and an expression of Fuchs (1964, as quoted by Monteith and Unsworth 1990, Chap. 9) for the drag coefficient. Nevertheless, other more accurate expressions for the settling velocity, like the Ferguson and Church (2004) with the coefficients $C1 = 18$ and $C2 = 0.4$, that reproduce the velocities of Rose et al. (1983b, Table 2), are simpler to handle. The equation of the settling velocity has been modified in this model. Sander et al. (2007) indicated that they have adopted the equation of Cheng (1997). Later Tromp-van Meerveld et al. (2008) affirmed that the equations of Dietrich (1982) and Ferguson and Church (2004) gave values of the settling velocity closer to their measured ones. The reduction of runoff due to the rainwater infiltration in the soil was determined by Walker et al. (2007) in laboratory experiments inserting the corresponding modifications in the Rose (1983a) model, or, what Jomaa et al. (2010) called, the Hairsine-Rose model. In the analysis of Tromp-van Meerveld et al. (2008) the values of the settling velocities of the different particle size intervals were optimized by fitting the predictions of their model to the measured data, but as mentioned earlier, these values could have been determined with some known equations. This result is encouraging due to the small influence of different factors on the settling of the particles discussed by Dey et al. (2019) as the particle shape, the hindered movement in the presence of other particles in the suspension (e.g. Chen et al. 2020), and the turbulence (Wang et al. 2018).

Using the data of the settling velocity of solid particles of Cheng (1997, Table 3) the relationship between the Best, X, and the Reynolds, R_e, numbers with the data of terminal velocity of rainfall of raindrops of Gunn and Kinzer (1949), were compared, Fig. 15.1. The two sets of data are very close.

In soils with coarse material and rock fragments in the surface, Jomaa et al. (2012a) found that the mass of particles loss by erosion was proportional to the fraction of the area exposed to the rain. This influence was integrated into the Hairsine-Rose model by Jomaa et al. (2012b), including the faster time to reach the steady state conditions in the rock fragments-covered than in the bare soils according to the laboratory results of Jomaa et al. (2013). These results were later generalized by Jomaa et al. (2017). The Hairsine-Rose erosion model has been successfully applied to the description of the sediment dynamics of soil erosion and sediment transport in heterogeneous surfaces by Heng et al. (2011).

The concept of sediment- and transport-limited processes, was introduced in the study of soil erosion by a seminal work by Meyer and Wischmeier (1969) where the contributions of the rainfall and runoff in the detachment of solid particles and their transport were separated in the simple case of a homogeneous slope. Nevertheless, it is rather hard to distinguish either pure detachment- or transport-limited processes as Pelletier (2012) pointed out. In a similar way, Sander et al. (2007) reanalyzed a previous work of Polyakov and Nearing (2003), indicating the relevance of the variable transport capacity in erosive processes.

15.3.2 A comparison of the full solution of surface flow equations with the kinematic wave approximation

Physically-based simulation of the surface run-off on a watershed involves overland and open channel flows (Eagleson 1970). A real watershed can be handled using a one-dimensional approach by approximating the real terrain by planes and channels with suitable connections among them. One of the earlier distributed flow models implemented in practice with this philosophy was the kinematic wave model, e.g., as in the KINEROS model (Woolhiser et al. 1990) based on the pioneering works by Liggett and Woolhiser (1967) and Woolhiser and Liggett (1967). The Rose et al. (1983a) model used a similar approach. The kinematic wave model produced a reasonable approximation to the full dynamic wave in relative steep terrain and permitted to simulate shocks using suitable techniques (Borah et al., 1980). With the increase in the computational power since Woolhiser and Liggett (1967), however, the solution of the full dynamic wave equations in a two-dimensional setting is a promising tool for the hydrodynamic modeling of run-off processes at the watershed scale (Kim et al., 2012). Earlier attempts were based on finite-difference techniques (Zhang and Cundy 1989), which are rather problematic for handling the wetting-drying processes involved in these natural phenomena, producing, additionally, mass conservation errors in some cases. The finite volume method is the dominant tool for modeling free surface flows (Toro 2001, Bradford and Sanders 2002, Begnudelli and Sanders 2006, 2007; Begnudelli et al., 2008). Among the advantages of this technique are its inherent mass-conservative properties, the robustness for the consideration of wetting and drying processes and the flexibility to simulate flows over complex domains. From this perspective, a real watershed is divided into connected cells where the two-dimensional shallow water equations are solved. Thus, earlier division into planes and channels with one-dimensional flow is replaced by a division in cells with two-dimensional flow.

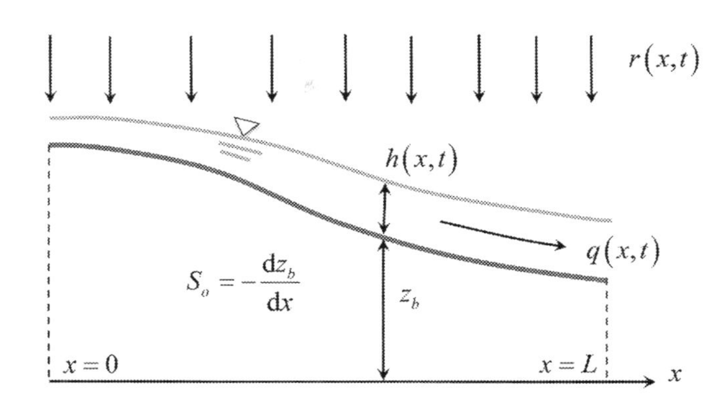

FIG. 15.2

Overland flow over a terrain receiving a net rainfall intensity *r(x, t)*.

In this section the basic features of the finite volume method for simulating run-off processes are presented for a one-dimensional setting taking as example the overland flow (Fig. 15.2). Although it contrasts to the current perspective of two-dimensional modeling at the watershed scale, extrapolation of a one-dimensional finite-volume solver to a two-dimensional setting is straightforward (Toro 2001), such that mastering the one-dimensional case is a prerequisite. Therefore, the basic features of the approach are described here in the one-dimensional setting.

15.3.2.1 Dynamic wave model

The dynamic wave equations are written in vector conservative form for a one-dimensional flow as (Toro 2001)

$$\frac{\partial \boldsymbol{U}}{\partial t} + \frac{\partial \boldsymbol{F}}{\partial x} = \boldsymbol{S} \tag{15.13}$$

where x is the space coordinate and t is time. Here $\boldsymbol{U}$ is the vector of the conserved variables, $\boldsymbol{F}$ the flux vector and $\boldsymbol{S}$ the source term vector, given for a rectangular channel by

$$\boldsymbol{U} = \begin{pmatrix} h \\ hU \end{pmatrix} \quad \boldsymbol{F} = \begin{pmatrix} hU \\ hU^2 + \frac{1}{2}gh^2 \end{pmatrix} \quad \boldsymbol{S} = \begin{pmatrix} r \\ -gh\frac{\partial z_b}{\partial x} - ghS_f \end{pmatrix} \tag{15.14}$$

with h as the water depth, U the depth-averaged velocity, r the net rainfall intensity, z_b the bed elevation, g the acceleration of the gravity and S_f the friction slope. Eq. (15.13) is integrated over a control volume in the x-t plane (Fig. 15.3), resulting in (Toro 2001, Castro-Orgaz and Hager 2019)

$$\boldsymbol{U}_i^{k+1} = \boldsymbol{U}_i^k - \frac{\Delta t}{\Delta x}\left(\boldsymbol{F}_{i+1/2} - \boldsymbol{F}_{i-1/2}\right) + \Delta t \boldsymbol{S}_i \tag{15.15}$$

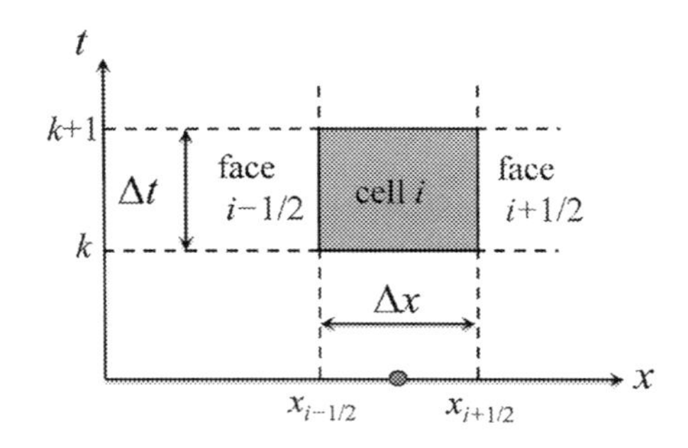

FIG. 15.3

Finite volume cell in the *x-t* plane.

Here $\boldsymbol{F}_{i+1/2}$ is the numerical flux across the interface between cells i and $i+1$, U_i is the spatial-average of the vector $\boldsymbol{U}$ at the corresponding time level, and $\boldsymbol{S}_i$ is the temporal-space average of the source term in the cell.

The model applied in this work is the MUSCL-Hancock finite volume solver, which is second-order accurate both in space and time. In-depth descriptions of each component of this numerical solver are available in Toro (2001) and Castro-Orgaz and Hager (2019).

For the simulation of overland flow (Fig. 15.2) the initial conditions are a dry terrain, $U(x, 0) = 0$ and $h(x, 0) = 0$ (Woolhiser and Liggett, 1967). The boundary conditions are settled in the model using ghost cells outside the computational domain. These cells are conveniently manipulated to produce the desired physical conditions at the boundaries. At the upstream boundary ($x = 0$) a zero flux condition is implemented in the ghost cell, and the water depth is computed by a zero-order extrapolation from the interior solution. At the downstream boundary ($x = L$) a transmitting or open section is implemented in the ghost cell to let the waves freely leave the computational domain. The intensity of the rain is instantaneously raised from zero to r at $t = 0$ in all the cells of the computational domain. This intensity is set back to zero at time t_r.

The MUSCL-Hancock solver reconstructs the solution within each cell using the MUSCL scheme, and the interface values obtained are evolved half the time step using Hancock's formula. The numerical flux is determined using an approximate solution of the Riemann problem resorting the HLL (Harten-Lax-van Leer) solver. The MUSCL-Hancock scheme permits an easy handling of dry terrain thanks to the dry-bed flux incorporation in the HLL Riemann solver. The effects of the source terms are incorporated in the solution using fractional steps. While the source term of the continuity equation originating from the rainfall requires no special care, the momentum equation needs careful attention for robust implementation of the topographic and friction source terms. The topographic source term is discretized using a well-balanced scheme to grant the C-property in full and partially wet cells, whereas the friction source term is treated implicitly for robustness of wet-dry front tracking, given the very shallow flows with high friction effects typical of overland flows.

The computational process is summarized as follows for a generic time loop:

1. Start at time level k with the cell-averaged values $\boldsymbol{U}_i$
2. Apply the CFL (Courant-Friedrichs-Lewy condition) to select a stable time step
3. Reconstruct the solution within each cell of the computational domain using the TVD MUSCL scheme applying a Minmod limiter, to regain second-order space accuracy
4. Evolve the interface values half the time step using Hancock's method to regain second-order time accuracy
5. Solve the Riemann problem at each interface using the evolved interface values with the HLL approximate Riemann solver
6. Apply the conservative equation to resolve the advection step for both the continuity and momentum equations

$$\boldsymbol{U}_i^{adv} = \boldsymbol{U}_i^k - \frac{\Delta t}{\Delta x}\left(\boldsymbol{F}_{i+\frac{1}{2}} - \boldsymbol{F}_{i-\frac{1}{2}}\right) \tag{15.16}$$

7. Include the effect of the bed-slope source term in the momentum equation using the values obtained at time k+1/2 with Hancock's method by a well-balanced topographic discretization
8. Include the effect of the rainfall source term in the continuity equation
9. Include the effect of friction in the momentum equation using an implicit scheme. In this work Manning's equation was adopted, thus only turbulent flows are accounted for.
10. Set boundary conditions for the new time loop
11. Go to step (1) and repeat the cycle until reaching the simulation ending time.

15.3.2.2 Test cases

15.3.2.2.1 Constant in time rainfall on a plane

Prior to applying the dynamic wave model for analyzing rainfall-runoff processes, the accuracy of the numerical solver was tested by adapting it to the solution of the kinematic wave equations for overland flow on a wide slope (Eagleson 1970), with Eq. (15.13)
with

$$\boldsymbol{U} = \begin{pmatrix} h \\ 0 \end{pmatrix} \quad \boldsymbol{F} = \begin{pmatrix} hU \\ 0 \end{pmatrix} \quad \boldsymbol{S} = \begin{pmatrix} r \\ S_0 - S_f \end{pmatrix} \tag{15.17}$$

where

$$S_f = \left(\frac{nU}{h^{2/3}}\right)^2 \tag{15.18}$$

with n as the Manning's roughness coefficient and S_0 the topographic slope. The MUSCL-Hancock finite volume method is thus only applied to the continuity equation, and the discharge is evaluated explicitly from Eq. (15.18) setting $S_f = S_0$.

The numerical solutions produced are to be compared with the analytical solution of the kinematic wave using the method of characteristics, extensively described in Jain (2001).

A test consisting in a plane of 24 m, $S_0 = 0.015$ and $n = 0.009$ m$^{-1/3}$s receiving a rainfall intensity of 1/12 cm s^{-1} was considered for rainfall durations $t_r = 10$ s, 23.96 s (concentration time) and 40 s. Computations were conducted in the numerical solver using 480 cells with CFL = 0.25, and the analytical and numerical results are displayed in Fig. 15.4, thereby showing the accuracy of the finite volume solver described herein.

The test conditions previously described corresponds to the classical experiments by Iwagaki (1955) in a rectangular flume of 19.6 cm width. Measurements of the water depth for t_r = 10 s, 20 s, 30 s and 40 s are plotted in Fig. 15.5, together with the kinematic wave solution for 40 s. Note that the kinematic wave produces poor results, thereby indicating the need to solve the full dynamic wave equations.

The dynamic wave model was run for this test and the results are displayed in Fig15.5. Computations were conducted in the numerical solver using 240 cells with CFL = 0.5. It can be observed that the dynamic wave produces an output in very good agreement with experiments, in contrast to the kinematic wave model. Note the significant underprediction of water depths by the kinematic wave solution for $t_r = 40$ s. The failure of the kinematic wave model was expected, given it is outside the range suggested by Woolhiser and Liggett (1967) for accuracy of kinematic wave solutions. The advantage of solving the full dynamic wave is that all flow cases are automatically accounted for without the need to pre-evaluate if a simplified solver is a suitable option.

15.3.2.2.2 Constant in time-variable in space rainfall on a cascade of planes

A second set of measurements by Iwagaki (1955) was conducted in a cascade of planes, each receiving a different rainfall intensity. Three consecutive reaches of

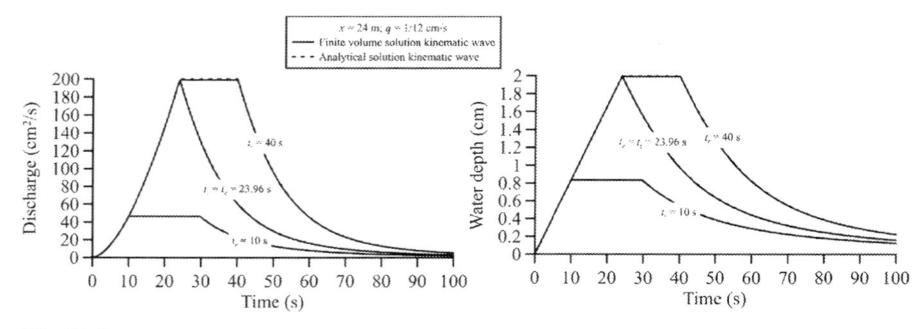

FIG. 15.4

Comparison of analytical (Eagleson 1970) and numerical solutions for the kinematic wave for a constant rainfall of 1/12 cm s^{-1} on a plane of slope $S_0 = 0.015$. In this particular simulation the hydraulic radius is taken as the water depth in the numerical solver.

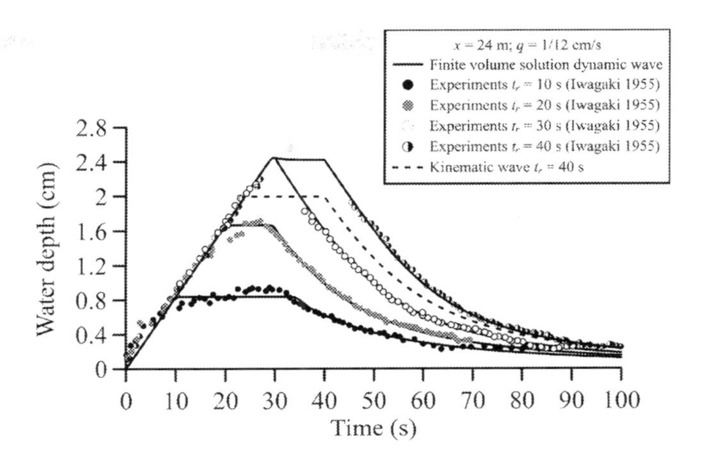

FIG. 15.5

Comparison of dynamic wave solution with experimental results for a constant rainfall of 1/12 cm s^{-1} on a plane of slope $S_0 = 0.015$.

slopes of 0.02, 0.015 and 0.01 in the experimental flume were used to emulate a cascade of planes in a watershed. Each reach is of 8 m in length, and the rainfall intensities are summarized in Table 15.1.

This experimental test is of notable interest, given the generation of complex runoff patterns. Further, the geometry is not standard, and the criteria developed by Woolhiser and Liggett (1967) cannot be applied in advance to infer if the kinematic wave is applicable. The dynamic wave model was applied to this experimental test for rainfall durations of $t_r = 10$, 20 and 30 s, and the computational results are displayed in Fig. 15.6 together with the measurements for the water depth and unit discharge. Computations were conducted in the numerical solver using 240 cells with CFL = 0.5. Note that the dynamic wave model produces a very good prediction of both variables at the end of the flume ($x = 24$ m) for all rainfall durations, albeit for $t_r = 10$ s there is a phase shift between computations and observations. Ascension and recession of the hydrographs are accurately predicted, as well of the peaks. Note formation of shocks, detected in the hydrographs by a sudden rise of the water depth and discharge (e.g. for $t_r = 10$ s).

The overland flow profiles are of considerable complexities as outlined in Fig. 15.7, where the spatial variation of the water depth is plotted at the instant of rainfall cut-off. Note the formation of continuous and discontinuous (shock) waves, both predicted without any special treatment by the finite volume solver. The wetting and drying of the terrain were simulated by the solver without any stability problem. The computational results displayed here in Fig. 15.7 are in excellent agreement with those reported by Iwagaki (1955).

Once the good performance of the dynamic wave model is verified with the aid of Fig. 15.6, the kinematic wave model was solved using the same solver for comparative purposes, and the results are displayed in Fig. 15.8.

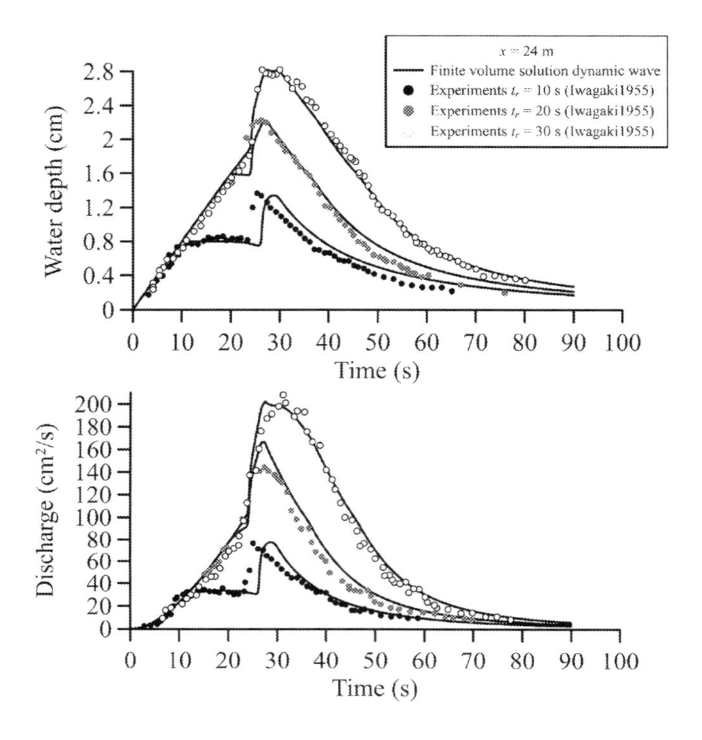

FIG. 15.6

Comparison of dynamic wave solution with experimental results for the cascade of planes summarized in Table 15.1.

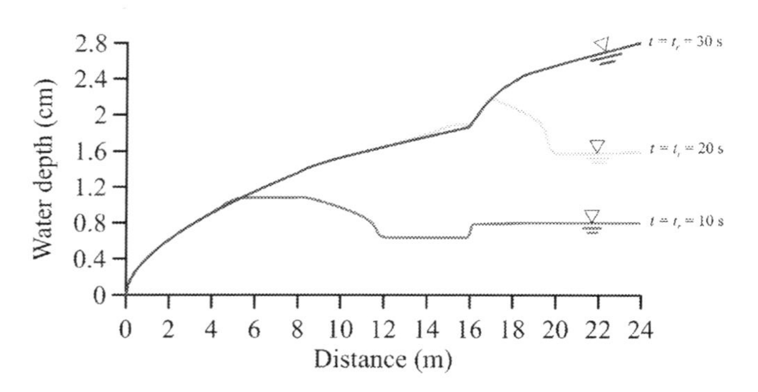

FIG. 15.7

Overland flow profiles for the cascade of planes described in Table 15.1, at the instant of rainfall cut-off.

The kinematic wave model produces poor rising branches of hydrograph for the discharge, but quite acceptable for the water depths. The computed hydrographs for $t_r = 40$ s are not good either for the water depth nor for the discharge. Note the significant underprediction of the maximum water depth for $t_r = 40$ s. Computational

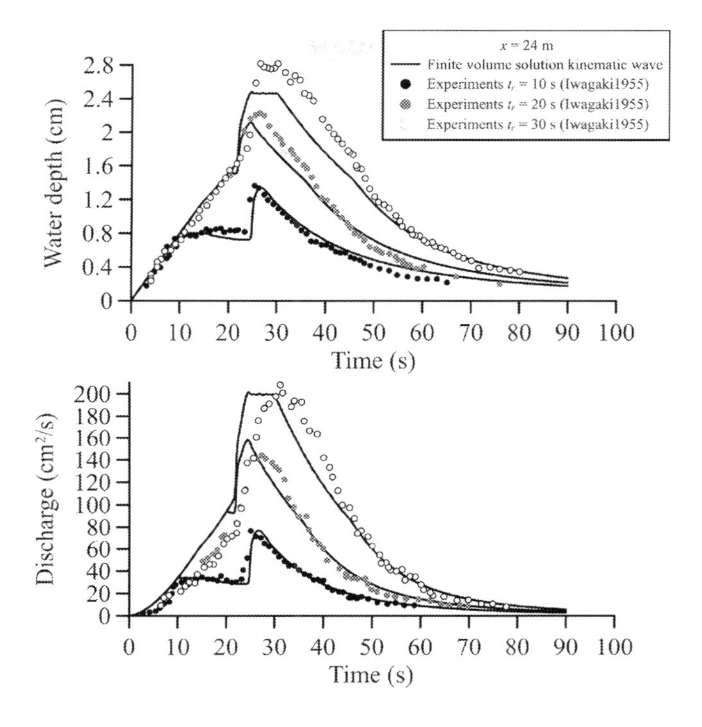

FIG. 15.8

Comparison of kinematic wave solution with experimental results for the cascade of planes summarized in Table 15.1.

results for t_r = 10 s are better than those for the dynamic wave, but this agreement seems to be pure fortuitous when the complete set of simulations is considered. These tests were previously simulated using the kinematic wave approach by Borah et al. (1980) applying an implicit finite-difference scheme and the method of characteristics incorporating shock fitting, and their results for the discharge prediction compared with the experiments by Iwagaki (1955). The results of the present finite volume simulations for the kinematic wave are better than those by Borah et al. (1980) in some cases, but the results of the dynamic wave solver are clearly superior. This test demonstrates that while for the simple case of constant rainfall on a slope it is possible to know in advance if the kinematic wave will work, it is not so the case for complex applications.

The dynamic wave model is able to reproduce complex flow patterns involving spatial variations of slopes and rainfall intensities, a situation where it is unknown in advance if the kinematic model is applicable, and, if applied in any case, results are shown to largely deviate from those of the dynamic wave model. Increase in the computational power is tending to replace progressively the kinematic wave by the dynamic wave model for the hydrodynamic modeling of watersheds.

15.4 Conclusions

Although the influence of the rainfall on the erosion and transport processes is very important for the rational conservation of natural resources such as soil and water, there are several aspects that demand further research.

The characterization of the erosive power of the rainfall, the mechanisms of the raindrop detachment, and the representation of the erosive phenomena, has progressed very much from the pioneer works of the middle of the last century, but the exacerbation of the meteorological processes and the increasing needs of food of our society require a more precise knowledge of those aspects.

References

Ahn, S., Doerr, S.H., Douglas, P., Bryant, R., Hamlet, C.A.E., McHale, G., Newton, M.I., Shirtcliffe, N.J., 2013. Effects of hydrophobicity on splash erosion of model soil particles by a single water drop impact, Earth Surf. Proc. Landf. 38 (11), 1225–1233.

Assouline, S., 2020. On the relationships between radar-reflectivity and rainfall rate and kinetic energy resulting from a Weibull drop size distribution. Water Resour. Res. 56 (e2020WR028156).

Barros, A.P., Prat, O.P., Testik, F.Y., 2010. Size distribution of raindrops. Nat. Phys. 6 (4), 232.

Beard, K.V, 1976. Terminal velocity and shape of cloud precipitation drops aloft. J. Atmos. Sci. 33 (5), 851–864.

Begnudelli, L., Sanders, B.F., 2006. Unstructured grid finite-volume algorithm for shallow-water flow and scalar transport with wetting and drying. J. Hydr. Eng. ASCE 132 (4), 371–384.

Begnudelli, L., Sanders, B.F., 2007. Conservative wetting and drying methodology for quadrilateral grid finite-volume models. J. Hydr. Eng. ASCE 133 (3), 312–322.

Begnudelli, L., Sanders, B.F., Bradford, S., 2008. Adaptive Godunov-based model for flood simulation. J. Hydr. Eng. ASCE 134 (6), 714–725.

Best, A.C., 1950. Empirical formulae for the terminal velocity of water drops falling through the atmosphere. Q. J. R. Meteorol. Soc. 76 (329), 302–311.

Beuselinck, L., Govers, G., Hairsine, P.B., Sander, G.C., Breynaert, M., 2002. The influence of rainfall on sediment transport by overland flows over areas of net deposition. J. Hydrol. 257 (1–4), 145–163.

Bielders, C.L., Baveye, P., Wilding, L.P., Drees, L.R., Valentin, C., 1996. Tillage-induced spatial distribution of surface crusts on a Sandy Paleustult from Togo. Soil Sci. Soc. Am. J. 60 (3), 843–855.

Borah, D.K., Prasad, S.N., Alonso, C.V., 1980. Kinematic wave routing incorporating shock fitting. Water Resour. Res. 16 (3), 529–541.

Bradford, S.F., Sanders, B.F., 2002. Finite-volume model for shallow-water flooding of arbitrary topography. J. Hydr. Eng. ASCE 128 (3), 289–298.

Bringi, V., Thurai, M., Baumgardner, D., 2018. Raindrop fall velocities from an optical array probe and 2-D video disdrometer. Atmosph. Meas. Techn. 11, 1377–1384.

Brown, L.C., Foster, G.R., 1987. Storm erosivity using idealized intensity distributions. Trans. ASAE 30 (2), 379–386.

Castro-Orgaz, O., Hager, W.H., 2019. Shallow Water Hydraulics. Springer-Verlag, Berlin.

Chen, X., Liu, Z., Chen, Y., Wang, H., 2020. Analytical expression for predicting the reduced settling velocity of small particles in turbulence. Environ. Fluid Mech. 20, 905–922.

Cheng, N.S., 1997. Simplified settling velocity formula for sediment particle. J. Hydr. Eng. ASCE 123 (2), 149–152.

Cornelis, W.M., Oltenfreiter, G., Gabriels, D., Hartmann, R., 2004. Splash–saltation of sand due to wind-driven rain: vertical deposition flux and sediment transport rate. Soil Sci. Soc. Am. J. 68 (1), 32–40.

Cugerone, K., de Michele, C., 2015. Johnson SB as general functional form for raindrop size distribution. Water Resour. Res. 51 (8), 6276–6289.

D'Adderio, L.P., Porcù, F., Tokay, A, 2018. Evolution of drop size distribution in natural rain. Atmos. Res. 200, (1 February 2018), 70–76.

Das, S.K., Simon, S., Kolte, Y.K., Krishna, U.V.M., Desphande, S.M., Hazra, A., 2020. Investigation of raindrops fall velocity during different Monsoon seasons over the Western Ghats. India, Earth Space Sci. 7. (e2019EA000956).

Dey, S., Ali, A.S., Padhi, E., 2019. Terminal fall velocity: the legacy of stokes from the perspective of fluvial hydraulics. Proc. R. Soc. A. 475 (2228).

Dietrich, W.E., 1982. Settling velocity of natural particles. Water Resour. Res. 18 (6), 1615–1626.

D'Odorico, P., Yoo, J.C., Over, T.M., 2001. An assessment of ENSO-induced patterns of rainfall erosivity in the Southwestern United States. J. Clim. 14 (21), 4230–4242.

Dunkerley, D., 2019. What does I30 tell us? An assessment using high-resolution rainfall event data from two Australian locations. Catena 180, (September 2019), 320–332.

Dunne, T., Malmon, D.V., Mudd, S.M., 2010. A rain splash transport equation assimilating field and laboratory measurements. J. Geophys Res. doi:10.1029/2009JF001302.

Eagleson, P.S., 1970. Dynamic Hydrology. McGraw-Hill, New York.

Ekern, P.C., 1951. Raindrop impact as the force initiating soil erosion. Soil Sci. Soc. Am. J. 15 (1), 7–10.

Ekern, P.C., Muckenhirn, R.J., 1948. Water drop impact as a force in transporting sand. Soil Sci. Soc. Am. Proc. 12 (C), 441–444.

Erpul, G., Gabriels, D., Norton, L.D., 2005. Sand detachment by wind-driven raindrops, Earth Surf. Proc. Landf. 30 (2), 241–250.

Erpul, G., Norton, L.D., Gabriels, D., 2002. Raindrop-induced and wind-driven soil particle transport. Catena 47 (1), 227–243.

Feingold, G., Levin, Z., 1986. The lognormal fit to raindrop spectra from frontal convective clouds in Israel. J. Climate Appl. Meteor. 25 (10), 1346–1363.

Ferguson, R.I., Church, M., 2004. A simple universal equation for grain settling velocity. J. Sedim. Res. 74 (6), 933–937.

Ferrier, B.S., 1994. A double-moment multiple-phase four-class bulk ice scheme. Part I: description. J. Atmos. Sci. 51 (2), 249–280.

Fister, W., Iserloh, T., Ries, J.B., Schmidt, R.-G., 2012. A portable wind and rainfall simulator for in situ soil erosion measurements. Catena 91, (April 2012), 72–84.

Furbish, D.J., Hammer, K.K., Schmeeckle, M.W., Borosund, M.N., Mudd, S.M., 2007. Rain splash of dry sand revealed by high-speed imaging and sticky paper splash targets. J. Geophys. Res. 112. doi:10.1029/2006JF000498.

Furbish, D.J., Childs, E.M., Haff, P.K., Schmeeckle, M.W., 2009. Rain splash of soil grains as a stochastic advection-dispersion process, with implications for desert plant-soil interactions and land-surface evolution. J. Geophys Res. doi:10.1029/2009JF001265.

Gabet, E.J., Dunne, T., 2003. Sediment detachment by rain power. Water Resour. Res. 39. doi:10.1029/2001WR000656.

Garcia, M.H., 2000. Discussion of "The legend of A.F. Shields. J. Hidr. Eng. ASCE 125, 718–719.

Ghadiri, H., Payne, D., 1981. Raindrop impact stresses. J. Soil Sci. 23 (1), 41–48.

Glass, L.J., Smerdon, E.T., 1967. Effect of rainfall on the velocity profile of shallow-channel flow. Trans. ASAE 10 (2) (336), 330–332.

Gossard, E.E., Strauch, R.G., Welsh, D.C., Matrosov, S.Y., 1992. Cloud layers, particle identification, and rain rate profiles from ZRVf measurements by clean air Doppler radars. J. Atmos. Oceanic. Technol. 9 (4), 108–119.

Gunn, R., Kinzer, G.D., 1949. The terminal velocity of fall for water droplets in stagnant air. J. Meteor. 6 (8), 243–248.

Hairsine, P.B., Rose, C.W., 1991. Rainfall detachment and deposition: sediment transport in the absence of flow-driven processes. Soil Sci. Soc. Am. J. 55 (2), 320–324.

Hairsine, P.B., Rose, C.W., 1992a. Modeling water erosion due to overland flow using physical principles: 1. Sheet flow. Water Resour. Res. 28 (1), 237–243.

Hairsine, P.B., Rose, C.W., 1992b. Modeling water erosion due to overland flow using physical principles: 2. Rill flow. Water Resour. Res. 28 (1), 245–250.

Hairsine, P.B., Sander, G.C., Rose, C.W., Parlange, J.Y., Hogarth, W.L., Lisle, I., Rouhipour, H., 1999. Unsteady soil erosion due to rainfall impact: a model of sediment sorting on the hillslope. J. Hydrol. 220 (3–4), 115–128.

Heng, B.C.P., Sander, G.C., Armstrong, A., Quinton, J.N., Chandler, J.H., Scott, C.F., 2011. Modeling the dynamics of soil erosion and size-selective sediment transport over nonuniform topography in flume-scale experiments. Water Resour. Res. 47. doi:10.1029/2010WR009375.

Hidy, G.M., 2019. Atmospheric aerosols: Some highlights and highlighters, 1950 to 2018. Aerosols Sci. Eng. 3, 1–20.

Hobbs, P.V., Rangno, A.L., 2004. Super-large raindrops. Geophys. Res. Lett. 31. doi:10.1029/2004GL020167.

Hudson, N.W., 1965. The Influence of Rainfall on the Mechanics of Soil Erosion with Particular Reference to Southern Rhodesia. M.Sc. Univ. Of Cape Town, South Africa.

Ignaccolo, M., de Michele, C., 2014. Phase space parameterization of rain: the inadequacy of Gamma distribution. J. Appl. Meteor. Climatol. 53 (2), 548–562.

Iwagaki Y. Fundamental studies on the runoff analysis by characteristics., Bull. 10, Disaster Prevention Research Institute, Kyoto University, Japan, 1955.

Jain S.C. Open channel flow, New York, 2001, John Wiley & Sons.

Izzard, C.F., 1946. Hydraulics of runoff from developed surface. Highw. Res. Bd. Proc. Annu. Meet 26, 129–150.

Jameson, A.R., Kostinski, A.B., 2001. What is a raindrop size distribution? Bull. Am. Meteor. Soc. 82 (6), 1169–1178.

Johnson, N.L., 1949. System of frequency curves generated by method of translation. Biometrika 36 (1/2), 149–176.

Jomaa, S., Barry, D.A., Brovelli, A., Sander, G.C., Parlange, J.Y., Heng, B.C.P., Tromp-van Meerveld, H.J., 2010. Effect of raindrop splash and transversal width on soil erosion: Laboratory flume experiments and analysis with the Hairsine–Rose model. J. Hydrol. 395 (1–2), 117–132.

Jomaa, S., Barry, D.A., Brovelli, A., Heng, B.C.P., Sander, G.C., Parlange, J.Y., Rose, C.W., 2012a. Rainsplash soil erosion estimation in the presence of rock fragments. Catena 92, (May 2012), 38–48.

Jomaa, S., Barry, D.A., Heng, B.C.P., Brovelli, A., Sander, G.C., Parlange, J.Y., 2012b. Influence of rock fragment coverage on soil erosion and hydrological response: Laboratory flume experiments and modeling. Water Resour. Res. 48. doi:10.1029/2011WR011225.

Jomaa, S., Barry, D.A., Heng, B.C.P., Brovelli, A., Sander, G.C., Parlange, J.Y., 2013. Effect of antecedent conditions and fixed rock fragment coverage on soil erosion dynamics through multiple rainfall events. J. Hydrol. 484, (25 March 2013), 115–127.

Jomaa, S., Barry, D.A., Rode, M., Sander, G.C., Parlange, J.Y., 2017. Linear scaling of precipitation-driven soil erosion in laboratory flumes. Catena 152 (5), 285–291.

Kim, J., Warnock, A., Ivanov, V.Y., Katopodes, N.D., 2012. Coupled modeling of hydrologic and hydrodynamic processes including overland and channel flow. Adv. Water Resour. 37, (March 2012), 104–126.

Kinnell, P.I.A., 1983. The effect of kinetic energy of excess rainfall on soil loss from non-vegetated plots. Aust. J. Soil Res. 21 (4), 445–453.

Kinnell, P.I.A., McGregor, K.C., Rosewell, C.J., 1994. The IXEA index as an alternative to the EI30 erosivity index. Trans. ASAE 37 (5), 1449–1456.

Larsen, M.L., Kostinski, A.B., Jameson, A.R., 2014. Further evidence for superterminal raindrops. Geophys. Res. Lett. 41 (19), 6914–6918.

Lavergnat, J., Golé, P., 1998. A stochastic raindrop time distribution model. J. Appl. Meteor. 37 (8), 805–818.

Laws, J.O., Parsons, D.A., 1943. The relation of raindrop-size to intensity. Eos. Trans. AGU 24 (2), 452–460.

Lee, G.W., Zawadzki, I., Szyrmer, W., Sempere-Torres, D., Uijlenhoet, R., 2004. A general approach to double-moment normalization of drop size distributions. J. Appl. Meteor. 43 (2), 264–281.

Leguédois, S., Planchon, O., Legout, C., Le Bissonnais, Y., 2005. Splash projection distance for aggregated soils: theory and experiment. Soil Sci. Soc. Am., J. 69 (1), 30–37.

Li, R.-M., Simons, D.B., Stevens, M.A., 1975. Nonlinear kinematic wave approximation for water routing. Water Resour. Res. 11 (2), 245–252.

Liggett, J.A., Woolhiser, D.A., 1967. Difference solutions of the shallow-water equations. J. Mech. Div. ASCE 93 (1), 39–71.

de Lima, J.L.M.P., van Dijk, P.M., Spaan, W.P., 1992. Splash-saltation transport under wind-driven rain. Soil Technol 5 (2), 151–166.

Loch, R.J., 1984. Field rainfall simulator studies in two clay soils of the Darling Downs, Queensland. III An evaluation of current methods for deriving soil erodibilities (K factors). Aust. J. Soil Res. 22 (4), 401–412.

Loch, R.J., Donnollan, T.E., 1983. Field rainfall simulator studies in two clay soils of the Darling Downs, Queensland.II Aggregate breakdown, sediment properties and soil erodibility. Aust. J. Soil Res. 21 (1), 47–58.

Lowell, C.J., Rose, C.W., 1988. Measurement of soil aggregate settling velocities. I. A modified bottom withdrawal tube method. Aust. J. Soil Res. 26 (1), 55–71.

Marshall, J.S., Palmer, W., 1948. The distribution of raindrops with size. J. Meteor. 5 (8), 165–166.

Marzen, M., Iserloh, T., Casper, M.C., Ries, J.B., 2015. Quantification of particle detachment by rain splash and wind-driven rain splash. Catena 127 (April 2015), 135–141.

Marzen, M., Iserloh, T., Lima, J.L.M.P.de, Ries, J.B., 2016. The effect of rain, wind-driven rain and wind on particle transport under controlled laboratory conditions. Catena 145 (October 2016), 47–55.

Marzen, M., Iserloh, T., Lima, J.L.M.P.de, Fister, W., Ries, J.B., 2017. Impact of severe rainstorms on soil erosion: Experimental evaluation of wind-driven rain and its implications for natural hazard management. Sci. Total Environ. 590-591 (15 July 2017), 502–513.

Mazurak, A.P., Mosher, P.N., 1968. Detachment of soil particles in simulated rainfall. Soil Sci. Soc. Am. J. 32 (5), 716–719.

Meyer, L.D., Wischmeier, W.H., 1969. Mathematical simulation of the process of soil erosion by water. Trans. ASAE 12 (762), 754–758.

Mitchell, D.L., 1996. Use of mass- and area-dimensional power laws for determining precipitation particles terminal velocities. J. Atmos. Sci. 53 (12), 1710–1723.

Monteith, J.L., Unsworth, M.H., 1990. Principles of Environmental Physics, 2nd edition. Edward Arnold, London.
Montero-Martínez, G., Kostinski, A.B., Shaw, R.A., García-García, F., 2009. Do all raindrops fall at terminal speed? Geophys. Res. Lett. 36. doi:10.1029/2008GL037111.
Moore, I.D., Burch, G.J., 1986. Sediment transport capacity of sheet and rill flow: application of unit stream power theory. Water Resour. Res. 22 (8), 1350–1360.
Moss, A.J., Walker, P.H., Hutka, J., 1979. Raindrop-stimulated transportation in shallow water flows: an experimental study. Sedim. Geol. 22 (3–4), 165–184.
Moss, A.J., 1988. Effects of flow-velocity variation on rain-driven transportation and the role of rain impact in the movement of solids. Aust. J. Soil Res. 26 (3), 443–450.
Moss, A.J., 1989. Impact droplets and the protection of soils by plant covers. Aust. J. Soil Res. 27 (1), 1–16.
Moss, A.J., 1991a. Rain-impact soil crust. I. Formation on a granite-derived soil. Aust. J. Soil Res 29 (2), 271–289.
Moss, A.J., 1991b. Rain-impact soil crust. II. Some effects of surface-slope, drop-size and soil variation. Aust. J. Soil Res 29 (2), 291–309.
Moss, A.J., Green, P., 1983. Movement of solids in air and water by raindrop impact: effects of drop-size and water-depth variations. Aust. J. Soil Res. 21 (3), 257–269.
Moss, A.J., Green, T.W., 1987. Erosive effects of the large water drops (gravity drops) that fall from plants. Aust. J. Soil Res. 25 (1), 9–20.
Mouzai, L., Bhouhadeff, M., 2003. Water drop erosivity: Effects on soil splash. J. Hydr. Res. 41 (1), 61–68.
Mualem, Y., Assouline, S., 1986. Mathematical model for rain drop distribution and rainfall kinetic energy. Trans. ASAE 29 (2), 494–500.
Nearing, M.A., Yin, S.Q., Borrelli, P., Polyakov, V.O., 2017. Rainfall erosivity: an historical review. Catena 157, 357–362.
Nissan, H., Toumi, R., 2013a. Dynamic simulation of rainfall kinetic energy flux in a cloud resolving model. Geophys. Res. Lett. 40, 3331–3336.
Nissan, H., Toumi, R., 2013b. On the impact of aerosols on soil erosion. Geophys. Res. Lett. 40 (22), 5994–5998.
Park, S.W., Mitchell, J.K., Bubenzer, G.D., 1983. Rainfall characteristics and their relation to splash erosion. Trans. ASAE 26 (3), 795–804.
Parsons, A.J., Abrahams, A.D., Simanton, J.R., 1992. Microtopography and soil-surface materials on semi-arid piedmont hillslopes, southern Arizona. J. Arid Environ. 22 (2), 107–115.
Pelletier, J.D., 2012. Fluvial and slope-wash erosion of soil-mantled landscapes: detachment- or transport-limited? Earth Surf. Proc. Landf. 37 (1), 37–51.
Peters, G., Fischer, B., Münster, H., Clemens, M., Wagner, A., 2005. Profiles of raindrop size distribution as retrieved by microrain radars. J. Appl. Meteor. 44 (12), 1930–1949.
Poesen, .J, 1985. An improved splash transport model. Z. Geomorphol. 29 (2), 193–211.
Poesen, J., 1986. Field measurement of splash erosion to validate a splash transport model. Z. Geomorphol. Supppl. 58, 81–91.
Polyakov, V.O., Nearing, M.A., 2003. Sediment transport in rill flow under deposition and detachment conditions. Catena 51 (1), 33–51.
Proffitt, A.P.B., Rose, C.W., Lovell, C.J., 1989. A comparison between modified splash-cup and flume techniques in differentiating between soil loss and detachability as a result of rainfall detachment and deposition. Aust. J. Soil Res. 27 (4), 759–777.
Proffitt, A.P.B., Rose, C.W., 1991a. Soil erosion processes: I. The relative importance of rainfall detachment and runoff entrainment. Aust. J. Soil Res. 29 (2), 671–683.

Proffitt, A.P.B., Rose, C.W., 1991b. Soil erosion processes: II. Settling velocity characteristics of eroded sediments. Aust. J. Soil Res. 29, 685–695.

Proffitt, A.P.B., Rose, C.W., Hairsine, P.B., 1991. Rainfall detachment and deposition: experiments with low slopes and significant water depths. Soil Sci. Soc. Am. J. 55 (2), 325–332.

Pruppacher, H.R., Klett, J.D., 1978. Microphysics of Clouds and Precipitation. Reidel, Dordrecht.

Ravi, S., Breshears, D.D., Huxman, T.E., D'Odorico, P., 2010. Land degradation in drylands: interactions among hydrologic–aeolian erosion and vegetation dynamics. Geomorphol 111 (3–4), 236–245.

Reeve, I.J., 1982. A splash transport model and its application to geomorphic measurement. Z. Geomorphol. 26, 55–71.

Renard, K.G., Foster, G.R., Weesies, G.A., McCool, D.K., Yoder, D.C., 1997. Predicting Soil Erosion by Water: A Guide to Conservation Planning with the Revised Universal Soil Loss Eq. (14.RUSLE). USDA Agriculture Handbook, 703 D.C, Washington, DC.

Rose, C.W., Williams, J.R., Sander, G.C., Barry, D.A., 1983a. A mathematical model of soil erosion and deposition processes: I. Theory for a plane land element. Soil Sci. Soc. Am. J. 47 (5), 991–995.

Rose, C.W., Williams, J.R., Sander, G.C., Barry, D.A., 1983b. A mathematical model of soil erosion and deposition processes: II. Application to data from an arid-zone catchment. Soil Sci. Soc. Am. J. 47 (5), 996–1000.

Salles, C., Creutin, J.D., Sempere-Torres, D., 1998. The optical spectropluviometer revisited. J. Atmos. Oceanic Technol. 15 (5), 1215–1222.

Salles, C., Poesen, J., 1999. Performance of an optical spectro pluviometer in measuring basic rain erosivity characteristics. J. Hydrol. 218 (3–4), 142–156.

Salles, C., Poesen, J., Govers, G. 2000. Statistical and physical analysis of soil detachment by raindrop impact: rain erosivity indices and threshold energy. Water Resour. Res. 36 (9), 2721–2729.

Salles, C., Poesen, J., Sempere-Torres, D., 2002. Kinetic energy of rain and its functional relationship with intensity. J. Hydrol. 257 (1–4), 256–270.

Sander, G.C., Parlange, J.Y., Barry, D.A., Parlange, M.B., Hogarth, W.L., 2007. Limitation of the transport capacity approach in sediment transport modeling. Water Resour. Res. 43. doi:10.1029/2006WR005177.

Sempere-Torres, D., Porrá, J.M., Creutin, J.D., 1994. A general formulation for raindrop size distribution. J. Appl. Meteor. 33 (12), 1494–1502.

Sempere-Torres, D., Porrá, J.M., Creutin, J.D., 1998. Experimental evidence of a general description for raindrop size distribution properties. J. Geophys. Res. 103 (D2), 1785–1797.

Shainberg, I., 1985. The effect of exchangeable sodium and electrolyte concentration in crust formation. Adv. Soil Sci. 1 (1), 101–122.

Shao, Y., Lu, H., 2000. A simple expression for wind erosion threshold friction velocity. J. Geophys. Res. 105 (22), 437–422, 443.

Skamarock, W.C., Klemp, J.B., Dudhia, J., Gill, D.O., Barker, D.M., Wang, W., Powers, J.G., 2005. A Description of the Advanced Research WRF Version 2 (No. NCAR/TN-468+STR). University Corporation for Atmospheric Research, Denver, CO. doi:10.5065/D6DZ069.

Sochan, A., Łagodowski, Z.A., Nieznaj, E., Beczek, M., Ryzak, M., Mazur, R., Bobrowski, A., Bieganowski, A., 2019. Splash of solid particles as a stochastic point process. J. Geophys. Res. Earth Surf. 124. https://doi.org/10.1029/2018JF004993.

Spilhaus, A.F., 1948. Raindrop size, shape, and falling speed. J. Meteor. 5 (3), 108–110.

Stanley, J.W., Parker, F.W., 1943. A study of new methods for size analysis of suspended sediment samplesRep. no. 7 of A Study of Methods Used in Measurement and Analysis of Sediment Loads in Streams. University of Iowa Iowa City, St. Paul U. S. Engineer District Sub-Office, Hydraulic Laboratory.

Terry, J.P., 1998. A rainsplash component analysis to define mechanisms of soil detachment and transportation. Aust. J. Soil Res. 36 (3), 525–542.

Terry, J.P., Shakesby, R.A., 1993. Soil hydrophobicity effects on rainsplash: simulated rainfall and photographic evidence, Earth Surf. Proc. Landf. 18 (6), 519–525.

Testik, F.Y., Barros, A.P., 2007. Toward elucidating the microstructure of warm rainfall: a survey. Rev. Geophys. 45. doi:10.1029/2005RG000182.

Testik, F.Y., Rahman, M.K., 2017. First in situ observations of binary raindrop collisions. Geophys. Res. Lett. 44 (2), 1175–1181.

Testud, J., Oury, S., Black, R.A., Amayenc, P., Dou, X., 2001. The concept of "normalized" distribution to describe raindrop spectra: a tool for cloud physics and cloud remote sensing. J. Appl. Meteor. 40 (6), 118–1140.

Thurai, M., Bringi, V.N., 2018. Application of the generalized gamma model to represent the full rain drop size distribution spectra. J. Appl. Meteor. Climatol. 57 (5), 1197–1208.

Toro, E.F., 2001. Shock-Capturing Methods for Free-Surface Shallow Flows. John Wiley & Sons, Singapore.

Tromp-van Meerveld, H.J., Parlange, J.-Y., Barry, D.A., Tromp, M.F., Sander, G.C., Walter, M.T., Parlange, M.B., 2008. Influence of sediment settling velocity on mechanistic soil erosion modeling. Water Resour. Res. 44. doi:10.1029/2007WR006361.

Ulbrich, C., 1983. Natural variations in the analytical form of the raindrop size distribution. J. Climate Appl. Meteor. 22 (10), 1764–1775.

Valentin, C., Bresson, L.M., 1992. Morphology, genesis and classification of surface crusts in loamy and sandy soils. Geoderma 55 (3–4), 225–245.

van Boxel, J., 1998. Numerical model for the fall speed of raindrops in a rainfall simulator. I.C.E. Special Report 1 (1), 77–85.

van der Meer, D., 2017. Impact on granular beds. Ann. Rev. Fluid Dyn. 49, 463–484.

van Dijk, A.I.J.M., Meesters, A.G.C.A., Bruijnzeel, L.A., 2002a. Exponential distribution theory and the interpretation of splash detachment and transport experiments. Soil Sci. Soc. Am. J. 66 (5), 1466–1474.

van Dijk, A.I.J.M., Bruijnzeel, L.A., Rosewell, C.J., 2002b. Rainfall intensity-kinetic energy relationship. A critical literature appraisal. J. Hydrol. 216 (1–4), 1–23.

Villermaux, E., Eloi, F., 2011. The distribution of raindrops speeds. Geophys. Res. Lett. 38. doi: 10.1029/2011GL048863.

Villermaux, E., Bossa, M., 2009. Single-drop fragmentation determines size distribution of raindrops. Nature Physics. doi:10.1038/NPHYS1340.

Walker, P.H., Hutka, J., Moss, A.J., KInnell, P.I.A., 1977. Use of a versatile experimental system for soil erosion studies. Soil Sci. Soc. Am. J. 41 (3), 610–612.

Walker, P.H., Kinnell, P.I.A., Green, P., 1978. Transport of a noncohesive sandy mixture in rainfall and runoff experiments. Soil Sci., Soc. Am. J. 42 (5), 781–793.

Walker, J.D., Walter, M.T., Parlange, J.Y., Rose, C.W., Tromp-van Meerveld, H.J., Gao, B., Cohen, A.M., 2007. Reduced raindrop-impact dr.iven soil erosion by infiltration. J. Hydrol. 342 (3–4), 331–335.

Wang, Y., Lam, K.M., Lu, Y., 2018. Settling velocity of fine heavy particles in turbulent open cannel flow. Phys. Fluids 30 (9), 095106. doi:10.1063/1.5046333.

Wischmeier, W.H., 1959. A rainfall erosion index for a universal soil-loss equation. Soil Sci. Soc. Am. J. 23 (3), 246–249.

Wischmeier, W.H., Smith, D.D., 1958. Rainfall energy and its relationship to soil loss. Trans. Am. Geophys. Union 39 (2), 285–291.

Wischmeier, W.H., Smith, D.D., 1965. Predicting rainfall erosion losses in the Eastern U.S.– a guide to conservation planning. Agricultural Handbook No. 282. USDA, ARS, Washington, DC.

Wischmeier, W.H., Smith, D.D., 1978. Predicting rainfall erosion losses. A guide to conservation planning. USDA Agriculture Handbook No. 537, ARS, USDA in cooperation with Purdue Agricultural Experiment Station, West Lafayette, IN.

Woolhiser, D.A., Liggett, J.A., 1967. Unsteady, one-dimensional flow over a plane-the rising hydrograph. Water Resour. Res. 3 (2), 753–771.

Woolhiser, D.A, Smith, R.E., Goodrich, D.C., 1990. KINEROS. A Kinematic Runoff and Erosion Model, Documentation and User Manual, ARS-77, U.S. Dept. of Agriculture, Agricultural Research Service, 130, Washington, DC.

Wright, A.C., 1986. A physically-based model of the dispersion of splash droplets ejected from a water drop impact. Earth Surf. Proc. Landf. 11 (4), 351–369.

Wright, A.C., 1987. A model of the redistribution of disaggregated soil particles by rainsplash, Earth Surf. Proc. Landf. 12 (6), 583–596.

Yang, C.T., 1973. Incipient motion and sediment transport. J. Hydr. Eng. ASCE 99 (10), 797–805.

Yarin, A.L., 2006. Drop impact dynamics: splashing, spreading, receding, bouncing…. Ann. Rev. Fluid Mech. 38, 159–192.

Yu, Y.S., Mcnown, J.S., 1964. Runoff from impervious surfaces. J. Hydr. Res. 2 (1), 3–24.

Zhang, W., Cundy, T.W., 1989. Modeling of two-dimensional overland flow. Water Resour. Res. 25 (9), 2019–2035.

CHAPTER

16 Rainfall and landslide initiation

Fausto Guzzetti[a], Stefano Luigi Gariano[b], Silvia Peruccacci[b], Maria Teresa Brunetti[b], Massimo Melillo[b]

[a]*Civil Protection Department, Office of the Prime Minister, Rome, Italy*
[b]*CNR IRPI - Italian National Research Council, Research Institute for the Geo-Hydrological Protection, Perugia, Italy*

16.1 Introduction

According to an estimate made by the United Nations Office for Disaster Risk Reduction, 91% of all disasters occurred worldwide from 1998 to 2017 were related to climatic and meteorological variables, in particular those caused by geo-hydrological hazards. Of all the reported disasters in the 20-year considered period, 5.4% were caused by landslides; although not too high, this percentage still concerns about five million people (Wallemacq and House, 2018).

A landslide is the movement of a mass of rock, debris, or earth down natural or engineered slopes (Cruden and Varnes, 1996; Hungr et al., 2013). Landslides involve different kinematic mechanisms, including flowing, sliding, toppling, falling, or spreading; moreover, many landslides evolve in different types of movements, simultaneously or in consecutive time period. According to their size, in particular the ratio of length to depth, landslides can be either shallow or deep-seated. Based on their velocity, mostly linked to their depth and the degree of strength loss during the failure, landslides can be classified from "extremely slow" (velocity less than 16 $mm \cdot y^{-1}$) to "extremely rapid" (more than 5 $m \cdot s^{-1}$) (Hungr et al., 2013). Overall, landslides are complex and diversified phenomena: their lifetime, triggering time, length, area, volume, velocity, affected area, number - span several orders of magnitude (Guzzetti et al., 2012).

Landslides are found in many areas of the world, contributing to the evolution of single slopes and entire landscapes. Because of their spread, they frequently pose a serious threat to population resulting in loss of life, injuries, and high economic and social damages (Nadim et al., 2006; Dowling and Santi, 2013; Pereira et al., 2015; Froude and Petley, 2018).

The stability/instability conditions of slopes are influenced by different phenomena, including hydrological processes, temperature changes, volcanic activity, earthquakes, and anthropogenic activities. Overall, all hydrologic processes including precipitation, snow and permafrost melting, water recharge into soils, lateral and vertical movement of water within the regolith, evapotranspiration, and water interception exert a large

RAINFALL. Modeling, Measurement and Applications. DOI: https://doi.org/10.1016/B978-0-12-822544-8.00012-3

influence on landslide initiation (Sidle and Ochiai, 2006). Among all, rainfall is by far the main trigger of landslides, because it affects most of the hydrological processes. Four rainfall parameters affect the generation of pore water pressure in unstable slopes, and consequently have a role in landslide initiation: (i) total amount of rainfall; (ii) short-term (mean or peak) rainfall intensity; (iii) antecedent rainfall; and (iv) rainfall duration (Sidle and Ochiai, 2006).

Rainfall infiltration causes temporary changes in groundwater dynamics reducing the stability conditions (Van Asch et al., 1999). Due to an increase in pore water pressure, the effective shear strength of the material decreases, becoming lower than a given value (Terzaghi, 1962), and a slope movement can initiate. Based on the intensity and duration of the rainfall, the combination of infiltration and runoff may cause several types of mass-movements at different depths.

The relative rates of all the above-mentioned hydrological processes, which are spatially and temporally distributed, may determine a transient level of groundwater in portions of the slope thus increasing the failure potential. Water filtration into soil and bedrock is controlled by their physical properties (Sidle and Ochiai, 2006). Groundwater may reach a given location within the slope through different paths and mechanisms, including: (i) surface flow (influenced by morphology); (ii) direct infiltration from the surface; (iii) flow within the soil mantle from upslope and side slopes; and (iv) seepage from the bedrock. These different paths can have different lengths and can be characterized by diverse hydraulic conductivity values: as a result, portions of the same rainfall event may reach a given site at different times, or different sites at the same time, so being responsible of different landslide activations/occurrences (Terranova et al., 2015). In addition, slope failures depend on soil initial conditions, slope angle, and soil mechanical parameters.

Given the numerous variables, and the different processes involved, investigating the relationships between rainfall and landslides is not trivial (Sidle and Ochiai, 2006; Guzzetti, 2015). Determining the rainfall conditions responsible for landslide occurrence, and predicting the possible occurrence of new phenomena are crucial tasks in geo-hydrological risk mitigation, and may contribute to save lives and properties. Petley (2012) estimated that about 90% of worldwide casualties due to slope failures can be attributed to rainfall-triggered landslides. Even worse, such natural phenomena have an increasing frequency in the last years, also due to global warming (Gariano and Guzzetti, 2016; Froude and Petley, 2018; Haque et al., 2019).

16.2 Modeling the relationships between rainfall and landslide

Two different approaches are generally adopted to model the relationships between rainfall and landslide occurrence/activation.

The first approach, hereafter referred to as "physically-based" (or "process-based", or "deterministic"), aims at understanding the physical laws controlling slope instability and modelling all (or at least most of) the effects of rainfall on slope in terms of overland flow, groundwater infiltration, pore pressure and related balance of shear stress and resistance (see e.g., Montgomery and Dietrich, 1994; Wilson and Wieczorek, 1995; Godt et al., 2008; Pisani et al., 2010). Usually, when dealing with this approach, numerical models (either complete or distributed) are employed, and a large amount of detailed data is required to model the slope in topographic, morphologic, hydrogeological, litho-structural, and geotechnical terms. Computation results in the calculation of a factor of safety, which expresses the ratio between the local resisting (R) and driving (S) forces, FS = R/S. Values of the factor smaller than 1.0 denote slope instability.

The second approach, hereafter referred to as "empirical" (or "statistical", or "probabilistic"), is based on the statistical analysis of past rainfall conditions that have (presumably) resulted in landslides, obtained by combining rainfall series with dates of occurrence/activation of slope movements, (see e.g., Campbell, 1975; Caine, 1980; Guzzetti et al., 2007; 2008, Brunetti et al., 2010). This approach needs less computational resources and time than the physically-based models. To apply this approach, empirical relations have to be determined by means of threshold values or functions, to distinguish among conditions that likely trigger or do not trigger a landslide. Different rainfall variables can be selected (Guzzetti et al., 2007, 2008; Segoni et al., 2018) including e.g., (i) the cumulated rainfall recorded in a given temporal interval (hours/days/months) before the landslide activation; and (ii) the duration, cumulated rainfall and/or mean intensity of a "triggering" rainfall event; or the average rainfall intensity in the same temporal window; or rainfall values normalized to some reference values (e.g., annual normal).

In the case of shallow landslides, empirical relationships between variables (e.g., triggering thresholds) can be derived from the correlations between the triggering rainfall (at daily, hourly or finer time resolution), corresponding to the initiation of the failure, and the rainfall cumulated over an antecedent period (usually, few days to two weeks before landslide initiation). Alternatively, empirical functions refer to relations between cumulated rainfall (in mm), rainfall mean intensity I (usually expressed in $mm \cdot h^{-1}$), and rainfall duration D (usually expressed in h). Antecedent rainfall can be also included in the calculations (Crozier, 1999). Larger amounts of antecedent rainfall should allow slope movements to be activated by less severe storms. Overall, for shallow phenomena, empirical approaches allow a relatively fast and reliable modeling. Empirical methods need statistically representative series of data and involve non-expensive computational requirements. In most of the cases, these methods result in the calculation of rainfall thresholds for the landslide initiation, which have become the most used tools to analyze the triggering conditions of slope failures and forecast the possible occurrence of a landslide or of a population of landslides in a given area.

On the other side, in case of deep-seated landslides, whose initiation mostly depends on groundwater dynamics, difficulties and limitations in using empirical modelling generally increase. Large landslides usually have complex and non-linear relationships with rainfall parameters, given that different hydrological mechanisms need to be considered (e.g., diverse groundwater aliquots may combine and reach the sliding surface).

16.3 The TRIGRS physically-based model

Among the physically-based models for the initiation of rainfall-induced landslides, here is briefly presented the TRIGRS (Transient Rainfall Infiltration and Grid-Based Regional Slope-Stability) model. TRIGRS is a code written in Fortran and released by the US Geological Survey (Baum et al., 2002; 2008), which is able to assess and predict the timing and spatial distribution of rainfall-induced shallow landslides in a given area. TRIGRS couples a grid-based, spatially distributed slope stability model with a rainfall infiltration model (Iverson, 2000), calculating the rainfall infiltration in the terrain and the stability conditions of the grid cells (Godt et al., 2008). The study area is discretized into a grid of regular cells (Fig. 16.1), based on a digital elevation model. Local terrain characteristics are used as an input for the solution of a system of equations whose output is the evolution of the Factor of Safety (FS) with time for different depths.

To model the landslide geometry (including the geometry of the topographic surface and the location of the failure surface) TRIGRS adopts a one-dimensional infinite slope approximation (Taylor, 1948) - which is rigorous but relatively simple, and is acceptable when the length of the slope is much larger than its thickness, i.e., in the case of shallow landslides. Within this approximation, in each cell the failure surface is assumed to be planar, at a fixed depth, parallel to the topographic surface, and of infinite extent (Fig. 16.1). The balance of the vertical component of gravity, against the resisting stress due to the basal Coulomb friction and the pore pressure (Richards, 1931) drives the stability of each cell. Forces acting on the sides of the sliding mass and inter-cell forces are neglected.

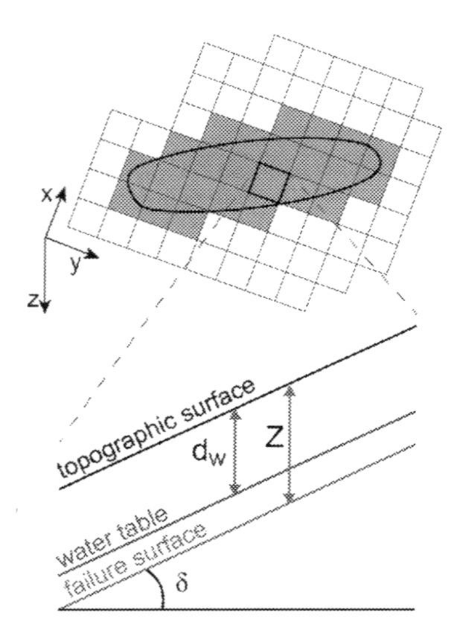

FIG. 16.1

Schematic representation of grid-based, one-dimensional infinite slope adopted in TRIGRS.

Along a potential failure surface located at depth Z (measured vertically from the topographic surface, Fig. 16.1), limit equilibrium conditions are reached (i.e., the failure is about to occur) when the mobilized shear stress equals the available shear strength. In terms of the *FS*, the ratio of the resisting *R* and the driving *S* forces is expressed as:

$$FS = \frac{R}{S} = \frac{tan\phi}{tan\delta} + \frac{c - \psi \gamma_w \tan \phi}{\gamma_s z \sin \delta \cos \delta} \tag{16.1}$$

where δ is the angle between the slope and the rupture plane, z is the slope-normalized coordinate ($z = Z / \cos \delta$), ϕ is the internal friction angle, c and γ_s are the cohesion and the soil unit weight of the material, γ_w is the groundwater unit weight, and ψ is the pressure head. Solution of Eq. (16.1) is made at discrete time steps and in the vertical coordinate.

TRIGRS requires both time-invariant and time-dependent input data (Raia et al., 2014). Time-invariant variables include: (i) hydro-mechanical properties of the slope material (i.e., γ_s, c, ϕ, water content, and saturated hydraulic conductivity), and (ii) geometrical characteristics of the sliding mass (i.e., δ and Z). Conversely, time-dependent information consists of the pressure head ψ (the pressure exerted by water on the sliding mass), which is a function of the depth of the water table d_w and is governed by the Richards equation (Richards, 1931). This non-linear partial differential equation does not have a closed-form analytical solution, and can be solved only by means of approximate solutions.

The original release of TRIGRS (Baum et al., 2002) calculates the stability conditions of individual grid cells in a given area, and models infiltration adopting the approach proposed by Iverson (2000) for one-dimensional vertical flow in isotropic, homogeneous materials, and for saturated conditions. In a second release of TRIGRS (Baum et al., 2008) also unsaturated soil conditions are included, considering the presence of a capillary fringe above the water table. Overall, four options for infiltration modeling are now available in TRIGRS (Tran et al., 2018): (1) saturated soil with infinite basal boundary depth, (2) saturated soil with finite basal boundary depth, (3) unsaturated soil with infinite basal boundary depth, and (4) unsaturated soil with finite basal boundary depth. To reduce the computational times of these four time-demanding execution modes of TRIGRS, an updated, parallel implementation of TRIGRS was issued by Alvioli and Baum (2016) to model time-varying rainfall inputs.

A common problem encountered in using TRIGRS to predict the occurrence of population of landslides over large areas is the difficulty of gathering sufficient and reliable information on the hydro-mechanical properties of the terrain. The adoption of literature values for the hydro-mechanical parameters may result in unrealistic modelling of the stability conditions of grid cells. To solve this problem diverse probabilistic approaches were proposed (e.g., TRIGRS-P by Raia et al., 2014 and PG_TRIGRS by Salciarini et al., 2017) and tested providing remarkable improvements in landslide prediction.

TRIGRS was also applied to determine the rainfall conditions related to landslide initiation and to define the corresponding physically-based rainfall thresholds at a regional scale (e.g., Salciarini et al., 2012; Peres and Cancelliere, 2014; Alvioli et al., 2014; 2018).

16.4 Rainfall thresholds for landslide initiation

A threshold can be defined as a condition, expressed in quantitative terms or through a mathematical law, whose overcoming results in a change of state of a system (White et al., 1996). With reference to landslides, a minimum threshold can represent the lower bound of known hydrological conditions (e.g., rainfall, infiltration, soil moisture) that resulted in landslides (Reichenbach et al., 1998). More in detail, a rainfall threshold defines the rainfall conditions that when reached or exceeded, are likely to trigger landslides (Guzzetti et al., 2008).

A threshold can be represented in a Cartesian plane as a lower boundary curve that delimits the portion of the space containing the hydrological (rainfall) conditions associated to known slope failures (Caine, 1980). A condition below the threshold roughly indicates slope stability, instead a condition above the threshold is likely associated to a slope failure (Fig. 16.2a). An upgrade of this approach is obtained by including in the analysis (and in the Cartesian plane) also those hydrological (rainfall) conditions known or presumably not associated to landslide occurrence (Fig. 16.2b). In this case, thresholds are defined as the best discriminator between triggering and presumably non-triggering conditions (Crozier, 1997). A further improvement consists in dividing the Cartesian plane in three sectors, by means of two thresholds (Crozier, 1997): a lower threshold, below which no landslides are expected, and an upper threshold, above which landslides always occur (Fig. 16.2c). Between the two curves, different probabilities of occurrence can be defined, taking into account uncertainties related to the partial knowledge of the physical process and to the incompleteness of the landslide database.

In the literature, methodological examples focus on two types of thresholds obtained for either single phenomena or for population of landslides (of a given type) occurred within a homogeneous geo-environmental region.

The concept of a minimum amount of rainfall necessary to trigger a landslide was firstly introduced by Endo (1969), and the first quantitative rainfall threshold for landslide initiation was proposed by Onodera et al. (1974). Subsequently, the two landmark articles about landslide-triggering rainfall thresholds were written by Campbell (1975) and Caine (1980). In particular, Campbell (1975) analyzed the rainfall conditions responsible for the initiation of soil slips in California in the period 1962–1971 and found that the cause of the failures was the combination of antecedent cumulated rainfall and event rainfall intensity. Caine (1980) studied 73 rainfall mean intensity (I) vs. rainfall duration (D) conditions that had resulted in shallow landslides and debris flows worldwide and proposed the first global threshold expressed by a power-law equation. Since those pioneering works – and despite some

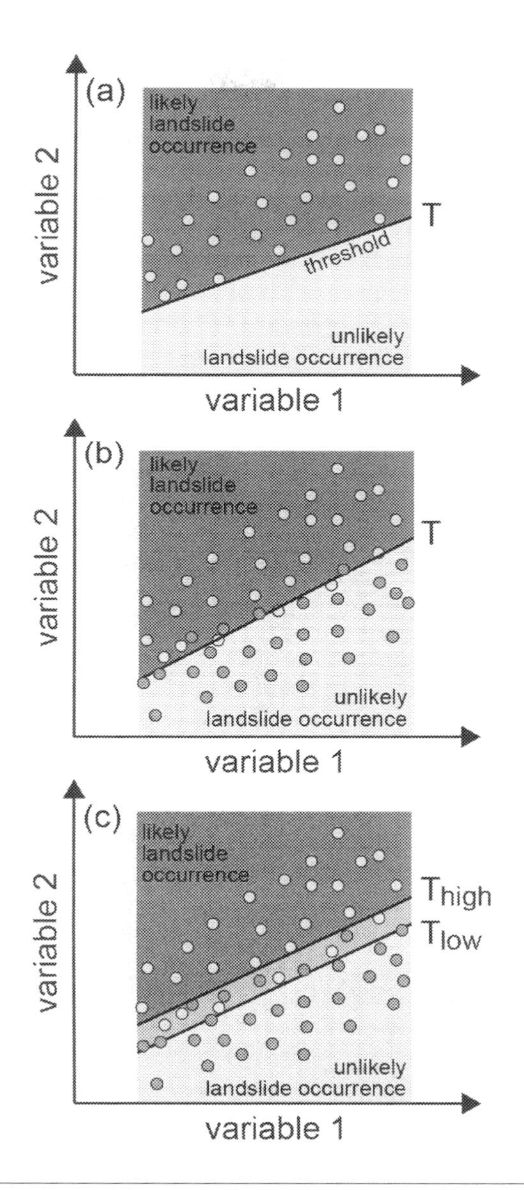

FIG. 16.2

Theoretical schemes of rainfall thresholds for landslide initiation.

criticisms (e.g., Bogaard and Greco, 2018) – rainfall thresholds have been widely used to model the relationship between rainfall and landslide initiation (De Vita et al., 1998; Reichenbach et al., 1998) at different geographical scales of analysis (global, regional, local), with a wide variety of rainfall parameters, in various physiographic settings, and for different landslide types (Guzzetti et al., 2007; 2008). Rainfall thresholds are defined using heuristic (e.g., Sengupta et al., 2010), statistical (e.g., Lagomarsino et al., 2015), probabilistic (e.g., Berti et al., 2012), and also physically-based methods (e.g., Salciarini et al., 2012); overall, statistical thresholds are the most adopted ones (Segoni et al., 2018).

Inspection of the literature (Guzzetti et al. 2007; 2008; Segoni et al. 2018) reveals that a very large number of meteorological and climate variables have been employed in the calculation of empirical rainfall thresholds. Both rainfall measurements obtained for specific rainfall events and antecedent rainfall condition are used. Additional hydrological variables (e.g., cumulated daily rainfall) can be combined with rainfall parameters. Among all, *ID* (mean rainfall intensity-rainfall event duration) thresholds are the most adopted type of thresholds (Segoni et al., 2018) although they have a few problems. From a theoretical point of view, the rainfall mean intensity is a function of the rainfall duration, therefore it is preferable to define *ED* (cumulated event rainfall–rainfall duration) thresholds, in which the two variables are not dependent on each other (Guzzetti, 2015; Gariano et al., 2020). Another reason in favor of *ED* thresholds is that, for operational landslide forecasting and early warning, measuring the cumulated rainfall over a given period is easier than calculating a derived amount such as the rainfall mean intensity.

Among the diverse methods adopted to calculate thresholds, it is worth mentioning the *frequentist* method proposed by Brunetti et al. (2010) and modified by Peruccacci et al. (2012), which was applied in several study areas worldwide (e.g., Peruccacci et al., 2017; Gariano et al., 2019; Jordanova et al., 2020; Melillo et al., 2020).

16.4.1 Frequentist method for the definition of rainfall thresholds

The *frequentist* method (Brunetti et al., 2010) is based on a frequency analysis of the empirical rainfall conditions that have resulted in known landslides in a given period and in a given area. The threshold curve is a power law relationship between cumulated event rainfall E (in mm) and the rainfall duration D (generally in h), according to the following equation:

$$E = (\alpha \pm \Delta\alpha)\, D^{(\gamma \pm \Delta\gamma)} \tag{16.2}$$

where α is the intercept (the scaling parameter) and γ is the slope (the scaling exponent) of the curve, and $\Delta\alpha$ and $\Delta\gamma$ are the uncertainties associated with α and γ, respectively. The method allows the calculation of objective and reproducible thresholds at different non-exceedance probabilities, and the uncertainties associated with the threshold parameters.

To establish reliable thresholds, a large amount of accurate information on rainfall conditions that triggered landslides in a given study area (generally collected in a catalogue) is needed. Information on landslide occurrence can be gathered from multiple sources, including physical or digital archives of national and local newspapers and technical reports. Rainfall data necessary to determine the rainfall amount responsible for the failures can be usually gathered from rain gauge networks. Radar (e.g., Marra et al., 2017) or satellite estimates (e.g., Nikolopoulos et al., 2017; Rossi et al., 2017; Brunetti et al., 2018) can also be used, especially in areas with low rain gauge density. It is worth pointing out that the adoption of remote sensing rainfall products is gaining consensus among the scientific community (e.g., Jia et al., 2020; Tang et al., 2020).

For each rainfall condition likely responsible for landslide initiation, the duration *D*, and the cumulated event rainfall *E* have to be calculated. To this aim, the definition of rainfall event (from which the triggering condition can be obtained) should be explained. In the literature, different criteria for the definition of the rainfall events or of the rainfall conditions responsible for landslides are proposed, sometimes ill-defined, poorly formalized or simply ambiguous. Recently, a few attempts were made to define procedures for a standardized, reproducible, and automatized calculation of landslide-triggering rainfall events. Melillo et al. (2015) proposed an automated algorithm for the objective and reproducible reconstruction of the rainfall conditions presumably responsible for the documented landslides. Starting from a continuous rainfall series, the separation of two consecutive rainfall events is based on the definition of a "no rain" time interval, which may depend on climatic and seasonal settings (e.g., it can be distinguished for a warm/dry and a cold/rainy season). Among all the reconstructed rainfall events, the selection of the conditions likely responsible for the triggering of known landslides goes through the assignment of a weight, which is a function of the cumulated rainfall, the mean rainfall intensity and the inverse distance between the landslide and the "representative" rain gauge (Melillo et al., 2016, 2018, 2020). Use of the algorithm accelerates greatly the compilation of large catalogues of rainfall events with landslides and reduces the uncertainty in the definition of landslide-triggering rainfall events.

In the process of filling the catalogue of rainfall conditions likely responsible of landslide triggering, several events are often discarded due to different reasons: (i) uncertain spatial location of the failure and/or unknown date of the landslide occurrence, (ii) unavailability or malfunctioning of the "representative" rain gauge (or grid cell, in case of remote rainfall estimates) used to reconstruct the amount of rainfall presumably responsible of the failure, (iii) unreliable or unavailable rainfall data.

Once the rainfall condition reconstruction is finished, the cumulated event rainfall – rainfall duration (*D,E*) pairs are plotted in a log-log chart (i.e., the empirical data are log-transformed). The distribution of the *DE* pairs is fitted (using a least square method) with the linear equation $\log(E) = \log(\alpha) + \gamma \cdot \log(D)$, which is entirely equivalent to the power law of Eq. (16.2) in linear coordinates (Fig. 16.3A). Next, for each (*D,E*) rainfall condition, the residuals, i.e., the differences between the logarithm of the cumulative event rainfall (*E*) and the corresponding value of the fit line are calculated (Fig. 16.3B). Then, the Probability Density Function (PDF) of the distribution of all differences is estimated through a Kernel Density Estimation (Silverman, 1986; Scott, 1992; Venables and Ripley; 2002), and the result is fitted (using a least square method) with a Gaussian function (Fig. 16.3C). Finally, thresholds corresponding to different non-exceedance probabilities are defined, based on the Gaussian distribution of the above-mentioned residuals: a threshold at a given non-exceedance probability is shifted from the fit line (50% non-exceedance probability) by the same distance δ between the mean of the PDF and the value which provides that fraction of the area

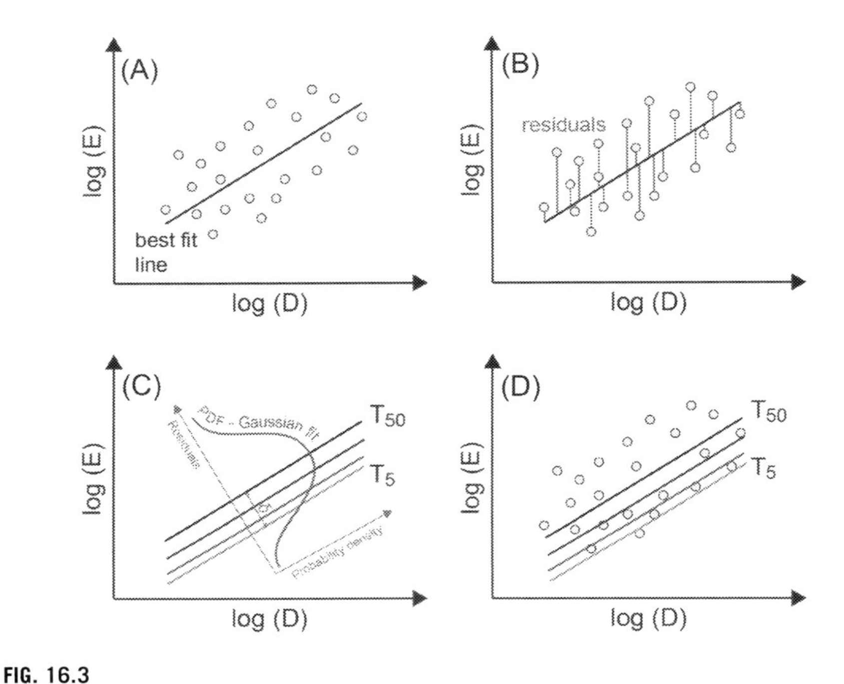

FIG. 16.3

Steps of the frequentist method for the calculation of rainfall thresholds for landslide initiation.

under the Gaussian curve (Fig. 16.3D). The distance δ between the mean value and a given probability value of the PDF is used to calculate the intercept of the given non-exceedance probability threshold curve. As an example, the 5% rainfall threshold T_5 is the curve (a line in the log-log- plot) parallel to the best-fit line T_{50} (with slope γ), having intercept $\alpha_5 = \alpha_{50} - \delta$. According to its definition, the thresholds at 5% non-exceedance probability should leave 5% of the empirical (D,E) pairs below the curve. Assuming that the catalogue of landslide-triggering rainfall conditions is sufficiently complete and representative for the study area, one can state that the probability of experiencing next landslides triggered by rainfall below T_5 threshold is less than 5%.

An improvement was made by Peruccacci et al. (2012) who proposed a method to estimate the uncertainty associated with the threshold parameters. In detail, threshold parameters are obtained from a large number of synthetic series of (D,E) pairs, randomly selected with replacement, generated by a bootstrap nonparametric statistical technique (Efron, 1979; Efron and Tibshirani, 1994) applied to the original (D,E) pairs. Therefore, α and γ in Eq. (16.2) are the mean values of the parameters, and $\Delta\alpha$ and $\Delta\gamma$ are their standard deviations.

Parameter uncertainties depend on the number and on the distribution and dispersion of the empirical data points in the DE domain. Peruccacci et al. (2012) found that a minimum number of rainfall conditions is needed to obtain stable mean values for the α and γ to have acceptable values for $\Delta\alpha$ and $\Delta\gamma$, and therefore to define reliable thresholds. The variation of the threshold parameters and their uncertainties as a function of the number of rainfall conditions can be easily modelled and therefore the minimum number can be obtained for any test site.

All the procedures described above (except for the part related to data collecting) was implemented in an algorithm and in a comprehensive software tool that makes the whole process objective, reproducible, automatic, and fast (Melillo et al., 2018).

16.4.2 Validation of rainfall thresholds

A proper validation of the predictive capability of the thresholds (using a dataset independent from the one used for their calculation) is mandatory, in particular if the thresholds are supposed to be used for operational landslide forecasting (Guzzetti et al., 2020). A threshold that is not properly validated may result in a high number of missed alarms (if the threshold is too high) or false alarms (if the threshold is too low). Gariano et al. (2015) proposed a quantitative procedure to validate frequentist thresholds, based on contingency tables and Receiver Operating Characteristic (ROC) analysis (Fawcett, 2006). The thresholds are considered as binary classifiers of the rainfall conditions that are likely (or not likely) to initiate landslides. The landslide initiation can be either true (T) or false (F), and the threshold predictions can be either positive (P, successful prediction) or negative (N, wrong prediction).

The procedure is composed of a number of steps. First, the catalogue of (D,E) pairs associated to landslide initiation is randomly divided into a calibration subset (e.g., 70%) and a validation subset (e.g., 30%). Second, for the whole investigated period, all the rainfall conditions that have presumably not triggered landslides are also included in the analysis. Third, the rainfall thresholds defined using the calibration subset are compared both with the landslide-triggering (D,E) pairs of the validation subset and with the rainfall (D,E) conditions that have not triggered landslides, allowing the calculation of a contingency table (Fig. 16.4A). The table includes: (i) true positives (TP) i.e., landslide-triggering (D,E) pairs located above the threshold; (ii) true negatives (TN) i.e., rainfall conditions not resulting in landslides located below the threshold; false positives (FP) i.e., rainfall conditions that have presumably not triggered landslides located above the threshold; and (iv) false negatives (FN) i.e., landslide-triggering (D,E) pairs located below the threshold. High thresholds might result in a high number of FN and a low number of TP. Conversely, low thresholds might produce many FP and a less TN.

Three skill scores can be calculated, namely: (i) the true positive rate, $TPR = TP/(TP + FN)$, representing the portion of landslides correctly predicted by the thresholds, i.e., the fraction of landslide-triggering (D,E) pairs above the threshold; (ii) the false positive rate, $FPR = FP/(FP + TN)$, that measures the landslides predicted but not occurred (i.e., the fraction of rainfall conditions without landslides above the threshold); and (iii) the Hanssen and Kuipers discriminant, $HK = TPR - FPR$, measuring the accuracy of the prediction of rainfall conditions with and without landslides.

Calculation of the contingency table and of the skill scores is repeated at any non-exceedance probability value. FPR and TPR skill scores, are then plotted in a

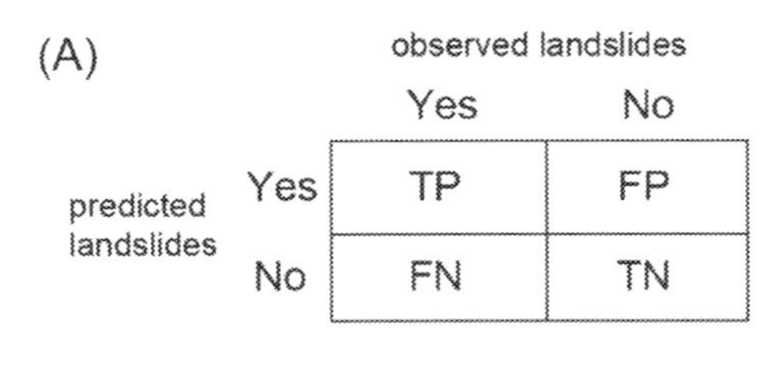

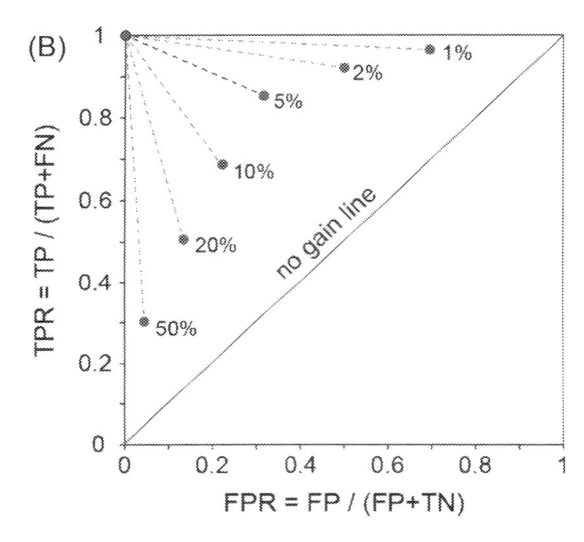

FIG. 16.4

(A) Contingency table showing the four possible outcomes for threshold validation: TP, true positive; TN, true negative; FP, false positive; FN, false negative. (B) ROC space, with hypothetical points associated to thresholds at different non-exceedance probabilities: TPR, true positive rate; FPR, false positive rate. The Euclidean distances between the points representing the threshold and the optimal point (TPR = 1, FPR = 0) are also shown.

Cartesian plane to represent the predicting capability of the thresholds: each FPR-TPR pair represents the prediction capability of a single threshold at a given non-exceedance probability (Fig. 16.4B). A random guess is defined as a point located along the diagonal from the bottom left to the top right corner of the plot (no gain line). Conversely, the point corresponding to the upper left corner of the plot represents the optimal performance (i.e., FN = FP = 0; TPR = 1 and FPR = 0). Generally, the closer is the point to the "optimal point" (i.e., the lower is the Euclidean distance of each point from the optimal point), the better is the prediction capability of the corresponding threshold.

When working in large areas (where the collection of full information on landslide occurrence is difficult), the validation procedure might be hampered by the lack of information on landslide occurrence. In particular, if many landslides have occurred and have not been reported, the number of TN and FP can be largely overestimated. Gariano et al. (2015) observed that even a small underestimation in the number of landslide initiations can result in a significant reduction of the threshold validation performance.

16.5 Operational prediction and forecasting of rainfall-induced landslides

Despite the difficulties to prove it theoretically, the assumption that rainfall-induced landslides can be predicted is raising consensus by the empirical evidence (Guzzetti et al., 2020). Nowadays, the prediction of rainfall-induced landslides is a relevant scientific and social topic, which can contribute to reduce the risk posed by landslides to population and properties. Overall, predicting where, when, how many, and how large landslides are expected in an area during a period of time is at the base of the landslide hazard assessment (Guzzetti et al., 2005).

When dealing with the prediction of a rainfall-induced landslide – as for other similar natural phenomena – one should assess the difference among "prediction", "forecasting", "nowcasting", and "hindcasting". The term "prediction" is used when referring to an estimate of an event happening in the future, the present or the past. A "forecast" can be referred as an estimate of the future state of a natural system obtained with a numerical model (Ramage, 1993). When this estimate is made in the past, e.g., for testing the performance of a model, this can be defined as a "hindcast". Lastly, a "nowcast" is a short-term forecast, typically up to six hours (WMO, 2017).

Physically-based and empirical methods are useful to assess the temporal probability of landslide occurrence, i.e., to predict when rainfall conditions are likely able to trigger one or landslides in a given study area. In particular, these methods can be applied for predicting the occurrence of population of landslides (instead of single phenomena).

The operational forecasting of population of rainfall-induced landslides over large areas is usually performed using landslide early warning systems (LEWSs), which can be defined as a "set of capacities needed to generate and disseminate timely and meaningful warning information to enable individuals, communities and organizations threatened by a hazard (one or more landslides) to act appropriately and in sufficient time to reduce the possibility of harm or loss" (UNISDR, 2009). Often, LEWSs are the most cost-effective mitigation measures and sometimes they are the only suitable option to manage the risk, particularly where structural measures cannot be implemented due to e.g., lack of resources, insufficient time to construct the structural measures, or unacceptable societal and economic impact of the measures (Glade and Nadim, 2014). Nowadays, LEWS operating around the globe are based on empirical rainfall thresholds and statistical analyses of rainfall variables; only a few systems rely on physically-based models (Piciullo et al., 2018; Segoni et al., 2018; Guzzetti et al., 2020). This finding deserves to be commented.

For operational landslide forecasting, it is preferable to use thresholds defined using information on landslides of the region where the LEWS operates, instead of thresholds defined at a coarser spatial scale (e.g., continental or global). However, given that a minimum number of events is necessary in order to obtain reliable thresholds, the definition of accurate rainfall thresholds for small geographical areas (where data could be scarce), remains difficult. For this reason, accurate thresholds defined for larger areas and characterized by a low uncertainty are preferable to thresholds defined for smaller areas but affected by a higher uncertainty. Moreover, the use of thresholds calculated using only extreme or even severe events, may result in a high number of missed alarms, and may compromise the prediction of numerous landslides in an area. Therefore, thresholds for operational landslide forecasting should be constructed using information that includes, as much as possible, all the rainfall conditions that likely initiated landslides in the area where the LEWS is going to operate (Peruccacci et al., 2017).

Overall, to obtain reliable thresholds and accurate warnings, a proper validation of the thresholds is mandatory before them being implemented in a LEWS (Piciullo et al., 2017; Segoni et al., 2018). To reduce threshold (and warning) uncertainties, rainfall data with the finest possible temporal resolution must be adopted (Marra, 2019; Gariano et al., 2020). Moreover, the temporal resolution of the prediction has to be the same used in the definition of the thresholds.

In many areas, the existing rain gauge networks are not adequate to allow reliable landslide forecasts (Lanza et al., 2022). Reasons for the inadequacy are manifolds (Gariano and Guzzetti, 2016), including that (i) in mountain or remote areas the sensor density is typically low or very low (Borga et al., 2008); (ii) the density of rain gauges is even lower at higher elevations, where many of the rainfall-induced landslides originate; (iii) rain gauge networks may not measure accurately high intensity, convective rainstorms, which are the main trigger of shallow and fast landslides (Borga et al. 2014); and (iv) existing networks of meteorological stations may be blind to snowfall and snowmelt events, particularly at low elevations. Finally, it could be surprising that, despite the large number of rain gauges worldwide, the area covered by all currently available stations (i.e., the surface area equivalent to all the orifice areas) is very small, equivalent to less than half a football field (Kidd et al., 2017). For all the above-mentioned reasons, the use of radar and satellite rainfall estimates (Borga et al., 2022; Kidd and Levizzani, 2022) is fostered and is increasingly adopted (and recommended) in operational landslide forecasting.

An often-neglected complication in the operational forecasting of rainfall-induced landslides is the effect of global warning, and the related ongoing and expected climate change. Rising of global temperature is unequivocal (IPCC, 2014) and is resulting in an increase of the frequency and the intensity of the rainfall events. In particular, the frequency and abundance of shallow, rapid and very rapid slope failures, which in many areas are the primary cause of landslide fatalities, is increasing (Gariano and Guzzetti, 2016; Haque et al., 2019). Such variations have to be taken into account in the design and management of new operational rainfall-induced LEWSs. A quantitative measure or estimate of how the climate, climate-related variables and their spatial and temporal variations, affect landslide processes at different temporal and spatial scales is mandatory for improving the reliability of new operational landslide forecasts.

16.6 Case study: frequentist thresholds for landslide initiation in Italy

The Italian territory is characterized by an abundance of landslides (Trigila et al., 2010, 2015) and a large physiographic variability. Therefore, also due to the high rain gauge density (Morbidelli et al., 2020), it constitutes a good case study to investigate the rainfall conditions that can result in landslides, and their variations in different environmental settings.

Using the above-defined methods and procedures, Peruccacci et al. (2017) calculated frequentist *ED* national thresholds for Italy, and 26 *ED* regional thresholds for environmental subdivisions of the Italian territory based on topography, lithology, land-use, land cover, climate, and meteorology. They used (i) hourly rainfall measurements captured by 2228 rain gauges and (ii) information (gathered mainly from national, regional and local newspapers and fire fighter reports) on the location and time of occurrence of 2819 landslides – mostly shallow – between 1996 and 2014. Combining these datasets, they obtained a catalogue of 2309 landslide-triggering rainfall conditions, which they used to calculate national thresholds at different non-exceedance probabilities. The Italian thresholds at 5% and 20% non-exceedance probabilities are shown in Fig. 16.5 – together with the 2309 (D,E) pairs used for their calculation – and have the following equations:

$$T_{5,IT}: E = (7.7 \pm 0.3) D^{(0.39 \pm 0.01)} \tag{16.3}$$

$$T_{20,IT}: E = (12.0 \pm 0.4) D^{(0.39 \pm 0.01)} \tag{16.4}$$

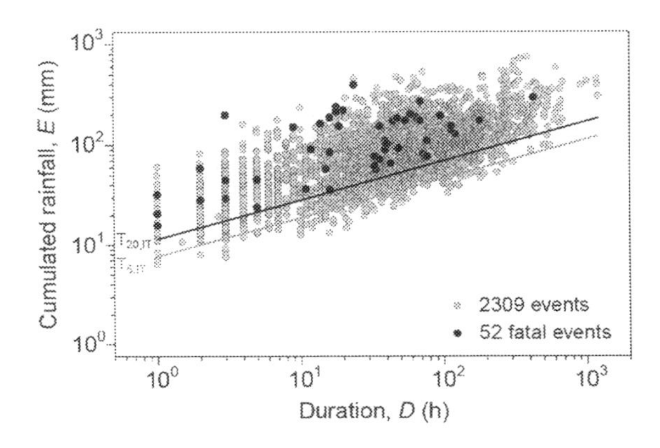

FIG. 16.5

Log-log chart showing 2309 (D, E) conditions responsible for the initiation of landslides in Italy in the period 1996 to 2014 (grey and black dots), and the corresponding 5% (T5, IT, grey curve) and 20% (T20, IT, black curve) ED thresholds. Black dots indicate a subset of 52 (D, E) pairs that initiated landslides with casualties in the same period. Modified after Peruccacci et al. (2017).

The $T_{5,IT}$ threshold can be interpreted as a minimum threshold for the Italian territory, and shows that a rainfall event with cumulated rainfall $E = 26.6$ mm in 24 h is expected to initiate shallow landslides in Italy. The $T_{20,IT}$ threshold is the lower boundary of the 52 rainfall conditions that have resulted in landslides with casualties in the investigated period in Italy (black dots in Fig. 16.5). Therefore, $T_{20,IT}$ threshold might be adopted to forecast the possible occurrence of fatal rainfall-induced shallow landslides in Italy.

The $T_{5,IT}$ threshold is implemented in SANF (an acronym for national early warning system for rainfall-induced landslides – *Sistema d'Allerta Nazionale per le Frane*), a pre-operational LEWS designed and managed by the Research Institute for Geo-hydrological Protection of the Italian National Research Council for the Italian National Civil Protection Department. The threshold is used to determine the probability of landslide occurrence for any given set of cumulated rainfall and rainfall duration value, in a range of duration from 1 to 1212 h (i.e., the range of duration of the (D,E) pairs used for threshold calculation). In the current version, SANF uses (i) hourly rainfall measurements provided by more than 2500 rain gauges available from a national network and various regional networks (average density of one station every ~120 km^2), (ii) quantitative rainfall forecasts provided by COSMO-I5 numerical weather model at 5 km × 5 km resolution updated twice daily, (iii) the above defined *ED* rainfall threshold, and (iv) a grid-based, synoptic-scale landslide susceptibility map to produce landslide nowcasts and forecasts for the whole Italian territory (Rossi et al., 2018; Guzzetti et al., 2020).

Peruccacci et al. (2017) investigated the variations in the rainfall conditions that can result in rainfall-induced landslides in Italy depending on different environmental conditions. As an example, Fig. 16.6 shows regional thresholds for different mean annual precipitation (MAP) regions in Italy. In particular, Fig. 16.6A portrays a subdivision of Italy into five mean annual precipitation (MAP) classes, modified after Desiato et al. (2014), and the geographical distribution of the slope failures in Italy (2819 white dots). Fig. 16.6B shows the (D,E) rainfall conditions that resulted in landslides in the four out of the five MAP regions, namely, low (LO, 800 mm < MAP ≤ 1200 mm, 1197 rainfall conditions), medium (ME, 1200 mm < MAP ≤ 1600 mm, 557 conditions), very low (VL, MAP ≤ 800 mm, 345 conditions), and high (HI, 1600 mm < MAP ≤ 2000 mm, 210 conditions) regions, with the corresponding 5% *ED* thresholds, $T_{5,LO}$, $T_{5,ME}$, $T_{5,VL}$, and $T_{5,HI}$. The four thresholds are also shown in Fig. 16.6C, in linear coordinates and in the reduced duration range $1\text{h} \le D \le 120$ h, with the shaded areas depicting the uncertainty associated to each threshold.

Inspection of Fig. 16.6 reveals that the thresholds become higher and steeper with increasing MAP values. This indicates that a larger amount of rainfall is needed to trigger landslides where the MAP is higher, confirming a landscape adaption to the average rainfall conditions (Peruccacci et al., 2017). Threshold uncertainties decrease as the number of rainfall conditions increase (Fig. 16.6C). In the case of the highest MAP class (>2000 mm), this number was not sufficient to define reliable thresholds. Ultimately, the thresholds related to the lowest two classes of MAP (VL: MAP ≤ 800 mm; LO: 800 mm < MAP ≤ 1200 mm) and the ones related to the

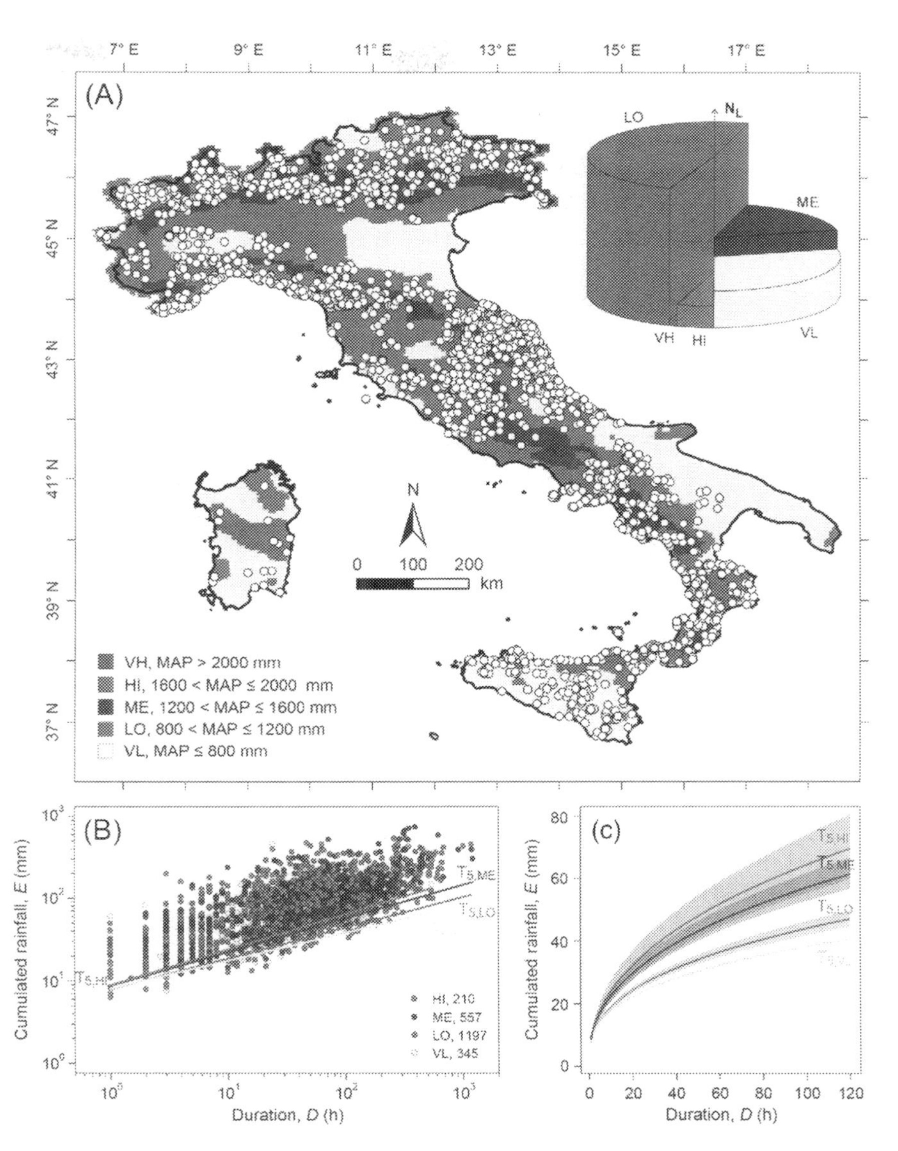

FIG. 16.6

(A) Map showing the distribution of 2820 rainfall-induced landslides over the Italian territory classified into five mean annual precipitation (MAP) classes; pie chart shows the percentage of the extent of the MAP regions (sector width) and the number of rainfall-induced landslides in each MAP region (NL, sector height). (B) Log-log chart showing 2309 (D, E) conditions, distinguished in the four main populated MAP classes, and the corresponding 5% thresholds. (C) The same thresholds, with associated uncertainty (shaded areas), reported in linear coordinates and in the range 1 h ≤ D ≤ 120 h. Modified after Peruccacci et al. (2017).

highest classes (ME: 1200 mm < MAP ≤ 1600 mm; HI: 1600 mm < MAP ≤ 2000 mm) can be separated and are statistically distinguishable. This indicates that the regional rainfall regime influences the triggering conditions for shallow landslides.

Among all the environmental subdivisions, the geographical distribution of the mean annual precipitation resulted in the most evident separation of the thresholds. In some cases, the low number of empirical points available in one of the subdivisions of the national territory did not allow the calculation of the threshold with acceptable uncertainties. Overall, for operational landslide forecasting, accurate thresholds defined for larger areas are preferable to uncertain, unreliable thresholds defined for smaller.

16.7 Conclusions

Despite difficulties and uncertainties, operational forecasting of rainfall-induced landslides and effective landslide early warning are feasible and can contribute to mitigate landslide risk and to reduce fatalities. Although there are still open questions, relevant developments in the definition of reliable methods for rainfall-induced landslide prediction have been made, including: (i) the definition of objective methods and automatic procedures; (ii) the adoption of multi-source rainfall data e.g., estimates provided by radars or satellites (as an alternative to unavailable or unreliable rain gauge measurements); (iii) the inclusion of other hydrogeological variables in the analysis e.g., soil moisture; (iv) the application of quantitative procedures to validate and to evaluate the performance of forecasting models; (v) the evaluation of diverse sources of uncertainty; (vi) the need to consider predisposing factors that may favor landslide initiation e.g., forest fires or cutting, agricultural changes; (vii) the definition of automatic procedures for landslide detection, particularly in data-scarce regions; (viii) the inclusion of the effect of climatic and environmental changes in the forecasting models; (ix) the implementation of the forecasting models into LEWSs, joining scientific and technical aspects.

Further research is desirable to refine all these aspects and to improve operational landslide forecasting. A better understanding of the complex relationships between rainfall and landslides and a better and more efficient landslide modelling and forecasting will be useful to provide decision-makers, media, and citizens reliable predictions in order to make appropriate decisions and to take effective actions to reduce landslide risk.

References

Alvioli, M., Baum, R.L., 2016. Parallelization of the TRIGRS model for rainfall-induced landslides using the message passing interface. Environ. Model Soft. 81, 122–135. doi:10.1016/j.envsoft.2016.04.002.

Alvioli, M., Guzzetti, F., Rossi, M., 2014. Scaling properties of rainfall induced landslides predicted by a physically based model. Geomorphology 213, 38–47. doi:10.1016/j.geomorph.2013.12.039.

Alvioli, M., Melillo, M., Guzzetti, F., Rossi, M., Palazzi, E., von Hardenberg, J., Brunetti, M.T., Peruccacci, S., 2018. Implications of climate change on landslide hazard in Central Italy. Sci. Tot. Environ. 630, 1528–1543. doi: 10.1016/j.scitotenv.2018.02.315.

Baum, R., Savage, W., Godt, J., 2002. TRIGRS—a Fortran program for transient rainfall infiltration and grid-based regional slope-stability analysis. USGS Open File Report 02–424.

Baum, R., Savage, W., Godt, J., 2008. TRIGRS—a Fortran program for transient rainfall infiltration and grid–based regional slope–stability analysis, version 2.0. USGS Open File Report 2008–1159.

Berti, M., Martina, M.L.V., Franceschini, S., Pignone, S., Simoni, A., Pizziolo, M., 2012. Probabilistic rainfall thresholds for landslide occurrence using a Bayesian approach. J. Geophys. Res. 117 (F4), 1–20, F04006. doi:10.1029/2012JF002367.

Bogaard, T., Greco, R., 2018. Invited perspectives: hydrological perspectives on precipitation intensity–duration thresholds for landslide initiation: proposing hydro-meteorological thresholds. Nat. Hazards Earth Syst. Sci. 18 (1), 31–39. doi:10.5194/nhess-18-31-2018.

Borga, M., Gaume, E., Creutin, J.D., Marchi, L., 2008. Surveying flash floods: gauging the ungauged extremes. Hydrol. Process 22 (18), 3883–3885. doi:10.1002/hyp.7111.

Borga, M., Marra, F., Gabella, M., 2022. Rainfall Estimation by Weather Radar. In: Morbidelli, R. (Ed.), Rainfall. Modeling, Measurement and Applications. Elsevier, Amsterdam, pp. 109–134. doi:10.1016/C2019-0-04937-0.

Borga, M., Stoffel, M., Marchi, L., Marra, F., Jakob, M., 2014. Hydrogeomorphic response to extreme rainfall in headwater systems: flash floods and debris flows. J. Hydrol. 518 (B), 194–205. doi:10.1016/j.jhydrol.2014.05.022.

Brunetti, M.T., Peruccacci, S., Rossi, M., Luciani, S., Valigi, D., Guzzetti, F., 2010. Rainfall thresholds for the possible occurrence of landslides in Italy. Nat. Hazards Earth Syst. Sci. 10 (3), 447–458. doi:10.5194/nhess-10-447-2010.

Brunetti, M.T., Melillo, M., Peruccacci, S., Ciabatta, L., Brocca, L., 2018. How far are we from the use of satellite rainfall products in landslide forecasting? Remote Sens. Environ. 210, 65–75. doi:10.1016/j.rse.2018.03.016.

Campbell, R.H., 1975. Soil slips, debris flows, and rainstorms in the Santa Monica Mountains and vicinity, Southern California. US Geological Survey Professional Paper, U.S. Government Printing Office, Washington, 851 (51).

Caine, N., 1980. The rainfall intensity-duration control of shallow landslides and debris flows. Geografiska Annal 62A, 23–27. doi:10.1080/04353676.1980.11879996.

Crozier, M.J., 1999. Prediction of rainfall-triggered landslides: a test of the antecedent water status model. Earth Surf. Process Landf. 24, 825–833.

Crozier, M.J., 1997. The climate-landslide couple: a southern hemisphere perspective. In: Matthews, J.A., Brunsden, D., Frenzel, B., Gläser, B., Weiß, M.M. (Eds.), Rapid Mass Movement as a Source of Climatic Evidence for the Holocene. Gustav Fischer, Stuttgart, pp. 333–354.

Cruden, D.M., Varnes, D.J., 1996. Landslide types and processes. In: Turner, A.K., Schuster, R.L. (Eds.), Landslides, Investigation and Mitigation, Special Report 247. Transportation Research Board, Washington D.C., pp. 36–75.

Desiato F., Fioravanti G., Fraschetti P., Perconti W., Piervitali E., 2014. Valori climatici normali di temperatura e precipitazione in Italia. ISPRA, Stato dell'Ambiente 55/2014, ISBN 978-88-448-0689-7, (in Italian).

De Vita, P., Reichenbach, P., Bathurst, J.C., Borga, M., Crozier, G.M., Glade, T., Guzzetti, F., Hansen, A., Wasowski, J., 1998. Rainfall-triggered landslides: a reference list. Environ. Geol. 35, 219–233.

Dowling, C.A., Santi, P.M., 2013. Debris flows and their toll on human life: a global analysis of debris-flow fatalities from 1950 to 2011. Nat. Hazards 71, 203–227. doi:10.1007/s11069-013-0907-4.

Efron, B., 1979. Bootstrap methods: another look at jackknife. Ann. Stat. 7, 1–26.

Efron, B., Tibshirani, R.J., 1994. An Introduction to the Bootstrap. Chapman and Hall, New York, p. 456.

Endo, T., 1969. Probable distribution of the amount of rainfall causing landslides, Annual Report 1968. Hokkaido Branch, For. Exp. Stn., 122–136.

Fawcett, T., 2006. An introduction to ROC analysis. Pattern Recogn. Lett. 27, 861–874. doi:10.1016/j.patrec.2005.10.010.

Froude, M.J., Petley, D.N., 2018. Global fatal landslide occurrence from 2004 to 2016. Nat. Hazards Earth Syst. Sci. 18 (3), 2161–2181. doi:10.5194/nhess-18-2161-2018.

Gariano, S.L., Guzzetti, F., 2016. Landslides in a changing climate. Earth-Sci. Rev. 162, 227–252. doi:10.1016/j.earscirev.2016.08.011.

Gariano, S.L., Brunetti, M.T., Iovine, G., Melillo, M., Peruccacci, S., Terranova, O., Vennari, C., Guzzetti, F., 2015. Calibration and validation of rainfall thresholds for shallow landslide forecasting in Sicily, southern Italy. Geomorphology 228 (1), 653–665. doi:10.1016/j.geomorph.2014.10.019.

Gariano, S.L., Sarkar, R., Dikshit, A., Dorji, K., Brunetti, M.T., Peruccacci, S., Melillo, M., 2019. Automatic calculation of rainfall thresholds for landslide occurrence in Chukha Dzongkhag, Bhutan. Bull. Eng. Geol. Environ. 78 (6), 4325–4332. doi:10.1007/s10064-018-1415-2.

Gariano, S.L., Melillo, M., Peruccacci, S., Brunetti, M.T., 2020. How much does the rainfall temporal resolution affect rainfall thresholds for landslide triggering? Nat. Hazards 100 (2), 655–670. doi:10.1007/s11069-019-03830-x.

Glade, T., Nadim, F., 2014. Early warning systems for natural hazards and risks. Nat. Hazards 70 (3), 1669–1671. doi:10.1007/s11069-013-1000-8.

Godt, J., Baum, R., Savage, W., Salciarini, D., Schulz, W., Harp, E., 2008. Transient deterministic shallow landslide modeling: requirements for susceptibility and hazard assessments in a GIS framework. Eng. Geol. 102 (3), 214–226. doi:10.1016/j.enggeo.2008.03.019.

Guzzetti, F., 2015. Forecasting Nat Hazards, performance of scientists, ethics, and the need for transparency. Toxicol. Environ. Chem. 98 (9), 1043–1059. doi:10.1080/02772248.2015.1030664.

Guzzetti, F., Reichenbach, P., Cardinali, M., Galli, M., Ardizzone, F., 2005. Probabilistic landslide hazard assessment at the basin scale. Geomorphology 72, 272–299. doi:10.1016/j.geomorph.2005.06.002.

Guzzetti, F., Peruccacci, S., Rossi, M., Stark, C.P., 2007. Rainfall thresholds for the initiation of landslides in central and southern Europe. Meteorog. Atmos. Phys. 98 (3), 239–267. doi:10.1007/s00703-007-0262-7.

Guzzetti, F., Peruccacci, S., Rossi, M., Stark, C.P., 2008. The rainfall intensity–duration control of shallow landslides and debris flows: an update. Landslides 5 (1), 3–17. doi:10.1007/s10346-007-0112-1.

Guzzetti, F., Mondini, A.C., Cardinali, M., Fiorucci, F., Santangelo, M., Chang, K.-T., 2012. Landslide inventory maps: new tools for an old problem. Earth-Sci. Rev. 112 (1–2), 42–66. doi:10.1016/j.earscirev.2012.02.001.

Guzzetti, F., Gariano, S.L., Peruccacci, S., Brunetti, M.T., Marchesini, I., Rossi, M., Melillo, M., 2020. Geographical landslide early warning systems. Earth-Sci. Rev. 200, 1–29, 102973. doi:10.1016/j.earscirev.2019.102973.

Haque, U., da Silva, P.F., Devoli, G., Pilz, J., Zhao, B., Khaloua, A., Wilopo, W., Andersen, P., Lu, P., Lee, J., Yamamoto, T., Keellings, D., Wu, J.-H., Glass, G.E., 2019. The human cost of global warming: deadly landslides and their triggers (1995–2014). Sci. Tot. Environ. 682, 673–684. doi:10.1016/j.scitotenv.2019.03.415.

Hungr, O., Leroueil, S., Picarelli, L., 2013. The Varnes classification of landslide types, an update. Landslides 11 (2), 167–194. doi:10.1007/s10346-013-0436-y.

IPCC: Climate Change 2014, Synthesis report. Contribution of Working Groups I, II and III to the Fifth Assessment Report of the Intergovernmental Panel on Climate Change, Geneva, Switzerland 151 p., 2014.

Iverson, R.M., 2000. Landslide triggering by rain infiltration. Water Resour. Res. 36 (7), 1897–1910.

Jia, G., Tang, Q., Xu, X., 2020. Evaluating the performances of satellite-based rainfall data for global rainfall-induced landslide warnings. Landslides 17 (2), 283–299. doi:10.1007/s10346-019-01277-6.

Jordanova, G., Gariano, S.L., Melillo, M., Peruccacci, S., Brunetti, M.T., Auflič, J.M., 2020. Determination of empirical rainfall thresholds for shallow landslides in Slovenia using an automatic tool,. Water 12 (5), 1–15, 1449. doi:10.3390/w12051449.

Kidd, C., Becker, A., Huffman, G.J., Muller, C.L., Joe, P., Skofronick-Jackson, G., Kirschbaum, D., 2017. So, how much of the Earth's surface is covered by rain gauges? Bull. Am. Meteorol. Soc. 98 (1), 69–78. doi:10.1175/BAMS-D-14-00283.1.

Kidd, C., Levizzani, V., 2022. Satellite Rainfall Estimation. In: Morbidelli, R. (Ed.), Rainfall. Modeling, Measurement and Applications. Elsevier, Amsterdam, pp. 135–170. doi:10.1016/C2019-0-04937-0.

Lagomarsino, D., Segoni, S., Rosi, A., Rossi, G., Battistini, A., Catani, F., Casagli, N., 2015. Quantitative comparison between two different methodologies to define rainfall thresholds for landslide forecasting. Nat. Hazards Earth Syst. Sci. 15 (10), 2413–2423. doi:10.5194/nhess-15-2413-2015.

Lanza, L.G., Cauteruccio, A., Stagnaro, M., 2022. Rain Gauge Measurements. In: Morbidelli, R. (Ed.), Rainfall. Modeling, Measurement and Applications. Elsevier, Amsterdam, pp. 77–108. doi:10.1016/C2019-0-04937-0.

Marra, F., 2019. Rainfall thresholds for landslide occurrence: systematic underestimation using coarse temporal resolution data. Nat. Hazards 95 (3), 883–890. doi:10.1007/s11069-018-3508-4.

Marra, F., Destro, E., Nikolopoulos, E.I., Zoccatelli, D., Creutin, J.D., Guzzetti, F., Borga, M., 2017. Impact of rainfall spatial aggregation on the identification of debris flow occurrence thresholds. Hydrol. Earth Syst. Sci. 21 (9), 4525–4532. doi:10.5194/hess-21-4525-2017.

Melillo, M., Brunetti, M.T., Peruccacci, S., Gariano, S.L., Guzzetti, F., 2015. An algorithm for the objective reconstruction of rainfall events responsible for landslides. Landslides 12 (2), 311–320. doi:10.1007/s10346-014-0471-3.

Melillo, M., Brunetti, M.T., Peruccacci, S., Gariano, S.L., Guzzetti, F., 2016. Rainfall thresholds for the possible landslide occurrence in Sicily (Southern Italy) based on the automatic reconstruction of rainfall events. Landslides 13 (1), 165–172. doi:10.1007/s10346-015-0630-1.

Melillo, M., Brunetti, M.T., Peruccacci, S., Gariano, S.L., Roccati, A., Guzzetti, F., 2018. A tool for the automatic calculation of rainfall thresholds for landslide occurrence. Environ. Modell. Softw. 105, 230–243. doi:10.1016/j.envsoft.2018.03.024.

Melillo, M., Gariano, S.L., Peruccacci, S., Sarro, R., Mateos, R.M., Brunetti, M.T., 2020. Rainfall and rockfalls in the Canary Islands: assessing a seasonal link. Nat. Hazards Earth Syst. Sci. 20 (8), 2307–2317. doi:10.5194/nhess-20-2307-2020.

Montgomery, D.R., Dietrich, W.E., 1994. A physically-based model for the topographic control on shallow landsliding. Water Resour. Res. 30, 1153–1171.

Morbidelli, R., García-Marín, A.M., Al Mamun, A., Atiqur, R.M., Ayuso- Muñoz, J.L., Taouti, M.B., Baranowski, P., Bellocchi, G., Sangüesa-Pool, C., Bennett, B., Oyunmunkh, B.,

Bonaccorso, B., Brocca, L., Caloiero, T., Caporali, E., Caracciolo, D., Casas-Castillo, M.C., Catalini, C.G., Custò, J., Dari, J., Diodato, N., Doesken, N., Dumitrescu, A., Estévez, J., Flammini, A., Fowler, H.J., Freni, G., Fusto, F., García-Barrón, L., Manea, A., Goenster-Jordan, S., Hinson, S., Kanecka-Geszke, E., Kanti Kar, K., Kasperska-Wołowicz, W., Krabbi, M., Krzyszczak, J., Llabrés-Brustenga, A., Ledesma, J.L.J., Liu, T., Lompi, M., Marsico, L., Mascaro, G., Moramarco, T., Newman, N., Orzan, A., Pampaloni, M., Pizarro-Tapia, R., Puentes Torres, A., Rashid, M.M., Rodríguez-Solà, R., Sepulveda Manzor, M., Siwek, K., Sousa, A., Timbadiya, P.V., Filippos, T., Vilcea, M.G., Viterbo, F., Yoo, C., Zeri, M., Zittis, F., Saltalippi, C., 2020. The history of rainfall data time-resolution in a wide variety of geographical areas. J. Hydrol. 590, 125528. doi:10.1016/j.jhydrol.2020.125258.

Nadim, F., Kjekstad, O., Peduzzi, P., Herold, C., Jaedicke, C., 2006. Global landslide and avalanche hotspots. Landslides 3 (2), 159–173. doi:10.1007/s10346-006-0036-1.

Nikolopoulos, E.I., Destro, E., Maggioni, V., Marra, F., Borga, M., 2017. Satellite rainfall estimates for debris flow prediction: an evaluation based on rainfall accumulation–duration thresholds. J. Hydrometeorol. 18 (8), 2207–2214. doi:10.1175/JHM-D-17-0052.1.

Onodera T., Yoshinaka R., Kazama H. Slope failures caused by heavy rainfall in Japan. In: Proceedings 2nd International Congress of the Int. Ass. Eng. Geol. San Paulo, 11, pp. 1–10, 1974.

Pereira, S., Zêzere, J.L., Quaresma, I., Santos, P.P., Santos, M., 2015. Mortality patterns of hydro-geomorphologic disasters. Risk Anal. 36 (6), 1188–1210. doi:10.1111/risa.12516.

Peres, D.J., Cancelliere, A., 2014. Derivation and evaluation of landslide-triggering thresholds by a Monte Carlo approach. Hydrol. Earth Syst. Sci. 18 (12), 4913–4931. doi:10.5194/hess-18-4913-2014.

Peruccacci, S., Brunetti, M.T., Luciani, S., Vennari, C., Guzzetti, F., 2012. Lithological and seasonal control of rainfall thresholds for the possible initiation of landslides in central Italy. Geomorphology 139–140, 79–90. doi:10.1016/j.geomorph.2011.10.005.

Peruccacci, S., Brunetti, M.T., Gariano, S.L., Melillo, M., Rossi, M., Guzzetti, F., 2017. Rainfall thresholds for possible landslide occurrence in Italy. Geomorphology 290, 39–57. doi:10.1016/j.geomorph.2017.03.031.

Petley, D., 2012. Global patterns of loss of life from landslides. Geology 40 (10), 927–930. doi:10.1130/G33217.1.

Piciullo, L., Calvello, M., Cepeda, J.M., 2018. Territorial early warning systems for rainfall-induced landslides. Earth-Sci. Rev. 179, 228–247. doi:10.1016/j.earscirev.2018.02.013.

Piciullo, L., Gariano, S.L., Melillo, M., Brunetti, M.T., Peruccacci, S., Guzzetti, F., Calvello, M., 2017. Definition and performance of a threshold-based regional early warning model for rainfall-induced landslides. Landslides 14 (3), 995–1008. doi:10.1007/s10346-016-0750-2.

Pisani, G., Castelli, M., Scavia, C., 2010. Hydrogeological model and hydraulic behaviour of a large landslide in the Italian Western Alps. Nat. Hazards Earth Syst. Sci. 10 (11), 2391–2406. doi:10.5194/nhess-10-2391-2010.

Ramage, C.S., 1993. Forecasting in meteorology. Bull. Am. Meteorol. Soc. 74, 1863–1871.

Raia, S., Alvioli, M., Rossi, M., Baum, R.L., Godt, J.W., Guzzetti, F., 2014. Improving predictive power of physically based rainfall-induced shallow landslide models: a probabilistic approach. Geosci. Model Dev. 7 (2), 495–514. doi:10.5194/gmd-7-495-2014.

Reichenbach, P., Cardinali, M., De Vita, P., Guzzetti, F., 1998. Regional hydrological thresholds for landslides and floods in the Tiber River basin (central Italy). Environ. Geol. 35 (2), 146–159. doi:10.1007/s002540050301.

Richards, L.A., 1931. Capillary conduction of liquids in porous mediums. Physics 1, 318–333.

Rossi, M., Luciani, S., Valigi, D., Kirschbaum, D., Brunetti, M.T., Peruccacci, S., Guzzetti, F., 2017. Statistical approaches for the definition of landslide rainfall thresholds and their uncertainty using rain gauge and satellite data. Geomorphology 285, 16–27. doi:10.1016/j.geomorph.2017.02.001.

Rossi, M., Marchesini, I., Tonelli, G., Peruccacci, S., Brunetti, M.T., Luciani, S., Ardizzone, F., Balducci, V., Bianchi, C., Cardinali, M., Fiorucci, F., Mondini, A.C., Reichenbach, P., Salvati, P., Santangelo, M., Guzzetti, F., 2018. TXT-tool 2.039-1.1 Italian national early warning system. In: Sassa, K., Guzzetti, F., Yamagishi, H., Arbanas, Ž., Casagli, N., McSaveney, M., Dang, K. (Eds.), Landslide Dynamics: ISDR-ICL Landslide Interactive Teaching Tools. Springer International Publishing, Cham, pp. 341–349.

Salciarini, D., Tamagnini, C., Conversini, P., Rapinesi, S., 2012. Spatially distributed rainfall thresholds for the initiation of shallow landslides. Nat. Hazards 61 (1), 229–245. doi:10.1007/s11069-011-9739-2.

Salciarini, D., Fanelli, G., Tamagnini, C., 2017. A probabilistic model for rainfall-induced shallow landslide prediction at the regional scale. Landslides 14 (5), 1731–1746. doi:10.1007/s10346-017-0812-0.

Scott, D.W., 1992. Multivariate Density Estimation. Theory, Practice and Visualization. Wiley, New York.

Segoni, S., Piciullo, L., Gariano, S.L., 2018. A review of the recent literature on rainfall thresholds for landslide occurrence. Landslides 15 (8), 1483–1501. doi:10.1007/s10346-018-0966-4.

Sengupta, A., Gupta, S., Anbarasu, K., 2010. Rainfall thresholds for the initiation of landslide at Lanta Khola in north Sikkim, India. Nat. Hazards 52 (1), 31–42. doi:10.1007/s11069-009-9352-9.

Sidle, R.C., Ochiai, H., 2006. Landslides: Processes, Prediction, and Land Use, 312.

Silverman, B.W., 1986. Density Estimation. Chapman and Hall, London.

Tang, G., Clark, M.P., Papalexiou, S.M., Ma, Z., Hong, Y., 2020. Have satellite precipitation products improved over last two decades? A comprehensive comparison of GPM IMERG with nine satellite and reanalysis datasets. Remote Sens. Environ. 240, 111697. doi:10.1016/j.rse.2020.111697.

Taylor, D.W, 1948. Fundamentals of Soil Mechanics. John Wiley & sons, Inc., New York.

Terranova, O.G., Gariano, S.L., Iaquinta, P., Iovine, G., 2015. ^{GA}SAKe: forecasting landslide activations by a genetic-algorithms-based hydrological model. Geosci. Model Develop. 8 (7), 1955–1978. doi:10.5194/gmd-8-1955-2015.

Terzaghi, K., 1962. Stability of steep slopes on hard unweathered rock. Geotechnique 12, 251–270.

Tran, T.V., Alvioli, M., Lee, G., H.U, An., 2018. Three-dimensional, time-dependent modeling of rainfall-induced landslides over a digital landscape: a case study. Landslides 15 (6), 1071–1084. doi:10.1007/s10346-017-0931-7.

Trigila, A., Iadanza, C., Spizzichino, D., 2010. Quality assessment of the Italian landslide inventory using GIS processing. Landslides 7 (4), 455–470. doi:10.1007/s10346-010-0213-0.

Trigila, A., Iadanza, C., Bussettini, M., Lastoria, B., Barbano, A., 2015. Dissesto idrogeologico in Italia: pericolosità e indicatori di rischio. Rapporto 2015. Istituto Superiore per la Protezione e la Ricerca Ambientale – ISPRA, Rapporti 233/2015, Rome, Italy, p. 162 (in Italian).

UNISDR, 2009. Terminology on Disaster Risk Reduction, 35, United Nations International Strategy for Disaster Reduction, Geneva, Switzerland, pp. 1–35.

Van Asch ThVJ, Buma J., Van Beek, L.P.H., 1999. A view on some hydrological triggering systems in landslides. Geomorphology 30 (1–2), 25–32. doi:10.1016/S0169-555X(99)00042-2.

Venables, W.N., Ripley, B.D., 2002. Modern Applied Statistics with S. Springer, New York.
Wallemacq, P., House, R., 2018. Economic Losses, Poverty & Disasters: 1998–2017. CRED, UNISDR, Geneva, Switzerland, p. 31.
White, I.D., Mottershead, D.N., Harrison, J.J., 1996. Environmental Systems, 2nd ed. Chapman & Hall, London, p. 616.
Wilson, R.C., Wieczorek, G.F., 1995. Rainfall thresholds for the initiation of debris flow at La Honda, CaliforniaEnviron. Eng. Geosci.1, 11–27.
WMO, 2017. Guidelines for Nowcasting Techniques. World Meteorological Organization, Geneva, Switzerland, pp. 1–68.

CHAPTER

Rainfall and droughts

17

Ashok Mishra[a], Ali Alnahit[b], Sourav Mukherjee[a]

[a]*Glenn Department of Civil Engineering, Clemson University, South Carolina, United States*

[b]*Department of Civil Engineering, King Saud University, Riyadh, Saudi Arabia*

17.1 Introduction

Drought is a natural and recurrent hydroclimatic phenomenon that occurs in all climatic zones (Mishra and Singh, 2010) with the potential to cause severe impact on society and the environment (Apurv et al., 2017; Han et al., 2019; Herrera-Estrada et al., 2019; Huang et al., 2017; Liu et al., 2019; Van Loon, 2015; Wu et al., 2017; Mishra et al., 2020). Drought differs significantly from other disasters (e.g., floods, tornados, and hurricanes) due to its longer duration and the challenges associated with capturing the onset and termination of the drought events.

The underlying causes and characteristics of drought can vary from region to region (Mishra and Singh, 2010). However, the most dominant driver that controls drought onset and evolution is the rainfall variability (Mishra and Singh, 2010). Rainfall or precipitation variability is usually spatially heterogeneous in nature, which controls the spatiotemporal evolution and propagation of droughts (Konapala et al., 2020). In combination with the rainfall variability, other hydro-meteorological factors such as temperature, background aridity, incoming radiation, and wind characteristics (e.g., magnitude and directions) can further contribute to the intensification and evolution of droughts (Mishra and Singh, 2010; Trenberth et al., 2014). These factors vary significantly across different spatiotemporal scales, and their relative contribution leads to a varying impact of drought on various stakeholders (Bachmair et al., 2016; Steinemann et al., 2015). These properties categorize drought as a complex phenomenon with the lack of a universal definition (Fang et al., 2019; Guo et al., 2019; Han et al., 2019; Mishra and Singh, 2010; Mukherjee et al., 2018.; Van Loon and Van Lanen, 2013; Wilhite, 2000). Four types of drought are generally identified: meteorological, agricultural, hydrological, and socioeconomic drought (Dai et al., 2020; Mishra and Singh, 2010; Mukherjee et al., 2018.; Wilhite and Glantz, 1985). Meteorological drought occurs when dry weather patterns dominate the area. In contrast, agricultural drought happens when there is a lack of sufficient moisture to meet the need of the crop and plant growth at a particular time. Hydrological drought refers to deficiencies in streams, reservoirs, and groundwater levels (usually following meteorological droughts). Socioeconomic drought is related to the supply and demand of goods with

RAINFALL. Modeling, Measurement and Applications. DOI: https://doi.org/10.1016/B978-0-12-822544-8.00004-4

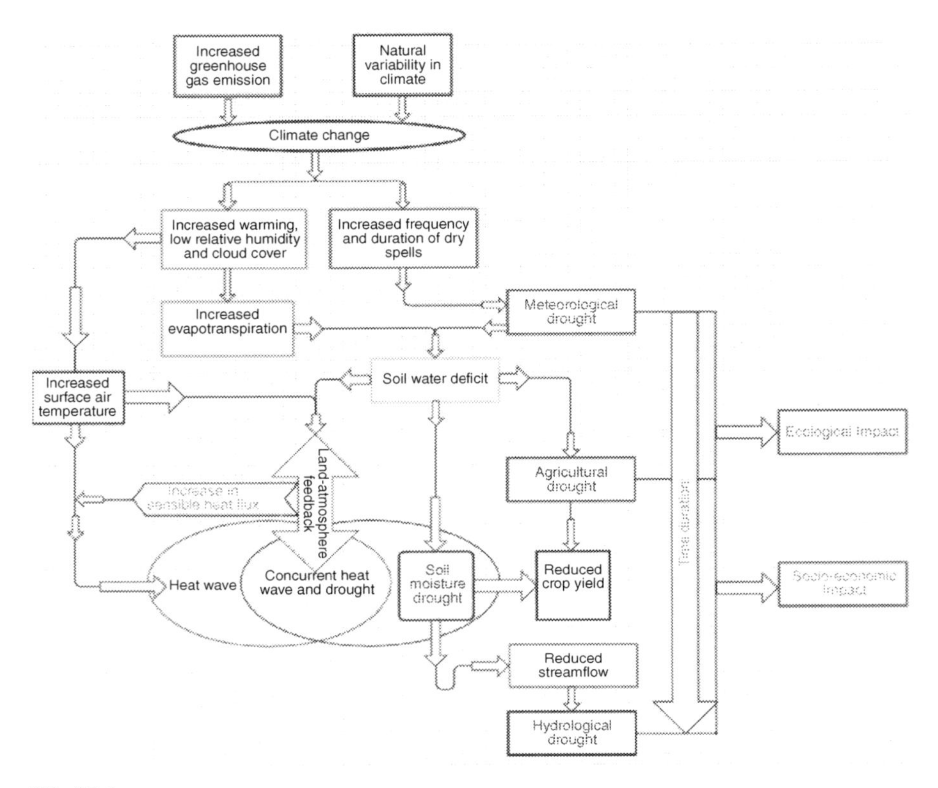

FIG. 17.1

Categories of drought and its development across the hydrological cycle. (Adapted from Mukherjee et al., 2018)

meteorological, agricultural, and hydrological drought elements. It occurs when the demand for an economic good exceeds the supply. This drought typology and the corresponding drought development phases are illustrated in Fig. 17.1.

Although these drought types differ significantly from one another, the root cause behind each of them stems from a common meteorological forcing, which is the persistent lack of rainfall (or dry spells) over a region (Fig. 17.1). Furthermore, drought events are further classified into different categories such as mild, moderate, severe, or extreme (Tallaksen and Van Lanen, 2004). These categories represent a progressively more severe impact on agricultural, industrial, and water supplies (Tallaksen and Van Lanen, 2004). Recent decades have witnessed massive increases in the frequency of persistent dry spells leading to severe-extreme drought events across different parts of the globe (Cindrić et al., 2010; Dai, 2011; Diffenbaugh et al., 2015; Kundzewicz et al., 2006; Mukherjee and Mishra, 2021). While some studies find a significant association between the dry spells and the natural climate variability (Li et al., 2016; Mukherjee and Mishra, 2021), most attribute the recent increases in dry spells to the increase in anthropogenic warming (Dai, 2011; Diffenbaugh et al., 2015; Mukherjee and Mishra, 2021).

The evaluation of drought conditions in a specific region (watershed) is the first step for planning water resources to prevent and mitigate the negative impacts of future drought occurrences (Mishra and Singh, 2010). The potential importance of rainfall in drought quantification is another crucial factor in planning and framing mitigation policies against the adversities of drought. To this end, several drought indices have been developed in the framework of evaluating the water supply deficit as a result of persistent rainfall (or precipitation) shortages (Hayes et al., 2007; Keyantash and Dracup, 2002; Zargar et al., 2011). More specifically, drought indices have been used as proxies to quantify drought severity by integrating data from one or several variables (e.g., precipitation and temperature) into a single numerical value. In addition to that, the nature of drought indices reflects different climatic conditions.

Among all drought indices proposed so far, the Standardized Precipitation Index (SPI), Standardized Precipitation-Evapotranspiration Index (SPEI), and the Palmer Drought Severity Index (PDSI) are the most used indicators. These indices rely on precipitation as the key variable and measure the deviation of actual precipitation from a historically established norm. Some of them even consider additional climatological variables such as temperature, evapotranspiration, or soil moisture. The reliability of drought analyses using these indices strongly depends on the primary data quality (e.g., precipitation) (Mukherjee et al., 2018; Mukherjee and Mishra, 2021). Therefore, in assessing dry (wet) periods, it is highly recommended to have long term dataset that is easy to access and can capture both dry and wet events uniformly across the whole studied area (Mishra and Singh, 2010; Mukherjee et al., 2018).

In this chapter, the importance of rainfall as a key component of drought assessment is highlighted by analyzing state-of-the-art rainfall-based drought indices in a systematic manner. The specific objectives of this chapter are: (a) to provide an overview of the most commonly used rainfall-based drought indices, the Standardized Precipitation Index (SPI), Standardized Precipitation Evapotranspiration Index (SPEI), the Palmer Drought Severity Index (PDSI), and (b) to compare SPI, SPEI, and PDSI in determining droughts across different spatial scales. The rest of the chapter is organized as follows. In Section 17.2, drought generating processes are discussed; Section 17.3 provides a review of the drought indices. Section 17.4 shows a case study to illustrate the use of drought indices, while conclusions are presented in Section 17.5.

17.2 Drought hydrology and generating processes

Droughts are generally driven by a period of low precipitation and high evapotranspiration. However, additional factors (e.g., watershed and geomorphic characteristics) can also play an important role in drought occurrence (Konapala and Mishra, 2020; Van Loon, 2015). The regional characteristics, such as soil type, topography, land use, land cover, underlying climate, groundwater system, and how groundwater is connected to the streamflow and neighboring regions can play an important role in the onset and propagation of drought. Overall, most droughts start with precipitation deficits but can be originated from multiple critical factors, such as seasons, antecedent conditions, modulations in the atmospheric demand, and the regional characteristics of the region (Fig. 17.1). The association between these factors and drought is particularly complex and needs to be assessed in the context of their cascading impacts. For example, precipitation deficits during the dry season commonly lead to a reduction in surface soil moisture, causing ecological droughts (Fig. 17.1). However, rainfall deficits in the dry season that sometimes leads to a reduction in groundwater levels may not always lead to hydrological droughts. Often, the rainfall deficits during the preceding wet season becomes a more dominant factor. Groundwater recharge during the dry season is naturally low; therefore, an extended period of dry spells in the preceding wet season exacerbates groundwater depletion in the subsequent dry season, which often propagates into causing a hydrological drought (Van Loon, 2015). In limited cases, however, hydrological drought can emerge even in the wet season due to continuous increases in evapotranspiration. Such conditions are often controlled by energy-limited conditions (in humid regions) and may lead to a significant reduction in soil moisture, which then lowers groundwater level and recharge (Condon et al., 2020).

Potential evapotranspiration (PET) is one of the most important components of drought generating processes. PET is defined as the maximum level of evapotranspiration that would occur if a sufficient water source is available. PET is triggered by high temperature, radiation, and dry winds. In most regions, PET exerts critical control on soil moisture variability and, therefore, has significant implications on vegetation health. However, such implications may vary spatially depending on the type of climate (dry or wet) of the region. In dry or arid regions, changes in evapotranspiration are limited by the water availability, meaning changes in evapotranspiration from plants is more sensitive to changes in soil moisture. On the other hand, in wet or humid regions, the incoming energy (such as temperature, incoming radiation coupled with dry winds) becomes more dominant factors in limiting evapotranspiration (Seneviratne et al., 2010). In addition, in dry regions, the higher than usual groundwater depletion rates override the implications from the lack of surface soil-moisture on vegetation health and agricultural yield. Plants often react to such high evaporative stress during the dry season by slowing down the evapotranspiration through modulation of stomatal resistance.

Droughts are manifests of abnormal hydrological anomalies that recur in time along the course of different hydrological processes (Mishra and Singh, 2010). Fig. 17.2 shows an example of how hydrological anomalies (wet and dry events) propagate through the hydrological system. Precipitation events are reflected immediately in the surface runoff. Moisture in the near surface of the soil also gets affected within two to three days. However, the propagation of the pulse depends on the depth until it reaches the groundwater aquifer. As the region is impacted by a drought event, soil moisture in the near-surface dries down fast over days to weeks, while moisture in the top few meters take weeks to months. Groundwater, on the other hand, does not get affected immediately; instead, it may take months to years to deplete the deep-water storages. It is also important to note that the speed of propagation of

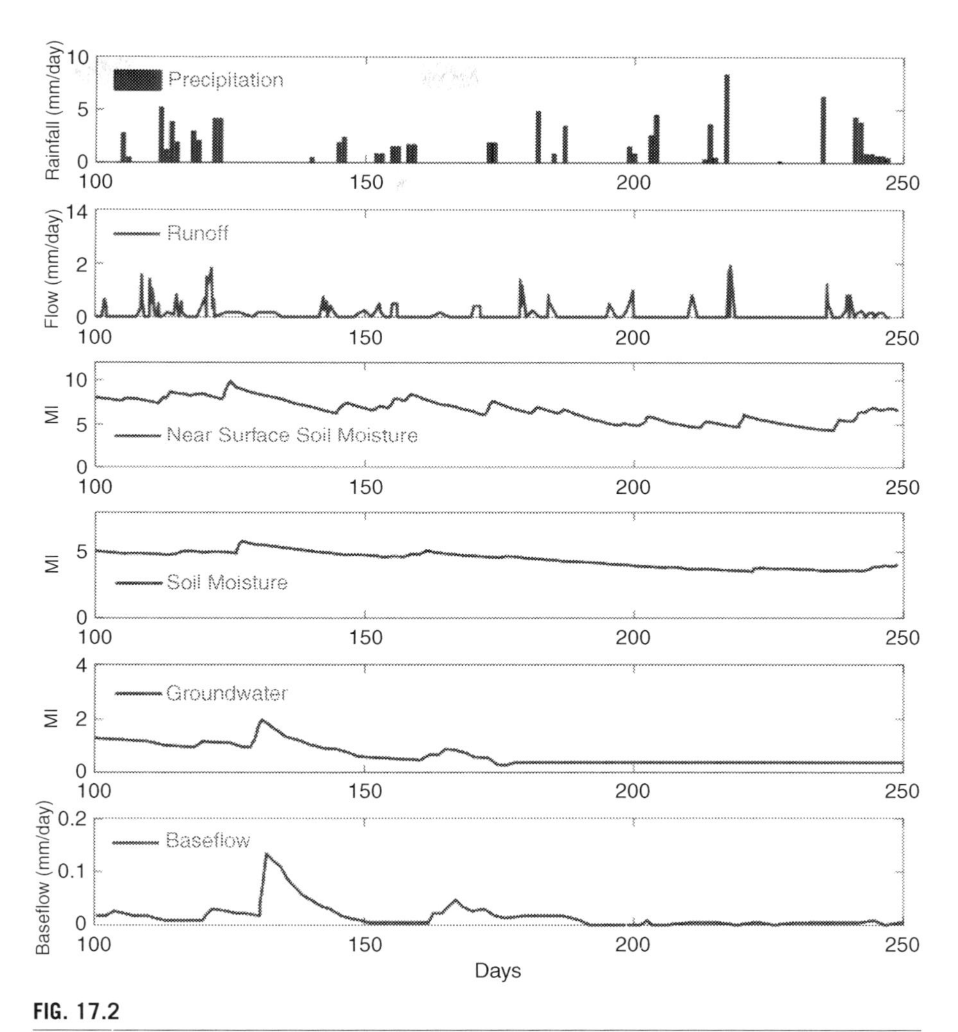

FIG. 17.2

The propagation of wet and dry events through the hydrological system. Adopted from Sheffield et al. (2012).

drought across different parts of the hydrological cycle depends on the climatic and watershed characteristics (Laaha and Blöschl, 2007).

17.3 Drought indices

Drought indices are proxies that are used to quantify droughts at a specific location by integrating climatic data such as precipitation and evapotranspiration into a single numerical value. Drought indices are useful in assessing the effect of drought and for defining different drought parameters (e.g., intensity, duration, and severity) on multiple timescales (in which the long-term data becomes essential). Drought analysis is usually performed best on a yearly to monthly time scale. The key advantage of using an annual time scale is that it captures useful information on droughts' regional behavior. On the other hand, the monthly time scale may be the best choice when monitoring the effects of drought episodes in situations related to agriculture, water supply, and groundwater abstractions (Panu and Sharma, 2002).

Many drought indices have been developed, evaluated, and implemented around the world (Hisdal et al., 2004; Mishra and Singh, 2010; Wanders et al., 2010). Based on previous studies, the proposed drought indices can be classified into two main groups. The first group of indices is based on addressing rainfall and precipitation deficits, while the second group of indices is related to soil moisture and water availability. Examples of precipitation-based drought indices include Standardized precipitation index (SPI), the Palmer drought severity index (PDSI), Standardized Precipitation-Evapotranspiration Index (SPEI). All these indices use precipitation as a single variable or in combination with other meteorological parameters to quantify droughts. For example, SPI uses only precipitation as an input parameter, PDSI uses temperature, precipitation, and soil moisture data to develop the index, while SPEI requires precipitation and potential evapotranspiration as key components. The following sections discuss the three indices, their usefulness, limitations, and intercomparison. These discussions are further summarized in Table 17.2.

17.3.1 Standardized precipitation index (SPI)

The Standardized Precipitation Index (SPI) is a wildly used drought index for quantifying and characterizing meteorological droughts. The SPI indicator was developed by McKee et al. (1993) and described in detail by Edwards et al. (1997). Using long-term historical precipitation records at a given location, SPI measures precipitation anomalies based on a comparison of observed total precipitation amounts for an accumulation period of interest (e.g., 1 month, 3 months, 6 months, etc.). The long-term record is fitted to a probability distribution (gamma distribution) and then transformed to a normal distribution (e.g., the mean SPI for that location with a period of zero). Since SPI values are in units of standard deviation, SPI can be used to compare precipitation anomalies for any geographic location and for any number of time scales. As such, for any given location, severe rainfall deficits are indicated as SPI below –2.0, while severe excess precipitation is indicated as SPI increases above 2.0 (Table 17.1). The flexibility of SPI lies in monitoring short-term water supplies, soil moisture, and long-term water resources (e.g., groundwater, streamflow, lake, and reservoir levels). In addition, a short time period (e.g., 1 month) may indicate soil moisture deficits resulting from low precipitation. In contrast, more extended time periods (e.g., 12 months) may highlight precipitation deficits, affecting streamflow, reservoirs, and groundwater levels (Hayes et al., 2007).

Several studies have highlighted the strength of the SPI. For example, Szalai and Szinell (2000) investigated the correlation of SPI with streamflow and groundwater

Table 17.1 Drought classification of SPI and SPEI.

Index value	Category
Value ≥ 2	Extremely wet
1.5 ≤ value <1.99	Moderately
1.0 ≤ value < 1.49	Slightly wet
-0.99 < value < 0.99	Near normal
-1.49 < value ≤ -1	Mild drought
-1.99 < value ≤-1.5	Moderate drought
Value ≤ -2	Extremely drought

levels at several stations in Hungary. For streamflow, the correlation strength was the highest using a 2-month time-scale, while for groundwater levels, the correlations were significant with different time scales. Szalai et al. (2000) highlighted that agricultural drought (based on the soil moisture content) was replicated best-using SPI on a 2–3 months scale. SPI was also used to examine different aspects of droughts. For example, SPI was used for forecasting (Belayneh et al., 2014; Kisi et al., 2019; Mishra et al., 2007; Mishra and Singh, 2009), frequency analysis (Bonaccorso et al., 2015; Hangshing and Dabral, 2018; Mishra and Singh, 2009),

Table 17.2 Summary of the commonly used drought indices.

Indices	Description	Strength	Weaknesses	reference
SPI	A simple method based on precipitation deficits over varying periods or timescales	Computed for flexible multiple timescales, provides early warning of drought and helps in assessing drought severity	Precipitation data are the only input data. SPI values based on long term precipitation may change. The long timescale (24 months) is not reliable	McKee et al. (1993)
SPEI	An improvement over the SPI by considering both precipitation and potential evapotranspiration	SPEI has been applied widely in many studies that investigated drought variability	Sensitive to the method to calculate potential evapotranspiration (PET)	Vicente Serrano et al. (2010)
PDSI	Calculated using precipitation, temperature, and soil moisture data. Soil moisture algorithm has been calibrated for relatively homogeneous regions	The first comprehensive drought index used widely to detect agricultural drought	PDSI can be slow to respond to develop and diminish droughts; it is also sensitive to precipitation and temperature.	Palmer (1965)

spatio-temporal analysis (Dabanlı et al., 2017; Haroon et al., 2016; Heddinghaus and Sabol, 1991; A K Mishra and Singh, 2009) and climate impact studies (Danandeh Mehr et al., 2020; Huang et al., 2016; Mishra and Singh, 2009). Besides that, the World Meteorological Organization employs SPI as the primary reference drought index (Hayes et al., 2002).

However, there are major challenges and limitations associated with SPI. One of the key limitations of SPI lies in the usage of precipitation data without considering evaporation demand. On the other hand, the length of record for precipitation and the nature of probability distribution plays an essential role for SPI. When precipitation data have similar gamma distributions for different periods, comparable and consistent results are observed for SPI values. However, when the distributions are different, SPI values show inconsistent results. It is highly recommended to be aware of the numerical differences in the SPI values, primarily when different lengths of record are used in interpreting and making decisions based on the SPI values. To illustrate, Wu et al. (2017) examined the impact of the length of precipitation on the SPI calculation by evaluating several parameters (e.g., correlation coefficients, the index of agreement, and dry/wet consistency events). The inconsistency in the SPI values was related to the changes in the gamma distribution parameters and scale parameters.

Furthermore, different probability distributions influence the SPI index values since the SPI index is based on the distribution fitting of precipitation time series. Commonly used distributions include gamma distribution (Edwards et al., 1997; McKee et al., 1993; Mishra and Singh, 2009), Pearson Type III distribution (Guttman, 1999), lognormal, and exponential distributions have been widely applied to simulations of precipitation distributions (Lloyd-Hughes and Saunders, 2002; Madsen and Rosbjerg, 1998; Todorovic and Woolhiser, 1976). Two problems can arise when fitting distribution for a given precipitation time series. When SPI values are computed for long time scales (longer than 24 months), fitting a distribution may be biased because of the limitation in the data length. This usually happens when quantifying droughts in watersheds with limited long-term precipitation data (Lloyd-Hughes and Saunders, 2002; Sönmez et al., 2005). Second, in dry climates, precipitation is seasonal, and zero precipitation values are common (e.g., there will be many zero precipitation values in a particular season). In this case, the computed SPI values at short time scales may not be normally distributed as the skewness is expected and limitation of the fitted gamma distribution. This may be led to high uncertainty when simulating precipitation distributions in dry regions using small datasets.

17.3.2 Standardized precipitation-evapotranspiration index (SPEI)

The standardized precipitation evapotranspiration index (SPEI) was proposed by Vicente-Serrano et al. (2010). SPEI improves the SPI by considering both precipitation and potential evapotranspiration (PET) as the key components of the water cycle. SPEI has been applied widely in many studies for investigating drought variability (Paulo et al., 2012; Potop, 2011; Sohn et al., 2013; Spinoni et al., 2013), drought reconstruction (Allen et al., 2010), drought atmospheric mechanisms (e.g., Boroneant et al., 2011; Manzano et al., 2019), climate change(Abiodun et al., 2013; Sohn et al., 2013; Wolf and Abatzoglou, 2011; Yu et al., 2014), and the identification of drought impacts on hydrological (Lorenzo-Lacruz et al., 2010; McEvoy et al., 2012), agricultural (Potop et al., 2012; Zarei and Moghimi, 2019), and ecological systems (Barbeta et al., 2013; Cavin et al., 2013; Deng et al., 2011; Martin-Benito et al., 2013; Toromani et al., 2011; Vicente-Serrano et al., 2012). In addition, the application of SPEI can also be found in developing drought monitoring systems (Fuchs et al., 2012). These previous studies unanimously reported that the SPEI shows a good correlation with hydrological and ecological variables compared to other drought indices.

The procedure for calculating the SPEI is similar to SPI. However, the calculations are performed using the potential monthly or weekly water deficit (i.e., the difference between precipitation and PET). The results of SPEI are aggregated over a given period of interest (e.g., 1 month, 3 months, 6 months etc.) and fitted to a probability distribution function. As the difference between precipitation and evapotranspiration can be negative, a three-parameter distribution is needed to model the deficit values. Vicente-Serrano et al. (2010) recommended the log-logistic distribution as it can fit extreme values better. The fitted cumulative probability density function is transformed to the standardized normal distribution. Positive values of SPEI indicate above-average moisture conditions (wet), while negative values indicate below normal (dry) conditions. The same scale used for categorizing SPI values can be used for SPEI (Table 17.1). The calculation of SPEI requires additional parameters (potential evapotranspiration, PET) as compared to the SPI. The PET values are computed indirectly from the measured meteorological variables (precipitation and temperature). There are several approaches available for calculating PET. The method proposed by Thornthwaite (1948) is relatively more straightforward as it requires only the monthly mean temperature and the latitude of the location of interest to compute PET (Hernandez and Uddameri, 2014).

17.3.3 Palmer drought severity index (PDSI)

The Palmer drought index (PDSI) was developed by Palmer (1965) to estimate the relative dryness based on precipitation and temperature. PDSI was the first comprehensive effort to assess the total moisture status of a region. The PDSI is calculated based on precipitation, temperature data, and soil water content. All terms of the water balance equation can be determined from these inputs, including evapotranspiration, soil recharge, runoff, and moisture loss from the surface layer. The calculation procedure of the PDSI has been discussed in many publications (Alley 1984; Karl, 1983; Karl, 1986). Some modified versions of PDSI have also been suggested. For example, The Palmer hydrological drought index (PHDI) was derived from the PDSI to quantify the long-term impact of drought on water supply systems (Karl, 1986). The modified Palmer drought severity index (WPLM) was proposed by the National Weather Service Climate Analysis Center for operational meteorological purposes (Heddinghaus and Sabol, 1991), modifying the original rules of accumulation during wet-dry spells.

Although the PDSI has several deficiencies, PDSI is still the most widely used regional drought index for quantifying and monitoring droughts (Mukherjee et al., 2018). PDSI has been used to investigate: the spatial and severity of various drought events (Karl and Quayle, 1981), spatial and temporal drought characteristics (Diaz, 1983; Karl and Koscielny, 1982; Soulé, 1993), periodic behavior of droughts (Rao and Padmanabhan, 1984), hydrologic trends, crop forecasts, and assess potential fire severity (Heddinghaus and Sabol, 1991), droughts over large geographic areas (Johnson and Kohne, 1993), the impact of natural variability and global warming on drought-related events (Mukherjee and Mishra, 2021), and drought forecasting (Kim and Valdés, 2003; Özger et al., 2009).

The Palmer indices are sensitive to both temperature and precipitation. (Guttman, 1991) performed a sensitivity analysis to assess the influence of temperature and precipitation anomalies on Palmer indices. It was observed that precipitation anomalies tend to dominate the change of PDSI in the cold season when evaporation is minimum. Additionally, the role of temperature in PDSI becomes more important in the warm season; however, the response of PDSI often lags behind the anomalies of temperature and precipitation by a few months (Karl, 1986). Furthermore, since PDSI depends on rainfall climatology (which is a function of time and fluctuates with temperature), PDSI can be equally affected by temperature and precipitation. This may complicate the usage of the index in interpreting precipitation anomalies and its application in inferring precipitation variations, particularly from reconstructed PDSI (Hu and Willson, 2000).

PDSI has been in use for a long time and has been tested and verified across many regions. PDSI is standardized and accounts for temperature and soil characteristics; therefore, robust comparison across different climatic zones is also possible. However, PDSI has several limitations. Although the time scale makes the PDSI a good indicator for agricultural droughts, it lacks the ability to robustly quantify the hydrologic droughts. In addition, the assumption that all precipitation is rainfall makes precipitation during winter months and at high elevations often uncertain in the framework of PSDI estimation. Furthermore, the PDSI assumes that runoff only occurs when all soil layers have become saturated, leading to underestimating the runoff. PDSI can even be slow to respond to developing and diminishing droughts (Hayes et al., 1999).

17.4 Case study: drought characterization using SPI, SPEI, and PDSI

17.4.1 Background

This section presents a comparative analysis between SPI, SPEI, and PDSI for two climate divisions located in Texas, in the South-Central region of the US between 100° W and 31° N (Fig. 17.3A). The climate of Texas is spatially diverse and varies

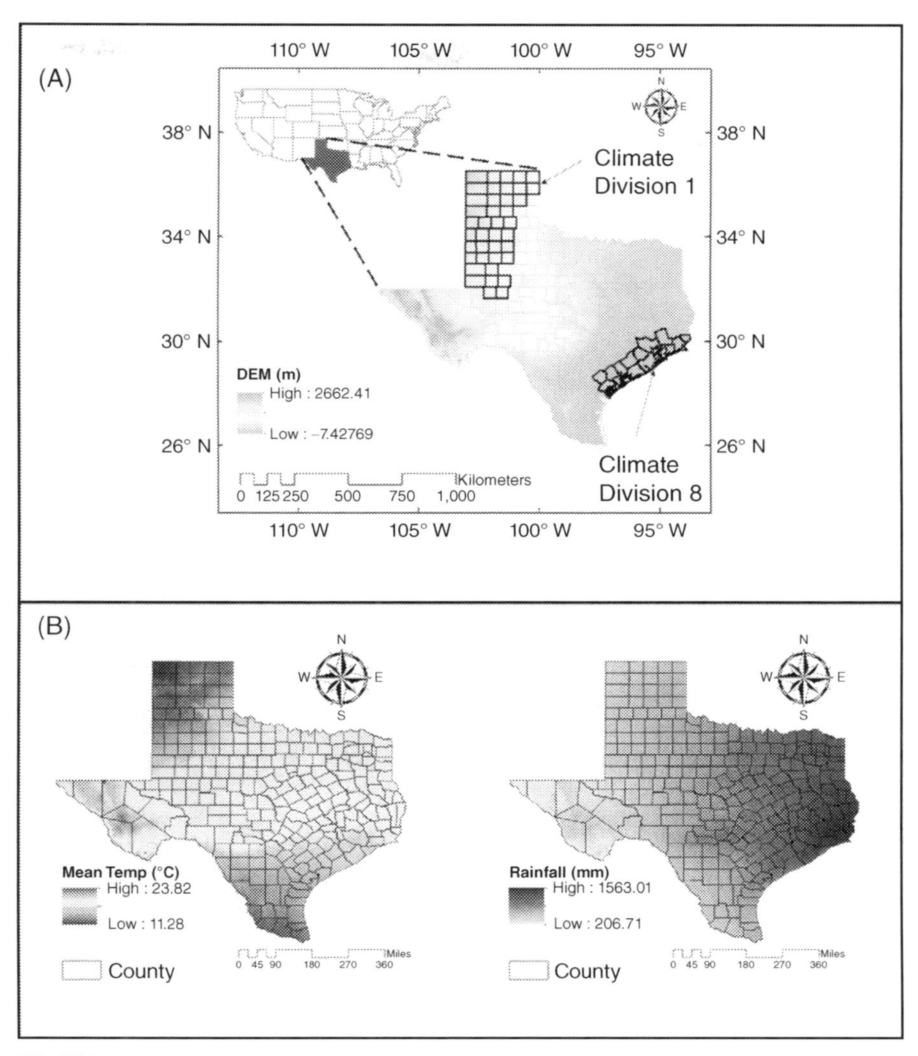

FIG. 17.3

(A) Digital elevation model (DEM) over the selected climatic subdivision located within the state of Texas. (B) Examples of climatic characteristics: mean annual temperature and mean annual rainfall over the states of Texas.

from arid in the west to humid in the east. There are ten distinct climatic regions, where each climatic region has different seasonal patterns. There is a clear trend that can be observed across the state (as shown in Fig. 17.3B). The mean normal precipitation decreases from the east (~1563 mm/year), influenced by the increase in the moisture supply from the Gulf of Mexico, to the west (~206 mm/year). The mean normal temperature decreases from the south to the north. Most of the rainfall in the state occurs in flash events (that produce large amounts of rainfall in a short duration). The climate of Texas is also impacted by extremes in temperature and precipitation, resulting in a high frequency of and floods. The rainfall associated with hurricanes is a characteristic of the Gulf Coast of Texas, while droughts occur throughout the state.

Texas has experienced frequent droughts; for example, two major drought events (2006 and 2011) resulted in billions of crop and livestock losses and massive wildfires (Combs, 2012). The 2011 drought event was one of the worst events recorded in the history of Texas. Almost 90% of the state experienced severe droughts (Combs, 2012). The economic impact is estimated at approximately 7.62 billion dollars (Fannin, 2012), which was mainly from agricultural losses. The record mean temperature was 2.9 °C higher than the normal mean temperature (Hoerling et al., 2013). The state's reservoir storage declined by almost 58 % in November 2011 (which was the lowest since 1978) (Combs, 2012). Watershed managers concluded that some regions, such as West Texas, maybe in the middle of new and more severe critical droughts events.

This section aims to characterize the historic droughts based on their duration and severity across two different climate subdivisions in Texas: Texas Climate Division 1- (High Plains) and Texas Climate Division 8 (Upper coast) (Fig. 17.3A). Three meteorological drought indices (SPI, SPEI, and PDSI) were evaluated over the selected subdivisions. The SPI and PDSI data were obtained from the dataset published in NOAA/National Centers for Environmental Information (NOAA/ NCEI), while SPEI was obtained from the global SPEI database. For SPI and SPEI, the 1 month, 3 months, 6 months, and 12 months timescales were selected for their relevance in capturing the short, medium, long term drought conditions (i.e., agricultural and meteorological drought).

17.4.2 Temporal evolution and characteristics of drought

Temporal analysis of meteorological droughts using SPI, SPEI, and PDSI is shown for climate subdivision 1 and subdivision 8 within Texas (Fig. 17.3A). The meteorological drought for both climate subdivisions is analyzed from 1960 to 2015. Drought assessment using SPI and SPEI is performed based on four-different time scales: 1, 3, 6, and 12 months. The temporal evolution of the SPI and SPEI for the selected subdivision is shown in Fig. 17.4 and Fig. 17.5, respectively. For time scales of 1-month and 3-month, the occurrence of both dry and wet periods is more frequent with no significant and notable trend (Fig. 17.4 and Fig. 17.5). On the other hand, for 6-months and 12-months, the historical drought events are apparent in the ascending order. The historical drought events are more prominent using the 6-month and 12-month accumulation periods for both indices. For subdivision 1 and subdivision 8, both SPI and SPEI detected the drought events of 1966, 1973, 1984, 2004, 2006, and 2011 with nearly equal magnitude.

Fig. 17.6 shows Pearson's correlation between SPI and SPEI obtained at 1-,3-,6-, and 12 months for both climate subdivisions from 1960 to 2015. Both observations

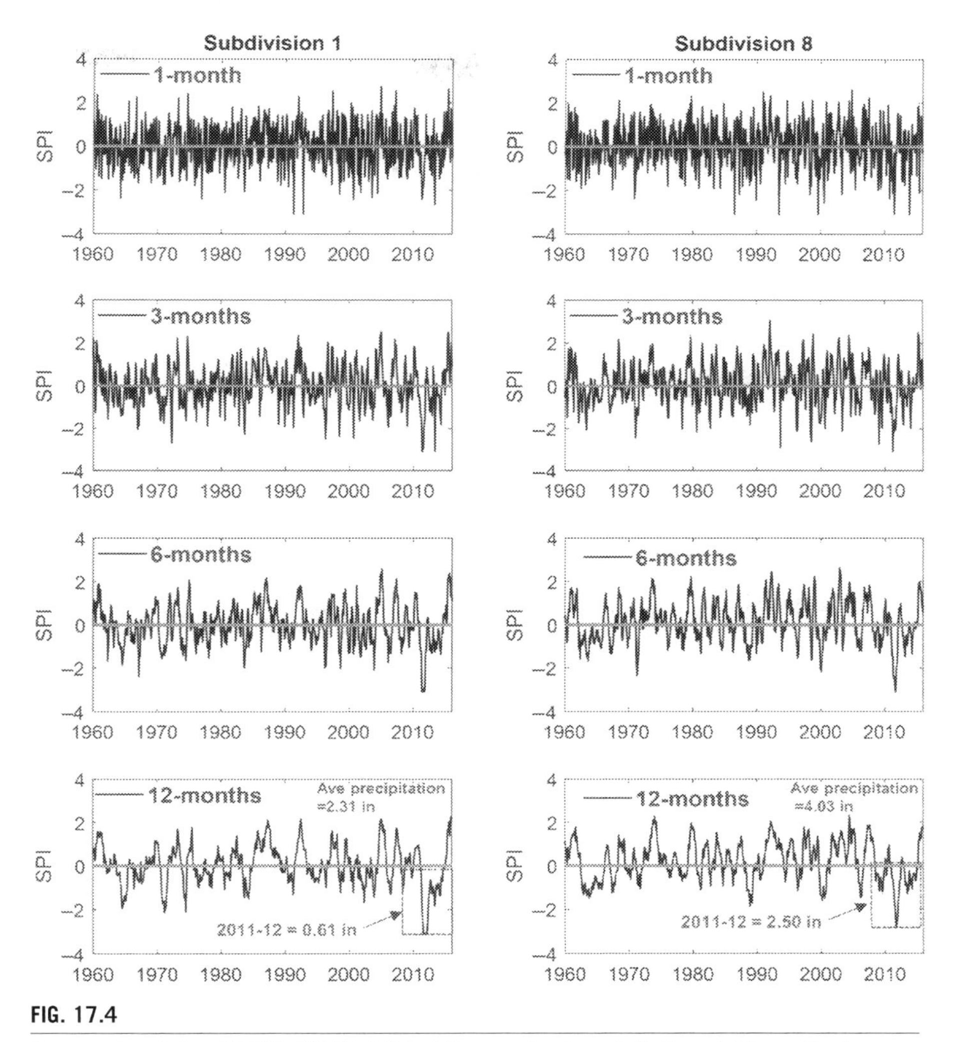

FIG. 17.4

Temporal evolution (1960–2015) of SPI at time-scales from 1,3, 6, and 12 months for climate subdivision 1 (left panel) and climate subdivision 8 (right panel).

showed a high correlation between SPI and SPEI. The correlation strength between SPI and SPEI was slightly stronger for subdivision 1 compared to subdivision 8 (as indicated by R^2 values). However, the number of droughts, mean duration, and mean severity is found to be different between the two indices (Table 17.3 and Table 17.4). For example, the total number of droughts events of severe droughts (<=-2.0) is higher in the case of SPI compared to SPEI. This may be because SPI is calculated only based on the precipitation data without the inclusion of potential evapotranspiration. Evapotranspiration plays a crucial role in propagating

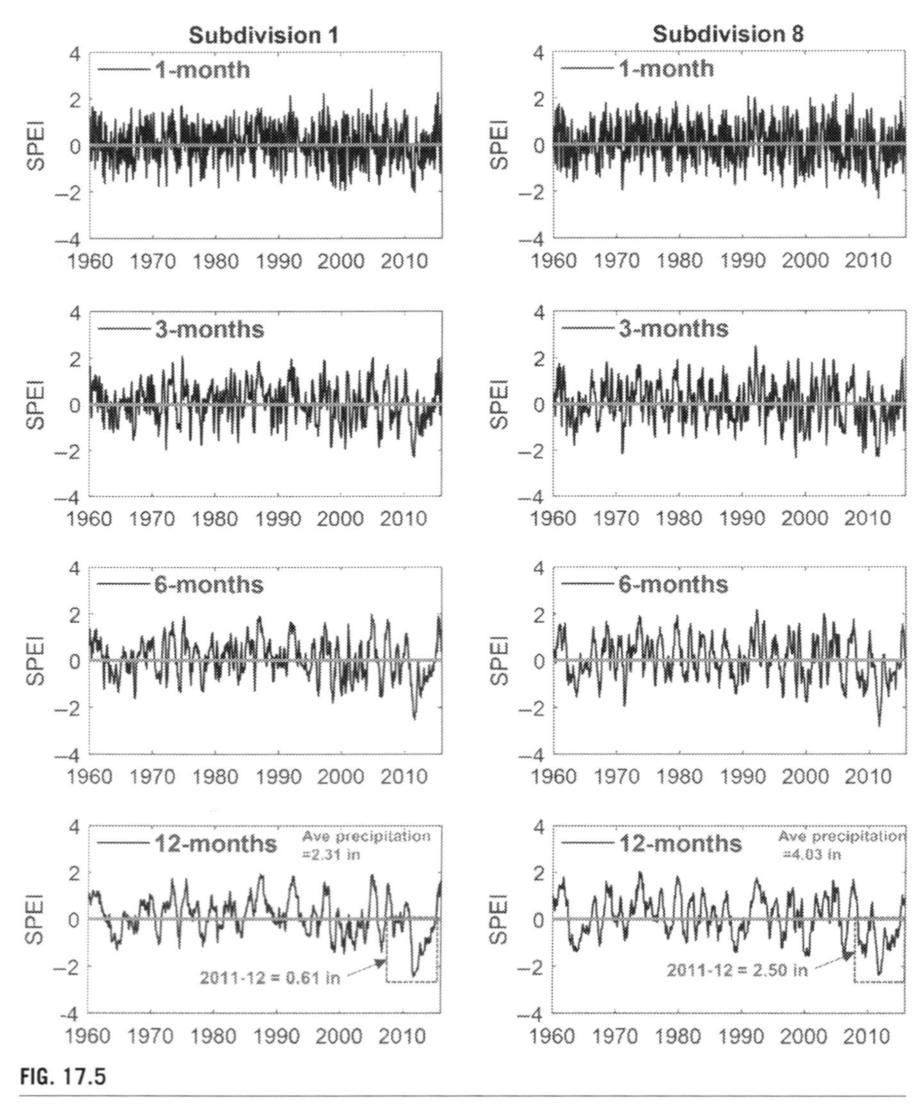

FIG. 17.5

Temporal evolution (1960–2015) of SPEI at time-scales from 1,3, 6, and 12 months for climate subdivision 1 (left panel) and climate subdivision 8 (right panel).

meteorological drought to hydrological drought (Van Loon, 2015). Therefore, the exclusion of potential evapotranspiration could overestimate the SPI, leading to an overestimation of drought severity. Byakatonda et al. (2018) observed a similar overestimation of SPI compared to SPEI in evaluating drought severity in Botswana located in Southern Africa. On the other hand, based on PDSI, both subdivisions have experienced many severe droughts events (<=-2.0) (accounted for at

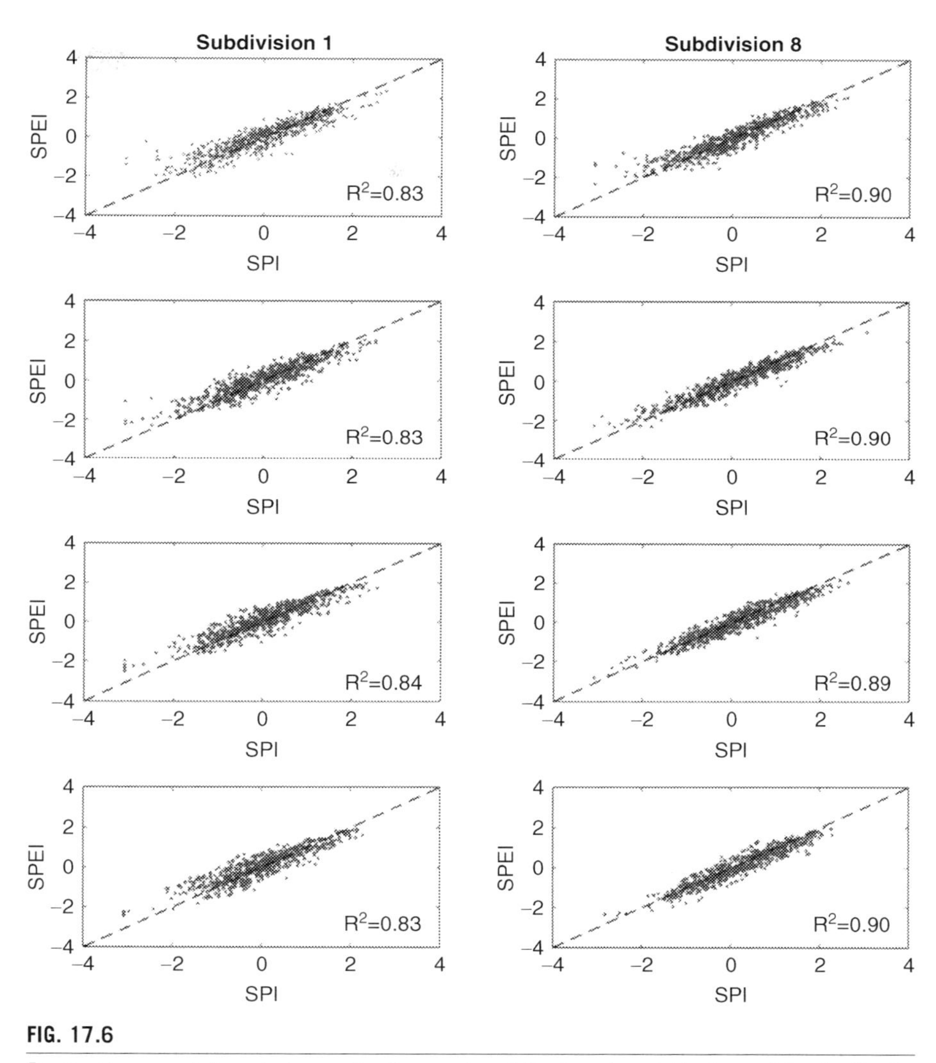

FIG. 17.6

Pearson's r correlations between the time series of the two drought indices of the 1-, 3-, 6-, and 12- month SPI and SPEI for subdivision 1 and subdivision 8.

least 19% of all wet and dry events from 1960 to 2015; Fig. 17.7). The frequency of severe drought increased up to 40% between 2000 to 2015. The total number of droughts events of severe droughts (<=-2.0) is higher in the case of PDSI compared to SPI and SPEI (Table 17.5). This may be because PDSI is more sensitive to variations in temperature and precipitation compared to SPI and SPEI due to the seasonal dependence of PDSI on temperature and precipitation (Guttman, 1991; Hoffmann et al., 2020; Mishra and Singh, 2010).

Table 17.3 Meteorological Drought characterization (severity and duration) using SPI for subdivision 1 and subdivision 8.

Region	Threshold	Accumulation period (months)	Total number of drought events	Mean Duration (months)	Mean Severity
Subdivision 1	-0.5 to -1.49	1	103	2	-1.55
		3	78	2	-1.86
		6	49	4	-3.40
		12	55	3	-2.34
	-1.5 to -1.99	1	25	1	-1.59
		3	16	2	-2.89
		6	8	2	-3.08
		12	9	3	-3.75
	<=-2.0	1	10	1	-3.58
		3	9	1	-3.46
		6	3	4	-9.17
		12	3	5	-12.89
Subdivision 8	-0.5 to -1.49	1	108	1	-1.11
		3	82	2	-1.57
		6	54	3	-2.62
		12	43	5	-3.21
	-1.5 to -1.99	1	29	1	-1.69
		3	24	1	-1.67
		6	12	2	-1.96
		12	7	2	-1.99
	<=-2.0	1	14	1	-2.57
		3	14	1	-2.67
		6	7	2	-3.53
		12	2	3	-6.50

17.5 Conclusion

Drought indices are commonly used to quantify drought and to help stakeholders to develop appropriate strategies to minimize the impact of droughts. A substantial number of drought indices are derived based on the combination of different meteorological forcing. These drought indices can be broadly categorized into two groups. While the first group of indices is based on addressing precipitation deficits, the second group of indices is focused on soil moisture and water availability. The aim of the chapter is to provide a systematic review of the state-of-the-art in drought assessment using the rainfall-based drought indices. Three most extensively used drought indices are selected to demonstrate the role of rainfall in drought quantification – (1) SPI that uses precipitation as the sole indicator variable, and (2) SPEI, and (3) PDSI, both of them use precipitation in combination with other meteorological

Table 17.4 Meteorological drought characterization (severity and duration) using SPEI for subdivision 1 and subdivision 8.

Region	Threshold	Accumulation period (months)	Total number of drought events	Mean Duration (months)	Mean Severity
Subdivision 1	-0.5 to -1.49	1	127	1	-1.26
		3	74	2	-2.06
		6	48	4	-3.06
		12	31	5	-4.88
	-1.5 to -1.99	1	17	1	-9.46
		3	15	1	-10.13
		6	12	2	-12.26
		12	6	2	-25.23
	<=-2.0	1	1	1	-2.32
		3	6	1	-2.58
		6	1	6	-14.5
		12	1	6	-13.78
Subdivision 8	-0.5 to -1.49	1	111	1	-1.27
		3	61	2	-2.16
		6	52	3	-2.58
		12	24	6	-5.37
	-1.5 to -1.99	1	22	1	-2.06
		3	14	1	-2.30
		6	8	2	-2.81
		12	3	2	-2.66
	<=-2.0	1	1	1	-2.02
		3	1	4	-8.71
		6	1	6	-13.87
		12	1	11	-25.03

parameters (such as temperature, potential evapotranspiration, and soil moisture) to quantify drought. The potential challenges and limitations of each of these indices are further emphasized.

Finally, a comparative analysis between these three indices (SPI, SPEI, and PDSI) is provided for two different climate divisions (subdivisions 1 and 8) of Texas using long-term (1960-2015) multi-scalar observations. All three indices identified the temporal variability of droughts and classified different types of droughts as indicated by the different timescales. SPEI captured more severe and moderate droughts under the study period of 1960–2015. The SPEI droughts occurred with a longer duration and increased magnitude. The PDSI was found to be most sensitive to variations in temperature and precipitation, which is strongly linked to the seasonal dependence of the PDSI on these meteorological variables.

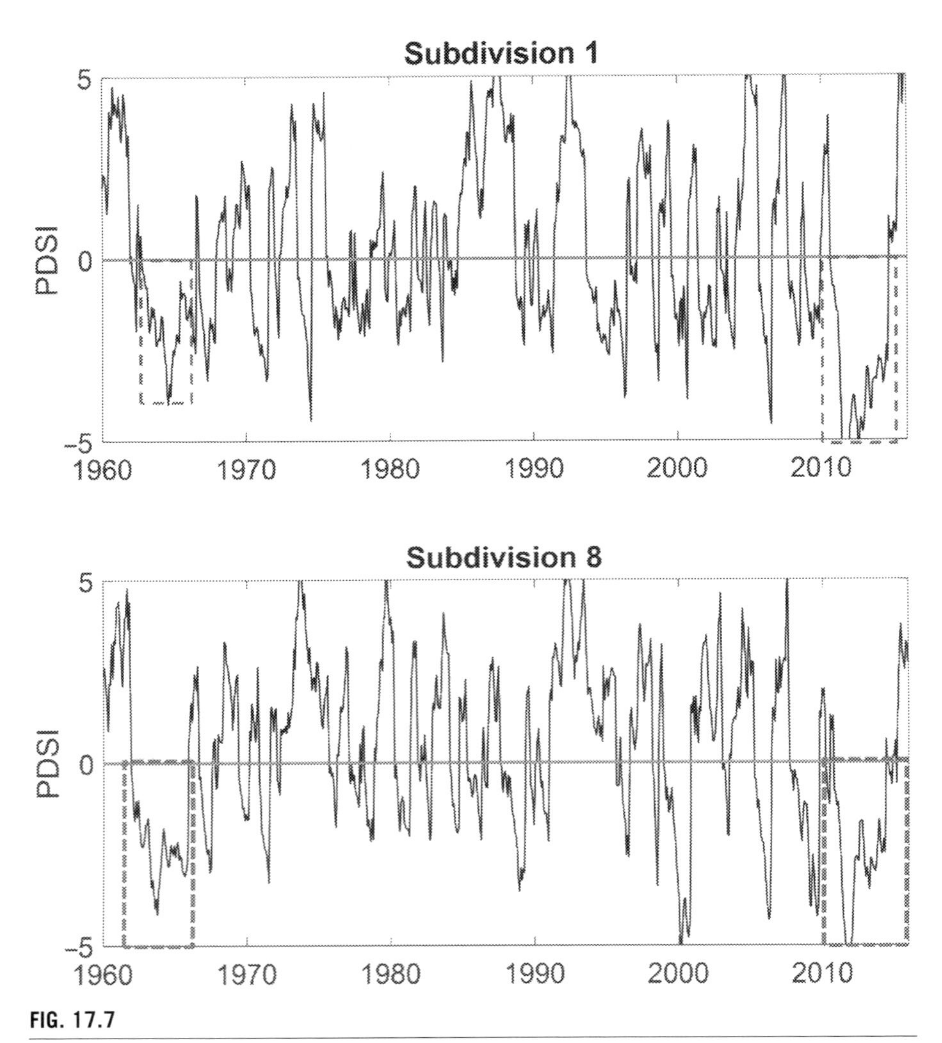

FIG. 17.7

Temporal evolution (1960–2015) of PDSI for climate subdivision 1 (upper panel) and climate subdivision 8 (bottom panel).

Overall, there are multiple challenges associated with the definition and quantification of drought. Although numerous drought definitions are available in the literature, only a handful of them recognizes droughts as different from the background aridity (Sherwood and Fu, 2014). In addition, drought has a varying and asymmetric impact on various stakeholders. For example, the reduced soil moisture can affect crop and plant growth, thereby initiating agricultural drought, but such conditions may not affect hydropower generation. In addition to that, there are multiple drivers of drought that further complicates the assessment of drought risk and decision making.

Table 17.5 Meteorological drought characterization (severity and duration) using PDSI for subdivision 1 and subdivision 8.

Region	Threshold	Total number of drought events	Mean duration (months)	Mean severity
Subdivision 1	-0.5 to -1.49	56	2	-2.50
	-1.5 to -1.99	41	1	-2.54
	<=-2.0	28	5	-15.06
Subdivision 8	-0.5 to -1.49	40	2	-2.10
	-1.5 to -1.99	30	2	-2.69
	<=-2.0	22	6	-18.26

These drivers may range from natural climate variability to the compounding impact from anthropogenic climate change and water availability and demand (Dai, 2011; Konapala et al., 2020; Van Loon and Van Lanen, 2013; Mukherjee et al., 2018; Fowler and Ali, 2022). Furthermore, high uncertainty associated with precipitation datasets makes rainfall-based drought assessment more challenging (Mukherjee et al., 2018). Thus, for a more reliable assessment, rainfall-based drought quantification should be more focused on choosing appropriate downscaling techniques, availability of high-quality precipitation data, estimation of potential evapotranspiration (PET), baseline period, non-stationary climate information, and anthropogenic forcing (Mukherjee et al., 2018).

References

Abiodun, B.J., Salami, A.T., Matthew, O.J., Odedokun, S., 2013. Potential impacts of afforestation on climate change and extreme events in Nigeria. Clim. Dyn. https://doi.org/10.1007/s00382-012-1523-9.

Allen, C.D., Macalady, A.K., Chenchouni, H., Bachelet, D., McDowell, N., Vennetier, M., et al., 2010. A global overview of drought and heat-induced tree mortality reveals emerging climate change risks for forests. For. Ecol. Manage. https://doi.org/10.1016/j.foreco.2009.09.001.

Alley, W.M., 1984. The Palmer drought severity index: limitations and assumptions. J. Appl. Meteorol. Climatol. 23, 1100–1109.

Apurv, T., Sivapalan, M., Cai, X., 2017. Understanding the role of climate characteristics in drought propagation. Water Resour. Res. 53, 9304–9329.

Bachmair, S., Stahl, K., Collins, K., Hannaford, J., Acreman, M., Svoboda, M., et al., 2016. Drought indicators revisited: the need for a wider consideration of environment and society. Wiley Interdiscip. Rev. Water 3, 516–536.

Barbeta, A., Ogaya, R., Peñuelas, J., 2013. Dampening effects of long-term experimental drought on growth and mortality rates of a Holm oak forest. Glob. Chang. Biol. 19, 3133–3144.

Belayneh, A., Adamowski, J., Khalil, B., Ozga-Zielinski, B., 2014. Long-term SPI drought forecasting in the Awash River Basin in Ethiopia using wavelet neural network and wavelet support vector regression models. J. Hydrol. 508, 418–429.

Bonaccorso, B., Cancelliere, A., Rossi, G., 2015. Probabilistic forecasting of drought class transitions in Sicily (Italy) using standardized precipitation index and North Atlantic oscillation index. J. Hydrol. 526, 136–150.

Boroneant, C., Ionita, M., Brunet, M., Rimbu, N, 2011. CLIVAR-SPAIN contributions: seasonal drought variability over the Iberian Peninsula and its relationship to global sea surface temperature and large scale atmospheric circulation. WCRP OSC Clim. Res. Serv. to Soc, 24–28.

Byakatonda, J., Parida, B.P., Moalafhi, D.B., Kenabatho, P.K., 2018. Analysis of long term drought severity characteristics and trends across semiarid Botswana using two drought indices. Atmos. Res. 213, 492–508.

Cavin, L., Mountford, E.P., Peterken, G.F., Jump, A.S., 2013. Extreme drought alters competitive dominance within and between tree species in a mixed forest stand. Funct. Ecol. 27, 1424–1435.

Cindrić, K., Pasarić, Z., Gajić-Čapka, M., 2010. Spatial and temporal analysis of dry spells in Croatia. Theor. Appl. Climatol. 102, 171–184.

Combs, S., 2012. The impact of the 2011 drought and beyond. Texas Comptrol. Public Accounts Spec. Rep. Publ:96–1704.

Condon, L.E., Atchley, A.L., Maxwell, R.M., 2020. Evapotranspiration depletes groundwater under warming over the contiguous United States. Nat. Commun. 11, 1–8.

Dabanlı, İ., Mishra, A.K., Şen, Z., 2017. Long-term spatio-temporal drought variability in Turkey. J. Hydrol. 552, 779–792.

Dai, A., 2011. Characteristics and trends in various forms of the Palmer Drought Severity Index during 1900–2008. J. Geophys. Res. Atmos. 116.

Dai, M., Huang, S., Huang, Q., Leng, G., Guo, Y., Wang, L., et al., 2020. Assessing agricultural drought risk and its dynamic evolution characteristics. Agric. Water Manag. 231, 106003.

Danandeh Mehr, A., Sorman, A.U., Kahya, E., Hesami Afshar, M., 2020. Climate change impacts on meteorological drought using SPI and SPEI: case study of Ankara, Turkey. Hydrol. Sci. J. 65, 254–268.

Deng, F., Chen, J.M., Dai, M., 2011. Recent global CO 2 flux inferred from atmospheric CO 2 observations and its regional analyses. Biogeosciences 8.

Diaz, H.F., 1983. Drought in the United States. J. Appl. Meteorol. Climatol. 22, 3–16.

Diffenbaugh, N.S., Swain, D.L., Touma, D, 2015. Anthropogenic warming has increased drought risk in California. Proc. Natl. Acad. Sci. 112, 3931–3936.

Edwards, C.D.C., McKee, T.B., Doesken, N.J., Kleist, J., 1997. Historical analysis of drought in the United States. 77th Conf. Clim. Var. 77th AMS Annu. Meet, 2–7.

Fang, W., Huang, S., Huang, Q., Huang, G., Wang, H., Leng, G., et al., 2019. Probabilistic assessment of remote sensing-based terrestrial vegetation vulnerability to drought stress of the Loess Plateau in China. Remote Sens. Environ. 232, 111290.

Fannin, B., 2012. Updated 2011 Texas agricultural drought losses total $7.62 billion. Agri. Life Today 21.

Fowler, H.J., Ali, H., 2022. Analysis of Extreme Rainfall Events Under The Climatic Change. In: Morbidelli, R. (Ed.), Rainfall. Modeling, Measurement and Applications. Elsevier, Amsterdam, pp. 307–326. doi:10.1016/C2019-0-04937-0.

Fuchs, B., Svoboda, M., Nothwehr, J., Poulsen, C., Sorensen, W., Guttman, N., 2012. A New National Drought Risk Atlas for the US from the National Drought Mitigation Center. Natl Drought Mitig Center, Univ Nebraska Lincoln, NE, USA.

Guo, Y., Huang, S., Huang, Q., Wang, H., Wang, L., Fang, W., 2019. Copulas-based bivariate socioeconomic drought dynamic risk assessment in a changing environment. J. Hydrol. 575, 1052–1064.

Guttman, N.B., 1999. Accepting the standardized precipitation index: a calculation algorithm 1. J. Am. Water Resour. Assoc. 35, 311–322.
Guttman, N.B., 1991. A sensitivity analysis of the palmer hydrologic drought index 1. J. Am. Water Resour. Assoc. 27, 797–807.
Han, Z., Huang, S., Huang, Q., Leng, G., Wang, H., Bai, Q., et al., 2019. Propagation dynamics from meteorological to groundwater drought and their possible influence factors. J. Hydrol. 578, 124102.
Hangshing, L., Dabral, P.P., 2018. Multivariate frequency analysis of meteorological drought using copula. Water Resour. Manag. 32, 1741–1758.
Haroon, M.A., Zhang, J., Yao, F., 2016. Drought monitoring and performance evaluation of MODIS-based drought severity index (DSI) over Pakistan. Nat. Hazards 84, 1349–1366.
Hayes, M.J., Alvord, C., Lowrey, J., 2007. Drought Indices. Intermountain West Climate Summary 3 (6), 2–6.
Hayes, M.J., Svoboda, M.D., Wiihite, D.A., Vanyarkho O, V., 1999. Monitoring the 1996 drought using the standardized precipitation index. Bull. Am. Meteorol. Soc. 80, 429–438.
Heddinghaus, T.R., Sabol, P., 1991. A review of the Palmer Drought Severity Index and where do we go from here. Proc. 7th Conf. Appl. Climatol., 242–246.
Hernandez, E.A., Uddameri, V., 2014. Standardized precipitation evaporation index (SPEI)-based drought assessment in semi-arid south Texas. Environ. Earth Sci. 71, 2491–2501.
Herrera-Estrada, J.E., Martinez, J.A., Dominguez, F., Findell, K.L., Wood, E.F., Sheffield, J., 2019. Reduced moisture transport linked to drought propagation across North America. Geophys. Res. Lett. 46, 5243–5253.
Hisdal, H., Tallaksen, L.M., Clausen, B., Peters, E., Gustard, A., VanLauen, H., 2004. Hydrological drought characteristics. Dev. Water Sci. 48, 139–198.
Hoerling, M., Kumar, A., Dole, R., Nielsen-Gammon, J.W., Eischeid, J., Perlwitz, J., et al., 2013. Anatomy of an extreme event. J. Clim. 26, 2811–2832.
Hoffmann, D., Gallant, A.J.E., Arblaster, J.M., 2020. Uncertainties in drought from index and data selection. J. Geophys. Res. Atmos. 125 (18), e2019JD031946.
Hu, Q., Willson, G.D., 2000. Effects of temperature anomalies on the Palmer Drought Severity Index in the central United States. Int. J. Climatol. A. J. R. Meteorol. Soc. 20, 1899–1911.
Huang, S., Huang, Q., Leng, G., Liu, S., 2016. A nonparametric multivariate standardized drought index for characterizing socioeconomic drought: A case study in the Heihe River Basin. J. Hydrol. 542, 875–883.
Huang, S., Li, P., Huang, Q., Leng, G., Hou, B., Ma, L., 2017. The propagation from meteorological to hydrological drought and its potential influence factors. J. Hydrol. 547, 184–195.
Johnson, W.K., Kohne, R.W, 1993. Susceptibility of reservoirs to drought using Palmer index. J. Water Resour. Plan. Manag. 119, 367–387.
Karl, T.R., 1986. The sensitivity of the Palmer Drought Severity Index and Palmer's Z-index to their calibration coefficients including potential evapotranspiration. J. Clim. Appl. Meteorol., 77–86.
Karl, T.R., 1983. Some spatial characteristics of drought duration in the United States. J. Appl. Meteorol. Climatol. 22, 1356–1366.
Karl, T.R., Koscielny, A.J., 1982. Drought in the united states: 1895–1981. J. Climatol. 2, 313–329.
Karl, T.R., Quayle, R.G., 1981. The 1980 summer heat wave and drought in historical perspective. Mon. Weather Rev. 109, 2055–2073.
Keyantash, J., Dracup, J.A., 2002. The quantification of drought: an evaluation of drought indices. Bull. Am. Meteorol. Soc. 83, 1167–1180.

Kim, T.-W., Valdés, J.B., 2003. Nonlinear model for drought forecasting based on a conjunction of wavelet transforms and neural networks. J. Hydrol. Eng. 8, 319–328.

Kisi, O., Gorgij, A.D., Zounemat-Kermani, M., Mahdavi-Meymand, A., Kim, S., 2019. Drought forecasting using novel heuristic methods in a semi-arid environment. J. Hydrol. 578, 124053.

Konapala, G., Mishra, A., 2020. Quantifying climate and catchment control on hydrological drought in the continental United States. Water Resour. Res. 56 (1), e2018WR024620.

Konapala, G., Mishra, A.K., Wada, Y., Mann, M.E., 2020. Climate change will affect global water availability through compounding changes in seasonal precipitation and evaporation. Nat. Commun. 11, 1–10.

Kundzewicz, Z.W., Radziejewski, M., Pinskwar, I., 2006. Precipitation extremes in the changing climate of Europe. Clim. Res. 31, 51–58.

Laaha, G., Blöschl, G., 2007. A national low flow estimation procedure for Austria. Hydrol. Sci. J. 52, 625–644.

Li, X., Meshgi, A., Babovic, V., 2016. Spatio-temporal variation of wet and dry spell characteristics of tropical precipitation in Singapore and its association with ENSO. Int. J. Climatol. 36, 4831–4846.

Liu, W., Sun, F., Sun, S., Guo, L., Wang, H., Cui, H., 2019. Multi-scale assessment of eco-hydrological resilience to drought in China over the last three decades. Sci. Total Environ. 672, 201–211.

Lloyd-Hughes, B., Saunders, M.A., 2002. A drought climatology for Europe. Int. J. Climatol. A. J. R. Meteorol. Soc. 22, 1571–1592.

Van Loon, A.F., 2015. Hydrological drought explained. Wiley Interdiscip. Rev. Water 2, 359–392.

Van Loon, A.F., Van Lanen, H.A.J., 2013. Making the distinction between water scarcity and drought using an observation-modeling framework. Water Resour. Res. 49, 1483–1502.

Lorenzo-Lacruz, J., Vicente-Serrano, S.M., López-Moreno, J.I., Beguería, S., García-Ruiz, J.M., Cuadrat, J.M., 2010. The impact of droughts and water management on various hydrological systems in the headwaters of the Tagus River (central Spain). J. Hydrol. 386, 13–26.

Madsen, H., Rosbjerg, D., 1998. 19: A regional Bayesian Method for Estimation of Extreme Streamflow Droughts, In: Parent, E., Bobée, B., Hubert, P., Miguel, J., Statistical and Bayesian Methods in Hydrological Science 327-340, Unesco Studies and Reports in Hydrology, Paris, France.

Manzano, A., Clemente, M.A., Morata, A., Luna, M.Y., Beguería, S., Vicente-Serrano, S.M., et al., 2019. Analysis of the atmospheric circulation pattern effects over SPEI drought index in Spain. Atmos. Res. 230, 104630.

Martin-Benito, D., Beeckman, H., 2013. Canellas: Influence of drought on tree rings and tracheid features of Pinus nigra and Pinus sylvestris in a mesic Mediterranean forest. Eur. J. For. Res. 132, 33–45.

McEvoy, D.J., Huntington, J.L., Abatzoglou, J.T., Edwards, L.M., 2012. An evaluation of multi-scalar drought indices in Nevada and Eastern California. Earth Interact. 16, 1–18.

McKee, T.B., Doesken, N.J., Kleist, J., 1993. The relationship of drought frequency and duration to time scales, Proc. 8th Conf. Appl. Climatol. Boston, vol. 17, 179–183.

Mishra, A.K., Desai, V.R., Singh, V.P., 2007. Drought forecasting using a hybrid stochastic and neural network model. J. Hydrol. Eng. 12, 626–638.

Mishra, A., Alnahit, A., Campbell, B., 2020. Impact of land uses, drought, flood, wildfire, and cascading events on water quality and microbial communities: a review and analysis. J. Hydrol., 125707.

Mishra, A.K., Singh, V.P., 2010. A review of drought concepts. J. Hydrol. 391, 202–216.
Mishra, A.K., Singh, V.P., 2009. Analysis of drought severity-area-frequency curves using a general circulation model and scenario uncertainty. J. Geophys. Res. Atmos. 114.
Mukherjee, S., Mishra, A., Trenberth, K.E., 2018. Climate change and drought: a perspective on drought indices. Curr. Clim. Chang. Reports 4, 145–163.
Mukherjee, S., Mishra, A.K., 2021. Increase in compound drought and heatwaves in a warming world. Geophys. Res. Lett. 48, e2020GL090617.
Özger, M., Mishra, A.K., Singh, V.P., 2009. Low frequency drought variability associated with climate indices. J. Hydrol. 364, 152–162.
Palmer, W.C., 1965. Meteorological Drought. vol. 45. US Department of Commerce, Weather Bureau, Washington D.C.
Panu, U.S., Sharma, T.C., 2002. Challenges in drought research: some perspectives and future directions. Hydrol. Sci. J. 47, S19–S30.
Paulo, A.A., Rosa, R.D., Pereira, L.S., 2012. Climate trends and behaviour of drought indices based on precipitation and evapotranspiration in Portugal. Nat. Hazards Earth Syst. Sci. 12, 1481–1491.
Potop, V., 2011. Evolution of drought severity and its impact on corn in the Republic of Moldova. Theor. Appl. Climatol. 105, 469–483.
Potop, V., Možný, M., Soukup, J., 2012. Drought evolution at various time scales in the lowland regions and their impact on vegetable crops in the Czech Republic. Agric. For. Meteorol. 156, 121–133.
Rao, A.R., Padmanabhan, G., 1984. Analysis and modeling of Palmer's drought index series. J. Hydrol. 68, 211–229.
Seneviratne, S.I., Corti, T., Davin, E.L., Hirschi, M., Jaeger, E.B., Lehner, I., et al., 2010. Investigating soil moisture–climate interactions in a changing climate: A review. Earth-Sci. Rev. 99, 125–161.
Sherwood, S., Fu, Q., 2014A drier future? Science (80-)343, 737–739.
Sohn, S., Ahn, J., Tam, C., 2013. Six month–lead downscaling prediction of winter to spring drought in South Korea based on a multimodel ensemble. Geophys. Res. Lett. 40, 579–583.
Sönmez, F.K., Koemuescue, A.U., Erkan, A., Turgu, E., 2005. An analysis of spatial and temporal dimension of drought vulnerability in Turkey using the standardized precipitation index. Nat. Hazards 35, 243–264.
Soulé, P.T., 1993. Hydrologic drought in the contiguous United States, 1900–1989: spatial patterns and multiple comparison of means. Geophys. Res. Lett. 20, 2367–2370.
Spinoni, J., Antofie, T., Barbosa, P., Bihari, Z., Lakatos, M., Szalai, S., et al., 2013. An overview of drought events in the Carpathian region in 1961-2010. Adv. Sci. Res. 10, 21.
Steinemann, A., Iacobellis, S.F., Cayan, D.R., 2015. Developing and evaluating drought indicators for decision-making. J. Hydrometeorol. 16, 1793–1803.
Szalai, S., Szinell, C., Zoboki, J., 2000. Drought Monitoring in Hungary. In: Wilhite, D.A., Sivakumar, M.V.K., Wood, D.A. (Eds.), Early Warning Systems for Drought Preparedness and Drought Management, WMO/TD n. 1037, Geneva, Swtizerland, 182–199.
Szalai, S., Szinell, C.S., 2000. Comparison of two drought indices for drought monitoring in Hungary—a case studyDrought drought Mitig. Eur. Springer, The Netherlands, pp. 161–166.
Tallaksen, L.M., Van Lanen, H.A.J., 2004. Hydrological Drought: Processes and Estimation Methods for Streamflow and Groundwater. Elsevier, Amsterdam, The Netherlands.

Thornthwaite, C.W., 1948. An approach toward a rational classification of climate. Geogr. Rev. 38, 55–94.
Todorovic, P., Woolhiser, D.A., 1976. Stochastic Structure of the Local Pattern of Precipitation. In: Shen, H.W. (Ed.), Stochastic Approaches to Water Resources, Vol. II, Fort Collins, 15.1-15.37, vol. 49, pp. 765-769.
Toromani, E., Sanxhaku, M., Pasho, E., 2011. Growth responses to climate and drought in silver fir (Abies alba) along an altitudinal gradient in southern Kosovo. Can. J. For. Res. 41, 1795–1807.
Trenberth, K.E., Dai, A., Van Der Schrier, G., Jones, P.D., Barichivich, J., Briffa, K.R., et al., 2014. Global warming and changes in drought. Nat. Clim. Chang. 4, 17–22.
Vicente-Serrano, S.M., Beguería, S., López-Moreno, J.I, 2010. A multi-scalar drought index sensitive to global warming: the standardized precipitation evapotranspiration index. J. Clim. 23, 1696–1718.
Vicente-Serrano, S.M., Beguería, S., Lorenzo-Lacruz, J., Camarero, J.J., López-Moreno, J.I., Azorin-Molina, C., et al., 2012. Performance of drought indices for ecological, agricultural, and hydrological applications. Earth Interact. 16, 1–27.
Wanders, N., Van Lanen, H.A.J., van Loon, A.F., 2010. Indicators for Drought Characterization on a Global Scale. Wageningen Universiteit, Wageningen, The Netherlands.
Wilhite, D.A., 2000. Drought as a Natural Hazard: Concepts and Definitions, In: Wilhite, D.A. (Ed.), Drought: A Global Assessment, vol. I, Routledge, London, United Kingdom, pp. 3–18.
Wilhite, D.A., Glantz, M.H., 1985. Understanding: the drought phenomenon: the role of definitions. Water Int. 10, 111–120.
Wolf, J.F., Abatzoglou, J., 2011. The suitability of drought metrics historically and under climate change scenarios, 47th Annu. Water Resour. Conf. Albuquerque, NM, 7–10.
Wu, J., Chen, X., Yao, H., Gao, L., Chen, Y., Liu, M., 2017. Non-linear relationship of hydrological drought responding to meteorological drought and impact of a large reservoir. J. Hydrol. 551, 495–507.
Yu, M., Li, Q., Hayes, M.J., Svoboda, M.D., Heim, R.R., 2014. Are droughts becoming more frequent or severe in China based on the standardized precipitation evapotranspiration index: 1951–2010? Int. J. Climatol. 34, 545–558.
Zarei, A.R., Moghimi, M.M., 2019. Modified version for SPEI to evaluate and modeling the agricultural drought severity. Int. J. Biometeorol. 63, 911–925.
Zargar, A., Sadiq, R., Naser, B., Khan, F.I., 2011. A review of drought indices. Environ. Rev. 19, 333–349.

Index

Page numbers followed by "*f*" and "*t*" indicate, figures and tables respectively.

W

X

Z

Printed in the United States
by Baker & Taylor Publisher Services